Springer Proceedings in Earth and Environmental Sciences

The series Springer Proceedings in Earth and Environmental Sciences publishes proceedings from scholarly meetings and workshops on all topics related to Environmental and Earth Sciences and related sciences. This series constitutes a comprehensive up-to-date source of reference on a field or subfield of relevance in Earth and Environmental Sciences. In addition to an overall evaluation of the interest, scientific quality, and timeliness of each proposal at the hands of the publisher, individual contributions are all refereed to the high quality standards of leading journals in the field. Thus, this series provides the research community with well-edited, authoritative reports on developments in the most exciting areas of environmental sciences, earth sciences and related fields.

Liudmila V. Radionova · Dmitrii V. Ulrikh
Editors

Advances in Ecology and Environmental Engineering

Proceedings of the 2nd International Conference on Ecology and Environmental Engineering (RusEcoCon-2024)

Springer

Editors
Liudmila V. Radionova
Department of Metallurgy
Moscow Polytechnic University
Moscow, Russia

Dmitrii V. Ulrikh
Institute of Architecture and Construction
South Ural State University (National Research University)
Chelyabinsk, Russia

ISSN 2524-342X ISSN 2524-3438 (electronic)
Springer Proceedings in Earth and Environmental Sciences
ISBN 978-3-031-64422-1 ISBN 978-3-031-64423-8 (eBook)
https://doi.org/10.1007/978-3-031-64423-8

This Springer imprint is published by the registered company Springer Nature Switzerland AG
The registered company address is: Gewerbestrasse 11, 6330 Cham, Switzerland

Preface

The International Russian Conference on Ecology and Environmental Engineering (RusEcoCon-2024) was organized by Moscow Polytechnic University, Moscow on March, 25–29, 2024.

The conference program encompassed a wide range of topics and was divided into four parts: Sustainable Urban Environmental Development; Resources and Environment Engineering; Energy Science, Green Technology and Eco-innovation; Technology Potential for Environmental Management 4.0.

Participants could take part in the conference as in a traditional face-to-face format and as format of video conference remotely.

The international program committee has selected a total of 48 papers for publishing in Springer Proceedings in Earth and Environmental Sciences (Springer International Publishing AG).

On behalf of the organizing committee, we express appreciation to our colleagues who participated in the review procedure of the papers and especially thank members of the international program committee, who helped us to organize this conference.

We express our gratitude to the participants for the active work at the conference sections and look forward to meeting at RusEcoCon-2025 next March in Sochi, Russia.

Moscow, Russia	Prof. Liudmila V. Radionova
Chelyabinsk, Russia	Prof. Dmitrii V. Ulrikh

Contents

Resources and Environment Engineering

Sustainable Urban Environmental Development

Environmental Characteristics of Imeretinka (Sochi) Coastal Hydraulic Structures

I. Kantarzhi

Abstract The paper overviews the design and use of the Imeretinka lowlands of Sochi hydraulic structures. These structures belong to the so-called 2014 Olympics coastal cluster. The main structures are the cargo port at the Mzymta River outflow with the annual cargo turnover of 5 million tons and the 5 km-long coastal protection dam from the port to Cape Konstantinovsky. Both structures were not initially planned and appeared at the design stage as a result of the insufficient previous research of natural processes. The article shows that the implemented seaport and coastal protection solutions do not ensure coastal stability. Hydro- and lithodynamic processes are investigated with mathematical modeling methods. Recent severe storms have significantly damaged the coastal structures, and the deterioration process may be expected will continue. The historical overview presented presents an important reference for similar problems.

Keywords Imeretinkanka Lowland · Coastal zone · Seaport · Coastal protection dam · Submarine Canyons · Construction history · Structural failures

1 Introduction

The so-called coastal infrastructure cluster of the 2014 XXII Olympic Winter Games was built in the Imeretinka Lowlands.

The causes for construction of the "heavy" structures in the Imeretinka Lowland were firstly, the decision to use the famous "Krasnaya Polyana" ski resort as a venue for the Games in Russia, and, secondly, the impossibility to place such structures on a mountainous landscape. However, this decision did not take into account the natural conditions of the Imeretinka Lowlands, which led to serious problems in the construction and operation of these structures.

I. Kantarzhi (✉)
Moscow State University of Civil Engineering, 26, Yaroslavskoe Shosse, Moscow 129337, Russia
e-mail: kantardgi@yandex.ru

L. V. Radionova and D. V. Ulrikh (eds.), *Advances in Ecology and Environmental Engineering*, Springer Proceedings in Earth and Environmental Sciences,
https://doi.org/10.1007/978-3-031-64423-8_1

1.1 Peculiarities of Natural Conditions [1]

The territory of the Imeretinka Lowland stretches along the Black Sea coast for 7 km, 0.8–2 km wide, and is a low seaside plain. The surface of the lowland is slightly elevated from 1.5 to 4 m on average above the sea level, and its significant areas are subject to waterlogging and bogging. The coastal zone is subject to abrasion processes with submarine canyons Novy, Konstantinovsky and a number of others developing in the coastal zone.

The total area of the Imeretinka Lowland is 1240 hectares (including the area of existing low-rise buildings—272 hectares, ponds—10 hectares and agricultural land—958 hectares before the construction of the 1014 Olympics structures).

The beaches of the Imeretinka Lowland are very dynamic as they lie in the vicinity of the outflows of the major coastal rivers, i.e., Mzymta and Psou, which are the main sources of river alluvium flow to the submarine slope. Waves move sand and gravel both in the coastal submarine slope and within the beach strip.

What is more, the beach dynamics is significantly influenced by the steepness of the coastal submarine slope, which varies within the area. The submarine slope of Imeretinka is shallow, with alternating high steepness areas (10–15°) where the tops of submarine canyons approach the shoreline (Fig. 1) and low steepness areas (3–5°) in the formed river alluvium cones areas at the river outflows.

The coastal submarine slope determines the wave action energy on the beach zone. With slopes over 10°, waves approach the shore with practically no change in parameters due to the collapse in the surf zone. The maximum storm wave runup is formed in those areas and, accordingly, the maximum power of beach material recycling is formed.

The beaches are mostly of full profile. The composition of the beach-forming material is sandy-gravel-pebble, with large pebbles and boulders occurring near the mouths of the Mzymta and Psou Rivers.

Construction on the existing weak Imeretinka soils required solving the soil reinforcement. It was proposed to "load" the ground with a layer of crushed stone up to

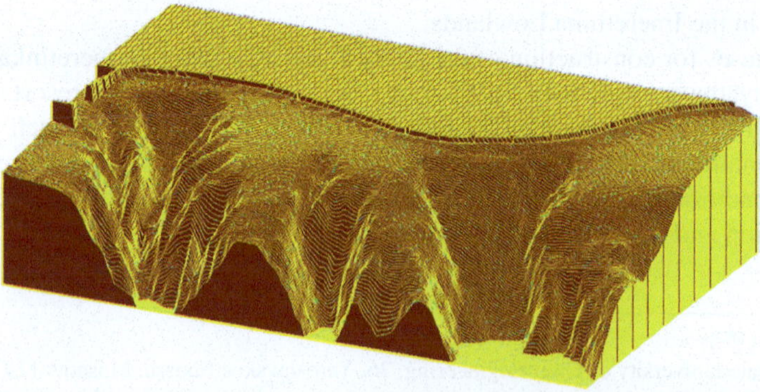

Fig. 1 Bottom relief of the Imeretinka Lowland with submarine canyons. Digital representation

Fig. 2 Seaport in Imeretinka at the mouth of the Mzymta River

3 m thick. This could have been accomplished by moving a significant amount of soil from the nearest quarry in Adygea in Russia to the site.

However, the Sochi conditions made the delivery of the required amount of crushed stone from Adygea to Imeretinka impossible. The currently existing one road and one railroad were not enough to organize such transportation without a significant damage to the tourist and social activity in the region.

The suggested alternative was to bring inert material from Turkey, Trabzon region by sea. To receive cargoes, it was decided to build a sea cargo port in Imeretinka, east of the mouth of the Mzymta River—Fig. 2.

The port is designed for the annual transshipment of up to 5 million tons of bulk and general cargo, including: cement (1 million tons p.a.), crushed stone, gravel, sand (3 million tons p.a.), rolled metal products (500 thousand tons p.a.), machinery (500 thousand tons p.a.). The port is all-year, 24/7.

It was planned that after the Olympic Games, the port would be modified from a cargo port to a yachting port. However, the relevant solutions were not developed in the project documentation.

1.2 Seaport

According to the design, the port layout along the shore (Fig. 2) eliminates the negative influences of submarine canyons on the stability of the barrier structures.

Structurally, the barrier breakwaters are made in the form of pile trestles with wave-dampening provided by the sketching of shaped massifs or concrete blocks.

The construction of the port revealed a new problem. The harbour barriers blocked the sediment flow of the longshore currents to the east and caused classic downstream scour. The downward scour began to form at the root of the eastern breakwater and moved eastwards along the shore. This caused the risk of a significant erosion of Imeretinka beaches.

The situation repeated the consequences of the downward scour that occurred after the construction of the cargo-passenger port in Sochi in the thirties of the XX century. As a result of that downstream scour, the central part of Sochi was virtually stripped off its beaches.

1.3 Coastal Protection Dam

The resulting coastal scour required the development of shore protection solutions. Before 1949, the beaches in Imeretinka were up to 60 m wide. After active excavation of the Mzymta River channel for construction purposes began, the beaches were reduced to 30–40 m. The attempts to construct coastal protection buoys or a coastal dike without repair fillings were unsuccessful and the structures collapsed.

A 5 km-long coastal protection dam was designed in the section from the seaport to Cape Konstantinovsky.

On the seaward side of the dam there is a 50-m sandy-pebble beach. From the shore, the beach is contiguous with a 1:5 slope protected by permeable reinforced concrete blocks (Fig. 3). The permeable slope is based on a thrust belt (pile foundation) with a foundation made of bored piles, 720 mm in diameter and 6.0 m in length, sunk with a 2.0 m spacing. The concrete slope is followed by an embankment with a wave protection parapet.

Thus, the design of the coastal cluster structures resulted in the need to develop two unplanned major hydraulic structures—the seaport and the coastal protection

Fig. 3 Appearance of the seaward face of the dam

dam. This the primary evidence that the regional coastal processes had not been studied to a sufficient degree prior to decision-making.

The expert assessment of the proposed coastal protection dam design allows for a conclusion that the beach within the dam shall be eroded as a result of downward scour. At the same time, a significant beach erosion should not be allowed, as in that case the inert material shall be washed out from under the slope, triggering its destruction.

Therefore, the most important question in the dam operation is to determine the extent of the beach replenishment repairs. Answering it required the following:

- To re-analyze the hydrometeorological conditions of the Imeretinka coastal zone over a long period of 25–30 years;
- To identify the most severe storms for the same period on embedded grids, from global—Black Sea to local—Imeretinka beach area, using wind-wave models;
- To verify the numerical wave modeling data with available in-situ measurement data and to set the correction factors;
- To model the propagation of the strong storm's waves from deep water to the coastal zone of the water area with submarine canyons, the modeling should be performed for natural conditions as well as for the port on place;
- To model the wave currents system caused by strong storms in natural conditions and in the presence of harbor barriers;
- To determine longshore sediment transport for sandy-pebble sediments at different sections of the dam for natural and harbor barrier conditions;
- To determine the amount of erosion and accumulation for the typical sections of the dam, and determine the amount of the average annual repair replenishment.

The listed problems were solved with numerical modeling methods [2] and the author's and open access models [3–13]. The team under the leadership of the author has accumulated many years of experience in using numerical and composite modeling methods in such problems [14–22].

The verification was based on the wave measurement data in the Gelendjik area and at the Roshydromet Sochi wave station [23].

2 Problems of Interaction Between the Structures and the Natural Processes

Figure 4 shows an illustrative numerical model results of a wave field in Imeretinka for a selected strong storm in January 1993. The significant wave height in this storm was 7.02 m. The wave calculation for the harbor scenario is shown.

The results presented in Fig. 4 show that the high waves do indeed travel through the submarine canyons to the shore. Thus, high waves are focused at the head of the south-west mole, what can be dangerous for the stability of the structure and for the mooring conditions in the port water area.

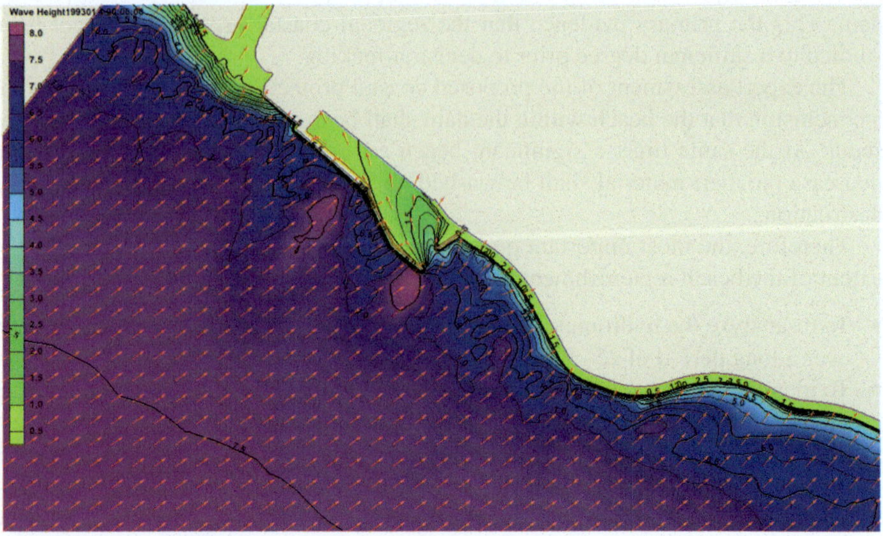

Fig. 4 Results of numerical modeling of waves in the Imeretinka water area

Figure 5 shows the alongshore sediment flow broken by the port facilities, which confirms the earlier analysis.

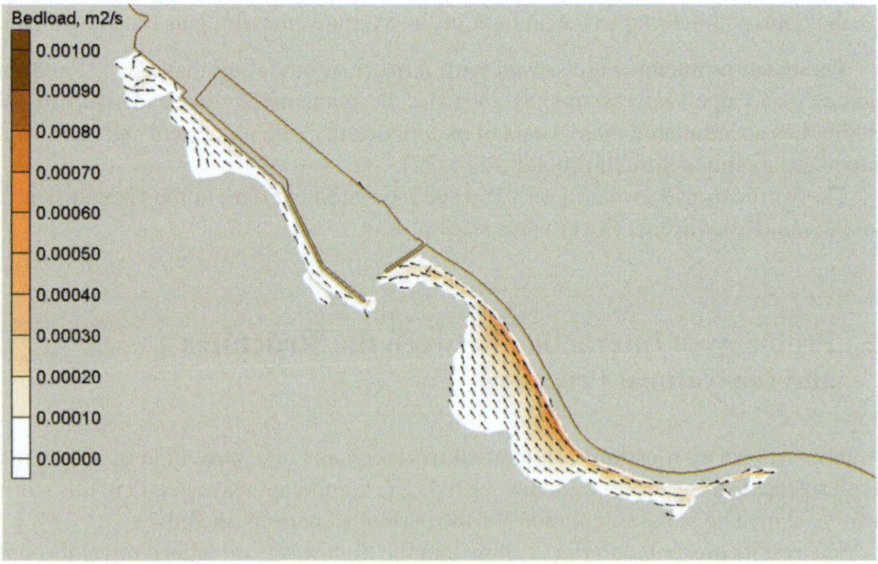

Fig. 5 Blocking along the shore flows of the entrained sediments by the port's protective structures. Simulation results

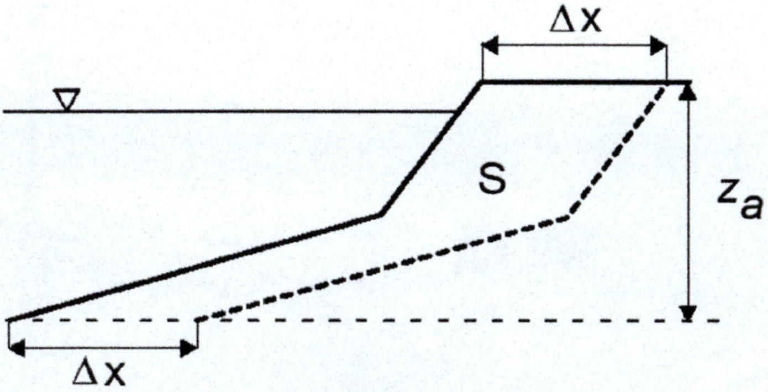

PK18:

$l \approx 400$ m $V = 7.7 + 12.5 \approx 20$ thousand m³/year

$S = 20 + 31 \approx 50$ m³m⁻¹year⁻¹. Δ x ≈5 m/year

PK22-PK28:

$l \approx 600$ m $V \approx 20$ thousand m³/year

$S \approx 30$ m³m⁻¹year⁻¹. Δ x ≈3 m/year

PK28:

$l \approx 400$ m $V \approx 17$ thousand m³/year

$S \approx 40$ m³m⁻¹year⁻¹. Δ x ≈4 m/year

Fig. 6 Calculating the dam scour volumes and the corresponding volumes of repair fill in the characteristic sections

Figure 6 shows the main final results of the coastal processes numerical modeling. The most dangerous sections of the shore are located in the vicinity of pickets PK18 and PK22, which drop directly into the canyons. Storms in these areas wash the sediment off the beach, transport it to the deeper areas which means it is irretrievably lost. The results of the calculations for these particular pickets are shown in Fig. 6.

Two sizes of beach material particles were used in the model, i.e., 40 and 80 mm. The strain distributions obtained for the two average diameters are similar. Although, the strain amplitudes for the larger sediment sizes are somewhat smaller.

The results obtained relate to the stabilization phase of the storm. The submarine part of the coastal slope undergoes the greatest changes. As concave profile is formed, the wave energy is the most uniformly dissipated along the profile. The eroded material is deposited mainly in the lower part of the slope, but a small part of it may reach the overwater beach.

On most profiles, the shore retreats to the landward side by several meters. The most "unfavorable" is PK18, where the landward drift is about 5–7 m per year.

As it was mentioned previously, the beach scour must be compensated by means of repair fillings. Otherwise, no dam and protected embankment stability can be guaranteed. This conclusion has been fully confirmed by the subsequent events. Despite that the years since 2014 have been relatively calm having no major storms,

Fig. 7 The Imeretinka dam and embankment destroyed by the storm on 27.11.2023

the downstream scour has slowly developed. The repair replenishment of the dam beach has not been carried out.

2.1 Storm of November 21, 2023

The first signs of failure of the reinforced concrete slope manifested themselves in 2017. And in 2023, after a relatively strong storm on November 21st, when the wave heights of about 4–5 m hit the coastal zone of Imeretinka, the beach was washed away and the reinforced concrete slope and embankment were significantly destroyed in a number of areas with the waves flow over the embankment, flooding the buildings of the Olympic Village (Fig. 7).

3 Conclusions

The main conclusion of the analysis made in the article is that the scientific research of the natural processes is crucial before making decisions regarding the hydraulic structures' layout and designs. It is not about engineering surveys, which, of course, are necessary and designed to obtain the initial data for research and design, but about the study of processes that determine the composition, layout and design of the hydraulic structures.

In the case of the structures in the Imeretinka Lowlands, such studies were not carried out to a sufficient extent. As a result, the design process was faced with the need to design new structures, caused by the errors made on the previous design stages. This is how the cargo seaport and the coastal protection dam appeared.

The main research method is mathematical modeling of processes, which allows considering the extreme design situations.

Acknowledgements The author acknowledges fruitful discussions with colleagues, and special technical assistance by my PhD students.

References

1. Kantardgi IG, Mordvincev KP (2011) Mathematical Modeling of the Coastal Processes to Support the Shore Protection of the Imeretinkansky Lowland. Int J for Computational Civil and Structural Eng. 7(2):64–71.
2. Kantardgi I, Zheleznyak M, Demchenko R, Dykyi P, Kivva S, Kolomiets P, Sorokin M (2014) Modelling of nonlinear hydrodynamics of the coastal areas of the Black Sea by the chain of the proprietary and open-source models. In: Proceedings of EGU. Vienna, April 27–May 02.
3. Holthuijsen LH (2007) Waves in oceanic and coastal waters. Cambridge University Press. Cambrige.
4. Booij N, Haagsma IJ, Holthuijsen L, Kieftenburg A, Ris R, van der Westhuysen A, Zijlema M SWAN Cycle III version 40.51. User manual. The Netherlands.
5. Hasselmann K, Barnett TP, Bouws E, Walden H (1973) Measurements of wind-wave growth and swell decay during the Joint North Sea Wave Project (JONSWAP). Ergnzungsheft zur Deutschen Hydrographischen Zeitschrift Reihe, no. 12, p. 95.
6. Aelbrecht D (1997) ARTEMIS 3.0: A finite element model for predicting wave agitation in coastal areas and harbours including dissipation. Computer modelling of seas and coastal regions III. Boston, pp 343–352.
7. Berkhoff JC (1972) Computation of combined refraction-diffraction. In: 13th Coastal Engineering Conference. New York, vol. 1.
8. Berkhoff JC (1976) Mathematical models for simple harmonic linear water waves: wave diffraction and refraction. Delft, Hydraulic Laboratory, p. 102.
9. Hervouet JM (1999) TELEMAC, a hydroinformatic system. La Houille Blanche, pp 3–4.
10. Kofoed-Hansen H, Sloth P, Sørensen OR, Fuchs J (2000) Combined numerical and physical modelling of seiching in exposed new marina. In: Proceedings of 27th International Conference of Coastal Engineering. Sydney Australia, 16–21 July, 2000.
11. Smit P, Stelling G, Zijlema M (2011) Assessment of non-hydrostatic wave-flow model SWASH for directionally spread waves propagating through a barred basin. In: Proceedings of ACOMEN 2011. Liege Belgium, 14–17 November, 2011.
12. Rijnsdorp DP, Smit PB, Zijlema M (2012) Non-hydrostatic modelling of infragravity waves using SWASH. In: Proceedings of 33rd International Conference on Coastal Engineering. Santander Spain, 1–7 July.
13. Zijlema M, Stelling G, Smit P (2011) SWASH: An operational public domain code for simulating wave fields and rapidly varied flows in coastal waters. Coastal Engineering 10(58):992–1012.
14. Hadla G, Kantarzhi I, Leont'yev I (2022) Modeling of Lithodynamic Processes in the Area of Sea Port. Advances in Science, Technology, and Innovation, pp 89–92. https://doi.org/10.1007/978-3-030-72547-1_20
15. Kantarzhi IG (2023) Forecast of the Dynamics of a Sandy Beach in Complexed Hydrodynamic Conditions, Oceanology 63(6):906–914.
16. Kantarzhi I, Zuev N, Shunko N, Zeleznyak M, Dikiy P, Sorokin M (2014) Numerical and physical modelling of the waves inside the new marina in Gelendjik (Black Sea). Application of physical modelling to port and coastal protection, Proceedings of 5th International Conference COASTLAB, Varna, Bulgaria.

17. Kantarzhi IG, Mordvintsev KP, Gogin AG (2019) Numerical analysis of the protection of a harbor against waves. Power Technology and Engineering 53(4):410–416.
18. Kantardgi I, Gogin A (2019) Dynamic Analysis of Moored Floating Objects In: 14th MEDCOAST Congress on Coastal and Marine Sciences, Engineering, Management and Conservation, MEDCOAST 2019. Marmaris Turkey, 2.
19. Gogin A, Kantarzhi I (2020) Numerical Study of the Floating Gravity Base Structure Mooring In: Lecture Notes in Civil Engineering 70:205–219.
20. Kantardgi IG, Gogin AG (2020) Modeling a berth with a moored ship under wind wave loads. Power Technology and Engineering 54(2):160–166.
21. Kantardgi IG, Gogin AG (2021) Submarine Pipeline Stability under Stream and Waves Action. Power Technology and Engineering 55(3):367–372.
22. Kantarzhi I, Gogin A (2022) Optimal Scale Modeling of Surf Zone Waves. Lecture Notes in Civil Engineering 170:259–270.
23. Kantarzhi I, Anshakov A, Gogin A (2021) Composite modelling of wind waves in designing of port hydraulic structures. In: Proceedings of the International Offshore and Polar Engineering Conference, Thessaloniki Greece.

Solving Some Environmental Problems Through Restoration, Plantation Building, and Recreational Woodland Management for the All-Around Sustainable Development of the Far Eastern Region of Russia

L. T. Krupskaya, M. Yu. Filatova, A. M. Orlov, A. V. Leonenko, and Y. G. Kocharyan

Abstract This article presents the results of multi-year research on solving some of the environmental problems in the mining territories of the Far Eastern Federal District (FEFD) by restoration, plantation building, and recreational woodland management for the sustainable development of the region. The rapid growth of forest and mining production, associated with forest cutting for the timber industry and gold mining in the past century, caused significant damage and environmental pollution leading to an environmental crisis that threatened not only human civilization but also the entire biosphere. However, identifying these trends did not require immediate action at the time, but now, according to N.V. Vasilyev et al. it is vital to realign the development strategies of the society and the region to shift the focus from immediate profits to long-term human development. Therefore, the goal of this research is to develop comprehensive proposals to reduce or eliminate the negative environmental impacts of the lands disturbed in the past century within placer gold mines to facilitate the all-around sustainable development of the region. The research focuses on forest ecosystems, technology, restored and unrestored plots, deforested areas, plantation forestry, and recreational woodland management. The geoenvironmental assessment of the region in question using the remote sensing of Earth, physical and chemical methods, and bioremediation helped define the current environmental situation in the region as critical. The authors made suggestions to

L. T. Krupskaya (✉) · M. Yu. Filatova · A. M. Orlov
Far Eastern Forestry Research Institute, 71, Street Volochaevskaya, Khabarovsk 680020, Russia
e-mail: ecologiya2010@yandex.ru

A. V. Leonenko
Mining Institute, 51, Street Turgenev, Khabarovsk 680000, Russia

Y. G. Kocharyan
Sochi State University, 94 Plastunskaya Street, Sochi 354003, Russia

L. V. Radionova and D. V. Ulrikh (eds.), *Advances in Ecology and Environmental Engineering*, Springer Proceedings in Earth and Environmental Sciences,
https://doi.org/10.1007/978-3-031-64423-8_2

solve some environmental problems. Their innovative nature is confirmed by the patents of the Russian Federation.

Keywords Forest resources · Technology · Restoration · Plantation · Recreational woodland management · Burnt areas · Restored and unrestored plots · Bioremediation

1 Introduction

The strategic goal of developing the Far East of Russia is anchoring the population in the region through economic development, creating a comfortable living environment, and catching up with the country's average social and economic development level in the far eastern regions of the Russian Federation. In the previous century, the rapid growth of industries, including timber and mining production associated with the deforestation of vast areas and mineral deposit development, caused significant environmental destruction and pollution leading to an environmental crisis that threatened human civilization and the entire biosphere. Identifying these trends did not require immediate action at the time, but now, according to Vasilyev et al. it is vital to realign the development strategies of the society and the region to shift the focus from immediate profits to long-term human development. Therefore, it is necessary to quickly restore the forests of the Far Eastregion to address the material needs of humans and recreate a favorable living condition in the lands disturbed by the previous economic activities through restoration, plantation building, and recreational woodland management. The works by Petrov et al. [1, 2] establish the efficiency of forest restoration in placer gold mines.. He claims that this is confirmed by the high survivability and preservation of forests over 5–11 years.

Drozdova et al. [3] presented the results of research on the prospects of the phytomining of gold from the washed sand from placer mines.

Rozinkina et al. [4] reviewed the problems of the classification of lands disturbed by mining and selecting optimal restoration methods based on the principles proposed by Kolesnikov who significantly developed the classification theory taking into account the type and rate of soil formation and demutation in disturbed lands depending on the geographic zoning of their location, substrate type, and the need for fertile earth mulching to provide the microbial flora. Rozinkina et al. [4] proposed a hierarchy of disturbed lands (with categories, classes, groups, and types).

The article of Belov et al. [5] is of special interest because it uses natural forests as the base for recreation plantings. The authors created a recreation plot with a water body and a forest in an industrialized part of the Urals. It provides comfort for the residents of Krasnogvardeysk. According to the authors, the same effect can be achieved by planting forest plantations. Apart from the forest plantation, there can be a water body represented by a pond or an abandoned red clay quarry used as a fire reservoir, which is very attractive for revelers (a combination of forest and water). The plot has to be cleared, and the timber produced during the clearing can be used for hard landscaping or fuel for legal fire pits. The amenities shall also include a recreation area near the water and a network of footpaths. Belov et al. [5] suggest using young coniferous trees, including the Siberian pine, as the base for the formation of multi-age plantations that are the most resilient against recreational loads. The researchers suggest using combined forest restoration with subordinate forest crops when the density of the parent forest stand and the number of young economic plants are low.

Zalesova et al. [6] suggest a landscape-spatial forest park structure facilitating the increased resilience and recreational attractiveness of the plantations. The researchers clarify the functional landscape-spatial zoning of forest parts and establish sustainable recreation-attractive planting based on it. Thus, Zalesova et al. [6] developed some requirements for landscape formation activities, including the following: Each of the landscaping plots needs to have the main and the accompanying plantation species, form, and structure; closed, semi-open, and open landscapes have to intermingle and, in an ideal case, they need to change one another every 100–300 m; it is necessary to consider the resilience of the plantation against the recreational loads; special attention must be paid to plots experiencing intensive digression and early degradation; it is necessary to maintain the optimal proportion of landscape types. The key landscaping activities aiming to transform forests into recreation areas include improvement cutting (landscape cutting), tree and shrub planting, and amenity improvement.

Shtukin et al. [7] present the results of the research carried out in Belarus, where a lot of attention is paid to the development of facilities for timber processing and plantation forestry that can cover the increasing paper wood procurement and reduce the lumbering load on natural forests. To avoid losses from timber production, the authors recommend considering different organizational and processing solutions for different stages of forest plantation lives. According to them, special project development must become the key stage of forest plantation building and plot arrangement. They believe that lands that used to be agricultural are the most interesting for plantation timber production, and the organization of this process can be heavily affected by their ownership types. The primary investment stage starts from the development of project documents to the beginning of improvement cutting.

A large group of Chinese researchers including Ding et al. [8] evaluated the impact of the transformation of pure plantations to mixed ones on the multifunctionality of the soil, stochastic processes, and stability of bacterial networks in subtropical southern China. They compared the research results from a pure plantation of Cunninghamia lanceolata and a mixed plantation of C. Lanceolata and Betula Luminifera. The authors concluded that the transformation of monoculture plantations to mixed ones improves soil fertility and has other benefits. Improved bacterial diversity. A more complex and stable network structure with more key taxons. Increased width of habitat niches and importance of random processes. Increased carbohydrate and energy metabolism. These benefits of mixed plantations help assess the environmental and economic consequences of forest restoration.

The research carried out by Luiz Marcelo Brum Rossi et al. [9] focused on the productivity and growth of forest vegetation in homogeneous experimental plantations in modified and abandoned areas of the Amazon region. The Amazon region has great opportunities for plantation in modified, abandoned, and degraded areas (due to agricultural activities and ranching). In Brazilian Amazonia, these lands make up about 40 million hectares. The size of deforested lands in Legal Amazonia reached 588,000 square kilometers before 2000 (INPE 2002), which is about 15% of the entire land area (MMA 2000). The restoration of forests is made more important by the lack of the most popular timber sorts, such as Ceiba pentandra, Schizolobium parahyba, and baboen. These trees cover over 50% of the plywood market and are only found in remote areas, which increases the location costs. Using abandoned plots to create forest plantations can reduce the costs of raw materials through the concentration of same-species plants or groups of plants closer to timber production centers. Researchers believe that the availability of plantation technology in degraded areas can help reduce the load on the tropical forests in the region, improve the productivity of plantations, and thus increase the profits of landowners through the sale of timber and seeds.

Riccioli et al. [10] attempted to calculate the recreational value of forests under different management modes. The researchers believe that forest degradation is a serious threat to the provision of ecosystemic services like timber production, biodiversity, and hydrogeological protection. The desolation of forests is one of the key reasons for their degradation in the Mediterraneanwhere the low added value of forestry affects economic sustainability. This problem calls for urgent restoration actions supported by the cost and benefit analysis covering the entire range of profitable forestry activities, including recreation. The impacts of forest management systems on timber production are well-studied, while the impact of the recreation value is not. Thus, the authors aim to demonstrate that various forest management modes result in different willingness to pay (WTP) for the maintenance of recreational forest use.

Musin et al. [11] note that the differentiated assessment of recreation potential criteria is a key base for the cadastral evaluation of lands that supports optimal decision-making concerning the improvement of the recreational properties of landscapes and ensures their protection and rational usage. This helped exclude some of the areas from recreational use, limit such use for the other group, and identify additional lease opportunities for the third group of forest lands. Thus, the authors suggested some redistribution options for the flow of revelers by creating forest parks based on the existing forest and forested landscapes. The researchers claim that popular recreational resources become the base of the economic development of regions. However, some aspects need to be adjusted, especially the rent prices. Currently, they are set without consideration of the environmental productivity of forest stands (they are the same for centuries-old pine forests near water and for homogeneous aspen woods). If the rent prices are differentiated depending on the environmental productivity, it will help increase the budget revenues by 1.5–2 times. If the value of economic productivity receives a correct monetary expression, market mechanisms can be used to establish the price of a favorable environment rather than expensive other random tools.

Thus, the preliminary and brief overview of literature and regulatory documents, as well as a search through patents show that the solving of environmental problems in the Far Eastern Federal District of Russiaby plantation building and recreational woodland management can be intensified. Currently, there is only limited experience in this area (Podkovykrin, Tikhmenev, Ivlev, Pugachev, Krupskaya, Kostenkov, Purtova, Filatova, etc.). Therefore, the goal of this research is to develop comprehensive proposals to reduce or eliminate the negative environmental impacts of the lands disturbed in the past century within placer gold mines to facilitate the sustainable development of the region.

The scientific significance of the research lies in the development of new, nature-like technologies, methods, and techniques to facilitate the all-around sustainable development of the Far Eastern region that would help reduce the industrial load on natural forests, increase the size of rented forest plots and revenues from woodland management, and create a comfortable environment for people's habitation and recreation. The practical significance of this project is obvious.

2 Methods and Materials

Methodologically, the research is based on Vernadsky's [12] theory of biosphere and noosphere, as well as the key provisions in the program and methods of studying technogenic biogeocenosis of Motorina and Kolesnikov [13]. This article uses the following set of key methods and techniques: systemic analysis and generalization of theoretical and experimental research, scientific forecasting, cartographic and mathematical modeling, zoning, and statistical processing of empirical data. The paper also used modern tools and traditional physical, chemical, and biological methods (Pollen Sterility and Growth Test systems), including microbiological research.

The length of the research placer is about 14 km with an average width of the industrial outline reaching 225 m (from 20 to 500 m). The placer development started in 1888. Technogenic formations (wastes) in the headwater are represented by hydraulic mining tailings, and by dredging tailings in the middle and lower reaches. The wastes include the gold-concentrating unit (GCU) wastes that are stored in piles within the settlement limits.

We used the data from the closed-down Kerbinsky gold mine to analyze the current state of the research placer area using the remote sensing of Earth with various spectral channel combinations of the space images from the Glovis Landsat 8 satellite that can be accessed freely through the NASA internet portal. We used the QGIS software to process the remote sensing data (RSD). To build cartographic models, we used the Ekolog 4 software. We loaded the images of the territories in question taken from 2014 to 2021 [14]. The RSD resources are objective, operative, and versatile, and they contain aggregate data on all landscape components. The data from the Landsat 8 satellite have a spatial expansion of 30 m and 9 spectral channels. For the inspected

Table 1 The NDVI scale for the study of vegetation development

Summer	NDVI value	Winter	NDVI value
Vegetation development		Coniferous forest density	
High	0.6–1	100%	0.8–0.9
Medium	0.3–0.6	90%	0.7–0.8
Low	0.2–0.3	80%	0.6–0.7
Lack of vegetation: water, mountains, fallow	−0.1–0.2	70%	0.5–0.6
–	–	60%	0.425–0.5
–	–	50%	0.4–0.425
–	–	40%	0.35–0.4
–	–	30%	0.3–0.35
–	–	20%	0.25–0.3
–	–	10%	0.2–0.25
–	–	Deciduous forest, snow, ice	−1–0.2

territories (dredging sites) of the closed-down Kerbinsky gold mine, we calculated the NDVI vegetation index used for the scientific analysis of satellite data [15, 16]. To analyze the phytomass in the research area, determined as the reflecting power of the plant leaf surface [17], we calculated the NDVI in the QGIS software. Using the bitmap calculator feature, we inserted the NDVI formula (3) [17]:

$$NDVI = \frac{NIR - R}{NIR + R} \tag{1}$$

where NIR (Channel 5) is the reflection factor in the near-infrared band; R—RED (Channel 4) is the reflection factor in the red band (the Table 1) [15].

The obtained index image was classified following the common scale shown in the Table 1.

3 Results and Discussion

The research conducted over multiple years (1990–2023) helped us identify the following environmental problems within the impact areas of mining facilities: 1. High waste production of mining facilities and intensive industrial pollution of the environment; 2. Low efficiency of extraction and utilization of minerals from the developed deposits; 3. Low level of green technology at mining facilities; 4. Problems associated with the development of technogenic deposits and tailing dumps; 5. The obsolete system of environmental requirements and the need for its reformation.

Increasing the rate of mining activities in the research area caused large-scale negative changes in all biosphere components. We established that the size of lands disturbed by mining activities is constantly increasing due to the development of technogenic deposits. For instance, from 2011 to 2023, it increased by over 5.5 times (from 492.4 hain 2011 to 2700 hain 2023). The gold concentrating unit of the former Kerbinsky gold mine, which could not handle the economic problems of the transition period, went bankrupt and shut down, presents the greatest risk and danger. The specifics of placer gold extraction and enrichment at this facility in the previous century included the extraction and processing of huge amounts of rocks and utilizing only a small fraction of it, while the rest accumulated as toxic industrial wastes stored in a specific location referred to as the gold concentrating unit. The economic activity in the research area in the previous century resulted in the formation of a technogenic system that combined natural and mining features. Shutting down the operation of the Kerbinsky mine created a big environmental problem: the industrial pollution of the surroundings with enrichment wastes promoted environment-related diseases among the residents of the Briakan mining town.

In terms of danger, the technogenic system combining natural and mining features can be classified as follows: Class 1: high environmental danger (gold concentration unit wastes, filter beds), where the Normalized Difference Vegetation Index (NDVI), a simple indicator of the amount of photosynthesizing biomass (also known as the vegetation index) is very low, from 0 to 0.1). Class 2: medium environmental danger (dredging sites, overburden and enclosing rock tailings), where the NDVI is also low, from 0 to 0.17, and the adjacent plots (where NDVI is a little higher, 0.3–0.6). The typical characteristics of Class 1 include pollutants belonging to danger classes 1 and 2 and featuring high migration ability, low protection of natural components, absence of protective drainage, and industrial facilities located in settlements. The industrial facilities presenting medium environmental danger (dredging sites, overburden and enclosing rock tailings) are characterized by storing toxic pollutants outside settlements.

Due to the rapid development of mining production in the previous century, there is a large ingress of chemicals in the environment, i.e. Environmental pollution. The industrial pollution of the environment has three main components: 1. Technospheric contamination sources that generate technogenic substances that pollute the environment; 2. Transit media that take, transform, and partially change technogenic substances; 3. Deposit environments—the biota and soil cover that accumulate pollutants.

At a closed-down mining facility, developed technogenic deposits and wastes generated during resource extraction present the greatest danger to the biota. The original vegetation of the gold mine plot was practically eliminated. The overgrowth is below satisfactory, with the dominance of weeds and pioneering species. The biodiversity and bioproductivity are significantly reduced. Note that the regeneration plant communities are formed by the surrounding landscape types,

The mineralogical analysis of washings (provided by Yatlukova) showed that the light fraction is represented by quartz and micaceous minerals. The heavy fraction is mainly composed of magnetic iron, hematite, and sulfides (pyrite, vermillion, and sometimes galena). Limonite after pyrite is widespread. We also observed ilmenite, zircon, chromite, grossularite, and anatase. Using spectral analysis, we found that some washing fractions had significant percentages of certain substances: Pb—0.2–0.3%, Cu—0.02%, Zn—0.03–0.06%, Sb—0.02%, As—0.1%, Sn—0.03%. To assess the wastes in the tailings of overburden and enclosing rocks, we determined the content of active forms of some chemical elements in them. We found a significant amount of toxic compounds of the following elements: Pb, Cu, As. Besides, the wastes are highly contaminated with technogenic and natural mercury (vermillion). Since the tailings and other wastes are not isolated from water systems, they harm the nearby ecosystems.

We believe that the placer gold development areas in the Far Eastern Federal District, where structural changes in industrial production are accompanied by a period of depression, poor environmental conditions, and the reduced quality of life of the residents, have to be redeveloped, including restoration and recreation activities or forest plantations. These areas can be classified as nature and mining systems.

The set of restoration activities has to provide a favorable habitat through the adjustment of atmospheric composition and cleanliness, feasible terrain formation, deceleration of erosion, controlling the water regimen, and other habitat parameters. Thus, the goal of restoration is to ensure the environmental safety of the mining facility territories.

Research on the restoration of the productivity of lands disturbed by mining activities has been carried out in the region in question for many years. Out signature technologies have been deployed, and their innovative features have been confirmed by the patents of the Russian Federation [18]. An example of this is using bioremediation, namely microorganisms, i.e. the inoculation of the substrate, toxic soils with a set of active adapted microorganisms, fungi, higher plants, and plant residues. Our research confirms the viability of using endomycorrhiza fungi and their repeated addition to the restored layer, as well as the mixing of this substrate with straw and mycelium (e.g. oyster mushroom), as well as biohumus, etc. At the same time, the environmental soil functions are restored, and a sustainable plant community is established in a very short term (one vegetation season), which is very important. This is largely promoted by microbial communities, mushroom substrates, and higher plants. This technology was used to restore about 18 hectares within the mining activity natural area created by the operation of the Kerbinsky gold mine. We used the following algorithm to rehabilitate the lands disturbed by mining in the Far Eastern region through forest restoration:

1. We developed a comprehensive territory rehabilitation plan where forest restoration is achieved by the restoration of the area's hydrology and hydrogeology and the recycling of all types of wastes including those of the timber industry;
2. We developed the procedures for forest restoration project activities;
3. The comprehensive plan and the procedures were approved on the regional level as regulatory documents (laws);
4. We suggested a method of selecting alternative uses for natural and mining territories, e.g. recreational purposes;

5. We rationalized the selection of plants for forest restoration;
6. We developed a flow diagram for forest restoration within one production process (extraction-overburden removal-waste disposal-restoration) in complete accordance with the Subsoil Law, the Forest Code of the Russian Federation, and the Land Code of the Russian Federation.

Currently, a complex plant community has formed at the location. This is confirmed by the inventory of the restored lands that showed that over the past 35 years from the restoration of the disturbed lands, a sustainable plant community has formed and the environmental functions of the soil on technogenic formations have been recovered due to the thick organ-accumulation layer (up to 5 cm) that formed during this period. The results of the environmental research on forest tree stand in restored and unrestored areas show that the productivity in the restored areas is significantly higher. Note that the new soil profile features a thick organ-accumulation layer that reflects a quality shift in some crucial environmental functions of the soil. We suggested criteria and indicators for the transfer of restored plantations to the forest reserves.

Thus, the performed assessment of the current state of forest ecosystems and environment within the impact area of the nature and mining technogenic system, including a set of geobotanical, soil, laboratory, forest management, environmental, and other research works providing data on the properties of disturbed land substrates of Class 1 (high environmental danger: gold concentration wastes, filler beds) and Class 2 (medium environmental danger: dredging sites, overburden and enclosing rock tailings), as well as adjacent areas, made it possible to perform restoration in 1988 using the bioremediation method. It was concluded that the forest communities newly established on the disturbed lands, that were designated as placer gold deposits in the previous century, shall be reclassified as forested lands suitable for various economic activities.

Note that the scale of forest reduction in the Far Eastern region has long been threatening. This reduction is mainly caused by human activities including mineral development and resource extraction that affect the entire biosphere but also timber procurement. Simultaneously, the demand for wood pulp and fuel, sawn timber, paper, etc. is constantly growing. Natural forests are cut down and it is necessary to quickly restore forests to satisfy the material needs of humans and create a favorable environmental situation in the areas previously disturbed by economic activities. The only way to preserve wild forests is to reduce the consumption of timber produced from them by reducing the consumption of sawn timber and paper and protecting the remaining natural forests from felling and transformation into agricultural lands, as well as conducting research on forest plantation building within nature and mining technogenic systems at previously restored sites.

The research was carried out on the plot of FEFRI in the Kavalerovskoye forest district of the Primorye territory, where young trees, including poplar, and shrubs were planted in the spring of 2022. The experiment used various bioactivators (huminol, amber acid, kornevin, and phototrophic bacteria). The inventory taking of the plantation in the autumn of 2023 showed the following: of all 188 trees and shrubs (larch and pine saplings, poplar scions of various kinds, elm-tree seedlings) planted with various bioactivators (kornevin, phototrophic bacteria, amber acid), 83% of the plantations survived, and 17% (32 items) died. Pine and spruce saplings showed good results. The best result in terms of diameter growth was achieved with kornevin, then phototrophic bacteria, and then amber acid. Sapling height showed another sequence: phototrophic bacteria came first, then kornevin, and then amber acid. The highest poplar scion rooting rate was observed for the Pioneer variety fertilized with huminol. Their height by the end of the first vegetation period reached 40 cm.

Thus, the novelty of this research lies in confirming the possibility of the successful growing of forest plantations on restored plots within nature and mining systems using bioactivators (biologically active substances). In the future, we are planning to continue working on this problem.

The research on the problem of the environmental and recreational potential of mining lands in the Far Eastern Federal District and Khabarovsk Krai in particular is currently at initial stages [19]. The problems of designing recreation activities in areas disturbed by mining companies have not been studied. There are no economic assessment methods for the ecosystem changes due to open-cast mining, and there are no scientific bases to optimize its impacts. There is practically no experience in comprehensive problem solving for the improvement of the environment and organization of recreational areas in mining lands in Russia as a whole, and the Far Eastern region in particular, including Khabarovsk Krai. Currently, there is an urgent need to solve specific problems of rational environmental management, such as the restoration and productivity recovery of ecosystems in technogenic landscapes of mining facilities so that they can be used for recreational purposes. We found that the problems of cultural, historical, industrial, environmental, and learning recreation, as well as organizing trips to old industrial monuments and using the developmental specifics of mining area societies for these purposes have not been studied. Doing this can give Khabarovsk Krai a boost in social and economic development, and the introduction of rational environmental management and recreational woodland management can help solve environmental problems.

Currently, the Far Eastern Federal District is experiencing an increased interest in recreational development that results in the design and operation of tourist routes and infrastructure facilities (e.g. Solnechny, Khabarovsk Krai). Solving this problem is required for the facilitation of the sustainable development of the region in terms of the rational development of its recreational resources while also preserving and restoring vulnerable ecosystems. Recreational research is useful for state-owned and private companies, environmental protection agencies, as well as researchers working with the environment and natural resource management.

The comparative analysis of the environmental conditions near the shut-down Kerbinsky gold mine shows that this facility is a perfect location for a recreational area within a mining development territory. Firstly, some environmental improvement activities have already been performed here (such as the restoration of disturbed lands after mining works), and secondly, the area in question has a great natural recreation potential, making it perfect for recreational development along with the adjacent plots. Besides, we assessed the climate factors and forest suitability which resulted in identifying four groups of parameters (attractiveness, sanitary and hygienic properties of the forest, comfort of the territory, and biogeocenosis sustainability) that helped characterize the forest resources of the recreation area in question as sufficiently suitable (Ir = 46.8 points), and recreations zones as partially suitable for recreation (Iz = 38 points). The parameters of forest recreation suitability (FRS) on various functional significance levels [20] should be evaluated for the organization of learning, walking, hiking, and picnicking recreation and partially for the organization of sports tourism.

We used the suggested indicators on two levels:

1. To assess a 3000-hectare Class 1 recreation area located 25 km away from Briakan. The territory was assessed as a single area comprising three allotments: S1 is 800 ha, S2 is 300 ha, and S3 is 1900 ha near the Kerbi River, Munali River, Dier River, Levy Dier River, Pravy Dier River, Maly Dier River, and Kondya river (the Fig. 1);
2. To assess a 10-hectare Class 2 recreation area near Briakan. This area includes a 1-ha restored plot and 9 hectares of forested land.

The suggested system includes a list of parameters that make it possible to select an optimal approach to the integral assessment under specific conditions. Each of the parameters was assessed in absolute units and points. Based on the permitted recreational loads and digression stages, we calculated the recreation capacity of the territory for two plots: Ez = 23 people per day, Er = 1920 people per day [20, 21].

During the zoning of the territory, a Class 1 recreation zone was established for recreational purposes, and we also identified three nature and mining zones (NMZ) that can be used for recreation (Classes 2 and 3) or cannot be used for this purpose. We believe that the Class 1 recreation area, as well as Classes 2 and 3 NMZ with low industrial impact levels, can be recommended for tourism. During stage 1, we suggest the following routes for the Class 1 recreation area:

1. climbing the Samotsvety summit; and
2. hiking to the Dier waterfalls.

It is required to observe the permitted recreation loads (PRL) for the routes to prevent their digression.

Thus, the forest resources of the Class 1 recreation area can be characterized as suitable for recreation, while those of the Class 2 recreation area are only partially suitable.

4 Conclusion

The rapid development of timber and mining industries in the previous century associated with tree cutting and the development of placer gold deposits in the Far Eastern Federal District and Khabarovsk Krai in particular caused significant damages and

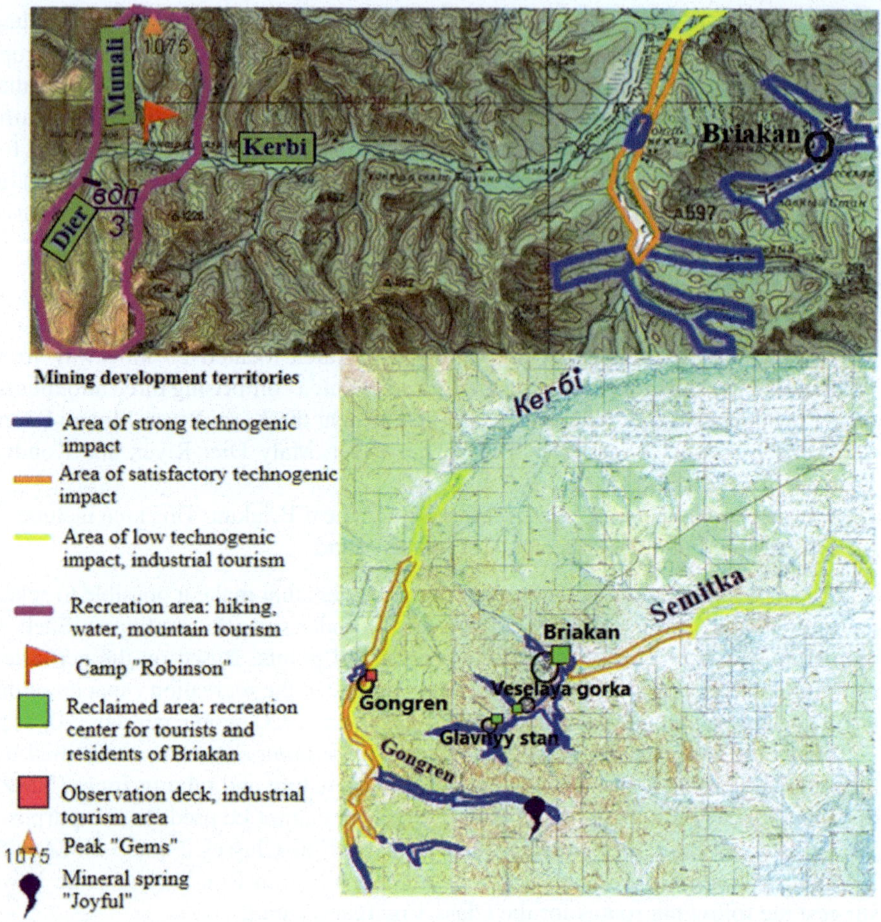

Fig. 1 The mining territories

technogenic pollution of the environment that resulted in an environmental crisis. In the region, a nature and mining technogenic system was formed that harms the environment and threatens the biosphere as a whole and its components.

The geoenvironmental assessment of the region in question using the remote sensing of Earth, physical and chemical methods, and bioremediation helped define the current environmental situation in the region as critical. The researchers developed proposals for the all-around sustainable development of the region designed to solve some of the environmental problems by restoration, plantation building, and recreational woodland management. Their novelty is confirmed by the patents of the Russian Federation.

Acknowledgements This research was carried out with funds from state assignment for applied research in 2023 and 2024 at plantation No. 3-I23

References

1. Petrov AI, Kotova VS, Osipenko RA, Zalesov SV (2023) Woodland management for the recovery of placer gold development sites. Forests of Russia and economy in them 2(85):16–23.
2. Petrov AI, Zalesov SV, Kotova VS (2023) The efficiency of growing common pine on dredging tailings. Siberian Journal of Forest Science 3:15–20.
3. Drozdova IV, Timofeeva SS, Timofeev SS (2023) The prospects of the phytomining of gold from the washed sands of placer mines. Mining Informational and Analytical Bulletin 5:99–111. doi:10.25018/0236_1493_2023_ 5_0_99.
4. Rozinkina EP, Bashegurov KA, Korchagin IE, Petrov AI, Morozov AE, Zalesov SV (2023) The methodological approaches to the restoration of disturbed lands. Relevant problems of the timber complex 63:71–75.
5. Belov LA, Bunkova NP, Zalesova ES (2020) Recreational plantation building. Bulletin of biotechnology 3(24):1–7.
6. Zalesova ES, Nesterova EN, Rubtsov LI, Vibe EP, Kartashova TYu (2017) Establishing sustainable recreation-compelling plantations in forest parks. Bulletin of biotechnology 2(12):7.
7. Shtukin SS, Volovich PI (2020) Organizational and technological stages, advantages, and drawbacks of plantation-based timber production. Proceedings of BSTU. Series 1: Forestry. Environmental management. Reprocessing of renewable resources 1(228):84–90.
8. Ding K, Zhang Yu, Wang Li, Ge Siyu, Zhang Yiman, Yang Qi, Huang Huahong, Tong Zaikang, Zhang Junhong (2023) Forest conversion from pure to mixed Cunninghamia lanceolata plantations enhances soil multifunctionality, stochastic processes, and stability of bacterial networks in subtropical southern China. Plant Soil. 488:411–429. https://doi.org/10.1007/s11104-023-05983-y.
9. Luiz Marcelo Brum Rossi, Celso Paulo de Azevedo, Cintia Rodrigues de Souza, Roberval M. B. de Lima. Potential forest species for plantations in Brazilian Amazonia. https://www.fao.org/3/xii/0537-b1.htm. Accessed 01.12.2023.
10. Riccioli F, Marone E, Boncinelli F, Tattoni C, Rocchini D, Fratini R (2019) The recreational value of forests under different management systems. New Forests, pp 345–360. https://doi.org/10.1007/s11056-018-9663-3
11. Musin HG, Denisov SV, Khalilov II, Minnichanov AR (2020) Differentiated assessment of the recreational potential of forests in the management of recreational forests. International Scientific-Practical Conference “Agriculture and Food Security: Technology, Innovation, Markets, Human Resources” (FIES 2019) 00222:5. https://doi.org/10.1051/bioconf/20201700222
12. Vernadsky VI (1989) Biosphere and noosphere. Nauka, Moscow, p 261.
13. Motorina LV (1978) The program and methodology of studying technogenic biogeocenoses. Nauka, Moscow, p 227.
14. Leonenko AV, Usikov VI, Krupskaya LT (2022) The remote analysis methodology for abandoned placer gold deposits (as exemplified by the Kerbinsky district). Izvestiya Tula State University. Sciences of Earth 1:98–113.
15. Lupyan EA, Konstantinova AM, Kashnitsky AV, Ermakov DM, Savorsky VP, Panova OYu, Bril AA (2022) The possibility of organizing a long-term remote monitoring of large industrial contamination sources to assess their impact on the environment. Current Problems In Remote Sensing Of The Earth From Space 19(1):193–213.
16. Cherepanov AS (2009) Spectral properties of vegetation and vegetation indices. In: Cherepanov AS, Druzhinina EG (ed) Geomatika 3:28–32.
17. Rouse R JWJr, Haas JH, Schell DA, Deering W (1974) Monitoring Vegetation Systems in the Great Plains with ERTS. Environmental Science.
18. Filatova MYu, Krupskaya LT, Bubnova MB, Leonenko AV (2023) The assessment of the environmental situation at gold mining facilities (to justify forest plantation building). Mining Informational and Analytical Bulletin 8:27–44.

19. Sapozhnikov AP, Glovatskaya OA (2005) On the environmental tourism. In: Sapozhnikov AP, Glovatskaya OA (ed) Geography and Natural Resources 1:158.
20. Sapozhnikov AP (1985) The problems of studying the recreational comfort of poorly-developed forest territories In: Sapozhnikov AP (ed) Lesnoye khozyaystvo 5:37.
21. Krupskaya LT, Leonenko AV, Morin VA, Gul LP, Orlov AM, Golubev DA, Onishchenko MS (2014) The environmental assessment of forest ecosystem in mining areas and evaluation of the functional economic use fitness of forests in Khabarovsk Krai. In: Problems of sustainable development in the management of the forests of Siberia and the Far East. The proceedings of the all-Russian conference dedicated to the 75th anniversary of the establishment of the Far Eastern Agricultural Research Institute with international participation, Khabarovsk, pp 435–442.

Dynamics of Green Areas in St. Petersburg and Adaptation to Climate Change

D. Alexeev, L. Dmitricheva, E. Mikhteeva, and A. Kanukhina

Abstract This paper explores the important role that the green infrastructure, i.e. the greenspace network, of a city can play in adapting for climate change. Significant differences in the overall greenness of Russia's largest urban agglomerations have been demonstrated. It uses the conurbation of Saint-Petersburg (Russia) as a case study site. The analysis of the dynamics of green spaces was carried out on the basis of satellite images of Landsat 7 and Landsat 8 for the period 2002–2018. A normalized vegetation index (NDVI) was used for quantification. Maps of the spatial distribution of NDVI were constructed over the observation period. Various climate change scenarios are also considered. Various risks for St. Petersburg in connection with possible climate changes are considered.

Keywords Climate change · Urban planning · Adaptation

1 Introduction

The problem of urban development is formulated as one of the seventeen goals of sustainable development for the near future. Namely, ensuring the openness, security, resilience and sustainability of cities and towns. The choice of a priority direction for improving the environmental situation in cities is largely determined by many factors. Among them are the complexity and eccentricity of the issues under consideration, the knowledge of the factors in the formation of the urban environment, the availability of reliable data on the ecological state and quality of the urban environment and the forecast of its possible changes under the influence of certain factors.

Among the whole complex of environmental problems of a modern city, landscaping of the urban area is in the first place. Green spaces, on the one hand, are an organic part of the planning structure of the city, and on the other hand, they perform

D. Alexeev (✉) · L. Dmitricheva · E. Mikhteeva · A. Kanukhina
Russian State Hydrometeorological University, 79, Voronegskaya, Saint-Petersburg 192007, Russia
e-mail: d.alekseev@rshu.ru

L. V. Radionova and D. V. Ulrikh (eds.), *Advances in Ecology and Environmental Engineering*, Springer Proceedings in Earth and Environmental Sciences,
https://doi.org/10.1007/978-3-031-64423-8_3

a number of important functions, for example, reducing the level of air pollution and the urban heat island [1].

According to the World Health Organization, the lack of physical activity, associated with poor access to recreational areas, may contribute to up to 3.3% of premature deaths. To ensure that urban populations have the opportunity for recreation, the availability of green spaces should be at 9.5 square meters per resident within a 15-min walking distance from their place of residence.

The study of the state of green spaces against the backdrop of rapid changes in the urban environment is an urgent task. The research study is focused on the St. Petersburg agglomeration. It is characterized by an extremely high rate of residential development for Russia. St. Petersburg is located on an area of approximately 8.8 thousand km^2; the population is about 6.0 million people: 5.23 million people in the city, 0.77 million people in the Leningrad region; population density is 682 people/km^2. Characteristically, the territory where 87% of the population of the agglomeration lives is 16% of its area. In accordance with the forecasts of urban development of the territory of the agglomeration, by 2030 the population within its boundaries will reach 7.3 million people, and by 2050—8.7 million people according to Committee information.

The purpose of the research work is to study the green spaces of the city of Saint-Petersburg according to satellite images. To achieve this goal, the following tasks were solved: analysis of the physical, geographical and climatological features of the research area; creation and processing of distribution maps of the normalized vegetation index for the study period; assessment of the dynamics of the green areas for various districts of Saint-Petersburg.

2 Materials and Methods

In the course of work, based on the data of multispectral satellite images Landsat 7 and Landsat 8 for the 2002 and 2018 years (Table 1), an analysis of the spatial distribution and assessment of the dynamics of green areas on the territory of St. Petersburg and its administrative-territorial units (intracity districts) was carried out. The work is carried out using two GIS packages: System for Automated Geoscientific Analyzes (SAGA GIS) and QGIS.

Table 1 Landsat 7 and Landsat 8 scene characterization

Scene ID	Satellite	Processing time	Scale (m)
LE07_L1TP_185018_20020711_20170129_01_T1	Landsat 7	2002–07–11/08:59:31	30
LC08_L1TP_185018_20180731_20180814_01_T1	Landsat 8	2018–07–31/09:10:24	30

NDVI (Normalized Difference Vegetation Index) is a normalized difference vegetation index, which was used to quantify the green areas of St. Petersburg. The index was described by John Rouse at el. [2] in 1973, is a simple quantitative indicator of phytomass:

$$NDVI = \frac{NIR - RED}{NIR + RED} \tag{1}$$

In this formula: RED is the reflection coefficient in the red spectral zone, NIR is the reflection coefficient in the near infrared spectral zone.

3 Results

According to the master plans of various Russian cities, the availability of green areas per capita varies from 4 to 135 m^2 (depending on the city and district). The lowest indicators of green space are in Chelyabinsk (2.1–5.8 m^2), Perm (4.0–10.0 m^2), Rostov-on-Don (6.7–10.0 m^2), and Volgograd (10.0 m^2). The maximum values are in Yekaterinburg (38.6 m^2) and Novosibirsk (88.0–135.0 m^2).

Considering the total area of all woody vegetation, the indicators in all cities are significantly higher. In Rostov-on-Don, Moscow, and St. Petersburg, the provision of woody vegetation per person is less than 100 m^2, in Volgograd, Perm, Samara, Chelyabinsk, Nizhny Novgorod, and Novosibirsk it is 100–200 m^2, while in Voronezh, Krasnoyarsk, Ufa, Omsk, Kazan, and Yekaterinburg it exceeds 200 m^2 [3].

The primary core of St. Petersburg green framework consists of specially protected natural areas. In St. Petersburg, there are numerous such areas, and they play a crucial role in preserving the city natural wealth and ensuring ecological sustainability. For example, swamps act as a natural filter, purifying the atmosphere from pollutants and providing clean drinking water. Moreover, many specially protected natural areas in St. Petersburg serve as migration sites for various animal species. Preserving these areas and integrating them into the city climate change adaptation system is an important task of sustainable development.

St. Petersburg has 17 regional specially protected natural territories, including 10 nature reserves and 7 natural monuments. The total area is 9212 hectares, accounting for 6.4% of the city square. Also St. Petersburg has federal dendrological parks and botanical gardens, such as the Botanical Garden of Peter the Great, the Botanical Garden of St. Petersburg State University, and the Botanical Garden of St. Petersburg State Forest Technical Academy named after S.M. Kirov.

In the literature, there are few assessments showing the proportion of a city's territory that should be covered by tree canopy to ensure its sustainability, and these assessments relate to specific effects of ecological services. For instance, in order to effectively mitigate the urban heat island effect, no less than 30% of the city's area should consist of unsealed surfaces (including "green roofs," "green walls," etc.), and their location in the central zone of the city with high building density is extremely important. In half of the cities considered above, these criteria are not met, not only in the largest cities located in forestless areas, as might be expected, but also in cities situated in relatively favorable conditions such as Kazan and Chelyabinsk.

The NDVI index is recommended for global monitoring of vegetation. Scale helps to compensate for changing lighting conditions, surface slope, exposure, and other external factors [4].

The following scale has been used for classification: NDVI = 0.49–1—high degree of vegetation development, good condition, NDVI = 0.38–0.48—medium degree of vegetation development, satisfactory condition, NDVI = 0.30–0.37—low degree of vegetation development, depressed state, NDVI = 0.18–0.29 open soils, very low degree of vegetation development, highly depressed state, NDVI = −0.25–0.17—artificial materials, NDV < −0.25—water bodies.

Maps of the spatial distribution of NDVI for 2002 and 2018 year were built based on this classification and satellite images. For various districts of St. Petersburg and the city as a whole, the area of green spaces was calculated according to the above classification (1).

As a result of a visual assessment of summer imagery data for 2002–2018, it was found that the distribution of relative phytomass within the urban area of St. Petersburg is extremely heterogeneous. The highest values of the index (> 0.49) are observed on the territory of the most remote areas from the administrative center of the city, which are located on the border with the Leningrad Region (Table 2).

According to the data obtained for 2002, the areas with the highest indicators of the area of the territory, characterized by the values of the NDVI index in the range of 0.49—1 (high degree of vegetation development, good condition) include such areas as Kurortny (137 km^2), Pushkinsky (136 km^2), Petrodvortsovy (74.1 km^2), Kolpinsky (56.4 km^2) districts. The minimum values of the area of the territory with index values within 0.49—1 are characterized by the Admiralteisky (0.23 km^2), Central (0,39 km^2) and Vasileostrovsky (0.68 km^2) districts.

The maximum values of the areas of the territory, characterized by the values of the NDVI index in the range of −0.25–0.17 (artificial materials), are observed in the following districts of the city: Nevsky (26.4 km^2), Primorsky (25.7 km^2), Moskovsky (24.3 km^2), Kirovsky (23.9 km^2), Vyborgsky (22.4 km^2).

In 2018 the areas with the highest indicators of the area of the territory, characterized by the values of the index in the range of 0.49–1 (high degree of vegetation development, good condition) include areas such as Kurortny (143 km^2), Pushkinsky (121 km^2), Petrodvortsovy (71.4 km^2) district. However, in the last two years (since 2002) there has been a decrease in area. The minimum areas of the territory with index values of 0.49–1 are characterized by the Admiralteisky (0.34 km^2), Central (0.52 km^2) and Vasileostrovsky (0.90 km^2) districts. The highest values of the areas of the territory, characterized by the values of the NDVI index −0.25–0.17 (artificial materials), are observed in Pushkin (35.6 km^2), Primorsky (34.5 km^2), Nevsky (33.5 km^2), Kolpinsky (33.1 km^2), Vyborgsky (31.7 km^2), Moskovsky (31.3 km^2) districts of the city. The maximum-recorded values of the area of objects "artificial materials" in 2002 did not exceed 27 km^2. The number of districts that have objects with an

Table 2 Distribution of the surface area (km^2) for various NDVI index intervals according to the classification (numerator—2018, denominator—2002)

District	Area, km^2				
	−0.25 – 0.17	0.18 – 0.29	0.3 – 0.37	0.38 – 0.48	0.49 – 1
St. Petersburg	*398* 291	*125* 125	*111* 119	*254* 288	*544* 620
Admiralteysky	*11.6* 11.5	*0.58* 0.79	*0.26* 0.32	*0.32* 0.35	*0.34* 0.23
Vasileostrovsky	*13.9* 11.2	*1.62* 2.10	*0.92* 0.99	*1.02* 0.78	*0.90* 0.68
Vyborgsky	*31.7* 22.4	*10.3* 10.7	*8.27* 9.13	*19.1* 24.1	*46.0* 50.6
Kalininsky	*21.1* 14.9	*5.76* 6.26	*3.74* 4.92	*4.54* 7.05	*4.71* 6.90
Kirovsky	*26.7* 23.9	*5.05* 6.17	*3.12* 3.93	*3.63* 4.35	*3.01* 3.34
Kolpinsky	*33.1* 13.9	*12.8* 7.12	*9.66* 6.75	*14.4* 19.8	*31.3* 56.5
Krasnogvardeysky	*26,9* 18.9	*4,65* 7.75	*6.18* 4.95	*7.68* 8.37	*9.60* 17.0
Krasnoselsky	*20.5* 12.4	*9.21* 9.20	*8.72* 8.27	*17.9* 19.5	*34.2* 42.4
Kronshtadtsky	*7.79* 8.01	*1.73* 2.22	*1.13* 1.54	*1.89* 2.50	*6.81* 4.47
Kurortny	*12.1* 11.5	*10.6* 9.90	*17.2* 23.8	*85.2* 91.5	*143* 137
Moskovsky	*31.3* 24.4	*7.63* 8.92	*6.22* 6.32	*12.4* 11.8	*15.9* 23.5
Nevsky	*33.6* 26.4	*8.51* 9.64	*5.04* 6.37	*5.55* 8.13	*5.49* 8.45
Petrogradsky	*11.5* 10.3	*1.46* 2.18	*0.87* 1.27	*1.24* 1.83	*1.93* 2.09
Petrodvortsovy	*12.5* 8.39	*7.52* 6.46	*6.99* 6.59	*14.0* 16.2	*71.5* 74.2
Primorsky	*34.6* 25.8	*10.2* 9.35	*7.56* 7.85	*13.4* 18.6	*43.5* 48.9
Pushkinsky	*35.7* 17.6	*20.8* 19.2	*20.5* 21.1	*46.1* 46.9	*121* 137
Frunzensky	*19.8* 16.1	*5.50* 6.16	*3.69* 4.03	*4.61* 5.66	*4.14* 6.05
Central	*13.5* 13.5	*0.70* 0.97	*0.43* 0.50	*0.53* 0.52	*0.52* 0.39

area of more than 30 km^2 related to "artificial materials" on their territory reached six by 2018.

The distribution of various surfaces in St. Petersburg in 2018 is characterized as follows: in percentage terms (excluding water bodies), the leaders are the objects "high degree of vegetation development, good condition"—38%, in second place with 29% are "artificial materials", "average degree of vegetation development, satisfactory condition" is 16%. "Low vegetation, depressed" and "Open soils, very low vegetation, severely depressed" score 8% and 9% respectively.

During the study period, a significant decrease in the NDVI indicator ranging from 0.49 to 1 was recorded in the Kolpinsky (Fig. 1), Krasnoselsky and Pushkinsky districts. A decrease in the range of 0.39–0.49 was recorded in the Kurortny, Primorsky, Pushkinsky districts. A change in the NDVI value of about 0.30–0.37 is observed in the Primorsky district. There is a relatively slight decrease in the areas of green spaces in Vyborgsky, Kalininsky, Kolpinsky, Krasnogvardeisky, Krasnoselsky, Kurortny, Moskovsky, Nevsky, Petrodvortsovoe. An increase in the NDVI index within values from 0.49 to 1 was recorded in the Kurortny, Primorsky, Petrodvortsovy, Moskovsky districts.

The increase in objects of the category "artificial materials" from 21 to 29% occurred over the period from 2002 to 2018. The category "average degree of development of vegetation, satisfactory condition" decreased by 3% of the total area of the city. The proportion of objects "high degree of development of vegetation, good condition" decreased by 5% compared to 43% in 2002.

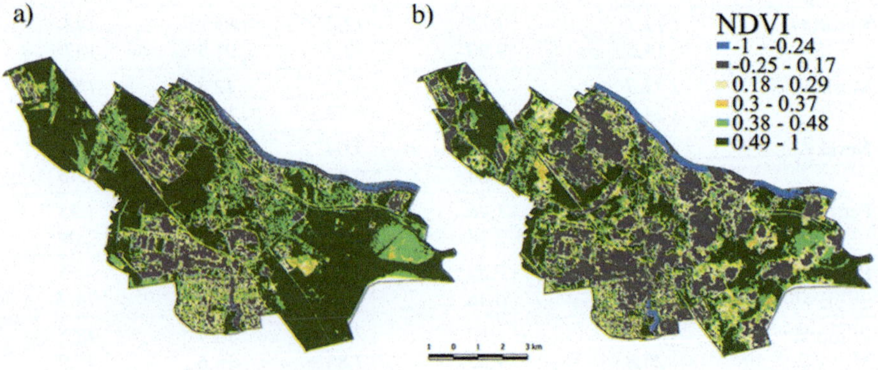

Fig. 1 An example of the spatial distribution of the NDVI index in the Kolpinsky district: **a** 2002, **b** 2018

4 Discussion

According to the estimates of the World Health Organization in Europe, climate change annually causes between 1 to 10% of deaths in the older age groups, and globally results in over 150,000 additional deaths and 5.5 million years of disability annually. This constitutes 0.3% of the total number of fatalities and 0.4% of the overall disability-adjusted life years. By the mid-twenty-first century, further increases in climate-related mortality are expected, by approximately 1–1.5%. The economic damage from additional mortality due to climate change worldwide varies widely, ranging from 6 to 88 billion dollars annually.

According to the assessments of Roshydromet, climate warming in Russia is occurring approximately 2.5 times more intensively than the global average: during the period 1976–2016, it is estimated as 0.45 °C per decade. The highest rate of annual average temperature increase is observed on the coast of the Arctic Ocean, especially in the Asian part of Russia (APR) (on the Taimyr Peninsula—more than + 0.9 °C per decade). In spring and autumn, the maximum warming is observed on the coast of the East Siberian Sea, and in winter—in the northwest of the European part of Russia (EPR). In summer, the fastest warming occurs in the south of the EPR. In spring, intensive warming is observed in Western Siberia (+0.74 °C per decade) and Central Siberia (+0.77 °C per decade). In Eastern Siberia, rapid warming occurs in spring (+0.72 °C per decade) and autumn (+0.74 °C per decade); in summer—in the Southern and Central Federal Districts (+0.76 °C per decade and + 0.70 °C per decade). The least warming on average per year is noted in the south of Western Siberia. Warming is observed in Russia's entire territory in summer and autumn; however, in Siberia (in the south in summer and in the center in autumn), growth is very weak in some places. For winter over the specified period, the trend on average across Russia is positive (0.32 °C per decade), but statistically insignificant. The increase in winter temperature ceased in the mid-1990s, after which a decline was observed; however, in the last five years, a trend towards an increase has emerged.

St. Petersburg has difficult climatic conditions hindering the development of green landscaping. The city is significantly affected by several climatic features, including high humidity, a prevailing wind regime, limited sunlight (approximately only 62 sunny days per year), and the influence of urban microclimatic conditions. Airborne particles such as dust and smoke diminish daytime solar radiation, impeding the cooling of the Earth's surface during the nighttime.

The temperature patterns in Saint Petersburg are primarily shaped by two factors: the radiation regime and atmospheric circulation. The arrival of Atlantic air masses, predominantly from South-Western and Western directions, typically brings windy and cloudy weather. On the other hand, the impact of the radiation factor becomes more pronounced during the formation of anticyclones, especially in clear and windless conditions.

The average annual temperature in Saint Petersburg according to long-term observations is 5.6 °C. The coldest months of the year are December and February, with average temperatures of −7.9 and −10.4 °C. The warmest month of the year is July with average daily temperature of 19.5 °C. The total change of mean annual air temperature in Saint Petersburg for the entire period of temperature measurement (from 1752) was 2 °C with the average change of 3.4 °C in winter and 0.5 °C in summer. However, over the last 30 years the average annual temperature increased by 1.7 °C: by 3.5 °C in winter and 1.5 °C in summer [5].

The twentieth century shows a clear trend to lower numbers of extremely cold days, amount of warm days indicates a much smaller change. The 1990s have been exceptionally warm but slightly exceed some earlier decades such as the 1930s [6].

The air temperature variation in Saint Petersburg exhibits a notable positive trend, influenced by both natural temperature variations and the emergence of a local "disturbance" in the thermal regime known as the Urban Heat Island. It's important to highlight that the majority of recorded minimum temperature values in Saint Petersburg date back to the 18th to the early twentieth century, whereas most of the maximum temperature values are observed in the beginning of the twenty-first century. Nevertheless, attributing the temperature increase solely to recent decades is challenging, as the Urban Heat Island phenomenon did not exist in the eighteenth-twentieth centuries.

Saint Petersburg experiences an average of 177 cloudy days per year, considering overall cloud cover. On sunny days, the average duration of sunshine varies from 10.1 h in June to 2 h in December. Saint Petersburg has an average of 177 cloudy days per year, in terms of the general cloud cover.

Since the annual amount of precipitation is about 200–250 mm and it exeeds the evaporation of moisture, Saint Petersburg is characterized by high humidity, about 80% (60–70% in summer, 83–88% in winter). The number of days with a relative humidity of at least 80% ranges from 140 to 155.

Considering its geographical location, Saint Petersburg is situated in a region characterized by excessive moisture. The average annual rainfall in the city over the past 30 years has been recorded at 653 mm. Cyclonic activity predominantly influences precipitation patterns in Saint Petersburg.

The distribution of precipitation is not uniform throughout the year, with the majority (67%) occurring during the warm period from April to October, reaching its peak in July and August. Conversely, only 33% of the precipitation falls during the cold period, with a minimum in February and March.

From December to February, solid precipitation is prevalent, while liquid precipitation dominates from May to October. Notably, the months of July and August experience exclusively liquid precipitation, and mixed precipitation, such as sleet or snow with rain, does not occur during this time.

In the northern part of the city and in the north of the suburban area, the annual rainfall is greater than in the central areas, by about 11% and by more than 20%, respectively. On the contrary, in the mouth of the Neva, the annual rainfall is lower than in the center by 5–7%. Even less precipitation falls on the coast of the Gulf of Finland (in the village of Lisy Nos, Petrodvorets, Strelna) and on the Islands (in Kronstadt), where the difference with the center reaches 8–9%.

The number of days with precipitation ≥ 0.1 mm ranges from 155–160 on the coast and Islands of the Gulf of Finland to 180–200 in the central and eastern parts of the city. Overall, the number of days with precipitation increased by 7.4 days over the period of 1936–2006, or by 32% from the long-term average for this period. One of the main characteristics of precipitation is its intensity. In the cold period of the year, their intensity is low, being 0.2–0.4 mm/h on average. In the summer months, the intensity increases to 1.1–1.3 mm/h due to heavy rainfall [5].

According to the Köppen climate classification [7] St. Petersburg is located in the Dfb (humid continental climate) zone. It is a moderately cold climate, with no dry season; in winter there is a stable snow cover; there are conditions for the growth of deciduous forests. The average air temperature of the coldest month does not exceed − 3 °C ($T_{min} \leq -3$ °C), the average monthly air temperature of the warmest month is less than 22 °C ($T_{max} < 22$ °C); while four months of the year the average air temperature exceeds 10 °C ($4\ T_{mon} \geq 10$ °C). The Cfb (oceanic) zone is characterised with a weaker variability of diurnal and annual air termperature. And possible shift from to Dfb to Cfb brings inreased precipitation amounts, cloudiness and higher relative humidity [8].

According to estimates of changes in the average annual and average monthly values of the average, maximum, minimum surface air temperature and precipitation for several scenarios of greenhouse gas emissions (B1, A1B and A2), the increase in the average surface air temperature by the end of the twenty-first century compared with the period 1971–2000 (5.4°C) will be: 2.8°C for scenario B1, 4.0°C for scenarios A1B and A2. Increase in precipitation compared to the period 1981–2010 will be: 228 mm for B1, 244 mm for A1B, 262 mm for A2 [6]. Under hard forecasts of global warming in the second half of the twenty-first century climatic conditions of St. Petersburg will become similar to the following cities: Birmingham (England), Vancouver (Canada), Hamburg (Germany), Glasgow (Scotland), Copenhagen (Denmark) [8].

The Earth's surface's ability to reflect incident radiation influences the thermal balance. The amount of solar energy that remains on Earth depends on the surface albedo. It is considered that the Earth reflects 37% of incoming radiation, but due to the greenhouse effect, this value may decrease, leading to an increase in surface temperature.

In the summer, engineering structures and road surfaces heat up under the influence of solar radiation, and after sunset they continue to be sources of radiation for a long time. The lower the albedo, the more the surface heats up. For example, the albedo of asphalt is 4%, red brick—10%. Trees with a large leaf area give the best effect in reducing temperature. The smaller the leaf, the less thermal energy the plant crown reflects. Plants with a large albedo provide good protection from thermal energy. Albedo of aspen is 61%, Norway maple is 50%.

In winter, the relationship between surface albedo and changes in average temperature in St. Petersburg is associated with the use of deicing materials during snowfall. The color of the melting portion of the snowpack changes from light to dark, more heat is absorbed, and the surface temperature rises. Temperature fluctuations depend on the surface area being treated with reagents, and as a consequence, the difference in temperature fluctuations across city areas can be explained by the diversity of the surface (park areas, water surface, where reagents are not used).

The increase in temperature in recent years has led to a sharp rise in the population of certain pests, which now reproduce several generations per year. Weakening of coniferous trees due to drought contributes to the reduced resilience of coniferous forests and the proliferation of pests. These factors were the cause of the massive die-off of coniferous forests in the Leningrad and Moscow regions from 2010 to 2016. In 2021, for the same reason, there was a recorded outbreak of the mass reproduction of the typographer bark beetle in spruce forests in the Leningrad region and Karelia. The rise in temperature, the likelihood of drought, as well as forest fires, leads to an increase in the duration of the wildfire season. The expansion of the area affected by forest fires will lead to a reduction in the biological diversity of forests, as it results in the decrease of coniferous forest areas and the increase in the area of non-forested clearings and dead plantations.

The correlation between the expansion of habitats and the increase in the population of dendrophagous insects with climate change is well-established. The warming climate has made it possible for the spread of certain exotic pests of woody plants. For example, the proliferation of the emerald ash borer has resulted in the complete devastation of ash trees in the European part of Russia. This pest has now emerged in St. Petersburg, marking the northern boundary of the ash tree's distribution.

Alobf with a possible increase in temperature, the most significant consequences of modern climate change for the agglomeration of St. Petersburg and the Leningrad region with a population of many millions is an increase in the number of "heat waves" [9]. According to the results obtained in the reduction of green areas, local overheating and increased impact of heat waves should be expected. The approaches designed for urban conditions presuppose that Green Infrastructure elements offer a variety of ecosystem services: mitigate the effects of climate change, reduce the "heat island" effect, regulate water flow, preserve biodiversity, contribute to food security, create favorable conditions for recreation, contribute to the formation of cultural identity [10].

5 Conclusion

As a result of a visual assessment of the summer images data for 2002 and 2018 it can be said that the distribution of the relative phytomass within the urban area of St Petersburg is extremely heterogeneous. The highest values of the index 0.49 are observed in the districts as far as possible from the administrative center of the city and located on the border with the Leningrad region. Period of 2002–2018 years is characterized by ultra-fast development of multi-storey buildings in St.Petersburg. It led to a sharp increase in the number and density of the population. This is accompanied by sharp decrease in the provision of the population with green spaces in the vast majority of residential areas. Along with possible climate change, sustainable urban planning and development of the St. Petersburg agglomeration is possible. But only if the factors of global and regional environmental changes are taken into account. The sharp decline in green infrastructure over the past decade, against the backdrop of global changes, could lead to more negative consequences. Reducing climate risks requires fundamental changes in information support, legislative backing, and project-planning decisions in the development of St. Petersburg. The expansion of specially protected areas is likely to play a key role in the sustainable development of St. Petersburg.

References

1. Nizamutdinov TI, Kolesnikova EV, Alexeev DK (2021) Influence of green spaces on the dynamics of air pollution in cities. PNRPU. Applied ecology. Urban development 1:58–73. https://doi.org/10.15593/2409-5125/2021.01.05
2. Rouse JW, Haas RH, Schell JA, Deering DW (1973) Monitoring vegetation systems in the Great Plains with ERTS. In: Third Earth Resources Technology Satellite–1 Symposium. Volume I: Technical Presentations NASA SP-351, Washington, D.C. NASA, pp 309–317.
3. Klimanova OA, Kolbowsky EYu, Illarionova OA (2018) The ecological framework of Russian major cities: spatial structure, territorial planning and main problems of development. Vestnik of Saint Petersburg University. Earth Sciences 63(2):127–146. https://doi.org/10.21638/11701/spbu07.2018.201
4. Lillesand TM, Kiefer RW, Chipman JW (2004) Remote Sensing and Image Interpretation. John Wiley and Sons, New Jersey.
5. Geological and Environmental Risks in St. Petersburg. Practical Recommendations for Climate Change Adaptation http://cliplive.infoeco.ru/. Accessed 15 Nov 2023.
6. Jones PD, Lister DH (2002) The daily temperature record for St. Petersburg (1743–1996). Climatic Change 53(1/3):253–267. https://doi.org/10.1023/A:1014918808741
7. Chen D, Chen HW (2013) Using the Köppen classification to quantify climate variation and change: An example for 1901-2010. Environmental Development 6:69–79.
8. Pavlovskii AA, Menzhulin GV (2018) Perspective estimates of change of natural climatic conditions of the territory of St. Petersburg in the XX century. Proceedings of the Russian State Hydrometeorological University. A theoretical research journal 51:44–57.
9. Malinin VN, Menzhulin GV, Pavlovskii AA (2016) Urban planning of St. Petersburg in the context of modern climate change. Proceedings of the Russian State Hydrometeorological University. A theoretical research journal 43:140–147.
10. Baro F, Chaparro L, Gomez-Baggethun E, Langemeyer J, Nowak DJ, Terradas J (2014) Contribution of ecosystem services to air quality and climate change mitigation policies: the case of urban forests in Barcelona, Spain. Ambio 4:466–479.

Development of a "Smart City" Model

E. A. Prelikova, N. R. Kukshin, and D. V. Nekrasov

Abstract The issue of sustainable development of cities and human settlements is of great importance. Attaining this goal can ensure a balance between the preservation of ecosystems and the development of infrastructure, comfort and well-being of people. Urban development in modern Russia is implemented as a rule by virtue of large urban agglomerations. New cities are almost not being built. The city's ability to transform to new realities is the key to its competitiveness and economic efficiency. However, this is not enough for the city to be considered "smart". It is very important for citizens to feel comfortable in the urban environment, that is, it is necessary to create favorable living conditions, attract new resources for the modernization of urban environment, use them competently, and create innovative jobs. The topicality of the research and the degree of its scientific development provide the basis for the definition of the scientific problem as the need to develop a "smart city" model. In the paper, the authors used methods of system, theoretical analysis, an intuitive-logical approach, a graphical method. The analytical approach allowed us to consider ways to develop smart city models, identify barriers and risks that militate against city smart transformation. The essence of the graphical method was the analytical process of studying the evolution of the city and the development of a "smart city" model.

Keywords Sustainable development · Smart city · Urban environment

1 Introduction

The age of technology has significantly changed people's views and demands concerning the surrounding reality. Information and telecommunication systems have transformed the system of "Human-Nature-Society" relations. Now there is no need to carry a public transport timetable with you, it is possible to order a taxi

E. A. Prelikova (✉) · N. R. Kukshin · D. V. Nekrasov
Southwest State University, 94, 50 Let Oktyabrya, Kursk 305040, Russia
e-mail: elena_prelikova@bk.ru

L. V. Radionova and D. V. Ulrikh (eds.), *Advances in Ecology and Environmental Engineering*, Springer Proceedings in Earth and Environmental Sciences,
https://doi.org/10.1007/978-3-031-64423-8_4

by pressing the phone button. Carsharing has appeared in major cities. Thanks to the public services website, access to many services has become as fast as possible. You can pay a fine, make an appointment with a doctor, request for a certificate and so on without leaving home. According to scientists, in 5–10 years, the predictive analytics system will independently make an appointment for a person for a medical examination according to the data collected by a fitness band on the wrist, based on the medical histories of patients with similar health data [1, 2].

It is obvious that a city should have favorable and safe living conditions for people, be comfortable, technological and, of course, "smart".

2 Relevance, Scientific Significance of the Issue and a Brief Literature Review

The relevance of the work is due to the need to provide people with comfortable and favorable living conditions, the lack of resources for the modernization of the urban environment and the creation of innovative jobs and the lack of technologies that contribute to the improvement of the urban environment and the sanitary and hygienic well-being of the population.

The theoretical foundations of the development of the "smart city" concept and the problems of transformation of the urban environment into a "smart" one are considered in the works of both Russian scientists (A. V. Tebekin, A. A. Egorova, D. E. Namiot, R. V. Gribkov, M. A. Mukhin, K. S. Petrov, E. S. Pityk, A. Yu. Chernoknizhnikova, A. V. Shustova, A. V. Artamonov, V. A. Molchanova) [3–8], and foreign scholars (N. A. Vukovic, V. A. Larionova, P. Morganti, A. Townsend, A. U. Berardi, R. M. Dangelico) [9–11].

The issues of organization of public services and amenities and their implementation are considered in the works of S. N. Bobylev, O. V. Kudryavtseva, S. V. Solovyova [12], D. A. Skorobogaty [13], V. V. Koneva, S. A. Khizhnyak [14], L. A. Pashkova [15], M. P. Starostina, O. A. Urzha [16], R. M. Nikaeva, X. M. Dikaeva [17], O. N. Gulik [18], E. A. Vetrova, O. A. Kulikova [19].

3 Statement of the Research Problem

The topicality of the research and the degree of its scientific development provide the basis for the definition of a scientific problem as the need to develop a "smart city" model. To solve this problem, the following tasks should be accomplished:

- theoretical and methodological substantiation of the "smart city" concept;
- studying the evolution of the "smart city" concept;
- identification of barriers and analysis of risks militating against city "smart" transformation.

4 Theoretical and Methodological Substantiation of the "Smart City" Concept

For the first time in the world, the term "Smart City" appeared at the end of 1960. The reason for this was the rapid development of computers and computer technologies [20]. In Russia, this term has been used since the late 1990s, when many people sought to move from rural areas to the city, thereby trying to improve their standard of living. However, the influx of a large number of people overloaded the urban infrastructure which was unable to cope with such a flow of new residents, and despite the expected "good" life, rural settlers felt uncomfortable.

Initially, the term "smart city" in this country was used only in terms of environmental friendliness and, as a rule, served as a synonym for the phrase "green technologies". In the early 2000s, the focus shifted to IT technologies, thanks to which the city was able to manage data, which ultimately improved the quality of life of the population, and the city itself was considered to be "smart".

Officially, the concept of creating a "smart city" was proclaimed by IBM in 2008, introducing the Smarter Planet model [21]. The meaning of this concept was as follows: every city dweller should feel comfortable. People should feel safe, secure and comfortable in everyday life, businesses should have all the conditions for effective development, and local authorities should have the ability to manage all urban processes.

Summarizing the above, we can present the concept of a "smart city" schematically. It is shown in Fig. 1.

Currently, the term "smart city" has several interpretations. Glazychev, a Russian scientist, researcher of design creativity and architectural heritage considers transformation into a "smart city", first of all, as an attempt to reform cities in accordance with the needs of modern society [22].

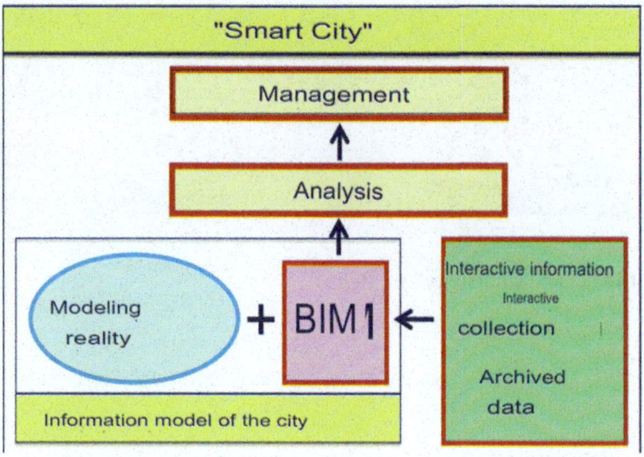

Fig. 1 The scheme of the "smart city" concept

Adam Greenfield, an American writer and urbanist, a fierce opponent of the concept of "Smart City" believes that the idea of transition to this concept is insufficiently elaborated, it lacks specifics. The problem people face when studying this issue is that there is not enough valid information on the transformation of the urban environment to create the whole image of a "smart city" [23].

The authors of the paper propose the following definition: "smart city" is an innovative city that implements a set of organizational and technical measures aimed at achieving the highest possible quality of service provision and sustainable use of resources, in order to create comfortable and favorable living conditions for both current and future generations.

Figure 2 shows the "smart city" model proposed by the author's team.

Human potential and social capital are the driving force of the city smart transformation. The age of the population and the level of education determines the susceptibility of citizens to "smart" technologies.

"Smart cities" have certain properties. These include mobility, socialization, virtualization, urbanization, and personification. In addition, smart cities must necessarily be comfortable to live in, have smart technologies at their disposal and be ready to modernize infrastructure. The properties of the "smart city" are shown in Fig. 3.

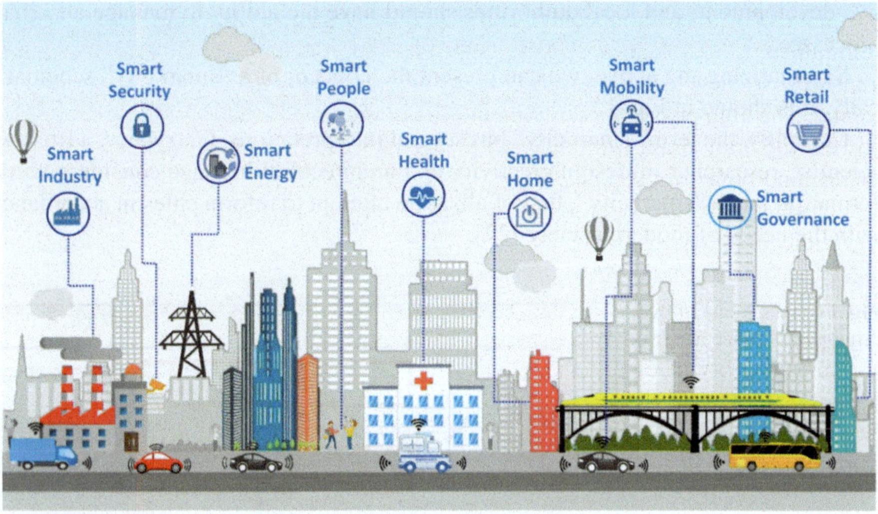

Fig. 2 A "smart city" model

Fig. 3 "Smart City" Properties

5 Studying the Evolution of the "Smart City" Concept

The task of building "smart" cities is quite relevant for our country, because since the dissolution of the Soviet Union, the engineering and utility infrastructure of most cities has been significantly worn out and outdated. It is extremely difficult to continue its development and try to adapt it to changing loads. An important role in the transition to a new technological economy is assigned to human resources and social capital. The high priority task facing the Government of the Russian Federation is to prevent the brain drain from the country, to create more comfortable and favorable living conditions for people in cities.

Let us consider the evolution of the "smart city" concept.

At the initial stage of the development of this concept, "smart city" 1.0 was presented as a technologically focused city. It was characterized by the availability of technologies that were used to improve the quality of life of the population, the city's resistance to technology-related risks and for the purpose of its manageability and sustainability. The key feature of this stage is the introduction of IT technologies. Electrification of urban facilities, use of semi-automatic systems and devices in the city infrastructure take place. The suppliers of technological solutions and services were most interested in the implementation of the "smart city" 1.0 concept.

At the second stage, Smart City 2.0, the city appears as a high-tech manageable city. The distinctive features of the city at this stage were the use of GIS technologies that allow optimizing solutions in the field of environmental protection, healthcare. 3G and 4G appeared. The initiators of this model of the city were local governments.

At the third stage, Smart City 3.0 is characterized as a highly intelligent city using technologies aimed at the development of entrepreneurship and social partnership. There is a digital transformation of all sectors of the city, it becomes possible to collect and process data in real time. The main participants in the implementation of the Smart City 3.0 stage are active and interested citizens [24].

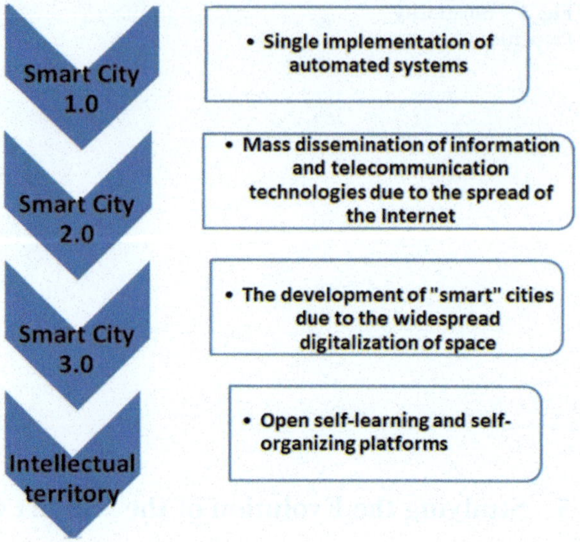

Fig. 4 Evolution of the "Smart City" concept

At the last stage, the "smart city" is transformed into an intellectual territory developed in all spheres and fields. Its borders expand significantly, which allows it to fit into the economy and politics of the state organically. Figure 4 shows the evolution of the "Smart City" concept.

The process of city "smart" transformation is the creation of a stable and solidary system for the development of the urban environment with the active involvement of all interested groups and actors into this process. As a rule, the request for "smart" transformation comes from caring citizens and local communities [25–27].

6 Identification of Barriers and Analysis of Risks Militating Against a City "Smart" Transformation

The main goal of the city "smart" transformation is to create motivation for the socio-economic development of the urban environment and improve the quality of life of the citizens. At the same time, innovative and modernizing development of the existing infrastructure of the city in accordance with the needs of modern society takes place.

As in any process, there are barriers and risks on the way to achieving "smart" transformation.

Barriers can be political and legal, economic, economic and social. They are described in more detail in Table 1.

The risks militating against the city "smart" transformation are divided into technological, social, political, legal and economic ones. In Table 2, you can read more about them.

Table 1 Barriers on the way of achieving a city "smart" transformation

Legal and political	Organizational	Economic	Social
Protectionism of domestic developers reduces motivation to compete with world leaders	Backwardness of the Russian system of codes and standards (especially in construction) in comparison with the world ones reduces the motivation to introduce smart technologies	Limited budgets of local authorities	Insufficient awareness of population about smart technologies and their capabilities
Excessive centralization of public administration reduces the initiative of cities	Low level of credibility among economic entities hinders the achievement of an agreement on the smart transformation of cities	Lack of business models for smart technology implementation	Low social level of the urban population often leads to the rejection of smart technologies (including vandalism and theft)
Political conditions are not conducive to long-term planning of smart development	Limited powers of local authorities	Low economic motivation of consumers to use smart technologies	The difficulty of involving citizens in smart city initiatives
The political situation limits the use of international experience	Excessive regulation of economic entities	Large project risks of long-term investment programs	Undeveloped methods of grassroots democracy
Attachment of management plans to political cycles and specific public officials	Lack of local managerial skills and qualified personnel	High maintenance and renewal costs, dependence on the suppliers of technological solutions	
	Municipal unitary enterprises have basically no motivation to introduce "smart" technologies	Excessive information security costs and risks for return on investment	

Despite the existing risks and barriers, the effect of creating a "smart city" will be great: the environmental burden will decrease by 20–40% (Fig. 5). In addition, the smart city will reduce costs in the following areas:

- traffic time will be reduced by 20%;
- the number of traffic accidents will decrease by 30% due to the traffic and transit control system;

Table 2 Risks on the way of the city "smart" transformation

Legal and political	Technological	Social	Economic
Risks of developing virtual crime	Risks of an increase in waste volumes due to the rapid obsolescence of equipment and its disposal	Digital divide: depriving part of the population of access to information services	Moving the shadow economy to the internet (via anonymous networks like tor)
Growth of political, social, economic Costs due to cyberattacks, data leakage	"junk information" - the volume of data transmitted over networks doubles every two years	Impossibility to express your opinion with the help of information and communication technologies	Cryptocurrency makes commodity-money exchange anonymous (ordering a hit, buying drugs and weapons, prostitution)
Use of personal data for criminal purposes	Risks of hacking smart city systems	Risks of discrimination and exclusion of certain categories of citizens from the process of consumption of public goods when using smart technologies	
Restrictions on human rights and freedoms when using digital identity	Risks of hacking personal (medical) smart devices	Psychological discomfort of a "smart" city	
The hazard of a one-time shutdown of all Public services in case of failures	Risks of inflating problems due to leaks due to the fact that the smart city network brings all personal data together	The problem of dilettantism in case of participation of the citizens in the city management	
Failures in copyright protection	Vulnerabilities in the software of smart home appliances lead to failures of service Due to virus attacks		

- up to 70% of energy and maintenance costs due to the use of energy-saving lamps and motion sensors;
- up to 30% of resources due to the creation of "smart" electricity and gas meters;
- up to 30% of the waste disposal costs due to the separate waste collection system;
- up to 20% of the cost of maintaining law enforcement and rescue services due to the video surveillance system.

Thus, the development of a "smart city" model will contribute to the development of comfortable, friendly human living environment. However, when developing a model, it is necessary to take into account the risks and barriers that militate against the transformation of the city into a "smart" one.

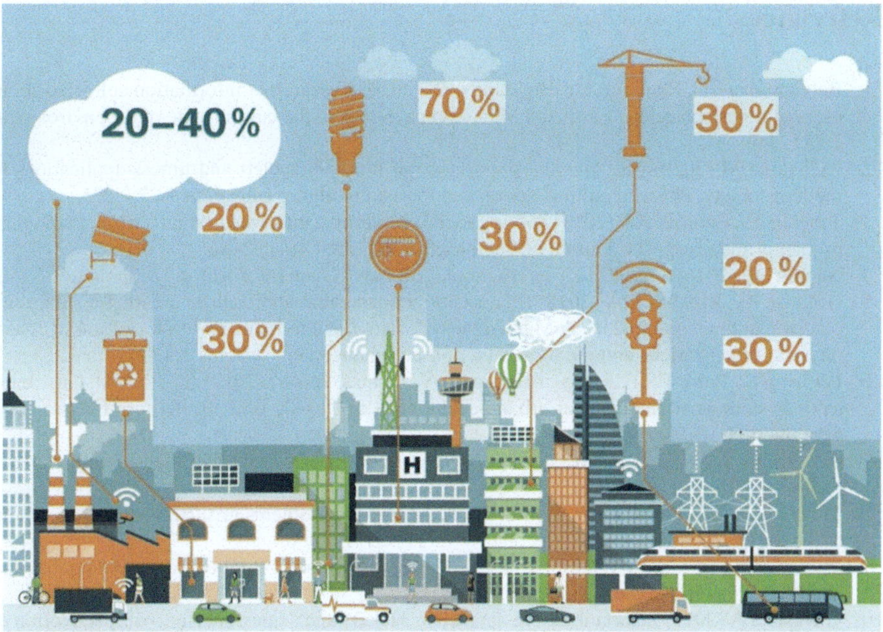

Fig. 5 The effect of the "Smart City"

7 Conclusion

City "smart" transformation supposes developing a relative and solidary system of relationships involving all interested actors and groups into the process. The creation of a "smart city" model implies, first of all, the sustainable development of the territory in accordance with the standards and principles of urbanism. This is necessary in order to form an innovative socio-economic development of the territory and improve the quality of life of the population. The tools for the city "smart" transformation are organizational, economic, technological and social development plans, an implementation strategy and an action plan developed for each group of subjects and actors of this process.

Acknowledgements The work was carried out within the framework of the development program of South-west state university project "Prioritet-2030"

References

1. Belyaninova EG, Ryabinin EV, Bogdanov PY (2020) Research of information collected about users of fitness bracelets. Earth and Man. Topical issues of the current state of the environment 1:247–249.
2. Kulbakina M (2020) The role and importance of fitness bracelets and fitness applications in modern society. Physical culture, sports and tourism in higher education 1:272–275.
3. Tebekin AV, Egorova AA (2019) Solving social problems of cities using smart city technologies: problems and prospects. Journal of Sociological Research 4(4):32–46.
4. Namiot DE (2016) Smart Cities. International Journal of Open Information Technologies 1:1–3.
5. Gribkov RV, Mukhin MA (2018) Smart City: a conceptual approach to creating a smart city platform at the municipal level. Management development in Industry 4.0: transition to cyber-physical organizations and the formation of their management systems 1:49–51.
6. Petrov KS, Pityk ES, Chernoknizhnikova AY, Shustova AV (2019) The Smart City Indicator System as an indicator of urban development in Russia. BST: Bulletin of Construction Machinery 7(1019):64–65.
7. Artamonov AV (2017) Smart City as a modern concept of city development. Design bureau 6:27–31.
8. Molchanova VA (2019) From a smart city to a fair city: problems of sustainable development in the digital economy. Creative economy 13(12):2371–2386.
9. Vukovic NA, Larionova VA, Morganti P (2021) Smart sustainable cities: smart approaches and analysis. Economy of Region 17(3):1004–1013.
10. Townsend A (2013) Smart cities: big data, civic hackers and a quest for a new utopia. Norton & Company, New York.
11. Albino V Umberto Berardi, Rosa Maria Dangelico (2015) Smart Cities: Definitions, Dimensions, Performance and Initiatives. J of Urban Technology 1:21–25.
12. Bobylev SN, Kudryavtseva OV, Solovyova SV (2014) Indicators of sustainable development for cities. The economy of the region 3(39):101–110.
13. Skorobogaty DA (2020) On the definition of the term “landscaping” in modern realities. A young scientist 42(332):136–138.
14. Koneva VV, Khizhnyak SA (2016) Development of measures for the improvement of the territory of the municipality. Infrastructure sectors of the economy: problems and prospects of development 12:43–49.
15. Pashkova LA (2014) Practical study of complex landscaping structures in the Belgorod region. Science Time 11:287–293.
16. Starostina MP, Urzha OA (2019) Management system of landscaping on the example of Sergiev Posad district. Materials of the Ivanovo readings 4(26):110–113.
17. Nikaeva RM, Dikaeva XM (2021) The management system of the improvement of the territory of the municipality. FSU Science 2(22):47–52.
18. Gulik ON (2018) Landscaping on the basis of recreation areas of higher educational institutions. Bulletin of Science 1(6):39–44.
19. Vetrova EA, Kulikova OA (2019) Recommendations for improving the landscaping of the territory (on the example of the urban settlement Udelnaya of the Ramenskoye municipal district of the Moscow region. Philosophy of social communications 1(46):82–84.
20. Nizamov RK (2018) Smart city. Smart City strategies. Digital documentaries for everyone 1:77–98.
21. Sizov YI, Medvedeva LN (2019) The development of an average city based on the concept: from a “smart house to a smart city”. Scientific works of the Free Economic Society of Russia 218(4):573–580.
22. Glazychev VL (2011) City without borders. Publishing House “Territory of the Future”, Moscow.
23. Grinfild A (2013) Against the smart city. Hachette Livre, London.

24. Akimova OE, Volkov SK, Kuzlaeva IM (2019) Unique advantages of using the concept of "smart city" in the context of the development of modern economic and strategic planning. Creative Economy 13(8):1521–1528.
25. 25 Prelikova EA, Belyakova OI, Grachev SV et al (2022) Development and implementation of a new format of eco shops "Without packaging – zero waste". IOP Conference Series: Earth and Environmental Science 1061(1):012042. https://doi.org/10.1088/1755-1315/1061/1/012042
26. Prelikova E, Timofeev G, Trifonov A et al (2023) Development of waste bins to collect cigarette butts as a way to ensure the safety and hygiene of the urban environment. AIP Conference Proceedings. 2910(1):020117. https://doi.org/10.1088/1755-1315/1061/1/012042
27. Prelikova E, Kukshin N, Kopylov M et al (2023) Development of automatic exchangers of small garbage for bird-seed and training of Corvidae family birds (the Crow family) to collect small rubbish. E3S Web of Conferences 411:02006.

Enhancing Intersection Capacity and Reducing Pollutant Emissions: Real-Time Analysis and Insights

V. Shepelev, A. Vorobyev, V. Mavrin, Z. Almetova, and U. Khassenov

Abstract Air containing harmful and dangerous substances to human health has long become an integral attribute of cities with developed infrastructure. The significant negative contribution is made by automobile transport. The situation is especially aggravated in the area of regulated intersections, where the driving modes of automobiles change sharply (braking, accelerating, decelerating, idling) and traffic jams often occur. Various measures are taken to improve the air quality, ranging from administrative measures to the use of new environmentally friendly materials. One way to improve the air quality in the intersection area is to increase the intersection capacity (IC), reducing the probability of deep congestion. This article proposes a methodology for analyzing the IC of a regulated intersection based on real-time data processed using neural network algorithms, as well as analyzing the emissions of pollutants (PM 2.5) when changing the queue structure of vehicles. The identified patterns of the passage sequence of different categories of vehicles with levels of pollutant emissions provide valuable information for making appropriate administrative decisions on traffic regulation to reduce the negative environmental impact.

Keywords Traffic capacity · Intelligent transportation systems · Regulated intersection · Emissions · Road traffic · PM 2.5

V. Shepelev (✉) · Z. Almetova
South Ural State University, 76, Prospekt Lenina, Chelyabinsk 454080, Russia
e-mail: shepelevd@susu.ru

A. Vorobyev
Moscow Automobile and Road Construction State Technical University, 64, Leningradsky Prospekt, Moscow 125319, Russia

V. Mavrin
Kazan Federal University, 10a, Syuyumbike Prosp, Naberezhnye Chelny 423800, Russia

U. Khassenov
NLC "Akhmet Baitursynuly Kostanay Regional University", 47, Baytursynov Street, Kostanai 110000, Kazakhstan

L. V. Radionova and D. V. Ulrikh (eds.), *Advances in Ecology and Environmental Engineering*, Springer Proceedings in Earth and Environmental Sciences,
https://doi.org/10.1007/978-3-031-64423-8_5

1 Introduction

The rapid development of the global economy has led to an increase in the number of vehicles, primarily in urban areas. The transportation sector makes a significant contribution to air pollution. The large number of vehicles on city roads has caused an increase in the number of road traffic accidents [1–3], the formation of deep traffic congestion [4], especially at intersections, which has resulted in increased pollutant emissions and decreased quality of life for people [5]. In recent decades, emissions have increased due to exponential growth in road traffic intensity [6–8]. Numerous studies have established a correlation between air quality and the incidence of diseases among residents [9–12].

There are many causes of traffic congestion (incorrect intersection geometry, weather conditions, road traffic accidents, presence of tram tracks, incorrect signal timing, etc.), one of which is low intersection capacity (IC). The IC of a regulated intersection is a controllable factor that can reduce the probability of traffic congestion situations [13, 14]. Fourati [15] assesses IC using automatically generated trajectories of probing vehicles. The development and widespread use of new technologies utilizing street surveillance cameras enables rapid and accurate determination of traffic parameters, vehicle types, their speeds, and other data [15–17]. To improve IC, the use of V2I (vehicle-to-infrastructure) in [18–20] has been proposed, forming platoons in [21], and optimizing traffic signal timings in [22, 23].

To increase the capacity of regulated intersections and, as a result, reduce the level of harmful emissions, the implementation of transport management technologies utilizing modern information and communication technologies is required. The combination of such technologies represents modern Intelligent Transportation Systems (ITS).

In order to choose the optimal tools to change the intersection capacity, it is necessary to conduct an analysis. Therefore, the aim of this research is to develop a methodology for analyzing the intersection capacity in conditions of mixed traffic and to identify patterns of how the queue of different types of vehicles affects the level of pollutant emissions.

2 Materials and Methods

In previous studies [24–26] a software system based on artificial intelligence was developed to collect data on traffic flows. Its objective is to provide an autonomous approach capable of assessing quantitative and qualitative parameters of road traffic (e.g., quantity, speed, trajectory, travel time, and vehicle idling), as well as emissions and pollutant concentrations. This system is based on the use of static cameras with a wide field of view that covers the entire physical territory of the intersection and adjacent roads. Several cameras provided by the company "Intersvyaz" with unrestricted access in the city of Chelyabinsk [27] were used to train the neural network.

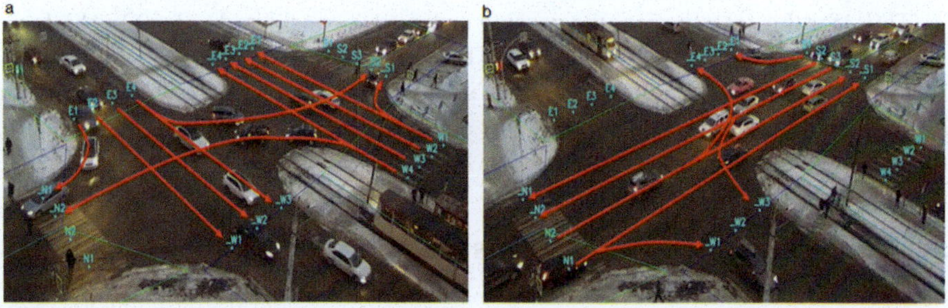

Fig. 1 Directions of movement during: **a** the first stroke of the traffic light; **b** the second stroke of the traffic light

The calculation of pollutant emissions in vehicle exhaust gases is performed using methodologies reflected in the national standard of the Russian Federation [28, 29].

To collect data on the structure of traffic flows and the emissions of pollutants from them, the intersection of Pobedy Avenue and Krasnoznamennaya Street in the city of Chelyabinsk (Fig. 1) was chosen. This particular intersection was selected due to the camera's sufficient field of view, allowing for the observation of multiple vehicle queues, a relatively short distance that ensures good visibility and enables accurate detection and classification of vehicles, as well as the absence of obstructions (such as poles or trees).

Video recordings were processed for 15 days during peak hours from 7:30 to 9:30 and from 17:00 to 19:00. In total, approximately 60 h of video data were analyzed.

Using the intersection's video footage, the traffic signal operation was measured. The operation of the signal consists of two main cycles:

1. Duration of the green signal is 51 s. The green signal is active for the following directions of movement: W1-_E1, W1-_S1, W2-_E2, W3-_E3, W4-_E4, W4-_N2, E1-_N1, E2-_W1, E3-_W2, E4-_W3, E4-_S2 (Fig. 1a).
2. Duration of the green signal is 37 s. The green signal is active for the following directions of movement: N1-_W1, N1-_S1, N2-_S2, N2-_E4, S1-_E1, S2-_N1, S3-_N2, S3-_W3 (Fig. 1b).

Based on the traffic signal data, the green cycle is determined when vehicles are allowed to move in the functional zone. The sequential number of each vehicle in its queue is also determined. Vehicles belong to the same queue if they are entering through the same lane.

As a result of collecting information on vehicle movement parameters at the intersection in all possible directions, a large dataset was obtained, consisting of 188,589 entries. It includes 15 days of observations, with two observation blocks per day. In total, there were 30 blocks, averaging about 6000 entries per block. For the analysis, only straight directions (from W to E and from E to W) were selected to exclude interferences from pedestrians and trams. The total number of selected entries was 45,297.

Thus, for this intersection, the passing capacity is considered to be 22 vehicles of the first category when moving straight. The maximum queue size that can fully pass through the intersection during the green signal of the traffic light is also assumed to be 22 vehicles.

3 Analysis of the Intersection's Capacity

The capacity of the intersection decreases when there is a non-passenger vehicle in the queue of passenger cars. As a result, the subsequent passenger vehicles in the queue will not move at the same speeds as they would in a queue consisting only of passenger vehicles.

It was decided to assess the impact of a non-passenger vehicle at different positions in the queue on the movement parameters of the entire queue. Based on this, we will calculate how the intersection capacity will decrease when vehicles of categories II (vans and minibusses up to 3.5 tons), III (trucks from 3.5 to 12 tons), and V (buses over 3.5 tons) are present in the queue.

The baseline value for assessing capacity will be the time chart of the intersection passage by the last passenger vehicle in the absence of other category vehicles in the queue.

For the first vehicle in the queue, the time it takes to reach the center of the measurement area is calculated using Eq. 1:

$$t_1 = \frac{2 \cdot S}{V_1}, \tag{1}$$

where *S*—distance to the center of the measurement area (20 m); V_1—the average vehicle speed in the measurement area (km/h).

For subsequent vehicles in the queue (21 cars), the additional time t_i is determined:

$$t_1 = \frac{2 \cdot (l_1 + L)}{V_{i-1} + V_i}, \tag{2}$$

where l_1—the average length of a passenger car (4.5 m); *L*—the average distance between cars in the queue (2.7 m); V_{i-1}—the average speed of the previous vehicle in the measurement area (km/h); *Vi*—the average speed of the vehicle in the measurement area (km/h).

According to the calculations, the total time for the last vehicle in the queue to pass through the intersection area is 15.1 s, which is significantly shorter than the green signal duration of 51 s. This can be attributed to the delay in the start of each vehicle in the queue. The averaged estimation of this delay is 1.7 s.

Similar calculations using Eqs. 1–2 were performed for queues with the first vehicle being of categories II, III, and V. Calculations were not conducted for category IV (trucks over 12 tons) due to insufficient data. For category II, records with queues

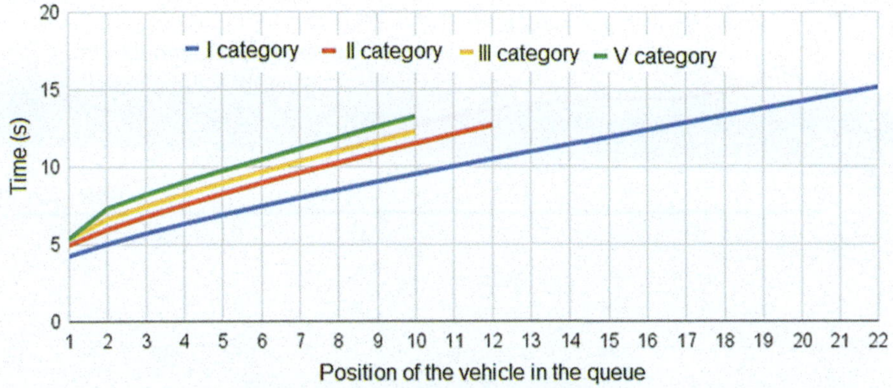

Fig. 2 Accumulation of total travel time in seconds for queues with vehicles of different categories in the first position

of 12 vehicles were available. For categories III and V, queues with 10 vehicles were considered. As a result, the total travel time for the last vehicle in the queue (*T*) is as follows:

- 12.7 s for category II, while the *T* for a queue of 12 passenger vehicles is 10.5 s;
- 12.3 s for category III while the *T* for a queue of 10 passenger vehicles is 9.5 s;
- 13.2 s for category V while the *T* for a queue of 10 passenger vehicles is 9.5 s.

Graphs of the accumulation of total travel time for a queue of passenger cars in the presence of one vehicle from other categories in the first place in the queue are shown in Fig. 2. The dynamics of accumulation is shown depending on the position of the vehicle in the queue of 10 and 12 cars.

Based on Eqs. 1–2, it is possible to calculate the passage time of the intersection for the last vehicle, depending on the position of vehicles of categories II, III, and V in the queue. Figure 3 shows the time for a queue of 12 vehicles for categories I and II. Figure 4 shows the time for a queue of 10 vehicles for categories I, III, and V. Values for category V at points 4, 9, and 10 were interpolated as there were no available data for them.

To determine the capacity of the intersection for a queue of vehicles with one vehicle of another category, extrapolation of the total travel time for a queue of 22 cars was carried out in the first place:

- a queue with category II vehicles passes the intersection in 55.5 s;
- a queue with category III vehicles passes the intersection in 57.1 s;
- a queue with category V vehicles passes the intersection in 59 s.

During the 51-s green signal cycle, a total of travels through:

- a total of 20 vehicles pass through with one vehicle of category II in the first position;

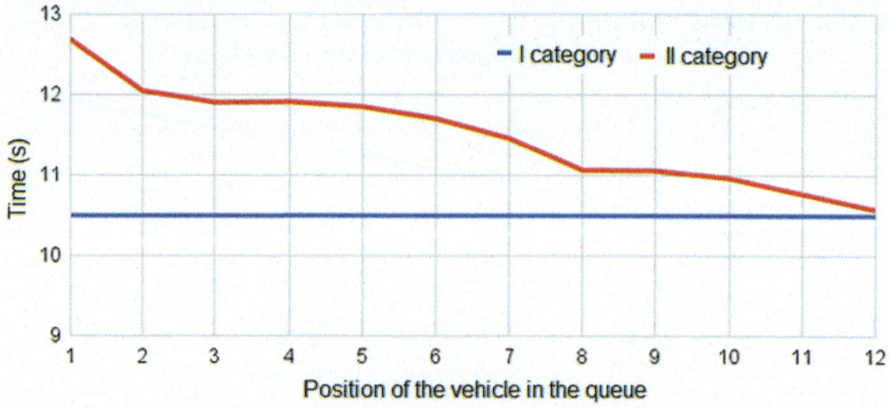

Fig. 3 Queue travel time with one vehicle of II category at different locations

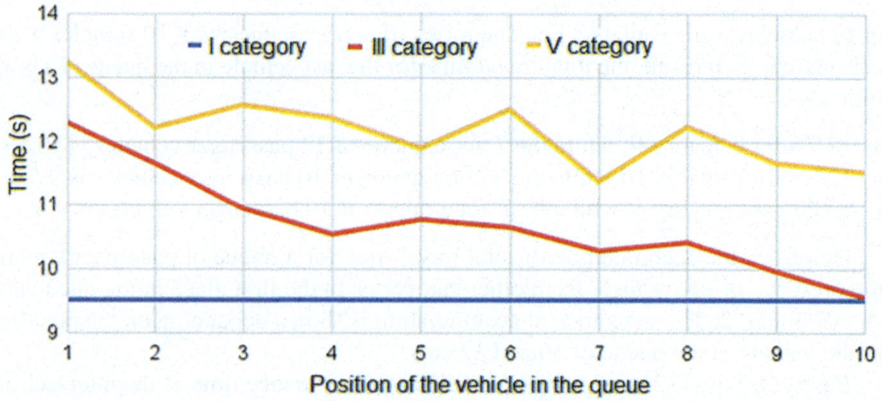

Fig. 4 Queue travel time with one vehicle of III and V categorys at different locations

- a total of 19 vehicles pass through with one vehicle of category III in the first position;
- a total of 18 vehicles pass through with one vehicle of category V in the first position.

4 Analysis of Pollutant Emissions When Changing the Queue Structure of Vehicles

Since microscopic suspended particles PM2.5 have the greatest impact on human health and these particles do not interact with other substances, an assessment analysis was conducted to evaluate the changes in their quantity when varying the structure of the traffic flow in the area of the examined regulated intersection [30].

Based on the obtained data, the amount of PM2.5 particle emissions per equivalent unit of a vehicle was calculated to account for different queue lengths simultaneously (Fig. 5).

As seen from the data, the emission levels from vehicles decrease as the position of the vehicle in the queue increases. This is due to the fact that the average speed of the traffic flow is lower when a non-passenger vehicle is closer to the front of the queue. A lower speed leads to an increase in the coefficient r_v in the emission calculation formula for moving vehicles. Since the speed can vary arbitrarily while the r_v coefficient is defined for a few values, a trend approximation of the dependence between the coefficient and speed was performed (Fig. 6). It is worth noting the high quality of the adopted approximation, supported by the value of the multiple correlation coefficient $R^2 = 0.9988$.

The previously obtained results only show emissions from a small section under consideration. Therefore, it is necessary to calculate the emissions from the entire flow starting from the moment the green signal of the traffic light is turned on. Based on data on vehicle sizes, distances between vehicles, delay before starting movement, and average speeds at the intersection, the amount of PM2.5 emissions (M) from the flow of passenger cars was calculated as follows:

- for a queue of 12 vehicles, $M = 0.0376309$ g;
- for a queue of 10 vehicles, $M = 0.0266996$ g.

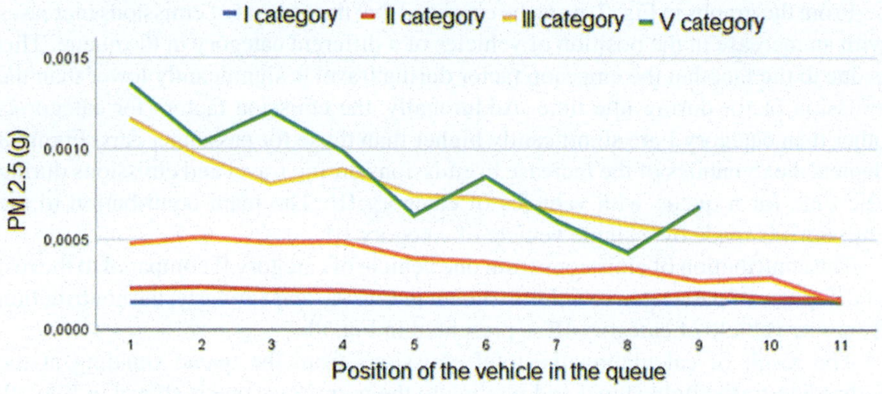

Fig. 5 Amount of PM2.5 emissions per given vehicle unit within the monitoring area

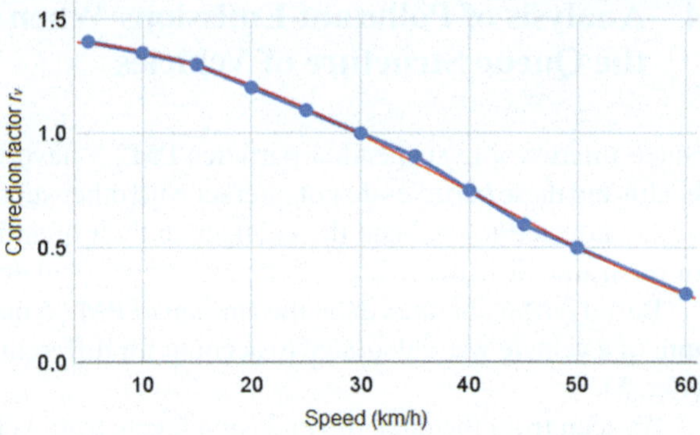

Fig. 6 Trend approximation for the r_v correction factor depending on the speed

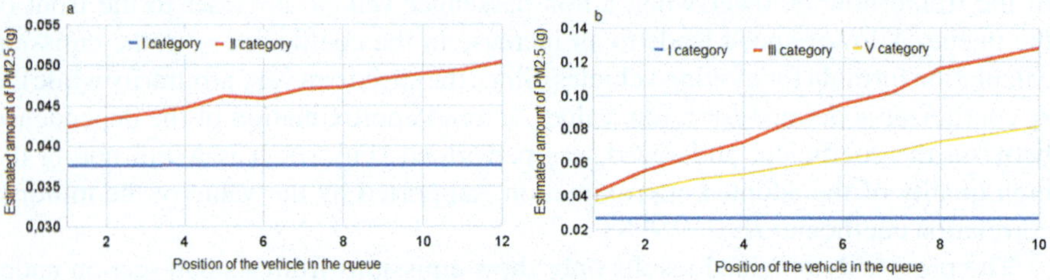

Fig. 7 Calculated amount of PM2.5 emissions from the flow of vehicles with **a** one vehicle of category II at different positions and **b** category III and V vehicles at different positions

Next, a queue with one category II vehicle was considered. In the available data, there were enough entries to analyze only queues with no more than 12 vehicles (Fig. 7a). For queues with one category III and V vehicle, queues with no more than 10 vehicles were considered from the available data (Fig. 7b). Category IV was not considered as there was insufficient data to identify patterns.

From the graphs in Fig. 7, it can be observed that the amount of emissions increases with an increase in the position of vehicles of a different category in the queue. This is due to the fact that the emission factor during travel is significantly lower than the emission factor during idle time. Additionally, the emission factors for categories other than category I are significantly higher than those for passenger cars. Figure 8 depicts the dynamics of the increase in emissions during travel and emissions during idle time for a queue with vehicles of category III. The main contribution to the emissions is made by a single vehicle of category III.

The contribution of emissions from one vehicle of category II compared to the rest of the queue of passenger vehicles is shown in Fig. 9a, and similarly, the contribution from one vehicle of category III is presented in Fig. 9b.

The result of calculating the total emissions from the queue standing at the forbidding traffic light signal and moving to the permissive one is shown in Fig. 10.

As seen from Fig. 10, emissions decrease as the position in the queue increases. This is because vehicles of categories other than category I, which produce more

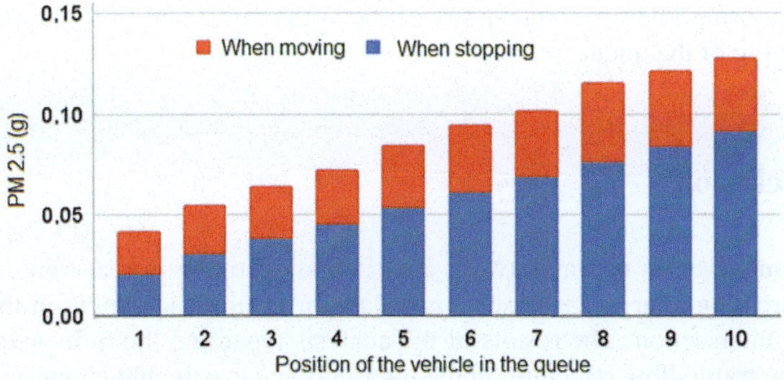

Fig. 8 The dynamics of the increase in emissions during travel and emissions during idle time for a queue with vehicles of category III, depending on the position

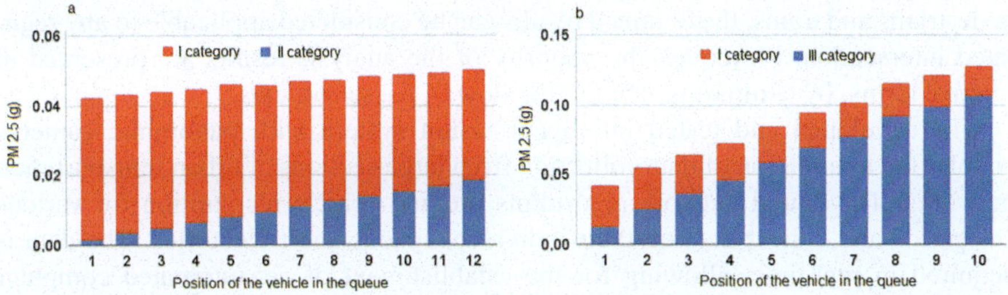

Fig. 9 The contribution of emissions from one vehicle compared to the rest of the queue of passenger vehicles, depending on the position in the queue **a** for category II vehicles and **b** for category III vehicles

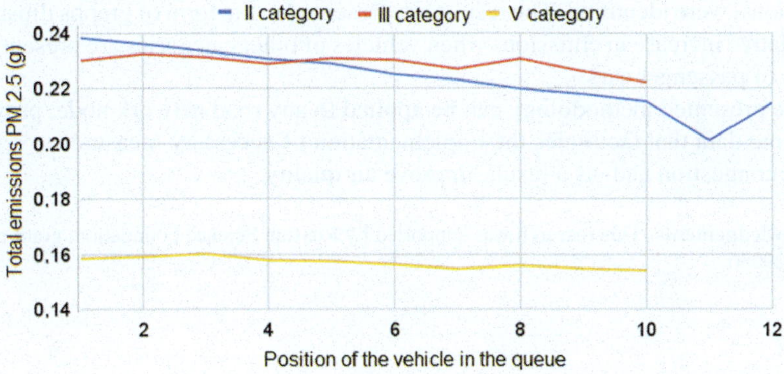

Fig. 10 Total emissions from a queue standing at a forbidding traffic light and moving to a permissive one

emissions during idle time, spend more time at the red signal of the traffic light at the beginning of the queue.

5 Conclusion

The system based on neural network algorithms, operating in real-time mode, has allowed creating a large database of raw data on vehicle movement in the area of a specific intersection. The results of the analysis regarding the influence of variations in the traffic flow structure on the intersection capacity and changes in PM2.5 particle emissions are valid only for this particular intersection. However, since only the movement of vehicles in the "straight" direction was considered, without taking into account the specifics of "right" and "left" turns, as well as interferences from pedestrians and trams, the obtained results can be considered applicable to any regulated intersection. Moreover, the majority of the analysis results are presented in relative terms, by comparing them to the flow of passenger cars.

The developed and tested intelligent vision system with panoramic cameras enables the automation of data collection from large sections of urban transportation networks. The neural network algorithms provide vehicle recognition for various categories and collect data on their individual speeds and trajectories. The data is acquired in real-time, allowing for the establishment of an automated computer system for continuous monitoring of the traffic situation on roads and estimation of emissions of various types of pollutants.

During the research, patterns related to the sequence of passage of different vehicle categories at regulated intersections, as well as the corresponding levels of pollutant emissions, were identified. The results are presented in the form of graphs illustrating the relative increase in emissions when vehicles of other categories are present in the queue of passenger cars.

The presented methodology can be applied to any road network node, providing real-time data that facilitates the implementation of necessary measures to regulate traffic congestion and, as a result, improve air quality.

Acknowledgements This research was supported by Russian Science Foundation, grant number 24-21-20086

References

1. Shesterov E, Mikhailov A (2017) Accident Rates at Signalized Intersections. Transportation Research Procedia 20:613–7. https://doi.org/10.1016/j.trpro.2017.01.099
2. Tian Z (2013) Speed-accident Relationship at Urban Signalized Intersections. Procedia - Social and Behavioral Sciences 96:1383–8. https://doi.org/10.1016/j.sbspro.2013.08.157
3. Al-Ghamdi AS (2003) Analysis of traffic accidents at urban intersections in Riyadh. Accident Analysis & Prevention 35(5):717–24. https://doi.org/10.1016/S0001-4575(02)00050-7

4. Luan S, Ke R, Huang Z, Ma X (2022) Traffic congestion propagation inference using dynamic Bayesian graph convolution network. Transportation Research Part C: Emerging Technologies 135:103526. https://doi.org/10.1016/j.trc.2021.103526
5. Yassin MF, Kellnerová R, Jaňour Z (2008) Impact of street intersections on air quality in an urban environment. Atmospheric Environment 42(20):4948–63. https://doi.org/10.1016/j.atmosenv.2008.02.019
6. Pandian S, Gokhale S, Ghoshal AK (2009) Evaluating effects of traffic and vehicle characteristics on vehicular emissions near traffic intersections. Transportation Research Part D: Transport and Environment 14(3):180–96. https://doi.org/10.1016/j.trd.2008.12.001
7. Adams MD, Kanaroglou PS (2016) Mapping real-time air pollution health risk for environmental management: Combining mobile and stationary air pollution monitoring with neural network models. Journal of Environmental Management 168:133–41. https://doi.org/10.1016/j.jenvman.2015.12.012
8. Popova I, Leonteva L, Danilov I, Marusin A, Marusin A, Makarova I (2021) Impact of vehicular pollution on the Arctic. Transportation Research Procedia 57:479–88. https://doi.org/10.1016/j.trpro.2021.09.075
9. Kaur S, Nieuwenhuijsen MJ, Colvile RN (2007) Fine particulate matter and carbon monoxide exposure concentrations in urban street transport microenvironments. Atmospheric Environment 41(23):4781–810. https://doi.org/10.1016/j.atmosenv.2007.02.002
10. Luo Z, Wang Y, Lv Z, He T, Zhao J, Wang Y et al (2022) Impacts of vehicle emission on air quality and human health in China. Science of The Total Environment 813:152655. https://doi.org/10.1016/j.scitotenv.2021.152655
11. Luo R, Zhang T, Wang L, Feng Y (2023) Emissions and mitigation potential of endocrine disruptors during outdoor exercise: Fate, transport, and implications for human health. Environmental Research 236:116575. https://doi.org/10.1016/j.envres.2023.116575
12. Mulholland E, Miller J, Bernard Y, Lee K, Rodríguez F (2022) The role of NOx emission reductions in Euro 7/VII vehicle emission standards to reduce adverse health impacts in the EU27 through 2050. Transportation Engineering 9:100133. https://doi.org/10.1016/j.treng.2022.100133
13. Verma JK, Ranga V (2021) Multi-Robot Coordination Analysis, Taxonomy, Challenges and Future Scope. J Intell Robot Syst 102(1):10. https://doi.org/10.1007/s10846-021-01378-2
14. Makarova I, Pashkevich A, Shubenkova K, Mukhametdinov E (2017) Ways to Increase Population Mobility through the Transition to Sustainable Transport. Procedia Engineering 187:756–62. https://doi.org/10.1016/j.proeng.2017.04.434
15. Fourati W, Friedrich B (2019) Trajectory-Based Measurement of Signalized Intersection Capacity. Transportation Research Record 2673(10):370–80. https://doi.org/10.1177/0361198119848412
16. Jeon H, Lee J, Sohn K (2018) Artificial intelligence for traffic signal control based solely on video images. Journal of Intelligent Transportation Systems 22(5):433–45. https://doi.org/10.1080/15472450.2017.1394192
17. Wang S, Patwary AUZ, Huang W, Lo HK (2022) A general framework for combining traffic flow models and Bayesian network for traffic parameters estimation. Transportation Research Part C: Emerging Technologies 139:103664. https://doi.org/10.1016/j.trc.2022.103664
18. Lin Y, Tiwari A, Fabien B, Ban X, Devasia S (2023) Increasing traffic capacity of mixed traffic at signalized traffic intersections using delayed self reinforcement. Transportation Research Part C: Emerging Technologies 157:104403. https://doi.org/10.1016/j.trc.2023.104403
19. Wu Y, Wang DZW, Zhu F (2022) Influence of CAVs platooning on intersection capacity under mixed traffic. Physica A: Statistical Mechanics and its Applications 593:126989. https://doi.org/10.1016/j.physa.2022.126989
20. Yang Q, Fu X (2024) An extended queueing model for vehicles at signalized intersections considering the platoon correlated arrivals. Physica A: Statistical Mechanics and its Applications 635:129483. https://doi.org/10.1016/j.physa.2023.129483
21. Timmerman RW, Boon MAA (2021) Platoon forming algorithms for intelligent street intersections. Transportmetrica A: Transport Science 17(3):278–307. https://doi.org/10.1080/23249935.2019.1692962

22. Boon M, Janssen G, Van Leeuwaarden J, Timmerman R (2023) Optimal capacity allocation for heavy-traffic fixed-cycle traffic-light queues and intersections. Transportation Research Part B: Methodological 167:79–98.https://doi.org/10.1016/j.trb.2022.11.010
23. Huang X, Wang H, Li Y, Huang L, Zhao H (2024) Reservation-based traffic signal control for mixed traffic flow at intersections. Physica A: Statistical Mechanics and its Applications 633:129426. https://doi.org/10.1016/j.physa.2023.129426
24. Shepelev V, Aliukov S, Nikolskaya K, Shabiev S (2020) The capacity of the road network: Data collection and statistical analysis of traffic characteristics. Energies 13(7)
25. Shepelev VD, Vorobyev AI, Shepeleva EV, Alferova ID, Golenyaev N, Yakupova G et al (2021) The Estimation of Traffic Flow Parameters based on Monitoring the Speed Values using Computer Vision. In: Berns K., Helfert M., Gusikhin O., (eds). International Conference on Vehicle Technology and Intelligent Transport Systems, VEHITS - Proceedings of Science and Technology Publications, Lda
26. Shepelev VD, Glushkov AI, Almetova ZV, Mavrin VG (2020) A study of the travel time of intersections by vehicles using computer vision. In: Berns K., Helfert M., Gusikhin O. (eds). VEHITS - Proc Int Conf Veh Technol Intell Transp Syst SciTePress. https://doi.org/10.3390/math11051144
27. Video surveillance. https://stream.is74.ru. Accessed 15 Dec 2023
28. On approval of the Methodology for determining emissions of pollutants into the atmospheric air from mobile sources for conducting summary calculations of atmospheric air pollution. Russian Federation (2019). https://www.garant.ru/products/ipo/prime/doc/73240708. Accessed 15 Dec 2023
29. On approval of methods for calculating the dispersion of emissions of harmful (polluting) substances in the atmospheric air. Russian Federation (2017). https://docs.cntd.ru/document/456074826. Accessed 15 Dec 2023
30. Kumar S, Ivanova O, Melyokhin A, Tiwari P (2023) Deep-learning-enabled multimodal data fusion for lung disease classification. Informatics in Medicine Unlocked 42:101367. https://doi.org/10.1016/j.imu.2023.101367

Functional and Ecological Approach to the Formation of a Coastal Binhai Park in Harbin

I. N. Maltseva, W. Sun, and N. N. Kaganovich

Abstract This article examines regional architectural and landscape traditions of China on the example of a conceptual project of a coastal park in the Qiunli district of the city Harbin. The methods, methodology and content of the centuries-old national landscape architecture that forms coastal parks are discussed. After analyzing the wishes, the functional scheme of the park has been developed for citizens of different age groups. Much attention is paid to the requirements of environmentalists for environmental protection and conservation untouched green areas and wetlands. The project of the park on the Sungari River is a gentrified area with a perfectly preserved landscape. After the project is implemented, the park will become a place of attraction for citizens and tourists, a business card of the Harbin.

Keywords Coastal park · National traditions · Environmental protection · Cultural heritage of China

1 Introduction

The globalization processes that have engulfed the world contribute to the complete erasure of "borders" between cities, countries, cultures, ethnicities. However, against the background of a cosmopolitan megapolis, people's need for spiritual culture has not diminished, and they are increasingly striving to live in a unique city. At present, humanity is coming to realize the value of individual cultures that have preserved their identity. Whether the urban landscape is distinctive or not, parks in every city of the world play a special role that cannot be ignored. Parks are not only works of landscape and architectural art in a city, but also an important platform for meeting the spiritual and cultural needs of the population, as well as a form of expression of unique national and regional architecture [1].

I. N. Maltseva (✉) · W. Sun · N. N. Kaganovich
Ural Federal University named after the First President of Russia B. N. Yeltsin, 19, Mira Street, Yekaterinburg 620002, Russia
e-mail: i.n.maltceva@urfu.ru

L. V. Radionova and D. V. Ulrikh (eds.), *Advances in Ecology and Environmental Engineering*, Springer Proceedings in Earth and Environmental Sciences,
https://doi.org/10.1007/978-3-031-64423-8_6

At the same time, with the continuous improvement of the quality of life, people's demands for habitat are also increasing. With the presence of a river in a city, the development of waterfront landscape becomes an inevitable trend, and waterfront parks naturally become a place of attraction for citizens to get in touch with water and nature [2].

Harbin is located on the banks of the beautiful Sungari River and is a truly unique and simply inimitable city. It is a city where completely different architectural styles and cultures of different nations coexist.

2 The Phenomenon of Harbin Architecture

China is an incredibly attractive and wonderful country with its own unique culture and particular architectural values [3], where the architecture of a specific region usually belongs to a unique cultural heritage, which has been carefully preserved since ancient times and up to now. Regional architecture, including landscaping, has a very big artistic significance and is a symbol of mature national Chinese culture. Using this heritage in creating a comfortable modern urban environment is the goal which many architects pursue [4].

Harbin is the center of politics, economy, culture and openness to the outside world in northeastern China, and a megacity with the largest area and a population of just over 10 million. Excessive urbanization has also caused many urban problems, such as traffic congestion, pollution, deteriorating air quality and so on, see Fig. 1. In recent years, Harbin has made active efforts to solve urban problems caused by the urbanization process and population growth.

Fig. 1 Harbin, China

A modern city, being a complex multifactor system, has another very important characteristic—its historical past—the time during which it existed and developed, formed its architectural and construction framework and urban infrastructure. It is as a result of the time factor that the city accumulates historical and architectural values and forms its socio-cultural component in the process of historical development. The time factor reflects not only the historical past, but also determines the urban planning potential, hence the strategies and prospects of urban development in the future [5].

Harbin is a city of diverse ethnic cultures, each with its own distinctive architectural features. The Han culture embodies traditional Chinese culture, the Manchu culture brings northern ethnic characteristics, and there is also a serious stylistic influence of Korean culture.

Due to special geographical factors, there was a Russian diaspora in Harbin, attracted by the construction of the Chinese Eastern Railway (CER). Russian way of life, Russian traditions and customs were formed on this territory, which certainly influenced the socio-cultural adaptation of Russians and the development of Russian architecture in China. Russians in Harbin not only preserved their national culture, spiritual values, language and mental environment, but also, in a different cultural environment, created a significant architectural heritage, made a great contribution to the development of Chinese culture [6]. Due to this, Harbin's urban architecture, ideas, lifestyle, language and art acquired the features of a foreign culture. For several decades, Harbin was considered a 'Russian island' in China [7]. This makes the image of the city more exotic, so it is sometimes referred to as 'Oriental Moscow', see Fig. 2.

This is what creates the unique essence of Harbin national character, and the identity of its architecture [8].

Fig. 2 Sofia square, Harbin, China

3 Basic Principles for the Formation of Waterfront Parks in China

The formation of a unified water-green framework of Harbin provides ecological balance of urban environment, presupposes the preservation of national traditions and the creation of significant natural areas in the structure of the waterfront that perform recreational and environmental functions. The integrated approach in modeling and structural transformation of coastal areas is based on the system of principles and methods of urban shorelines organization. Among the main principles of waterfront zones formation, the following are proposed.

The direct mapping method refers to a way to accurately and clearly reproduce elements of regional architecture without interference or additional artistic treatment [8]. This principle implies maximizing the inclusion of natural components into the structure of designed coastal spaces and respecting the existing natural resources. This is the basis for increasing the role of the natural framework in infilling waterfront spaces, which consists in the preferential selection of nature components for the purposes of eco restoration and the creation of new urban waterfronts [9]. For parks, this means the use of landscape design, composed of indigenous flowers, plants, bushes, trees, rock slides, hills, rivers, and ponds.

Most Chinese gardens are designed in this way. For example, the well-known Suzhou Gardens and Hangzhou Parks. They are called a Chinese paradise because of their natural features: they are surrounded by a series of lakes and are rich in virtually untouched nature, see Fig. 3.

Fig. 3 The humble administrator's garden has appeared in Suzhou, China

Fig. 4 Suzhou Gradens—the humble administrator's Garden, Suzhou, China

In these gardens, indigenous plants are combined with the existing natural landscape: mountains, rocks, water surface of ponds and lakes, creating organic Chinese-style parks and forming a unique style of a garden system with national architectural features [10].

The concept of the Suzhou Garden continues the ideas of Chinese parks, see Fig. 4. The park-garden design does not destroy the natural landscape; it respects nature and combines regional landscape architecture to form a unique example of a Chinese garden.

The principle of aesthetic harmonization refers to the use of indirect, euphemistic and ambiguous techniques to fully illuminate regional park architecture and planning. This principle focuses on enhancing the human-perceived visual qualities of the waterfront environment to achieve compositional unity of buildings, green space systems, visual orientation aids, small forms, etc. The use of this principle allows to ensure historical continuity and preservation of the 'spirit of place'.

Landscape architects of the park in Harbin resort to the simultaneous use of several elements of European style, which very carefully fit into the Chinese classical design. They contrast with each other and, at the same time, complement each other [11]. In line with classical Chinese traditions, three small bridges in the classical European style have been built over a small river in Zhaolin Park with untouched local nature, which reflects the harmonious fusion of traditional Chinese landscape architecture and European style, the convergence of multi-ethnic cultures in the park see Fig. 5 [12].

Fig. 5 Zhaolin Park, Harbin, China

4 Binhai Waterfront Park Project

Binhai Modern Multifunctional Park is located on the banks of the Sungari River in the Qunli District of Harbin. During the survey of the land plot, it was found that its area is 67159 square meters, the width of the river bank is from 12 to 79 m, and the total length along the river is 1222 m. The majority of the land plot consists mainly of agricultural land and the terrain has a slight slope. There is a complete lack of landscaping on the site, which provides an opportunity for a complete and conflict-free design of the area. There are scattered forests and densely growing bamboos on both sides of the river. The original vegetation on the site has been partially preserved after the park was laid out. Near the Sungari River, there is convenient transportation infrastructure, with the Yangmingtan Bridge crossing the river. The current appearance of the bridge is unremarkable and the cladding on the west side of the bridge is in the form of a rigid retaining wall, which is not aesthetically pleasing. In the new project, the retaining wall and the area surrounding the bridge will be designed using natural materials such as natural stone and wood, and landscaping to perceive the architectural environment as an extension of the natural environment, see Fig. 6.

4.1 File Naming and Delivery

The project concept is to create a multifunctional public space that is interconnected with the existing urban planning context as an element of the socio-cultural cluster forming in the area.

Fig. 6 Landscaping of the Yangmingtan Bridge area

The project objectives are to change the image of the city, preserve Chinese traditional architecture, create a tourist attraction, improve the living environment of residents, and preserve the green zone.

The environmental sustainability of the concept is aimed at creating 'conditions for sustainable urban development, for high, ecologically sound quality of waterfront areas, for restoring the city's water façade and maintaining ecological balance'. Green infrastructure of the urban waterfront, flood control, air purification, rainwater harvesting, and water conservation provide quality living conditions for urban residents.

The practicality of the concept lies in the feasibility of combining green areas with surrounding urban areas to meet the needs of city dwellers in recreation and sports, as well as maximizing the efficient and rational use of coastal resources.

The aesthetic component of the project is that the green corridor organized in the central part of the city becomes a symbol of Harbin's architectural image. At the same time, the park green zone reflects traditional landscape architecture, forms a regional landmark and is a visiting card of the city.

4.2 Functional Zoning of the Area

The project is "people-oriented" and meets the requirements of modern architecture. A social survey revealed that the city residents lack sports and children's playgrounds, long bicycle and jogging paths, quiet and aesthetic recreation areas in a green zone with clean air. Urban development of coastal zones is inextricably linked to the environmental safety of the city. The formation of a unified water-green framework of the city as the main planning means of ensuring the ecological balance of the urban environment involves the preservation and creation of significant natural areas in the structure of waterfronts, which perform recreational, environmental, as well as health improvement functions. The project takes into account the requirements of ecologists: a modern park should not damage the surrounding pristine nature, but should maximally preserve the original natural appearance of wetlands.

Thus, the design project was developed jointly by architects, urbanists and residents of the surrounding neighborhoods.

Taking into account the wishes of citizens of different ages from the surrounding residential neighborhoods, the park was divided into four functional zones: natural landscape, contemplative, cultural and exhibition, and sports, see Fig. 7. With the help of cultural and creative elements and landscape sketches, unique major landscape nodes such as park entrances, recreation plazas, rain gardens and open lawns are created to realize the creative heritage and continuation of the historical context in the cultural and exhibition zone it is planned to present ice sculpture art in winter.

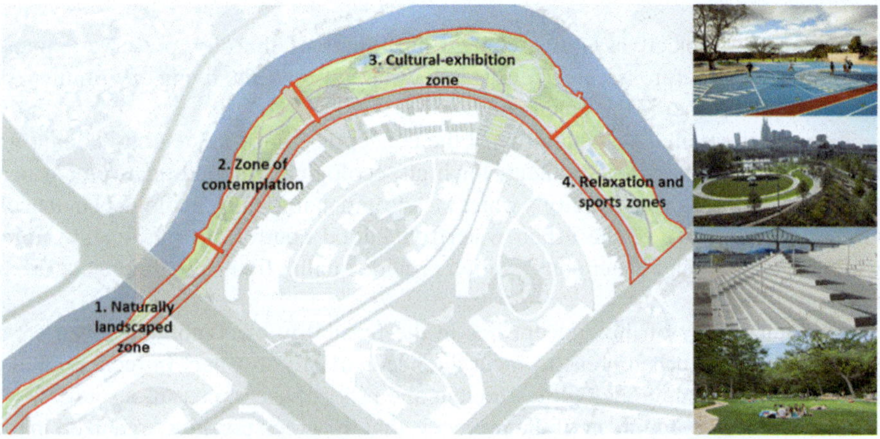

Fig. 7 Functional zoning of the waterfront park

Fig. 8 Venues and terraces for quiet relaxation for citizens

4.3 Landscape Architecture with Regional Peculiarities

The landscape design of the park is mainly developed by directly borrowing the traditions and elements of Chinese landscape architecture. In designing the park, first of all, the relief of the terrain—the slope of the waterfront—was taken into account. The topography allowed the design of a platform and viewing terraces, so that people can enjoy the natural scenery and attractions of the park, directly at the entrance to the park area. The project utilized the natural areas of the land plot to create quiet recreation areas, sports fields, which are becoming increasingly sought after by Harbin residents, see Fig. 8.

While providing the basic functions of the park, this project respects the local ecological environment and reflects the unique natural landscape of the Sungari River. The wetland part of the park has been beautified, with convenient and safe paths created so that visitors can contemplate the beautiful and diverse vegetation and wetland inhabitants, see Fig. 9. The project aims to achieve the goals of the China Environmental Management and Cultural Heritage Committee [13].

4.4 The Regional Character of the Park's Architecture

The design of landscape elements and small architectural forms, as mentioned earlier, should reflect Harbin's regional culture. Benches, street lamps and road signs inside the park are designed in the style of the original hunting and fishing culture of the

Fig. 9 Wetlands area

ancient city. Small architectural forms are inspired by Russian and European architecture to express the characteristics of Harbin's multicultural integration. Sculptures and murals are inspired by the local culture of unique ice and snow sculptures to fully demonstrate the architectural identity of the northeastern province of China [14]. The combination of different styles in the landscape design of the waterfront park is necessary to represent the multi-culture of the region, increase the interest of tourists and citizens in the park, and preserve the original traditions of the region.

4.5 The Use of Indigenous Plants

The selection of plants in the park takes into account regional characteristics and utilizes native plants from the cold northeastern region of China. The soil in the lowland areas is saturated with water and partially hypoxic, so the plants planted in the park were carefully selected according to the plant classification for water-saturated and normal areas of the park [15, 16].

In areas with water systems, seedlings that have the ability to tolerate overwatering and can grow normally after brief immersion in water were designed. Examples include metasequoia, pool cedar, river willow, maple poplar, liquidambar, and begonia. In the plots without water, plants common for this region were chosen for landscaping: ginkgo, eucalyptus, peach tree, cherry blossom, magnolia, camellia, see Fig. 10.

Seasonal changes were taken into account, plant diversity was increased and year-round green plants were included. The landscape design includes an organic combination of plants to significantly increase the decorativeness and color diversity of individual areas of the park. For example, a combination of evergreen plants (such as red pine, eucalyptus) plus ornamental plants (carnation and cloves) and deciduous plants (such as white birch or magnolia) creates a great effect [17, 18]. Many interesting compositions have been developed, taking into account the changing topography of trees, flowering shrubs, stone slides and water bodies.

Fig. 10 Green areas of the park: in water saturated areas (left), in normal areas (right)

5 Conclusion

In modern society, cities are devoid of regional architectural features and ecological systems have been destroyed. Binhai Nature Park on the banks of the Sungari River will become a necessary and important part of Harbin's comfortable urban environment. There, when planning open spaces of embankments, natural green areas will be preserved as much as possible, and at the same time platforms for active and quiet recreation of citizens will be included. In conclusion, with the development of megacities, the rich traditional Chinese park culture is being lost, so when designing gardens and parks, especially waterfront parks, it is necessary to integrate regional and national architectural and landscape traditions into design solutions. Modern parks should continue and develop the special cultural heritage of Chinese regions to become places of attraction and pride for city dwellers, areas for recreation, sports and entertainment.

References

1. Junfang Xie. Helen Woolley, Binyi Liu et al (2021) Overview of urban planning policy and urban green space system at a national level in China. In: IOP Conference Series Earth and Environmental Science 349(1):012021. https://doi.org/10.1088/1755-1315/349/1/012021
2. Marshall R (2001) Waterfronts in post-industrial cities. Spon Press, London.

3. Wen Xinyu., Qin Xiaoyan (2016) Harbin: A Unique City Where Different Cultures Meet. In: China Today. http://www.chinatoday.com.cn/chinese/mlzg/mlcs/201603/t20160329_800053394.html. Accessed 29 Dec 2022.
4. Razin AD, Cui Wenhe (2021) A Study on Regional Design of Coastal Landscape Planning in China. Sciences of Europe 64:16–17. https://cyberleninka.ru/article/n/issledovanie-regionalnogo-dizayna-pribrezhnogo-landshaftnogo-planirovaniya-v-kitae. Accessed 19 Feb 2022.
5. Komarova KA (2023) Aktualnye problemy ekspluatatsii i prisposobleniya pamyatnikov arkhitektury (Actual problems of operation and adaptation of architectural monuments). J Young Scientist 13(460):36–40.
6. Li YQ (2009) Creation of Regional Landscape in New Rural Construction. J Southwest Agricultural University J 7:86–89.
7. Krotova MV (2019) Fenomen Harbina kak polifonichnogo goroda (The Phenomenon of Harbin as a Polyphonic City). J Discussion Platform "Reflections of Historian" 3(2):18–37. https://doi.org/10.23859/2587-8344-2019-3-2-6
8. Guo Lijuan (2014) Excavation of regional culture to express regional characteristics. J Shanxi Architecture 34:206–214.
9. Voronina ON (2005) Park kak zona obshchestvennoy rekreatsii. Rezhimy funktsionirovaniya. Prioritet (The park as a public recreation area. Operating modes. Priorities). In: Velikie reki 2005: General reports, abstracts of the "Great rivers 2005" international congress reports, Nizhny Novgorod State University of Architecture and Civil Engineering, N. Novgorod, 17–20 May, pp 34–36.
10. Votinov VA (2013) Sovremennyye priyemy formirovaniya obshchestvennykh prostranstv v istoricheskoy srede goroda (Modern techniques for the formation of public spaces in the historical environment of the city). Bulletin of BSTU named after. V. G. Shukhova 5:11–14.
11. Wang Yi (2017) Analysis of the Russian Architectural Style in Harbin. J Building materials and finishing 39:47–52.
12. Xianglin Yu (2002) Russian Overseas Chinese Culture in Harbin during the Republic of China. J of Beijing University of Science and Technology: Social Sciences 39:7–15.
13. Che Yuning (2010) Building Harbin's Urban Landscape with Regional Cultural Features. J Art History 3:88–96.
14. Savelev MV, Kiseleva DA, Bondar NV et al (2019) Printsipy formirovaniya gorodskikh obshchestvennykh rekreatsionnykh zon naberezhnykh territoriy (The principles of the organization of public recreational areas in the city waterfront territories). Bulletin of Tomsk State University. Cultural studies and art history 33:173–187. https://doi.org/10.17223/22220836/33/15
15. Ma Yang, Li Pengbo (2013) Expression of the features of regional culture in the sketches of the urban landscape. J Modern Horticulture 10:52–54.
16. Razin AD, Cui Wenhe (2021) Issledovaniye regional'nogo dizayna pribrezhnogo landshaftnogo planirovaniya v kitaye (A study of the regional design of coastal landscape planning in China). J Sciences of Europe 64:16–17. https://cyberleninka.ru/article/n/issledovanie-regionalnogo-dizayna-pribrezhnogo-landshaftnogo-planirovaniya-v-kitae. Accessed 19 Feb 2023.
17. Yang Baojun Dong Ke (2007) Discussion of coastal urban design. J of Architecture 7:7–10.
18. Yao Yongzheng (2009) Landscape of a Chinese garden. China Forestry Press, Beijing, p 156.

Creating Sustainable Urban Transportation Systems Through Innovative Traffic Management Strategies

O. Fadina, Z. Almetova, O. Ivanova, A. Vorobyev, and G. Assanova

Abstract In recent years, research on urban microclimates has become increasingly important, drawing attention from both governments and scientists. Continuous monitoring and management of urban air quality are crucial for ensuring the health and well-being of city residents. In this study, an assessment of pollutant emissions at an intersection in Chelyabinsk city was conducted using the AIMS-Eco monitoring system. This system utilizes real-time video processing with the YOLOv4 neural network, which collects information on traffic movement and air quality. Based on data analysis, a mathematical model for dynamic regulation of the traffic flow speed in ensuring uninterrupted passage through a regulated intersection by group vehicles was developed. The evaluation of harmful emissions, specifically carbon dioxide, was performed. The model takes into account factors such as the number of vehicles in the queue, the length of the section, acceleration and speed of the traffic flow, and waiting times. The mathematical modeling demonstrated that by optimizing the speed regime of vehicles, it is possible to reduce emissions by up to 30%. Therefore, the establishment of sustainable urban transportation systems, aided by innovative traffic management strategies, plays a significant role in improving the lives of city residents and preserving the environment.

Keywords Queue of vehicles · Harmful emissions · Traffic flow · Neural network · Sustainable transport

O. Fadina (✉) · Z. Almetova · O. Ivanova
South Ural State University, 76, Prospekt Lenina,, Chelyabinsk 454080, Russia
e-mail: fadinaos@susu.ru

A. Vorobyev
Moscow Automobile and Road Construction State Technical University, 64, Leningradsky Prospekt, Moscow 125319, Russia

G. Assanova
NLC, "Akhmet Baitursynuly Kostanay Regional University", 47, Baytursynov Street, Kostanai 110000, Kazakhstan

L. V. Radionova and D. V. Ulrikh (eds.), *Advances in Ecology and Environmental Engineering*, Springer Proceedings in Earth and Environmental Sciences,
https://doi.org/10.1007/978-3-031-64423-8_7

1 Introduction

Modern society is characterized by urbanization and increasing urban density. This leads to a high number of vehicles and pedestrians in cities. While this lifestyle brings economic benefits, it also creates new environmental problems associated with transport emissions. Therefore, city authorities need to address these issues.

Various approaches to traffic management and the impact of vehicle speed, acceleration, and queue length on pollutant emissions are discussed in scientific papers. In the study [1], a similar method has been developed that utilizes existing video surveillance systems for real-time monitoring of traffic conditions and precise measurement of the composition and structure of non-platoon traffic flow. Paper [2] analyzes the mechanisms governing pollutant dispersion in urban street canyons and explores strategies for air pollution control near roadways. In [3], the levels and distribution of pollutants are analyzed in a specific area. Authors [4] have developed a model for air pollution dispersion in urban areas (UTAP). Researchers [5] have investigated the airflow distribution around buildings resulting from tests conducted in an aerodynamic tunnel. Paper [6] focuses on microscopic-scale traffic measurements combined with microscopic-scale emission modeling for a large study area. Authors in [7] present a model that combines macroscopic traffic modeling and emissions tailored for a capital district. This model allows estimating hourly emission variation on the road network level. Paper [8] presents an analysis of emission characteristics during vehicle movements. Authors [9] have created an innovative model for vehicle emissions, providing a balance between calculation accuracy and speed. Highly accurate CFD models have been developed by researchers, representing a comprehensive urban environment with well-designed buildings and trees in the target area [10]. In this work [11], the development and effectiveness of a new simulation tool for vehicle transport are described. The created model accurately predicts fuel consumption, NOX, and CO2 emissions with high temporal and spatial resolution. Research on mixed traffic management using Connected Vehicle Systems (CVS) and Cooperative Automated Vehicle Systems (CAVS) to enhance control efficiency has also been explored [12]. An assessment of the impact of connected and automated vehicles on fuel consumption and emissions in a mixed flow on a high-speed highway was conducted [13]. These studies examined the influence of CVS and CAVS on urban traffic management efficiency and their contribution to fuel consumption and emissions reduction. In a study conducted by the authors [14], the effects of speed limit and traffic signal control on emissions of harmful substances such as NOx and PMx were investigated. The study [15] examined the impact of vehicle speed on fuel consumption, acceleration, and emissions of gas pollutants like NOx, CO2, and CO. Research has been conducted to identify areas with the highest air pollution levels resulting from gas emissions by studying traffic flows [16].

In our previous studies, we conducted a comparative assessment of emissions from road traffic at urban intersections. We also examined the details of the traffic patterns at these intersections [17], using neural networks to forecast the volume of emissions associated with traffic flow [18].

2 The Survey of Traffic Flow Intensity

To determine the quantitative movement of traffic at the intersection of Komsomolsky Avenue and Voroshilova Street in Chelyabinsk, a series of measures were undertaken to study the intensity of traffic flow. The intensity of vehicle movement at the intersection is defined as the equivalent hourly flow rate, calculated as the ratio of the number of vehicles to the time interval over which this number of vehicles passes through a specific section of the street or road [19]. The survey of traffic flow intensity was conducted every 20 min throughout the day, taking into account the following types of vehicles, to obtain the necessary data for calculating the maximum instantaneous and aggregate emissions from motor vehicles (According to the Order of the Ministry of Natural Resources of Russia dated 27.11.2019 No. 804 and GOST R 56,162–2019 [20]) (Table 1).

The captured images of the traffic flow from a camera were processed and merged into digital streams using a trained convolutional neural network (CNN).

Figure 1 shows the boundaries of the zone for measuring traffic parameters (Komsomolsky Avenue—Voroshilov Street, Chelyabinsk).

In order to make it easier for visual analysis, due to the large amount of data, this study presents hourly indicators of traffic parameters (Fig. 2).

Based on the hourly daily intensity (Fig. 2), a morning peak hour from 7:00 to 10:00 and an evening peak hour from 16:00 to 20:00 were identified. These data are necessary to determine the off-peak hours, as the study was conducted for this time of day. During off-peak hours, there are many business trips as well as personal movements of the population. Therefore, it is important to ensure comfortable movement without delays and with the least air pollution for road users during this time.

In off-peak hours at this intersection, the average number of individual vehicles and the amount of CO emissions generated by these vehicles was determined (Fig. 3).

Table 1 Categories of vehicles

Vehicle category	Vehicle
I	Passenger cars
II	Vans and Minibuses up to 3.5 tons (cargo-passenger vans with seating capacity of up to nine, including the driver's seat, and light vans, both passenger and cargo vehicles with low carrying capacity, regardless of the presence or absence of trailers, including camper vans and recreational vehicles)
III	Trucks from 3.5 to 12 tons (two-axle and three-axle trucks without trailers)
IV	Trucks over 12 tons (four-axle trucks, as well as trucks with one or more trailers; tractor units with semi-trailers and one or more trailers; tractor units without trailers and semi-trailers; and specialized vehicles such as self-propelled road rollers, bulldozers, cranes, and military-style tanker trucks)
V	Buses over 3.5 tons (city buses, intercity buses)

Fig. 1 Image of the intersection under examination (Komsomolsky Avenue—Voroshilov Street, Chelyabinsk)

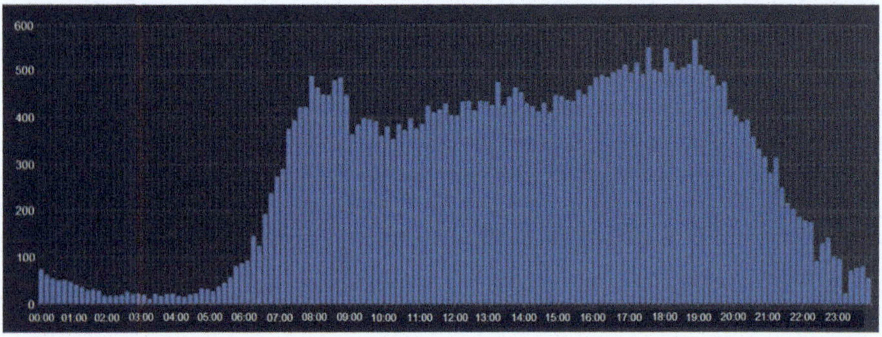

Fig. 2 The daily intensity of the traffic flow at the intersection (units/20 min)

Correct, the amount of emissions will increase depending on the idle time of individual vehicles while waiting for a green signal at the traffic light.

On Fig. 4, we can see that buses and trolleybuses have the highest idle time on the studied intersection—18 s. Passenger cars have an idle time of 11 s. The cumulative idle time per day is 120 h.

The speed of vehicles also affects the amount of emissions. The lowest amount of emissions is formed at a speed of 60 km/h (Table 2).

On Fig. 5, we can see that passenger and commercial vehicles have the highest speed at the intersection, approximately 27 km/h, while trolleybuses have the lowest speed at 17 km/h. From this data, we can conclude that the intersection has a low speed, which increases the amount of emissions, as shown in Table 2.

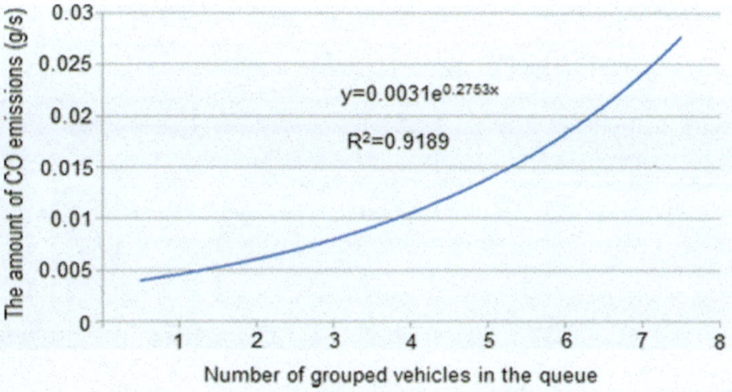

Fig. 3 The amount of CO emissions (g/s) depending on the number of grouped vehicles in the queue

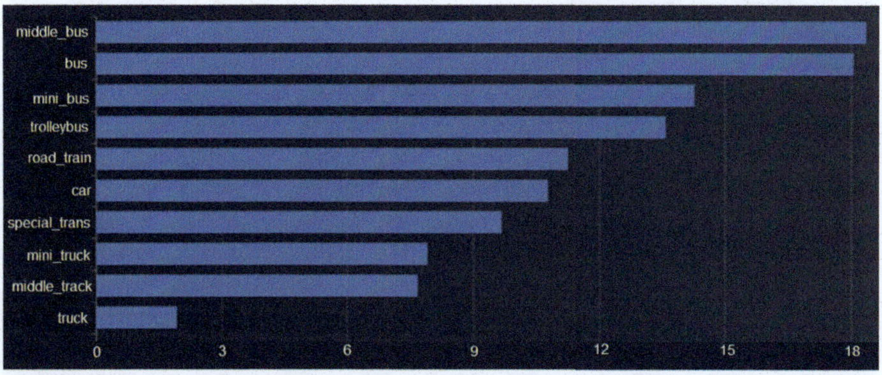

Fig. 4 The average idle time per day

Table 2 Values of the coefficient $r_{vk.l}$ taking into account changes in emissions depending on the average vehicle speed [20]

	Travel speed *V* (km/h)					
V	30	35	40	45	50	60
$r_{vk.l}$	1.00	0.90	0.75	0.65	0.50	0.30

The amount of emissions also depends on the category of vehicles. Figure 6a represents the structure of the traffic flow and Fig. 6b shows the sources of CO emissions at the intersection per day.

From Fig. 6, we can conclude that the highest amount of CO is generated by passenger vehicles, as well as buses and category II trucks, which make up only 8% of the entire traffic flow.

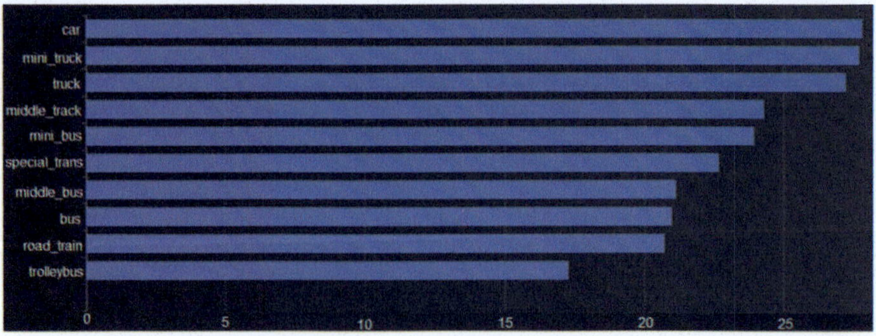

Fig. 5 The average speed of vehicles per day (km/h)

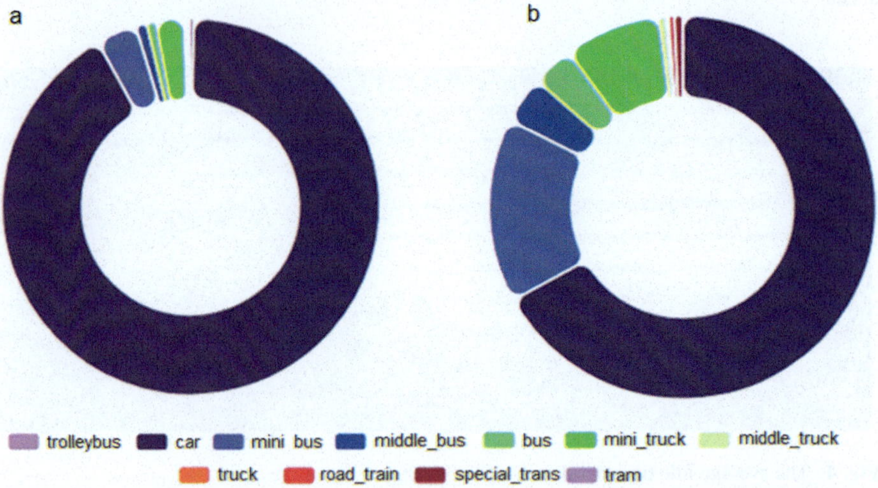

Fig. 6 **a** Structure of the traffic flow; **b** Sources of CO emissions (g)

The analysis of the data has shown that it is important to prioritize the passage of public transport at this intersection in order to adhere to the schedule and reduce idle time, thus increasing the average speed.

3 The Method of Calculating the Time of Non-Stop Travel of a Section of the Road Network

To ensure uninterrupted passage between two consecutive nodes of the road network, the following times need to be calculated:

- The time required for the first vehicle to cross the length of the intersection (the distance from the stop line to the far boundary of the crossed carriageway) after the green signal of the traffic light. This time depends on the length of the intersection and the driver's reaction time.
- The time taken by the first vehicle, after crossing the far boundary of the crossed carriageway, to reach the maximum allowed speed (60 km/h) of the traffic flow at a distance from the next intersection. This time depends on the speed at which the vehicle left the far boundary of the crossed carriageway.
- The time required for the first vehicle to change its speed from the maximum to the recommended speed indicated by a dynamic sign.
- The time taken by the first vehicle from reaching the recommended speed to reaching the stop line of the next intersection, in the absence of any queue of individual vehicles.

To determine the time interval within which the car will reach the stop line of the next intersection from the moment the green light signal is turned on at the previous one (t_p), the following formula can be used:

$$t_p = \left(t_i^{sl} + t_i^{\text{int}} + t_i^r\right) + \frac{\left(v_i^m - v_i^{\text{int}}\right)}{a_i} + \frac{S_{ds}}{v_i^m} + \frac{(S_{uds} - S_v - S_{ds})}{v_i^m} \tag{1}$$

where: $t^{sl}{}_{\text{i}}$ is the time taken for the i-th vehicle to reach the stop line from the moment the traffic signal turns green (s); $t^{int}{}_i$ is the time taken for the i-th vehicle to traverse the length of the intersection (s); $t^r{}_i$ is the driver's reaction time for the 1st vehicle (delay in starting the first car) (s); $v^m{}_i$ is the maximum allowed speed for the i-th vehicle between intersections (m/s); $v^{int}{}_i$ is the speed attained by the i-th vehicle at the far boundary of the crossed roadway (m/s); a_i is the acceleration of the i-th vehicle between intersections (m/s^2); S_{ds} is the distance to the dynamic sign from the point where the i-th vehicle reached $v^m{}_i$ (m); S_{uds} is the distance from the far boundary of the crossed roadway to the stop line of the next intersection (m); S_v is the distance required for the vehicle to accelerate from its current speed at the far boundary of the crossed roadway to the maximum allowed speed (m).

The time taken by a vehicle to cross an intersection depends on the acceleration at which the i-th vehicle crosses the intersection, it is determined by:

$$t_i^{\text{int}} = \sqrt{\frac{2 \cdot S_{\text{int}}}{a_i^{\text{int}}}} \tag{2}$$

where S_{int} is the length of the intersection (m); $a^{int}{}_i$ is the acceleration of the i-th vehicle while crossing the intersection (m/s^2).

The distance required for a vehicle to accelerate from the initial velocity of the vehicle at the far boundary of the road being crossed to the maximum permitted speed is determined by:

$$S_v = \frac{\left(v_i^m\right)^2 - \left(v_i^{\text{int}}\right)^2}{2 \cdot a_i} \quad (3)$$

To calculate the time it takes for a vehicle to reach the stop line of the next intersection on a green signal, it is necessary to consider the fact that by the time the vehicle approaches the stop line, the traffic light should already be in the green signal mode. Otherwise, the vehicle will have to engage in emergency braking and come to a stop, and then resume movement.

According to [21], the restrictions on t_p should be as follows:

$$20s \leq t_p \leq 120s \quad (4)$$

4 The Method of Calculating the Speed Limit of a Dynamic Sign

Continuous movement of group vehicles as they approach the next intersection is proposed to be ensured by reducing the speed in advance to the required level. This value should be set as a recommended speed compared to the average flow speed between the nodes of the road network and indicated on the dynamic road sign "6.2".

Thus, it is assumed that the dynamic road sign will display a recommended speed for the traffic flow, at which the passage of the next node of the road network will occur without any forced stops.

In cases where non-group vehicles have accumulated at the next intersection during the red signal phase, or not all vehicles in the queue have been able to leave the intersection during the green signal phase, it will be necessary to adjust the speed between the intersections for uninterrupted flow of traffic from the previous intersection. This adjusted speed will be indicated on the dynamic sign "6.2".

The recommended speed for the dynamic sign, in the presence of non-group vehicles at the next intersection, depends on the time it takes for the queue to clear.

$$V_{ds} = \frac{(S_{uds} - S_v - S_{ds})}{\frac{(S_{uds} - S_v - S_{ds})}{v_i^m} + \sum t_i^{sl}} \quad (5)$$

where $\sum {t^{sl}}_i$ refers to the total time it takes for the non-group vehicles in the queue to clear (s).

The time it takes for the non-group vehicle queue to clear depends on the number of vehicles in the queue and their category. The limitations for the recommended speed include:

$$V_{\min} \leq V_{DS} \leq V_{\max} \quad (6)$$

According to [21], the maximum allowed speed in urban areas is 60 km/h, and there is no specific minimum threshold set. However, considering driving psychology, this value should not be lower than 20–30 km/h.

5 Investigation of the Effect of Speed

Based on the proposed methodology for calculating the speed regime for the dynamic sign, calculations of the dynamic speed have been performed for traffic flow on sections of different lengths, depending on the queue of non-group vehicles (Fig. 7), considering different initial speeds at the far boundary of conflicting directions of the lead vehicle.

Thus, if we consider the influence of the lead vehicle's initial speeds on a single section (Fig. 7), it is evident that the higher the initial speed (i.e., the speed at which the vehicle leaves the intersection zone), the smaller the magnitude of the change in dynamic speed required for uninterrupted passage of the node in the road network.

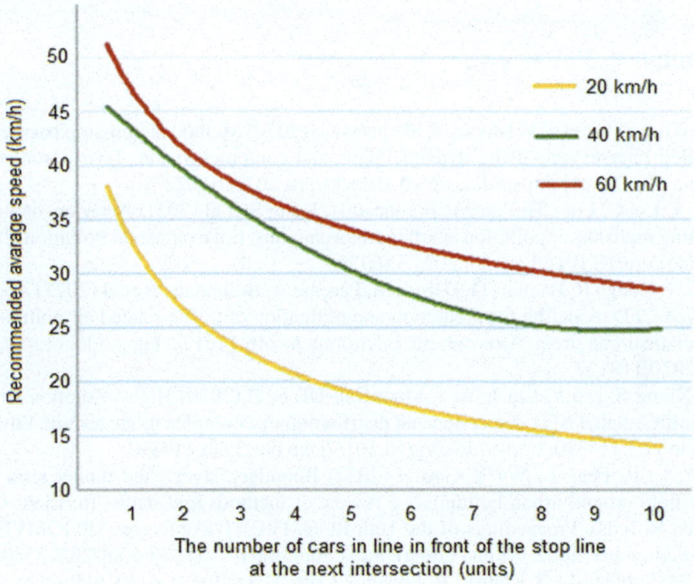

Fig. 7 The recommended speed depends on the initial speed and the number of cars in the queue

6 Conclusions

The analysis of the data in this study demonstrates the need for measures to minimize emissions at the surveyed intersection. Stopping vehicles at red traffic lights and delays in movement, as well as low speeds, are among the primary factors contributing to harmful emissions in urban areas. A significant amount of gas emissions (67%) is produced by passenger cars, with small capacity buses contributing 20%.

The mathematical model for dynamic traffic speed regulation can help reduce vehicle delays and increase traffic flow speed. By utilizing calculations based on this model, waiting times for vehicles at traffic lights can be reduced by 20%, the average speed of vehicle movement can be increased by 10–15%, and the air quality in the city can be improved by 30%. This optimization will facilitate smoother traffic flow, enhance the efficiency of urban transportation, and create a safer and environmentally cleaner environment.

Therefore, the use of real-time intelligent transportation systems offers the opportunity for authorities to address specific issues contributing to air quality deterioration in the city while establishing sustainable urban transportation systems.

References

1. Gorodokin V, Shepelev V, Buyvol P, Shepeleva E (2020) Method of non-stop passage of signal-controlled intersections using dynamic signs and computer vision. Transportation Research Procedia 50:174–81. https://doi.org/10.1016/j.trpro.2020.10.022
2. Huang Y, Lei C, Liu CH, Perez P, Forehead H, Kong S et al (2021) A review of strategies for mitigating roadside air pollution in urban street canyons. Environmental Pollution. 280:116971. https://doi.org/10.1016/j.envpol.2021.116971
3. Lauriks T, Longo R, Baetens D, Derudi M, Parente A, Bellemans A et al (2021) Application of Improved CFD modeling for prediction and mitigation of traffic-related air pollution hotspots in a realistic urban street. Atmospheric Environment 246:118127. https://doi.org/10.1016/j.atmosenv.2020.118127
4. Fu X, Xiang S, Liu Y, Liu J, Yu J, Mauzerall DL et al (2020) High-resolution simulation of local traffic-related NOx dispersion and distribution in a complex urban terrain. Environmental Pollution 263:114390. https://doi.org/10.1016/j.envpol.2020.114390
5. Zhao Y, Li R, Feng L, Wu Y, Gao N (2022) Boundary layer wind tunnel tests of outdoor airflow field around urban buildings: A review of methods and status. In: Li A, Olofsson T, Kosonen R, (eds). Proceedings of the 16th ROOMVENT Conference (ROOMVENT 2022), 16–19 September, 2022, Xi'an (China). https://doi.org/10.1051/e3sconf/202235604031
6. Nyhan M, Sobolevsky S, Kang C, Robinson P, Corti A, Szell M et al (2016) Predicting vehicular emissions in high spatial resolution using pervasively measured transportation data and microscopic emissions model. Atmospheric Environment 140:352–63. https://doi.org/10.1016/j.atmosenv.2016.06.018
7. Rodriguez-Rey D, Guevara M, Linares MP, Casanovas J, Salmerón J, Soret A et al (2021) A coupled macroscopic traffic and pollutant emission modelling system for Barcelona. Transportation Research Part D: Transport and Environment 92:102725. https://doi.org/10.1016/j.trd.2021.102725

8. Wu Y, Song G, Yu L (2014) Sensitive analysis of emission rates in MOVES for developing site-specific emission database. Transportation Research Part D: Transport and Environment 32:193–206. https://doi.org/10.1016/j.trd.2014.07.009
9. Jamshidnejad A, Papamichail I, Papageorgiou M, De Schutter B (2017) A mesoscopic integrated urban traffic flow-emission model. Transportation Research Part C: Emerging Technologies 75:45–83. https://doi.org/10.1016/j.trc.2016.11.024
10. Brozovsky J, Simonsen A, Gaitani N (2021) Validation of a CFD model for the evaluation of urban microclimate at high latitudes: A case study in Trondheim, Norway. Building and Environment 205:108175. https://doi.org/10.1016/j.buildenv.2021.108175
11. Smit R (2013) Development and performance of a new vehicle emissions and fuel consumption software (PΔP) with a high resolution in time and space. Atmospheric Pollution Research 4(3):336–45. https://doi.org/10.5094/APR.2013.038
12. Li J, Yu C, Shen Z, Su Z, Ma W (2023) A survey on urban traffic control under mixed traffic environment with connected automated vehicles. Transportation Research Part C: Emerging Technologies 154:104258. https://doi.org/10.1016/j.trc.2023.104258
13. Yao Z, Wang Y, Liu B, Zhao B, Jiang Y (2021) Fuel consumption and transportation emissions evaluation of mixed traffic flow with connected automated vehicles and human-driven vehicles on expressway. Energy 230:120766. https://doi.org/10.1016/j.energy.2021.120766
14. Rhode J, Wagner P, Ziemke T (2022) The effect of speed limits and traffic signal control on emissions. Procedia Computer Science 201:568–73. :https://doi.org/10.1016/j.procs.2022.03.073
15. Coman G, Burciu MS, Baroiu N (2019) Vehicles emissions under different driving conditions in urban areas. Rev Chim 70(2):438–41. https://doi.org/10.37358/RC.19.2.6930
16. Ambroziak T, Jachimowski R, Pyza D, Szczepański E (2014) Analysis of the traffic stream distribution in terms of identification of areas with the highest exhaust pollution. AoT 32(4):7–16. https://doi.org/10.5604/08669546.1146993
17. Shepelev V, Glushkov A, Fadina O, Gritsenko A (2022) Comparative Evaluation of Road Vehicle Emissions at Urban Intersections with Detailed Traffic Dynamics. Mathematics 10(11):1887. https://doi.org/10.3390/math10111887
18. Shepelev V, Glushkov A, Slobodin I, Balfaqih M (2023) Studying the relationship between the traffic flow structure, the traffic capacity of intersections, and vehicle-related emissions. Mathematics 11(16):3591. https://doi.org/10.3390/math11163591
19. Transportation Research Board, National Academies of Sciences, Engineering, and Medicine. Highway Capacity Manual 7th Edition: A Guide for Multimodal Mobility Analysis. Washington, D.C.: National Academies Press, 2022. https://doi.org/10.17226/26432
20. Order of the Ministry of Natural Resources of the Russian Federation No. 804 (11.27.2019). On approval of the methodology for determining emissions of pollutants into the atmospheric air from mobile sources for conducting summary calculations of atmospheric air pollution. https://www.garant.ru/products/ipo/prime/doc/7324070
21. Traffic regulations of Russia (2023). https://www.pdd24.com

Real-Time Modeling of Traffic Flow and Emissions for Enhancing Urban Air Quality

V. Shepelev, S. Zhankaziev, L. Fatikhova, I. Alferova, and M. Aitbayev

Abstract Reducing emissions intensity associated with road traffic is of paramount importance and requires careful investigation to improve air quality. Various models are employed to assess air pollution, providing quantitative evaluations to optimize traffic management. Intelligent driver models account for the human factor to represent real-world behavior. Real-time modeling using video surveillance cameras enables precise monitoring of traffic flows. Neural networks process high-speed data for comprehensive analysis of traffic movement and emission levels. This study focuses on modeling traffic flows with different vehicle categories approaching regulated intersections and calculates emissions of CO, NOx, PM2.5, and PM10. The model results demonstrate good correlation with laboratory measurements. The methodology enables effective implementation of measures within intelligent transportation systems, combating congestion, and improving urban air quality.

Keywords Intelligent driver model · Intelligent transportation systems

---

V. Shepelev (✉) · I. Alferova
South Ural State University, 76, Prospekt Lenina, Chelyabinsk 454080, Russia
e-mail: shepelevd@susu.ru

S. Zhankaziev
Moscow Automobile and Road Construction State Technical University, 64, Leningradsky Prospekt, Moscow 125319, Russia

L. Fatikhova
Kazan Federal University, 18, Kremlyovskaya Street, Kazan 420008, Russia

M. Aitbayev
Kostanay Socio-Technical University named after Academician Zulkharnai Aldamzhar, 27, Kobylandy Batyr Avenue, Kostanay 110000, Kazakhstan

L. V. Radionova and D. V. Ulrikh (eds.), *Advances in Ecology and Environmental Engineering*, Springer Proceedings in Earth and Environmental Sciences,
https://doi.org/10.1007/978-3-031-64423-8_8

1 Introduction

Reducing and managing the intensity of emissions associated with road traffic is an important task that requires detailed examination. Mitigating traffic congestion has a positive impact on air quality [1–3]. Numerous models assist in evaluating emission hotspots from vehicles, providing quantitative assessments of air quality necessary for optimizing traffic management efficiency.

Urban traffic flows (TF) are highly diverse and consist of various types of vehicles (VT) with different dynamic characteristics and dimensions [4, 5]. TF is inherently unpredictable and subject to various influences, some of which are beyond human control. Therefore, different approaches are used to model TF dynamics [6–8]. The intelligent driver model is an approach that accounts for human factors, such as reaction time, anticipation, and errors, to represent real-world traffic behavior [9–11]. This model offers several advantages as it incorporates human factors that influence the nature of real-world traffic movement.

Currently, models that receive real-time data [12] and video surveillance cameras [13–15] are widely used for analyzing traffic conditions and monitoring traffic flows. Continuous video streaming provides immediate data on vehicle movement parameters, which can be effectively and accurately processed using neural networks. The large datasets generated allow for the analysis of traffic situations, the calculation of pollutant emissions from vehicles, and the identification of hidden patterns between traffic parameters and pollutant concentrations. The resulting current information facilitates the implementation of necessary measures in intelligent transportation systems to tackle congestion and improve urban air quality.

The actual state of air quality at the regulated intersection is a complex task with multiple different aspects [16, 17]. The goal of this study is to model the traffic flow based on real-time data and regulatory documents [18, 19] in order to calculate the quantity of harmful substances emitted by vehicles at the regulated intersection and to identify the relationship between the position of a vehicle in the queue and the quantity of emitted harmful substances.

In this article, based on previous studies [13, 17], where the work of a trained neural network was presented, the traffic flow was simulated with different types of vehicles approaching the regulated intersection, and the quantities of harmful substances, such as CO, NOx, PM 2.5, and PM 10, were calculated. The calculated data presented here show a good correlation with the data obtained from real-world measurements.

2 Materials and Methods

A developed AI-based software system was used to collect data on traffic flow [20]. Its objective is an autonomous approach capable of assessing quantitative and qualitative parameters of road traffic, including vehicle quantity, speed, trajectory, travel time, and idle time, as well as emissions and pollutant concentrations. For real-time object recognition tasks, the YOLOv4 neural network demonstrated the best performance [21, 22]. The system relies on the use of static cameras with a wide field of view, providing visibility across the entire physical area of the intersection and the adjacent roads. Several cameras belonging to the "Intersvyaz" company with open access in the city of Chelyabinsk were used to train the neural network [23]. The computation of traffic flow parameters is presented in [24, 25].

For additional analysis of the impact of traffic flow structure on PM2.5 emissions, a simulation program was developed using the Python programming language. Visualization was implemented using the OpenCV library.

The program is based on the intelligent driver model, which utilizes intuitive behavioral parameters like desired speed, acceleration, and comfortable deceleration. The model calculates the acceleration of the vehicle (*a*).

$$a = k \cdot a_{\max} \tag{1}$$

where a_{max}—maximum possible acceleration (m/s^2); *k*—the coefficient calculated Eq. 2:

$$k = 1 - \left(\frac{V_c}{V_{\max}}\right)^q - \left(\frac{S}{S_{\min}}\right)^2 \tag{2}$$

where Vc—current vehicle speed (m/c); V_{max}—maximum possible speed (m/s); q—the coefficient of driving aggressiveness; S_{min}—minimum distance to the vehicle in front (m); S—the dynamic desired gap, calculated Eq. 3:

$$S = S_{\min} + V_c \cdot T + \frac{V_c \cdot dV}{2 \cdot \sqrt{a_{\max} \cdot |a_{\min}|}} \tag{3}$$

where T—the safe time headway (s); dV—the difference between the speeds of the current and the vehicle ahead (m/s); a_{min}—the minimum possible acceleration during braking (m/s^2).

To select the parameter q, its effect on the coefficient k was analyzed. The dependence of the parameter k on q is shown in Fig. 1.

The following parameters q were adopted for different vehicle categories: for I (passenger cars)—10, for III (trucks from 3.5 to 12 tons) and for V (buses over 3.5 tons)—4. Queues of 10 vehicles with one car of category III or V were studied. The simulation starts with the following initially set parameters:

- the number of vehicles in front of the stop line—0;
- distance from the stop line to the first approaching vehicle—100 m;
- the speed of vehicles approaching the intersection—50 km/h;
- the number of vehicles approaching the intersection—10;
- the number of vehicles of category III or V in the queue approaching the intersection—1;

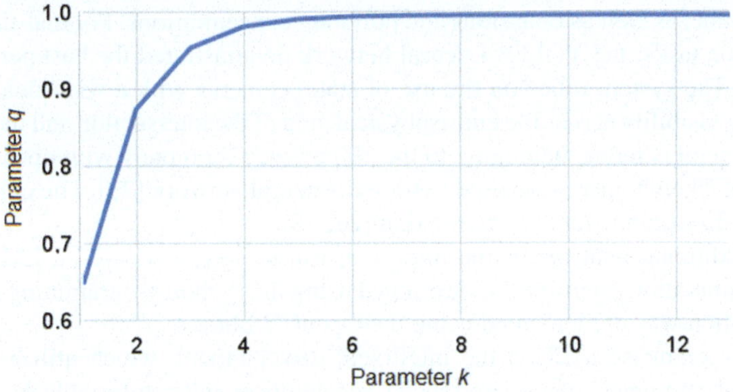

Fig. 1 The dependence of the parameter k on q

Restrictions are set for passenger vehicles:

- vehicle length (4.5 ± 0.5 m);
- minimum distance to the next vehicle (2.7 ± 0.5 m);
- maximum acceleration (2.1 ± 0.2 m/s^2);
- acceleration during braking (-3 ± 0.2 m/s^2);
- start delay (1.7 s).

Restrictions are set for vehicles of category III:

- vehicle length (7 ± 0.5 m);
- minimum distance to the next vehicle (2.7 ± 0.5 m);
- maximum acceleration (1.26 ± 0.2 m/s^2);
- acceleration during braking (−3 ± 0.2 m/s^2);
- start delay (1.7 s).

Restrictions are set for vehicles of category V:

- vehicle length (12 ± 1 m);
- minimum distance to the next vehicle (2.7 ± 0.5 m);
- maximum acceleration (1.2 ± 0.2 m/s^2);
- acceleration during braking (−3 ± 0.2 m/s^2);
- start delay (1.7 c).

Calculations are performed 25 times per second. Each time, starting with the first vehicle, the acceleration, velocity, and coordinates are computed. For the first vehicle, the acceleration is assumed to be at its maximum. Additionally, when starting from a standstill, subsequent vehicles do not immediately begin moving but instead experience a delay (a characteristic of each vehicle).

The queue of vehicles approaches the intersection during the red signal of the traffic light and stops before the stop line. After the green signal is activated, the vehicles start moving. The average speed and emissions are calculated within a specified zone throughout the entire simulation time (174–100 m before the stop line

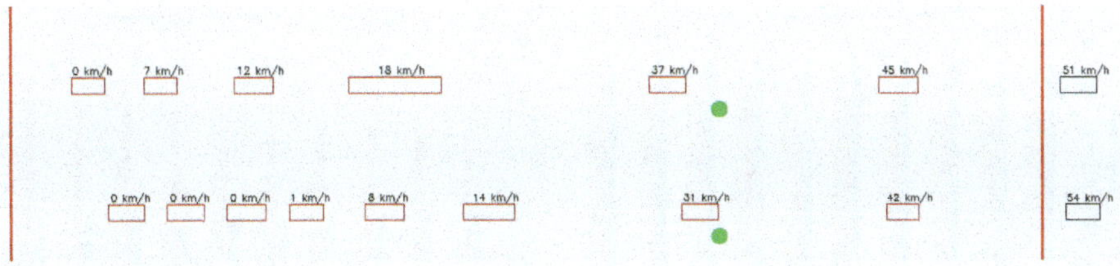

Fig. 2 Examples of configuration modeling with TC III and V categories

and 74 m after it). To calculate the average speed, intervals where the vehicle speed was zero are not considered. An example simulation is shown in Fig. 2.

Table 1 shows the simulation data of a queue with one vehicle of a given category in all positions. Each configuration was simulated 10 times, and the average value of all attempts was calculated as a result.

Figure 3a shows the relationship between the average speed of all vehicles and the position of the investigated category vehicles in the queue. The presence of category

Table 1 Simulation data of a queue with one vehicle of a given category in all positions

Category	Number in the queue	Average TF rate (km/h)	Total number of emissions (g)
I		27.31	0.08611
III	1	25.06	0.23317
III	2	24.97	0.24295
III	3	25.26	0.25133
III	4	25.53	0.25960
III	5	25.77	0.26784
III	6	26.03	0.27616
III	7	26.30	0.28460
III	8	26.54	0.29258
III	9	26.78	0.30056
III	10	27.03	0.30854
V	1	25.38	0.16443
V	2	25.15	0.16333
V	3	25.44	0.16716
V	4	25.67	0.17080
V	5	25.89	0.17450
V	6	26.10	0.17815
V	7	26.34	0.18198
V	8	26.58	0.18563
V	9	26.80	0.18912
V	10	27.05	0.19284

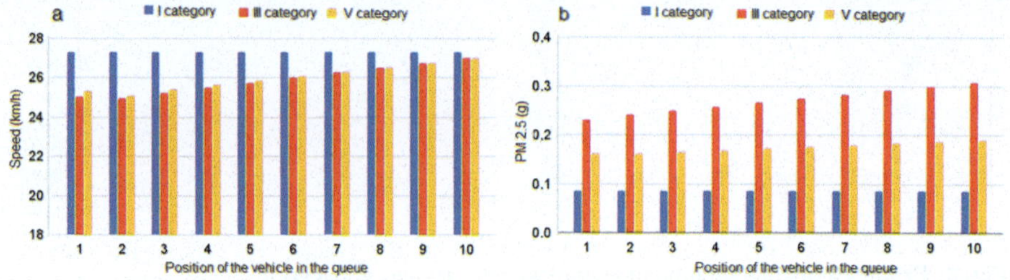

Fig. 3 Dependence on the position of the vehicle of the studied category in the queue **a** the average speed of all vehicles; **b** PM2.5 emissions

III and V vehicles in the queue slows down the traffic flow compared to a queue consisting only of passenger vehicles. In particular, vehicles at the beginning of the queue experience greater deceleration since, during the start of the green signal of the traffic light, the speed of the entire queue is determined by the leading vehicle. The difference in speed impact between category III and V vehicles is not significant.

Figure 3b illustrates the relationship between PM2.5 emissions and the position of the investigated category vehicles in the queue. It is evident that category III vehicles have a significant influence on emissions, as the coefficients used to calculate emissions for category III vehicles are approximately twice as large as those for category V. The further back in the queue category III and V vehicles are, the higher the emissions generated by the queue since the idle time of heavy vehicles increases, and they contribute the most to environmental pollution.

3 Forecasting the Number of Vehicles and RM 2.5

The first developed neural network predicted the quantity of vehicles. The best configuration was selected as a model with 3 layers and 300 neurons in each layer. The test results of this model for two consecutive time intervals within an 8-h period are presented in Fig. 4.

The comparison of the predicted and actual traffic flow in terms of the number of vehicles accumulated over 48 twenty-minute intervals (totaling 16 h) showed an increase in the predicted values from 42,345 to 44,120 units, which amounts to only a 4.2% deviation.

The second developed neural network predicted the amount of PM2.5 emissions. The best configuration was selected as a model with 4 layers and 600 neurons in each layer. The test results of this configuration for a duration of 24 h (72 intervals of 20 min each) are shown in Fig. 5.

The comparison of the cumulative emissions of PM2.5 particulate matter over a 24-h period is presented in Fig. 6. The calculation showed a decrease in the predicted daily values from 477.5 g to 448 g, which amounts to only a 6.2% deviation from the actual values.

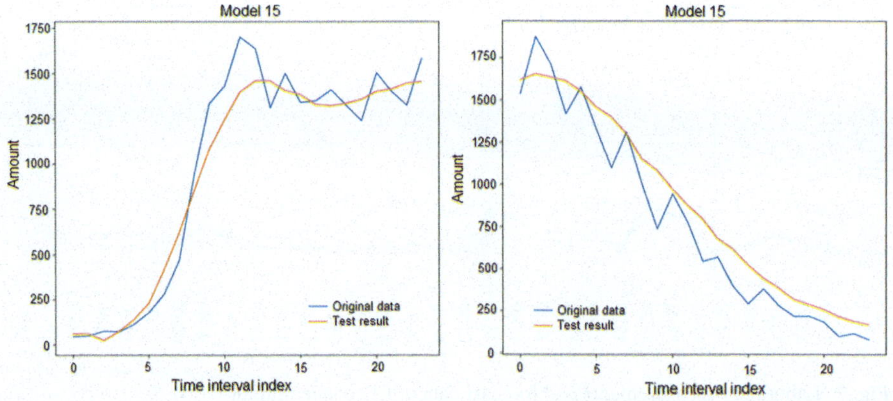

Fig. 4 The original and predicted by the neural network number of vehicles over two time periods

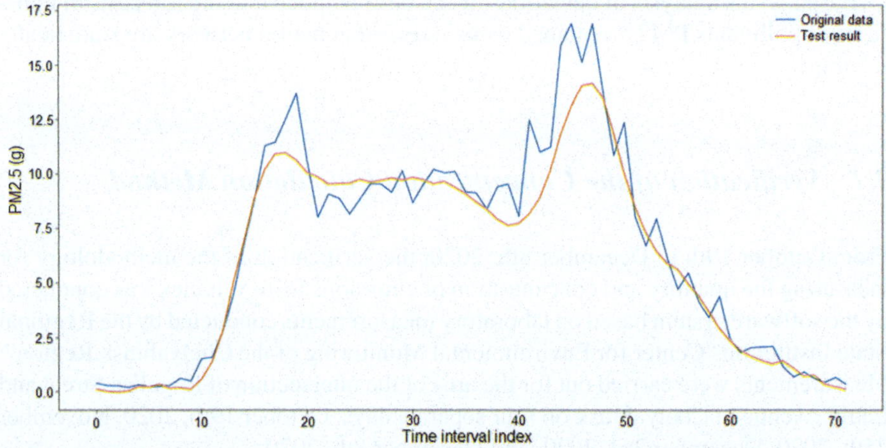

Fig. 5 Measured PM2.5 emissions over 24 h and PM2.5 emissions forecasted by the neural network

Fig. 6 Actual and forecasted cumulative PM2.5 emissions

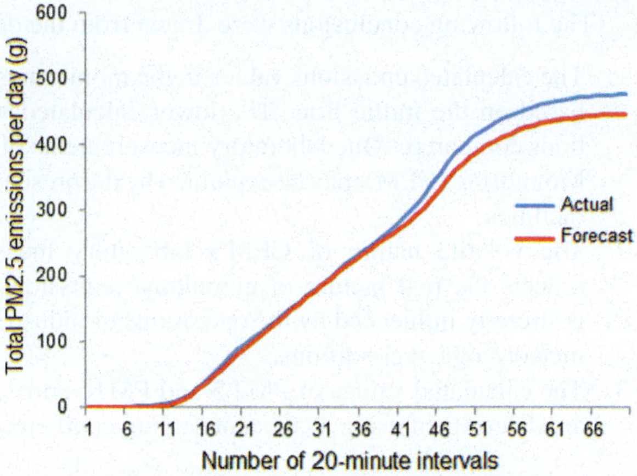

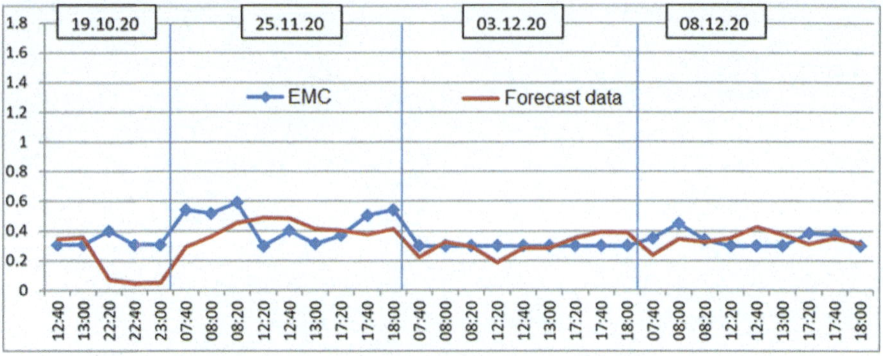

Fig. 7 Laboratory measurements and forecast data of CO concentration

Based on the analysis of deviations, it can be concluded that the forecasts of emissions of pollutants PM2.5 obtained using a recurrent neural network are statistically reliable.

3.1 Verification of the Concentration Calculation Method

From October 19th to December 8th, 2020, the verification of the methodology for measuring the quantity and concentration of emissions from vehicles was conducted by the software system based on laboratory measurements conducted by the Regional State Institution "Center for Environmental Monitoring of the Chelyabinsk Region". Measurements were carried out for the area of the intersection of Engel'sa Street and Lenin Avenue in Chelyabinsk on four separate days: October 19th, 2020, November 25th, 2020, December 3rd, 2020, and December 8th, 2020.

The results of the laboratory and computational measurements are presented in Figs. 7, 8, 9 and 10.

The following conclusions were drawn from the data analysis:

1. The calculated emissions values in the monitoring system are determined solely based on the traffic flow. The lower calculated values of emission concentrations compared to the laboratory measurements of the Center for Environmental Monitoring (CEM) may be explained by the presence of emissions from industrial facilities.
2. The volatile nature of CEM's laboratory measurements over several days reflects the real picture of cumulative emissions of particulate matter, which is strongly influenced by the operations of industrial facilities and unpredictable meteorological conditions.
3. The calculated values of PM2.5 and PM10 emissions in the monitoring system are determined using a constant scaling coefficient.

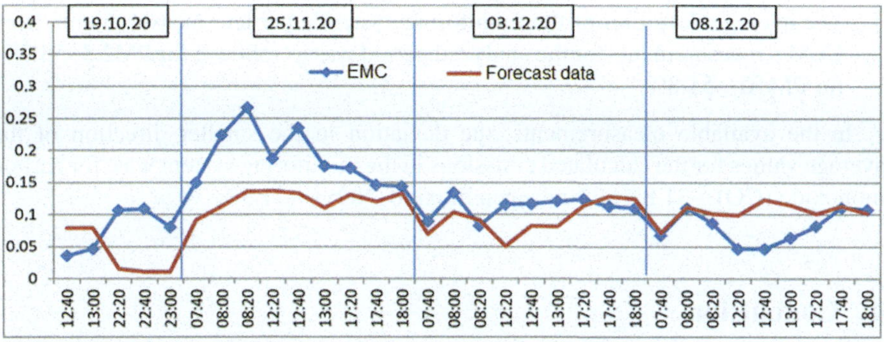

Fig. 8 Laboratory measurements and forecast data of NOx concentration

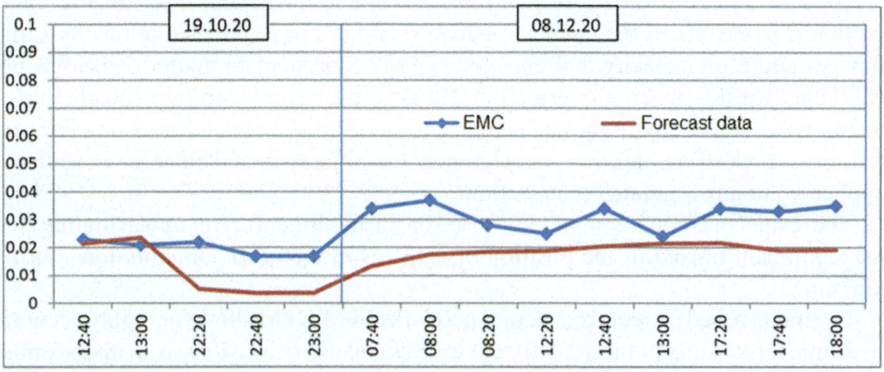

Fig. 9 Laboratory measurements and forecast data of PM2.5 concentration

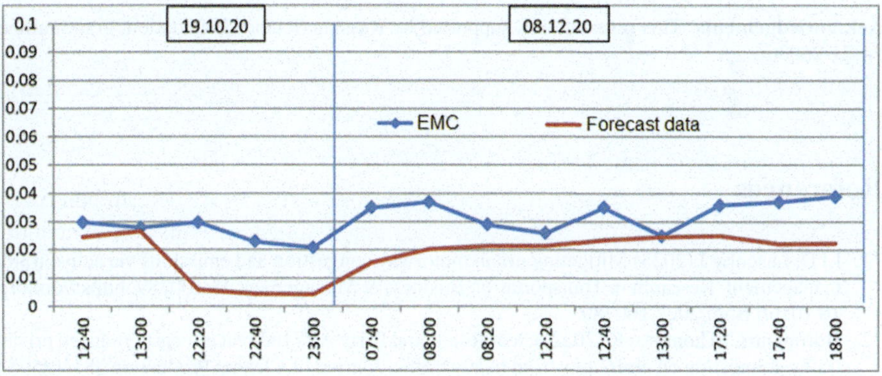

Fig. 10 Laboratory measurements and forecast data of PM10 concentration

4. The average deviation of the monitoring system's particle measurements from CEM's measurements for the analyzed period was as follows: for PM2.5–53.9%, for PM10—51.8%.

In the available measurements, the deviation in the smaller direction of the average values for the calculated emissions in the monitoring system was: for carbon monoxide (CO)—31.8%, for nitrogen dioxide (NOx)—21.5%.

4 Conclusion

The neural network-based system operating in real-time mode has enabled the creation of a large database of raw data on vehicle movement at a specific intersection. The results of the conducted analysis on the impact of variations in traffic flow structure on capacity and changes in PM2.5 particulate matter emissions are valid only for this specific intersection. However, since only straight vehicle movements were considered, without taking into account the peculiarities of right and left turns, pedestrian and tram interference, the obtained results can be considered applicable to any regulated intersection.

The results of computer modeling based on the intelligent driver model highlighted the significant impact of the position of vehicles in the queue on emissions during idle time.

Programs based on recurrent neural networks like LSTM allow for highly accurate real-time forecasting of the quantity of vehicles and their emissions at an intersection. By combining monitoring and prediction capabilities, this hybrid software system provides the ability to assess road conditions and manage environmental risks.

Acknowledgements This research was supported by Russian Science Foundation, grant number 24-21-20086

References

1. Li D, Lasenby J (2023) Mitigating urban motorway congestion and emissions via active traffic management. Research in Transportation Business & Management 48:100789. https://doi.org/10.1016/j.rtbm.2022.100789
2. Hoffmann C, Thommes K (2022) Clear Roads and Dirty Air? Indirect effects of reduced private traffic congestion on emissions from heavy traffic. Journal of Cleaner Production 381:135053. https://doi.org/10.1016/j.jclepro.2022.135053
3. Makarova I, Gabsalikhova L, Gritsenko A (2020) Improvement of environmental compliance of urban transport system through enlarging fleet of gas-engine municipal machinery. Transportation Research Procedia 50:405–13. https://doi.org/10.1016/j.trpro.2020.10.048
4. Pan Y, Wu Y, Xu L, Xia C, Olson DL (2024) The impacts of connected autonomous vehicles on mixed traffic flow: A comprehensive review. Physica A: Statistical Mechanics and its Applications 635:129454. https://doi.org/10.1016/j.physa.2023.129454

5. Makarova I, Khabibullin R, Shubenkova K, Boyko A (2016) Ensuring sustainability of the city transportation system: problems and solutions. In: Kortov S, Terlyga N, Balk I (eds). Proceedings of E3S Web of Conferences, Ekaterinburg, Russia, 6:02004. https://doi.org/10.1051/e3sconf/20160602004
6. Li X, Li Z, Han X, Dai S (2008) Jamming transition in extended cooperative driving lattice hydrodynamic models including backward-looking effect on traffic flow. Int J Mod Phys C 19(07):1113–27. https://doi.org/10.1142/S0129183108012698
7. Ge HX, Dai SQ, Dong LY, Xue Y (2004) Stabilization effect of traffic flow in an extended car-following model based on an intelligent transportation system application. Phys Rev E. 70(6):066134. https://doi.org/10.1103/PhysRevE.70.066134
8. Nagatani T, Nakanishi K (1998) Delay effect on phase transitions in traffic dynamics. Phys Rev E. 57(6):6415–21. https://doi.org/10.1103/PhysRevE.57.6415
9. Xing Y, Lv C, Cao D, Velenis E (2021) Multi-scale driver behavior modeling based on deep spatial-temporal representation for intelligent vehicles. Transportation Research Part C: Emerging Technologies 130:103288. https://doi.org/10.1016/j.trc.2021.103288
10. Derbel O, Peter T, Zebiri H, Mourllion B, Basset M (2013) Modified Intelligent Driver Model for driver safety and traffic stability improvement. IFAC Proceedings 46(21):744–9. https://doi.org/10.3182/20130904-4-JP-2042.00132
11. Xin Q, Fu R, Yuan W, Liu Q, Yu S (2018) Predictive intelligent driver model for eco-driving using upcoming traffic signal information. Physica A: Statistical Mechanics and its Applications 508:806–23. https://doi.org/10.1016/j.physa.2018.05.138
12. Kheder MQ, Mohammed AA (2023) Real-time traffic monitoring system using IoT-aided robotics and deep learning techniques. Kuwait Journal of Science 100153. https://doi.org/10.1016/j.kjs.2023.10.017
13. Shepelev V, Aliukov S, Nikolskaya K, Das A, Slobodin I (2020) The Use of Multi-Sensor Video Surveillance System to Assess the Capacity of the Road Network. Transp Telecommun 21(1):15–31. https://doi.org/10.2478/ttj-2020-0002
14. Barbosa R, Ogobuchi OD, Joy OO, Saadi M, Rosa RL, Otaibi SA et al (2023) IoT based real-time traffic monitoring system using images sensors by sparse deep learning algorithm. Computer Communications 210:321–30. https://doi.org/10.1016/j.comcom.2023.08.007
15. Mahmoud N, Abdel-Aty M, Cai Q, Yuan J (2022) Estimating cycle-level real-time traffic movements at signalized intersections. Journal of Intelligent Transportation Systems 26(4):400–19. https://doi.org/10.1080/15472450.2021.1890072
16. Fan J, Li A, Ilahi A, Gao K (2023) Emission impacts of left-turn lane on light-heavy-duty mixed traffic in signalized intersections: Optimization and empirical analysis. Heliyon 9(5):e16260. https://doi.org/10.1016/j.heliyon.2023.e16260
17. Shepelev V, Slobodin I, Almetova Z, Nevolin D, Shvecov A (2022) A Hybrid Traffic Forecasting Model for Urban Environments Based on Convolutional and Recurrent Neural Networks. In: Khomenko A, Kovalev I, Voroshilova A (eds), Krasnoyarsk. https://doi.org/10.1016/j.trpro.2023.02.059
18. On approval of methods for calculating the dispersion of emissions of harmful (polluting) substances in the atmospheric air. Russian Federation (2017). https://docs.cntd.ru/document/456074826
19. On approval of the Methodology for determining emissions of pollutants into the atmospheric air from mobile sources for conducting summary calculations of atmospheric air pollution, Russian Federation (2019). https://www.garant.ru/products/ipo/prime/doc/73240708
20. AIMS eco. https://aims.susu.ru/demo
21. Shepelev V, Glushkov A, Vorobyev A (2023) Using Computer Vision to Analyze the Sequence of Vehicles Passing Through Regulated Intersections. In: Proc - Int Russian Smart Ind Conf, SmartIndustryCon. https://doi.org/10.1109/SmartIndustryCon57312.2023.10110803
22. Glushkov AI, Shepelev VD, Shepelev SD, Magdin KA, Slobodin I, Burzev A, et al (2021) Monitoring of Transport Flow Emissions based on the Use of Convolutional Neural Networks. In: Berns K., Helfert M., Gusikhin O.(eds) Proccedings of International Conference on Vehicle Technology and Intelligent Transport Systems, 2021

23. Video surveillance. https://stream.is74.ru/
24. Khazukov K, Shepelev V, Karpeta T, Shabiev S, Slobodin I, Charbadze I (2020) Real-time monitoring of traffic parameters. J Big Data 7(1). https://doi.org/10.1186/s40537-020-00358-x
25. Fedorov A, Nikolskaia K, Ivanov S, Shepelev V, Minbaleev A (2019) Traffic flow estimation with data from a video surveillance camera. J Big Data 6(1). https://doi.org/10.1186/s40537-019-0234-z

Resources and Environment Engineering

Forestry Operations as a Management System for New Man-Made Forest Plantings in the Far Eastern Federal District

M. Yu Filatova, L. T. Krupskaya, A. M. Orlov, and A. V. Leonenko

Abstract The relevance of the issues to be solved is determined by the recently significantly aggravated problem of supplying forest industry resources to certain regions of Russia, including the Far Eastern Federal District (FEFD). The ecologically and economically available potential of wood resources is no longer sufficient for annual procurement and the heavily depleted forest fund. Not only the financial issues of creating forest plantings but also many organizational, legal, forestry, and environmental ones should be resolved. Under the conditions of the FEFD, the theoretical framework should be developed for the accelerated cultivation of commercial timber with predetermined specifications through reclamation and the creation of commercial timber plantings, where the role of forestry operations is very large. As a result, the more than a century-accumulated experience in studying the development of sustainable forest management principles by domestic foresters may form the ground for solving current problems, including increasing the efficiency of forestry operations. In this regard, the research objective was to justify the forestry operations of the new commercial forestry plantings in the Far Eastern Federal District. Natural-mining man-made systems served as the study object. The practice of performing some forestry operations in the FEFD has been analyzed, involving firefighting, forest health monitoring, and environmental reclamation by bioremediation that contribute to the reforestation and protection of forest plantings and their sanitary and hygienic properties. The authors have developed techniques, the novelty of which is supported by RF Patents.

Keywords Forest plantings · Forestry operations · Reclamation · Fire hazard · Forest health monitoring · Bioactivator · Poplar cuttings

M. Yu Filatova · L. T. Krupskaya (✉) · A. M. Orlov
Far Eastern Forestry Research Institute, 71, St. Volochaevskaya, Khabarovsk 680020, Russia
e-mail: ecologiya2010@yandex.ru

A. V. Leonenko
Mining Institute, 51, St. Turgenev, Khabarovsk 680000, Russia

L. V. Radionova and D. V. Ulrikh (eds.), *Advances in Ecology and Environmental Engineering*, Springer Proceedings in Earth and Environmental Sciences,
https://doi.org/10.1007/978-3-031-64423-8_9

1 Introduction

Forestry planting is the key way for global forestry to meet the demands of the wood processing industry. The Russian Federation, characterized by the idea of the inexhaustibility of its forest resources, has resumed the policy of preserving natural forests by creating forest plantings only recently due to the continuous reduction of forests available for exploitation by their environmental and economic parameters. In this regard, many foreign and Russian scientists consider it expedient to develop the forestry planting system with increased productivity and a shortened cutting turnover. E.g., Tyurin et al. [1] reviewed the practice of forestry planting in various natural areas of the world. They showed its potential for accelerated forest growing and production of commercial wood with the required specifications. It has been established that forestry planting is most effective when intensifying technological methods and using machinery at all stages of forest growing. An average annual increase of 7–30 m^3/ha in forest plantings was observed on a global scale. Due to those measures, wood procurement in natural forests is reduced by an average of 26%, which contributes to the preservation of biodiversity, carbon sequestration, and the preservation and restoration of environmentally favorable living conditions.

Prokal et al. [2] have developed a project for creating multi-purpose forest plantings by silvicultural methods that will ensure average intensification of forest management and help reduce the time to grow forest plantings. This eliminates the need for replacing available forest plantings that poorly perform their ecological functions and improves the structure, fire safety, and health of forests in territorial entities in general.

Panichev et al. [3] have found that in the European part of Russia, 1.5–2.5 times more products can be obtained from 1 ha of commercial timber plantings. According to them, the Voronezh VNIILGISbiotech has identified clones of poplar, aspen, and Karelian birch as target species for creating commercial plantings. The obtained result for recording the black poplar stock per 1 ha ranged from 550 to 1350 m^3. Similar high results were obtained for aspen, aspen-poplar hybrids, and Karelian birch. The use of these species will speed up obtaining wood for the timber processing industry by 1.5–2.5 times.

Gizaw Ebissa et al. [4] have analyzed forest plantings grown in Ethiopia for carbon sequestration, roundwood procurement, and wood processing. The study results have shown that in the 10th and 15th years of planting growth, the most feasible is wood production, which allows for obtaining a supplementary product (lumber, raw materials for the pulp and paper industry, furniture, parquet, etc.), in contrast to two other goals. In Ethiopia, carbon sequestration plantings require the use of exotic tree species ensuring stable and high biomass growth and the implementation of biomass-increasing techniques. Unlike carbon sequestration plantings, roundwood ones should focus on wood volume and quality. The authors have concluded that the costs of managing forests for obtaining carbon credits and producing logs are almost the same but significantly higher than those required to procure raw materials for the timber processing industry.

The study by Augustynczik et al. [5] was aimed at assessing several strategies for integrating forest management operations in forest planting. They have developed logging planning models according to the most effective spatial pattern of forest plantings, considering compliance with the biodiversity condition when planting them. Thus, effective structural pattern of trees at the forest planting planning stage allows for reducing the movement of machinery and the costs of building and maintaining the road network during further timber procurement.

Hughes et al. [6] in their paper examined the quality of water in a stream in the upper reaches of a Pinus radiata forest planting after forestry operations. The most significant effect was obtained from tree planting and thinning over four years with an increase in nitrate nitrogen (NO_3) and total nitrogen (N) concentrations. According to the authors, this was caused by a change in physical conditions (e.g., increased stream scumming, expansion of nitrogen-fixing weeds, and reduced spatial distribution of seepage wetlands) due to changes in land use. In turn, the increased concentration of nitrogen nitrate and total nitrogen after thinning shows that nitrogen absorption by growing trees is significant. Over the entire observation period, their concentration increased fourfold (from 400 to 1590 $\mu g\ L^{-1}$). An increase in concentration was also observed during the intermediate felling and cutting down of plantings, but it was short-term, less than 12 months.

Thus, the analysis, summarization, and systematization of literature data have shown the need to continuously search for a compromise between wood procurement and the preservation of biodiversity and the environment as a whole to maintain a balance in the region under study. This can only be achieved with a comprehensive understanding of the environmental impact of forestry planting.

In this regard, the research objective was to justify the forestry operations of the new commercial forestry plantings in the Far Eastern Federal District. According to the objective, the following tasks have been identified: (1) Summarizing domestic and foreign practices in studying the described problem; (2) Identifying the most significant threats to forest plantings in the FEFD; (3) Developing proposals for improving forestry operations under the conditions of forest plantings considering the specifics of the Far Eastern region.

The scientific novelty of the research results is in the assessment of the forest fund state for the region under study, the identification of the most significant threats to forest plantings, and the justification of forestry operations in commercial forest plantings in the Far Eastern Federal District.

2 Methods and Materials

In the study, standard methodological approaches were used: silvicultural, forest taxation, geoecological, and statistical.

The following techniques were used for scientific justification of forestry operations: theoretical analysis of domestic and foreign literature; historical method, and scientific forecasting, as well as empirical, structural, functional, and system analyses.

In the study, legal and literary data sources were used, including databases of the International Group of Experts on Climate Change, the results of research and development work in the field of climate change [7], Fundamentals of State Policy in the Field of Use, Protection, and Regeneration of Forests in the Russian Federation for the Period up to 2030 [8], the Paris Agreement adopted at the 21st session of the Conference of the Parties to the UN Framework Convention on Climate Change in Paris on December 12, 2015, and other regulatory legal documents [9].

In the study, scientific, statistical, reporting, and other industry-related materials were used, containing relevant information on the specifics, conditions, and dynamics of forest lands in the region under study, forest management, wood processing, protection, and regeneration of forests. The research was based on the Forest Plans of the constituent entities of the Far Eastern Federal District of the Russian Federation, the State Forest Register (SFR) materials, industry-related regulatory legal acts on the use, regeneration, and protection of forests, and thematic literary sources.

3 Results and Discussion

In recent years, the problem of wood procurement for the forestry industry and farms in certain regions of Russia, including the Far Eastern Federal District (FEFD), has aggravated. The transport-accessible forest fund for environmentally and economically efficient wood procurement is heavily depleted here.

Forestry planting in Russia has a long history. Thus, already in 1980, by s Resolution No. 95 of the USSR State Planning Committee dated April 30, 1980, the scope of work on planting (including 1990 and subsequent years) was determined at 35 thousand ha/year, and the planned wood procurement volume, starting from 2030, at least 18–20 mln m^3/year.

According to the UN FAO, as of 2000, about 187 mln ha of forest plantings were created globally, which is relatively little with a total forested land area of 4.2 billion ha. However, in individual countries, plantings play a very important role [10].

In the FEFD, forest plantings may give impetus to the entire forest sector for the transition to intensive innovative development since it requires the integration of science and practice and the use of the latest technologies for growing and processing wood and advanced achievements in genetics and selection. As a result, the industrial load on natural and hard-to-reach mountain forests exercising various protective functions should be reduced [11, 12]

In this regard, in recent years, creating commercial forest plantings has been resumed at the state level, but many issues of their cultivation and subsequent maintenance have not been resolved.

In our opinion, some negative consequences of creating commercial forest plantings can be avoided at the stage of their arrangement. Along with solving direct economic problems, they should contribute to the protection, restoration and preservation of adjacent natural forests and should not negatively affect them. It is important to ensure the possibility of movement (migration) of fauna species between forest formations, arrange various water protection bands, and create plantings of different ages, species composition, and cutting turnover.

To more effectively use the potential of forest plantings, not only the degree of compliance with the growing conditions but also the goals of growing a given forest planting should be considered. They should not further lead to soil degradation, deterioration in water quality, changes in water levels, outbreaks of insect pests, the spread of fires, etc.

The aforementioned allows for identifying criteria and indicators for sustainable management of forests in general and commercial timber plantings in particular [13].

General criteria for the arrangement of forestry as part of national criteria and indicators for sustainable forest management are:

1. The optimal level of forest cover and the target forest fund structure can be maintained through:
   - a system of regulating the economic activity on the forest planting territory;
   - effective pattern and fragmentation of cutting areas in the elementary basins of the planting;
   - balance of average growth and cutting of wood resources;
   - a system for long-term assessment of the forest planting conditions based on a complex of forest inventory and monitoring operations.
2. Preservation and maintenance of forest stand productivity:
   - resource-saving logging technologies;
   - an effective forest planting territory development scheme;
   - effective forest fire prevention measures [14, 15].
3. Preservation and maintenance of water protection and habitat-forming functions of the forest through:
   - effective arrangement and fragmentation of protective bands of certain parameters in the structure of catch basins on the forest planting territory;
   - continuous monitoring of the parameters of water runoffs (including solid ones).
4. Preservation and maintenance of the biological diversity of the surrounding forests and the planting itself through maintaining the composition and structure of its forest stands.
5. Maintaining an acceptable sanitary condition through:
   - identifying the causes and monitoring the forest stand condition;
   - planning and implementing measures to prevent harmful impacts;
   - prevention of diseases and mass outbreaks of harmful insects;
   - timely use of dead planting resources.
6. Maintaining the socio-economic forest functions through:
   - creating conditions for socioeconomically effective production development based on the integrated and multi-purpose use of forest resources and the formation of progressive green approaches to the arrangement of forest management on the forest planting territory through a system of socioeconomic incentives;

- implementing the sustainable development principle, multi-purpose, non-exhaustive use of forest resources, and greening of forest planting management;
- the use of modern progressive forms of management;
- improving the regulatory framework and inculcating civil and professional legal consciousness;
- the desire of specific forest management entities to participate in national and international processes of environmental protection and careful treatment of forest resources.

The quality characteristics of forests, including man-made plantings, can only be maintained and their conditions improved through the use of forestry operations. This is a set of organizational and technical forest cultivation, preservation, protection, and reforestation measures aimed at improving the sustainability, productivity, habitat-forming, environmental, sanitary, hygienic, and health-promoting properties of forests.

Given the extreme fire danger and flammability of managed FEFD forests, forest planting fire protection measures are of the most importance and need. About 78% of its forest fund belongs to the highest classes of natural fire danger. This is mainly caused by periodically recurring extreme dry seasons, an abundance of combustible materials, mountainous terrain, and harsh wind conditions. In recent years, catastrophic fires in the FEFD recur every 6–8 years [16]. Herewith, they have long-term environmental and economic consequences negatively affecting the entire environment.

According to the established forest fire risk zoning [17], the FEFD with a total area of forest fund lands amounting to about 40% of this area of the Russian Federation, includes 4 forest fire risk zones, of which the most complex is the 373,082.3 thousand ha zone of exclusive satellite identification and predominantly ground fire suppression (75.1%). According to statistical data, in the FEFD, predominantly ground fires occur—68%, and crown fires make up 32%. 90% of all forest fires are caused by humans and only 10%—by thunderstorms.

The forest fire safety measures include: (1) prevention of forest fires; (2) monitoring of forest fire danger and forest fires; (3) development and approval of forest fire suppression plans [15].

Since 1997, the Forest Fire Remote Monitoring Information System of the Ministry of Natural Resources of the Russian Federation (RMIS MNRRF) has been operating in Russia [18]. The key tasks of the system are collecting information on fires in protected and unprotected areas and the environmental conditions and promptly providing data required for making management decisions and assessing the forest fire consequences.

Preventing forest fires involves the fire-preventing arrangement of forests and the provision of forest fire protection and extinguishing means. Fire protection measures for forest plantings include:

1. Construction, reconstruction, and operation of forest roads designed to protect forests from fires. In the case of creating forest plantings, this point is most effective;

2. Construction, reconstruction, and operation of pads for helicopters used for aerial forest protection. Since under the FEFD conditions, only small forest areas can be used for planting, arranging helicopter pads for each planting is infeasible. It would be expedient to arrange forest plantings in proximity to available pads;
3. Laying fire-fighting mineralized bands;
4. Construction, reconstruction, and operation of fire monitoring points (towers, masts, pavilions, etc.) and firefighting equipment allocation points; arranging fire reservoirs and access points to firewater sources;
5. Preventive controlled burning of brushwood, forest litter, dry grass, and other forest combustible materials; other measures.
6. Fire danger monitoring;
7. One of the most effective measures to prevent forest fires is restricting and prohibiting people's access to the forest (barriers, posts, patrols, patrol groups), which is easily feasible in commercial forest plantings.

To improve the efficiency of measures to prevent and combat forest fires, it is proposed to use an improved formula for calculating a complex Forest Fire Danger Index depending on weather conditions by introducing a correction factor for precipitation into the existing formula, due to which it takes [19] the following form:

$$CI = \sum_{1}^{n} t^{o}\left(t^{o} - \tau^{o}\right) - (k\alpha)$$

According to the formula, the complex index for the current day is determined as the sum of the products of air temperature (t^o) by the difference between this temperature and the dew point (τ^o) of that day for the number of days (n) after the last rain minus the precipitation correction factor ($k\alpha$), considering their daily amount (α) in mm. Thus, the use of the precipitation correction factor allows for determining the *CI* at its different possible values before precipitation and at different precipitation amounts.

According to the current methodology, when precipitation is more than 3 mm/day, the complex indicator is reset to zero, regardless of its current value, i.e., it is unreasonably adopted that precipitation of more than 3 mm completely reduces the forest fire danger, and therefore, forest sunburns and fires are impossible. This approach is invalid since small precipitation (4–6 mm) cannot eliminate the forest fire danger accumulated, e.g., during a month-long drought. Thus, a closed canopy of dark coniferous timber can retain precipitation of up to 5 mm, and to stop forest fires after a drought lasting 12–15 days, precipitation of at least 25 mm is required. Such a methodological approach to calculating the complex indicator cannot serve as a reliable basis for regulating the work of forest fire services. The formula should allow for evaluating the complex indicator at different possible drought levels and precipitation amounts. Previous studies by the Far East Forestry Research Institute Federal Budgetary Institution (FEFRI FBI) established that continuous (widespread) precipitation of 1 mm per day reduces the complex indicator by $k = 400$ U/mm, and local precipitation by $k = 300$ U/mm. Thus, the use of the precipitation correction factor allows for determining the *CI* at its different possible values before precipitation and at different precipitation amounts. The use of a new technique for calculating the complex indicator will allow for objectively assessing the fire danger and reasonably extend the fire hazard period and emergency regime in specific forest areas and districts.

The choice of the species composition of a forest planting should consider that in the FEFD, forest formations with potentially heavy fire consequences include cedar, pine (especially young forests), spruce-fir, oak, ash-elm, and cedar-elfin. More fireproof and less fire-hazardous are larch, birch, and aspen forests.

Arranging protective fire-resistant bands (FRBs) of a certain silvicultural-pyrological structure is inexpedient due to the small forest planting areas. In turn, commercial timber planting may play the role of a firebreak, subject to the proper arrangement and planning of the tree planting scheme and schedule. Thus, larch bands in the coniferous-deciduous forest zone acquire fire protection properties for 4–5 years, and in the taiga zone—for 5–7 years. Poplar bands start functioning a year earlier and can perform firefighting functions for 40 years without significant economic measures.

Forests are protected from harmful organisms (viable plants of any species, varieties, or biological types, animals, or pathogens of any type that may harm forests and forest resources) according to Articles 60.1–60.11 of the Forest Code of the Russian Federation in force.

Forest protection is aimed at identifying harmful organisms in forests, preventing their spread, and in case of outbreaks, their elimination. It is regulated by regulatory legal acts and comprises the implementation of sanitary safety measures and the elimination of outbreaks of harmful insects, which include [20] forest protection zoning; state forest health monitoring; forest health examinations; preventing the spread of harmful organisms; other forest sanitary safety measures taken according to the forest plans of the constituent entities of the Russian Federation, forestry regulations of forest districts and parks, and forest development projects.

Forest protection zoning identifies zones of weak, medium, and strong forest health threats, as well as those requiring the implementation of ground and (or) remote state forest health monitoring means and the performance of forest health studies.

In 2004–2006, the FEFRI jointly with the Pacific Institute of Geography of the Far Eastern Branch of the Russian Academy of Sciences and the Forest Institute of the Siberian Branch of the Russian Academy of Sciences developed forest maps of pathological regions of the Khabarovsk and Primorsky Territories [21] and intended to perform forest protection zoning for the FEFD forests, including the spruce forests of Sikhote-Alin and the Amur region. However, forest protection zoning of controlled forests still is not being performed in the FEFD forest areas under study.

Thus, to ensure the maintenance of quality parameters, improve the condition of forests, and prevent forest degradation in the FEFD, it is expedient to arrange forest health monitoring as a system of real-term and continuous control over the manifestation, spread, and development of foci of entomo- and phytopathology, the forest conditions, and the damage to forests under the impact of natural and anthropogenic factors.

The Far East taiga and Amur-Primorsky coniferous-deciduous forest areas have a high probability of mass reproduction of some most dangerous needle- and leaf-eating insects—Lepidopterans capable of causing defoliation and further drying out of coniferous species, and weakening and reduced fruitification of deciduous species for 1–3 years over large areas. The Siberian silk moth (Dendrolimus superans sibiricus Tschetv.), the tussock moth (Calliteara abietis Den. and Schif.) accompanying it in outbreaks, and, to a lesser extent and in some years, the black arches (Lymantria monacha L.) pose the threat of drying out of mature cedar, fir, and spruce trees in coniferous-deciduous forests. The Asian gypsy moth (Lymantria dispar praeterae Kard.), and in the south of the Primorsky Territory, also the pink gypsy moth (Lymantria mathura aurora Butl.) damage many deciduous species, primarily oak. Other lepidoptera such as the lackey moth (Malacosoma neustria testacea Motsc.) and the spring complex of phyllophages composed of geometer moths and leaf-rolling moths are less significant.

Currently, a system for monitoring the Far Eastern populations of the listed predominant pests has been developed using forest health zoning and regional methodologies for forest health examination of foci have been drawn up, natural factors of natural mortality have been identified and assessed, which allows for forecasting their condition and the threat of damage to plantings.

The efficiency of protective measures is ensured by the timeliness, scope, and quality of forest health monitoring and examinations. Creating forest plantings should consider that forest districts with large nut-harvesting zones in the FEFD should be treated as the zone of a strong forest health threat. In the Khabarovsk Territory, the degree of forest health threat should be elevated from weak to medium. This covers the Northern and Sovetskoye forest districts, as well as those with walnut-harvesting zones or genetic cedar reserves (Gurskoye, Khorskoye, Avanskoye, Nanaiskoye, Mukhenskoye).

A way to improve the efficiency of forestry operations is reproducing the productivity of areas that have been subject to negative anthropogenic impact in the last century through reclamation and phytomelioration measures and the further creation of commercial timber plantings.

The development of mineral raw materials in the south of the FEFD (Amur Region, Khabarovsk and Primorsky Territories) disturbed over 800 thousand ha of land, most of which was withdrawn from the forest land fund. Large areas of valley forests have been destroyed. To reduce the negative impact of disturbed lands, their timely technical and biological reclamation after mining is required.

The technical reclamation stage involves the following types of work:

1. Coarse and finishing land leveling, and its restoration;
2. Anti-erosion treatment of the territory, incl. slopes;
3. Reclamation of toxic rocks;
4. Laying potentially fertile rocks (where appropriate);
5. Applying humic compounds to the surface of potentially fertile rocks;

At the biological stage, the productivity of disturbed lands is restored. This includes a set of agrotechnical and phytomeliorative measures aimed at restoring flora and fauna and creating sustainably functioning biocenoses.

The basic type of biological stage of the reclaimed area development should be perennial grassing and transplanting of trees and shrubs. In the scope of biological reclamation, special attention should be paid to the enrichment of the substrate with organic matter. In this regard, the authors recommended using the potential of biological systems (bioremediation), phototrophic bacteria, humic compounds, and other bioactivators (RF Patents, 2013, 2015, 2022, etc.) [22]. The proper choice of shrub, tree, and perennial grass species from a significant number of those suitable for creating targeted commercial forest plantings is of great importance as a way to most effectively meet both the economic and environmental needs of the region. Thereat, it is recommended to use the most fire-resistant plantings—poplar and larch [15]. Creating forest plantings will also contribute to the consolidation of moving sands, afforestation of small slopes (no more than 10^0), the creation of shelterbelts and water- and snow-regulating plantings, ensuring favorable conditions for human habitat, etc.

In this regard, we experimented with the rooting of hybrid varieties of the Voronezh VNIILGISbiotech poplar and Gmelin larch using bioactivators for accelerated growth of cuttings at the FEFRI's site in the village of Nekrasovka, the Khabarovsk Territory, in the nursery forest of Rastim Les LLC and on the reclaimed site of the Kavalerovsky forestry of the Primorsky Territory. At the experimental sites, poplar cuttings of 10 varieties (Pioneer, Volosistoplodny, Chinese, Sakrau, Versiya, E.S., etc.) and Gmelin larch cuttings were planted using Rostivin, Succinic Acid, Huminol, and Florentine Water. Stable rooting results were obtained for a single growing season. The average growth of poplar and larch cuttings was, respectively, 70 and 14 cm. An application for an invention was filed.

4 Conclusion

In recent years, the problem of supplying forest industry resources to certain regions of Russia, including the Far Eastern Federal District (FEFD), has significantly aggravated. The ecologically and economically available potential of wood resources is no longer sufficient for annual procurement and the heavily depleted forest fund. Under the conditions of the FEFD, the theoretical framework should be developed for the accelerated cultivation of commercial timber with predetermined specifications through reclamation and the creation of commercial timber plantings, where the role of forestry operations is very large. In this regard, domestic and foreign experience in creating commercial forest plantings and improving the efficiency of forestry operations is summarized, the role of which is great. Natural-mining man-made systems served as the study object. In the study, firefighting, forest health monitoring, and environmental reclamation by bioremediation have been analyzed that contribute to the reforestation and protection of forest plantings and their sanitary and hygienic properties. The authors have developed bioremediation techniques, the novelty of which is supported by RF Patents.

Acknowledgements The study was performed at the expense of the state assignment for applied scientific research for 2023 and 2024, No. 3-I23, Plantings

References

1. Tyurin DS, Danilov DA, Zaitsev DA, Shchekalev RV (2023) Forestry Planting in World Practice. In: Pamfilov EA (ed) Current Problems of the Forestry Complex. Coll. of sc. papers, BGITU, Bryansk, p 321
2. Proka IYu, Babynin SN (2018) Creating Multi-Purpose Forest Plantings Using Silvicultural Methods. Bulletin of the Kazan State Agrarian University 13(4.51):77–82. https://doi.org/10.12737/article_5c3de3887620f5.55458216
3. Panichev G, Zenina N, Rusin N Forest Plantings. Living Forest Internet Magazine. https://givoyles.ru/articles/nauka/lesnye-plantacii/#VoronezhExperience. Accessed 10 Dec 2023
4. Gizaw Ebissa, Aramde Fetene, Hayal Desta (2023) Comparative Analysis of Managing Plantation Forests: The Case of Keeping Plantation Forests for Carbon Credit and Industrial Profits in Oromia Region, Ethiopia. Heliyon, vol. 9, issue 4. https://doi.org/10.1016/j.heliyon.2023.e15151
5. Augustynczik ALD, Arce JE, Silva ACL (2016) Aggregating Forest Harvesting Activities in Forest Plantations Through Integer Linear Programming and Goal Programming. Journal of Forest Economics. Elsevier 24(C):72–81
6. Hughes AO, Quinn JM (2019) The Effect of Forestry Management Activities on Stream Water Quality Within a Headwater Plantation Pinus Radiata Forest. Forest Ecology and Management 439:41–54
7. Morin VA, Gul LP et al (2017) Assessing the Impact of Forestry Operations on the Carbon Cycle in Managed Forests of the Far Eastern Taiga and Amur-Primorsky Coniferous-Deciduous Forest Areas: research report (final). FEFRI FBU; lead. L.T. Krupskaya; executive, Khabarovsk, p 233
8. Order of the Government of the Russian Federation No. 1724-r dated September 26, 2013, On the Fundamentals of State Policy in the Field of Use, Protection, and Restoration of Forests in the Russian Federation for the Period up to 2030. https://www.garant.ru/products/ipo/prime/doc/70361820/. Accessed 15 Dec 2023
9. Decree of the Government of the Russian Federation No. 318 dated April 15, 2014 (as amended on October 18, 2021) On Approval of the Forestry Development State Program of the Russian Federation. https://base.garant.ru/70644228/. Accessed 10 Dec 2023
10. Shutov IV (ed) (2007) Forestry Planting. Polytechnic Publishing House. Univ., St. Petersburg, p 366
11. Alekseenko Ayu, Nikitenko EA (2017) Prospects for Creating Forest Plantings in the Far East. Forestry Bullten 4:13–18
12. Pokrovskaya AA (2009) Scots Pine (Pinus Sylvestvis) as a Basis for Planting Afforestation. Issues of Green Planting Management in Khabarovsk: materials of the 4th city scientific and practical conference, Khabarovsk, November 25, 2009, pp 96–97
13. Order of Rosleskhoz No. 21 dated 02/05/1998, On Approval of the Criteria and Indicators of Sustainable Forest Management in the Russian Federation. Electron. Ref. Legal System. Access mode: https://docs.cntd.ru/document/901706209, free. Title from the screen. Lang. Rus. Accessed 10 Dec 2023

14. Decree of the Government of the Russian Federation No. 1614 dated October 7, 2020, On Approval of Fire Safety Rules in Forests. ConsultantPlus: Official website of ConsultantPlus. Electron. Ref. Legal System. https://online.consultant.ru/riv/cgi/online.cgi?req=doc&base=LAW&n=364560&dst=1000000001&cacheid=D566D4D2552FB39D5CA1FE5B1CACCFDF&mode=splus&rnd=tyZ7Fw#ihe02zT4YCKXQCgC1. Accessed 10 Dec 2023
15. Order of Rosleskhoz No. 174 dated 04/27/2012, On Approval of the Standards for Fire Protection of Forests (Registered with the Ministry of Justice of Russia on 06/07/2012, No. 24488). https://docs.cntd.ru/document/902348623. Accessed 12 Dec 2023
16. Control over Forest fires at the Ecoregional Level. Materials of the International Scientific and Practical Seminar. Aleks Publishing House, Moscow, 2004, p 208
17. Annex to Order of Rosleskhoz No. 22 dated January 26, 2022, On Establishing Forest Fire Zoning and Invalidating Order of the Federal Forestry Agency No. 753 dated 08/05/2020. https://docs.cntd.ru/document/350105395?marker=6560IO, free. Title from the screen. Lang. Rus. Accessed 12 Dec 2023
18. Belyaev AI, Ershov DV, Lukyan EA et al (2004) National System for Collecting, Processing, and Analyzing Data on Natural Fires and Its Connection to International and Regional Data Networks. Materials of the International Scientific and Practical Seminar. Aleks Publishing House, Moscow, pp 156–166
19. Sheshukov MA, Kovalev AP, Pozdnyakova VV (2018) Methodological Recommendations for Supplementary Measures to Improve the Efficiency of Implementing a Special Fire Regime and Emergency Regime in the Forests of the Far Eastern Federal District. Publishing House of FEFRI FBU, Khabarovsk, p 29
20. Order of the Ministry of Natural Resources of Russia dated No. 912 dated November 9, 2020 (as amended on May 12, 2022), On Approval of the Rules for Implementing Measures to Prevent the Spread of Pests. https://online.consultant.ru/riv/cgi/online.cgi?from=371458-31&req=doc&rnd=IRZ8YA&base=LAW&n=425581#fzf12zTuwefUl9PD, free. Title from the screen. Lang. Rus. Accessed 14 Dec 2023
21. Yurchenko GI, Baranchikov YuN, Korets MA (2005) Forest Pathological Areas of the Khabarovsk Territory (Maps). ILSORAN, Krasnoyarsk
22. Krupskaya LT, Kovalev AP, Shugaley IV, Orlov AM, Filatova MYu (2023) Justification of Biological Methods for Restoring the Protective Functions of Man-Made Waste Contaminated Forest Areas to Create Forest Plantings. Ecological Chemistry 32(1):28–40

Research on the Self-healing of Biota in Natural Ecosystems Disturbed by Mining Enterprises

Ju Galchenko and Ju Ozaryan

Abstract This article presents data from a study regarding the influence of stone mining processes built upon granodiorite fine earth deposits on the natural restoration of vegetation. Throughout the process of conducting field research, a pattern was found regarding the distance of the transfer of seeds of primary phytocenoses from primary phytocenoses of a natural forest. The successive series of of nodal phytocenoses self-regeneration on juvenile surfaces upon new technogenic formations are largely determined by the intensity of soil formation due to the dry plant mass of herbaceous plants. Our studies show that natural restoration of soil and vegetation coverage occurs more intensively in the presence of a developed plant community found near man-made constructions. The mine's surface and surrounding areas are partially and unevenly covered with vegetation, in contrast to other observations made throughout the study. Based on the results of this study, the reclamation of lands disturbed by mining should be carried out with consideration for the natural restoration of the native and technical environment.

Keywords Natural restoration · Technogenic deposit · Self-healing · Revitalization · Reclamation · Disturbed lands · Succession

1 Introduction

The modern state of scientific and technical development determines a swift increase in the consumption of natural resources. The extraction of useful deposits favorably affects economic and social spheres. But like any technogenic intervention,

---

J. Galchenko
Institute of problems of integrated development of subsurface resources RAS Academician N.V. Melnikov, 4, Kryukovsky Tup., Moscow 111020, Russia

J. Ozaryan (✉)
Mining Institute FEB RAS, 51, Turgeneva Street, Khabarovsk 680000, Russia
e-mail: ozaryanigd@gmail.com

L. V. Radionova and D. V. Ulrikh (eds.), *Advances in Ecology and Environmental Engineering*, Springer Proceedings in Earth and Environmental Sciences,
https://doi.org/10.1007/978-3-031-64423-8_10

subsoil use produces a negative effect on the components of the lithosphere, atmosphere, and biosphere. With the extraction of primary forest territories, the absence of ecotone prompts animal migration and, accordingly, a decrease in biodiversity in the surrounding territories.

Mining production has a powerful and complex impact on all components of the environment, leading to its transformation, and often to its complete degradation [1, 2]. This is considered one of the most destructive types of economic activity, causing deterioration in the functioning of the natural ecosystem.

The air in our environment is exposed to pollution due to the radical transformation of the lithogenic base with soil coverage is transformed and polluted and flora and fauna oppressed and exterminated; landscapes are modified with relief from pollution only occurring on surface and underground water [3–5].

To accelerate the processes of natural biota restoration, it is necessary to identify the patterns of formation of secondary phytocenoses—considering to what extent external biotic factors influence the processes, accelerate them, and ensure positive dynamics. The Vegetation Index (NDVI), a measure of the amount of photosynthetically active biomass, is designed to measure the environmental and climatic characteristics of vegetation. The dependence of the index on productivity (temporary changes), quality, and quantity of biomass have been established, with soil moisture at various depths and varied organic matter saturation, also including a volume of impurities, height, and elemental composition of snow-covered grounds.

The geological nature of the location of ore deposits leads to the fact that mining enterprises under construction and operation require a man-made invasion of the natural biota of the Earth. That is why the degree of ecological danger of inevitable changes in this biota determines almost all prospects for the existent possibilities of natural biotope preservation over vast territories.

Most of Russia's resource potential is concentrated in zones of seasonal permafrost. Its development is determined by a high level of technogenic action, and natural restoration occurs extremely slowly. Severe natural and climatic conditions require a special approach to the rational organization of the technological process of the mining complex, the activities of which are characterized by significant environmental risks with global consequences for low productivity and unstable biogeocenosis of the permafrost zone. The large-scale impact of the mining complex does not always make it possible to give a qualitative assessment of the state of natural objects in the field, especially when the main reserves of the minerals raw materials are concentrated in the zone of seasonal and permafrost.

For mining enterprises, the lifespans are limited by the exhaustibility of balance reserves. The area affected by natural biota due to transit technogenic factors is many times greater than the area occupied by the enterprise itself. Creating conditions for the successful self-restoration of phytocenosis of natural ecosystems is possible only by studying the features and patterns of restoration succession and the direction development of de-mutation processes of regeneration for tree species in various parts of the studied territory in the post-exploitation period.

2 Methodology and Objects of Research

These new approaches were built on principles of sustainable development, which are interpreted in our country as a strategy for the sustainable development of the environment and society [2, 6].

Considering that the natural-technical system was formed through man-made impacts and includes a technological component, components in our ecosystem surround the territory where man-made objects are located and include the mining complex. The parameters of the technological process, as well as natural and climatic characteristics, including precipitation, humidity, air, and soil temperature, were studied.

The study combines field observations and remote sensing data, as well as laboratory analysis, to assess the natural recovery processes of biota components and the environmental impacts of mining. Different natural-technical systems were selected for the study. The features of the development of secondary plant communities in the study area were studied.

The restoration of plant communities was also studied in areas where industrial activity continued 5, 10, 15, 25, and 40 or more years ago. A field study of the site limitation was carried out, along with a geobotanical description of the surrounding forest area, as well as its site limitation. The size of the training areas is 2 × 50 m. The distance between the profiles is 2.5–4 km. The areas were laid out on dumps formed at the deposits of building stones and connected with the impact of an underground mine with a capacity of more than 1 million tons of ore.

3 Results and Discussion

Natural restoration has been considered an effective method of revegetation [7]. This method, based on self-design theory, involved creating physical and chemical conditions that favor the growth of desired species in suitable habitats. Compared to cultivated plant communities, phytocenoses that followed natural restoration were more likely to approximate native plant communities [8, 9] with more biodiverse species and abundance [10, 11]. According to the findings of this research, the channel pebbled beach and the second bottom of the valley were planted by Populus maximovichzii, Henri, Chosenia macrolepis, Turaz, Fraxinus mandshurica, Rupr, Aser mono, Rupr, Acer mandshuricum Maxim, Suringa amurensis, Rupr, Alnus hirsuta, Turcz, and some other arboreal species [11, 12].

Moreover, the natural restoration of vegetation usually leads to the enrichment of soil with organic matter [12]. Extreme conditions of natural-technical systems influence natural recovery. A small part of the research concerns the structure of phytocenoses [7, 8, 10–12], which is important for identifying the mechanisms of vegetation restoration.

The background plant community characteristic of the study area on both slopes of the valley was represented by a three-tier coniferous-deciduous forest. In the process of studying the established sites, the succession of herbaceous, shrub, and tree species was studied. The secondary phytocenosis is founded with the participation of Fraxinus mandshurica, Rupr, Acer mandshuricum Maxim and Aser mono, Rupr, Populus maximovichzii, Henri, Chosenia macrolepis, Turaz, in the synusia of the edificator they enter the background [2, 10].

The influence of the mining and industrial complex on the components of the ecosystems of adjacent territories located in the river valley determined the formation of a new plant complex, different in composition from the primary one (Fig. 1).

The species composition significantly depends on its location relative to the slope and its steepness. The presence of various types of birch was revealed in both the upper and lower parts of the slope. It was noted that the lower part of the slope is more densely populated by various representatives of the flora, forming a deciduous forest in which the species of the original phytocenosis are recorded.

The forest-forming species here were Betula mandshurica Na Kai, Betula costata, Trautv, Populus tremula, L., Tilia amurensis, Kom, Aser mono, Rupr, Acer mandshuricum Maxim. During a survey of the territory, single specimens of Abies holophulla, Maxi Pinus koraiensis, and Sieb were found. et Zucc.

The differences in the restoration of herbaceous and woody plants were also documented. The revitalization of the wood species zone located between the primary forest and lands disturbed during the production of mineral raw materials has several notable features.

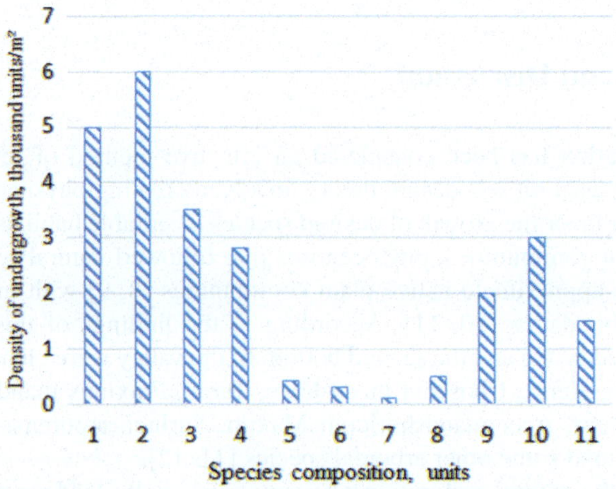

Fig. 1 Density of undergrowth species of secondary phytocenosis along the slopes of the valley. 1. Asermono (5 thousand pieces/ha), 2.Asermandshurica Maxima (6 thousand pieces/ha), 3.Acerpseudosie boldianum, Kom. (3.5 thousand pieces/ha), 4.Tiliaamurensis, Kom. (2.8 thousand pieces/ha), 5. Betulamandshurica NaKai, 6. Populustremula, L., 7. Pinuskoraiensis, Sieb. etZucc, 8. Abiesholophulla, Maxim, 9. Kalopānax septemlōbus, 10. Micromelesalnifolia, 11. Betula schmidtii

The formed boundary of the transition zone between the two studied zones is pronounced and is clearly distinguished in the process of studying objects, it is also visible when observing the space shots. All this indicates that in the future, after the removal of technogenic and anthropogenic loads, successional development will most likely go in the direction of the formation of this territory of multi-species plantations with a predominance of deciduous species. That is, changes in the primary biota caused by the long-term existence of mining enterprises have become largely irreversible.

In the studied territory, the industrial development zone is located where all the primary forests are surrounded by the transition zone close to the technogenically broken (Fig. 2).

The zone of complete destruction of the initial biota during the construction and operation of a mining enterprise includes two components: a land allotment, within the boundaries of the entire production complex, and a system for storing solid mining waste or temporary storage of various industrial products.

The size of the industrial site of an enterprise depends on its annual productivity (Fig. 3).

A detailed analysis, including under field conditions, showed that the qualitative and quantitative composition of secondary plant communities differs significantly from the primary one and is mainly represented by herbaceous and shrub vegetation. On the study sites, it was discovered that there were traces of ungulates identified due to the young plant growth being an excellent food supplier.

As a result of the analysis of remote sensing data of natural and technical systems with the identification of the boundaries of man-made formations, it was established that, under equal natural and climatic conditions, the restoration of vegetation in

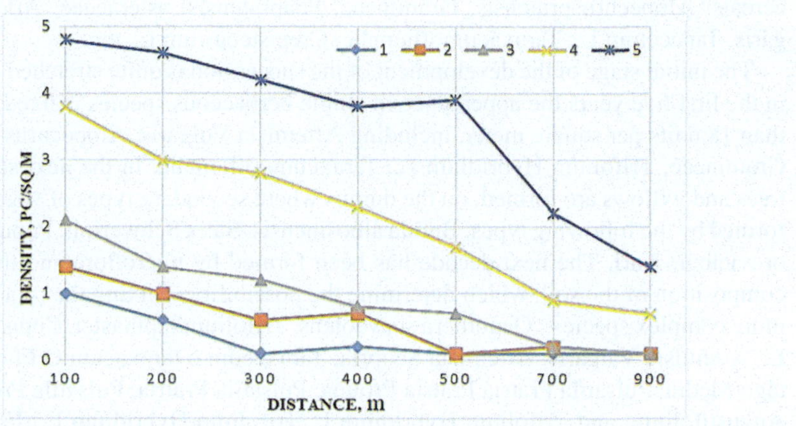

Fig. 2 Dynamics of the main forest-forming species undergrowth density within the zone of the ecotone between the standard and second-growth forest communities (1—Pinus koraiensis, Sieb. et Zucc; 2—Jurglans manshuriga Maxim.; 3—Abies holophulla, Maxim.; 4—Acer; 5—Betula costata, Trautv.)

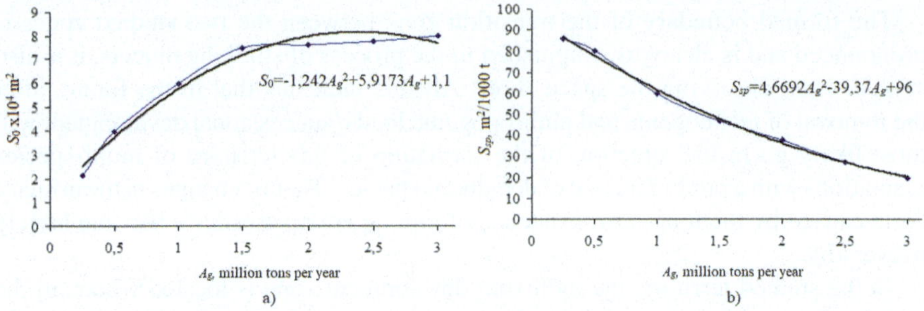

Fig. 3 Influence of the annual productivity of the mine (Ag) on the total (**a**) and specific (**b**) area of biota destruction (So and Ssp) during the creation of the industrial infrastructure of the mining enterprise

areas where mining has ceased is more intensive than in the dumps located next to the production complex.

The processes of revitalization of plant communities were studied on dumps formed during the extraction of fine earth granodiorite, located near the Bolshekhehtsirsky Nature Reserve. Various types of broad-leaved, coniferous-deciduous, and dark-coniferous forests are widespread. The composition of the vascular plants of the reserve includes more than 1000 units, which contributes to the acceleration. As a result of the studies of physical–chemical characteristics and agrotechnical properties, the overburdened rocks of the Korfovo stone quarry are potentially fertile.

It was also found that 1–2 years after the formation of the dump, pioneer herbaceous vegetation appeared on its surface, which are represented by herbs of the following families: legumes: Trifolium hybridium L., Melilotus albus Medik; cereals: Alopecuruspratensis, Gramineae, Poapraténsis; asteraceae: Artemísiavulgáris, Tanacetum L., Taraxácumofficinále, Crepistectorum L., etc.

The initial stage of the development of the succession is quite stretched over time; in the first five years the appearance of single herbaceous species is fixed, not more than 18 units per square meter, including Artemísia Vulgáris, Alopecurus Pratensis, Gramineae, Trifolium Hybridium L., Taraxcum Officinále. In the next stage, birch trees and willows are formed, on the dumps where secondary types of vegetation are formed by the following types: Betula albosinensis, Salix Schwerinii, Populus maxyorovichzii, Anri. The next decade has been formed by microflora and the mineral composition of the soil, which determine the possibility of fixing the root system of more complex species.: Oenothera suaveolens, Trifolium lupinaster, Cepis Tectorum L., Artemísia Vulgáris, Artemísia Scopria, Taraxacum o taraxacum O Fficele, Plantágo Média, Pulsatíla Pratria Pratria Prouna, Pulsatila Pratria, Pulsatila Prouna, Pra, Angustifolium, and Trifolium Hybridium L. Trifolium Hybridium L. (Fig. 4). The development of phytocenosis favorably affects the balance of the nitrogen in soil, which stimulates humusonacotics.

The analysis of the qualitative characteristics of the technogenic substrate showed a low content of humus and microelements in it. These results have a positive effect

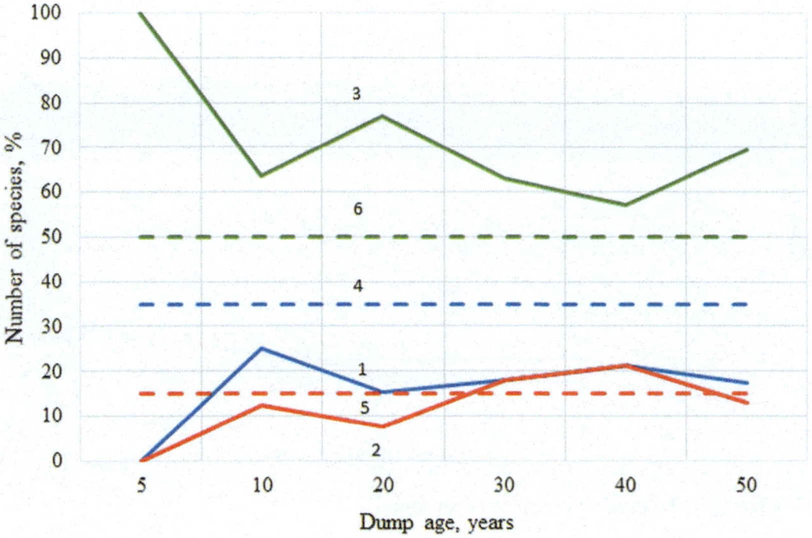

Fig. 4 Change in the coenotic structure of secondary plant communities over time (1—tree species, 2—shrub species, 3—herbaceous species, 4–6—background indicators, respectively)

on plant growth and development. Soils formed around the influence of mineral raw materials are mixtures of overburdened rocks and processing waste, the physical and mechanical composition is like a rocky-sand mixture. Based on the above, the data obviously shows that moisture is not retained in the substrate; no hydrological regime is present.

The number of identified herbaceous species on the studied sites is 46 species, but on the dumps where the vegetation has been formed for more than 45 years, the share of herbaceous species has been reduced by half. Instead, 16 wood and 7 shrub species appear (Fig. 5).

Such characteristics of self-organized vegetation processes can be explained by the low content of humus in the substance and its unfavorable physical and mechanical features. The soils consist mainly of rocky and sandy mixtures, due to the lack of moisture in the substance, and this leads to the absence of the water regime necessary for vegetation.

Key plant communities have been identified based on the conducted analysis, and the key succession series of the plant community self-regeneration at rock spoil heaps of the Korfovsky Rock Querry was produced (Fig. 6).

A general idea of the directions of development of succession is given by studying the regeneration of tree species in various parts of the studied area. In most of its parts, the structure of the undergrowth is quite consistent with the known features of digressive-demutational changes that occur after the disturbance of primary coniferous-deciduous forest plantations. The maples that regenerate better than others are mono (5 thousand pieces/ha), green bark (6 thousand pieces/ha), false Siebold (3.5 thousand pieces/ha), Amur linden (2.8 thousand pieces/ha). /ha). Coniferous

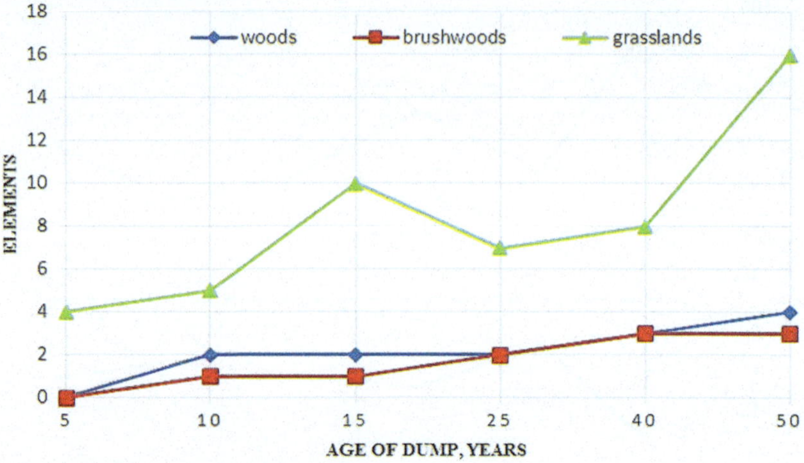

Fig. 5 Change in the number of species over time

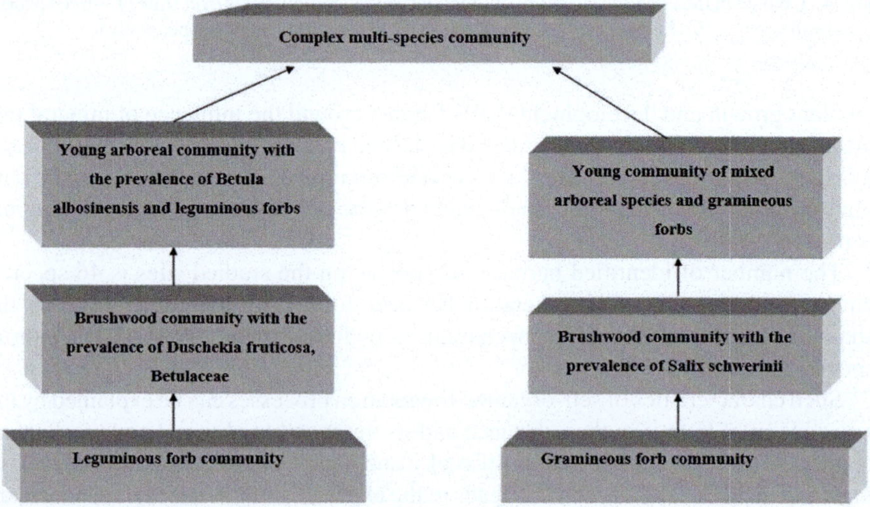

Fig. 6 Succession seres of the plant community self-regeneration at rock spoil heaps

undergrowth is also practically absent. All this indicates that in the future, after the removal of technogenic and anthropogenic loads, successional development will most likely go in the direction of the formation of this territory of multi-species plantations with a predominance of deciduous species (primarily maples and linden). That is if the changes in the primary biota caused by the long-term existence of a mining enterprise have become irreversible.

The process of regeneration of tree species in the ecotone zone between secondary forest plantations and areas of primary forest preserved in hard-to-reach places looks completely different. Due to the technogenic reasons for its formation, the boundary between the mature forest stand of these two communities is expressed quite clearly and for this component of the phytocenosis there seems to be no zone of transition of properties. However, there is a visible concern with the undergrowth of the main forest-forming species. Around all preserved areas of the primary biota of this ecosystem, a stripe (ecotone) is marked, within which, along with the renewal of species of the forest, the secondary community, the main forest-forming species of the reference community also appear.

The intensity of this process for all these species decreases with the distance from the mother plants, depending on the mechanisms of seed transfer (wind in fir, birch, maples, and zoogenic in cedar and walnut). The spread of the latter two listed is also hampered by the fact that their seeds are eaten by animals and birds, and most importantly, collected by humans. Moreover, as the area of primary forest plantations in the zone of anthropogenic impact decreases, the intensity of this form of anthropogenic load increases rapidly.

As part of the studies, the main criteria for evaluating the restoration of vegetation were the quantitative and qualitative composition of phytocenosis, the speed of restoration of vegetation cover, and the vegetation index. The studies made it possible to conclude that the restoration of the studied objects of the Khabarovsk Territory occurs, but not in full. For further clarification of the results, additional expeditionary studies will be carried out.

The main parameters have been established that affect the natural restoration of vegetation:

1. Natural climatic conditions.
2. A method of extraction and processing of raw materials.
3. Mineral raw material types.
4. The intensity of the technogenic load.
5. The physical and mechanical properties of soil and the shape of the dump.
6. The presence in the immediate vicinity of developed phytocenosis.

Apparently, there is no regular pattern in the changing stages of the succession sere. The appearance of some or other species in the plant community depends on the nature of the transfer of seeds taking place in a particular season of the year, otherwise, it requires other specific conditions of the natural environment. Nevertheless, special mention should be made of the total absence of natural regeneration of coniferous species, which indicates that the regeneration succession process under these conditions has passed the point of bifurcation at a very early stage of its development [4].

It also has been established that the active resumption of biota components occurs in areas with low population density and low development of the territories, when anthropogenic effects are limited only by the activities of the mining and industrial complex, since there are no artificially created plant communities of agricultural and forestry purposes, and there are no kind of any other industrial enterprises.

4 Conclusions

1. The objects under the study have a powerful potential for natural restoration, which must be used when developing reclamation measures.
2. The processes of vegetation restoration and their dynamics have been identified. However, the structure, species composition, and other characteristics of secondary vegetation differ significantly from the original ones.
3. Restoration of soil and vegetation covered in the studied areas, according to forecast estimates, ranges from 10 to 45 years. This is the time during which the area will be covered with vegetation, but its qualitative and quantitative composition will differ significantly from the original.
4. The conducted studies indicate that the possibility of natural restoration and development of plant communities remains possible even while maintaining the existing technogenic load.

A complex multi-species community regenerates through the development of a key succession sere, but because of the bifurcation of the process natural regeneration of the initial biota does not happen at the local succession level.

References

1. Postawa A, Motyka J (2019) Selected trace elements and metals in groundwater within Permian sediments near Olkusz (Zn-Pb ore mining region, S Poland). Environmental Science and Pollution Research 26:34–43
2. Jahanshahi B, Zare M (2015) Assessment of heavy metals pollution in groundwater of Golgohar iron ore mine area, Iran. Environmental Earth Sciences 74(1):505–520
3. Manakov YuA, Kupriyanov AN (2009) Criteria for diagnosing the primary stages of succession on Kuzbass dumps. Mining Information and Analytical Bulletin 7:186–193
4. Bogorodskaya AV, Trefilova OV, Shishikin AS (2014) Processes of primary soil formation in technogenic ecosystems on the dumps of the Borodino brown coal deposit (eastern part of KATEK). Bulletin of Tomsk State University 382:214–220
5. Klaas GJN, van Lagen B, Buurman P (2001) Composition of plant tissues and soil organic matter in the first stages of a vegetation succession. Geoderma 100:1–24
6. Galchenko YuP, Bubnova MB, Ozaryan YuA (2013) Environmental Aspects of Development of Environmental-and-Technical Systems of Exploitation of Mineral Resources in Conditions of the Far East. Environmental Systems and Tools 7:25–32
7. Ahmed Abdelaal, Mohamed Sultan, Mohamed Elhebiry et al (2021) Integrated studies to identify site-specific parameters for environmentally benign mining operations: A case study from the Sukari Gold Mine, Egypt, Science of The Total Environment, vol. 750. https://doi.org/10.1016/j.scitotenv.2020.141654
8. Galchenko Ju, Ozaryan Ju (2018) Method of quantitative assessment of the regularities of natural restoration of biota in zones of technogenic disturbance by extractive enterprises. Problems of Complex Development of Georesources. electronic resource. "E3S Web of Conferences", 04006. https://doi.org/10.1051/e3sconf/20185604006
9. Joy B Zedler (2000) Progress in wetland restoration ecology. Trends in Ecology & Evolution 15(10):402–407. https://doi.org/10.1016/S0169-5347(00)01959-5

10. Karel Prach, Petr Pyšek (2001) Using spontaneous succession for restoration of human-disturbed habitats: Experience from Central Europe. Ecological Engineering 17(1):55–62. https://doi.org/10.1016/S0925-8574(00)00132-4
11. Trubetskoi KN, Galchenko YuP (2015) Geoecology of the development of the Earth's interior and ecogeotechnology for the development of deposits. Nauchtekhlitisdat, p 360
12. Trubetskoi KN, Galchenko YuP, Burtsev LI (2003) Environmental Problems of Exploitation of Mineral Resources under Sustainable Development of the Environment and Society. Nauchtekhlitisdat, p 262

Changes in the Structure of Coniferous Forests Within the Broad-Leaved Coniferous Forest Zone in PreUrals (Russia) Based on Data from 50-Year Direct Observations

A. Davydychev and G. Zaitsev

Abstract The growth and development of individual trees as well as the interactions between various tree species determine the structure and composition of forests. Evaluating the processes of forest formation is made possible by examining the dynamics of alterations in the composition and structure of forests. To evaluate the changes in forest structure and composition over a 50-year period, five types of coniferous forests (spruce, fir, spruce-fir, pine, and larch forests) within the broad-leaved coniferous forest zone in the PreUrals were studied. Data from the 1971 forest inventory (for fir forests in 1984) and the 2020 forest inventory were used as the basis for this investigation. Three permanent trial plots per type of coniferous forest totalling fifteen were laid out. Over a 50-year period, conifer forests have grown and developed successfully. Coniferous forests are undergoing changes in composition and structure. New spruce and fir generations have arisen, allowing stands in all types of coniferous forests to develop sustainably. The diameter of the trees differs significantly. In pine and larch forests, pine and larch trees are gradually replaced by spruce. This is because spruce can grow under various environmental conditions. A high threshold of sustainability is indicated by the overall pattern of tree difference in coniferous forests for the viability of natural coniferous forest ecosystems.

Keywords Coniferous forests · Tree diameter · Stand dynamics · Forest formation

A. Davydychev · G. Zaitsev (✉)
Ufa Institute of Biology, Ufa Federal Research Centre of the Russian Academy of Sciences, 69, Pr. Oktyabrya, Ufa 450054, Russia
e-mail: forestry@mail.ru

L. V. Radionova and D. V. Ulrikh (eds.), *Advances in Ecology and Environmental Engineering*, Springer Proceedings in Earth and Environmental Sciences,
https://doi.org/10.1007/978-3-031-64423-8_11

1 Introduction

The biosphere functions of forest ecosystems, their sustainability, the intensity of biocoenotic processes, and the efficiency of producing and depositing organic matter are all influenced by the structure and composition of forests. [1, 2]. Forest structure and composition refers to the overall sum of the variability of stand characteristics (tree height and diameter, age, stand density, etc.), the distribution of trees based on these characteristics, and the interactions among these characteristics in the dynamics of the forest community [3]. The distribution of different tree species by diameter is commonly used to describe the structure of forests. A tree diameter distribution series thoroughly characterises the processes of tree growth, differentiation, and forest mortality.

Numerous publications have recently been published that cover the theoretical and methodological aspects of forest structure research. These publications also include partial and generalised models of the species-wise structure of individual stands and their aggregates, accounting for factors such as age, origin, and other characteristics, across a range of geographic regions [4, 5]. Nevertheless, there are not many researches on how the structure of forests changes over a long period of time [6, 7]. One way to identify the primary factors driving the formation of forest communities throughout time is to analyse data from direct long-term research conducted in permanent study areas. This fact has determined the relevance and novelty of our study. The aim of this study was to study the dynamics of the structure of coniferous forests within the broad-leaved coniferous forest zone in PreUrals (Ufa Plateau, Russia).

2 Materials and Methods

2.1 Site Descriptions

The research locations were mature natural conifer forests on the Ufa Plateau (PreUrals), which are in the water-protection zone of the Pavlovka Reservoir (Ufa River). The coniferous forests studied here are at the eastern boundary of the distribution of East European boreo-nemoral forests. The existence of a mosaic vegetation cover (owing to various environmental factors) generates a unique research area. These forests have developed over a long period of time (beginning with the Holocene) under a variety of distinct forest conditions, each with a varying degree of impact from the main environmental elements [8]. During the process of forming forests, the co-growth of coniferous and broad-leaved tree species near the edge of their geographical range results in complex competitive dynamics. The spread of perennially frozen soils, which are unique to the PreUrals, vertical zonation, and altitude differentiation in the distribution of climatic indicators (temperature inversion) are other environmental factors that impact the creation of forests in the region

[8]. Additionally, human activity (felling operations) has had a significant impact on the current condition of coniferous forests [9]. Previously, a full description of the natural and climatic circumstances, as well as a complete characterisation of the research area's forests, has been provided [10, 11].

The most prevalent types of coniferous forests in the area were used to investigate the dynamics of forest composition. Spruce forests (with Siberian spruce (*Picea obovata* Ledeb.)), with the grassy layer dominated by green mosses and small ferns. These forests thrive on steep, shady slopes with humus-carbonate mountain-forest soil derived on loamy, large-block rubble, and eluvio-deluvia limestone deposits. Larch forests (with Sukachev larch (*Larix sukaczewii* Dyl.)), where sphagnum and green mosses predominate in the grassy layer. These forests are located on north-facing slopes where peat-humus mountain forest soils, which are created by layers of rubble-clay eluvio-deluvia and large blocks of gravel and limestone eluvia, are continuously wetted. The soil is perennially frozen. Pine forests (with Scots pine (*Pinus sylvestris* L.)) are associated with insolated, steep slopes, where various species of sedge and green mosses predominate in the grassy layer. The soil of these forests is a humus–carbonate mountain-forest, formed by rubble-clay limestone deluvia. Fir forests (with Siberian fir (*Abies sibirica* Ledeb.)), where goutweed, wood sorrel, and horsetail thrive in the grassy layer. Growing on the light-grey mountain-forest soil developed on clay deluvia, fir forests may be found on the bottom and lower portions of slopes at all exposures. Fir-spruce forests, dominated by big ferns and goutweed in the herbaceous layer. These forests grow on the tops of slopes and on broad plateaus where flaggy-rock limestone eluvia formed the grey mountain-forest soils.

2.2 Data Collection and Analysis

Based on standard method guidelines [12, 13], the forest inventory was made. In 1971, three trial plots (a total of fifteen) were laid out for each type of coniferous forest. In fir forests, trial plots were laid out in 1983. We used the most widely practised trial plots, i.e., 0.25 ha trial plots (50 × 50 m). [14, 15]. The key characteristics of the tree stand, including the tree's age, height (H), and diameter at breast height (DBH), were measured. Measurements were taken for every tree in the trial plots. The widely used measurement of the total basal area of all trees was used to calculate the stand density [15]. Growth class (also called bonitet) is an inventory feature that describes the potential productivity of a stand and the growth rate of trees. Based on the average age and height of the trees as well as the stand origin (vegetative or seminal), we defined growth classes using the Orlov table [16]. In 2020, an iterative forest inventory was conducted on the same trial plots. All test data were statistically processed [17], and all calculations (descriptive statistics) were performed using MS Excel and Statistica 8.0 software.

3 Results and Discussion

Research indicates that the growth class, stand density, and composition of spruce forests have remained essentially unchanged over the past 50 years (Table 1). There has been an increase in the growing stock, average diameter, and average height. In the composition of forests, the proportion of trees with small trunk diameters has notable declined, whereas the proportion of trees with large diameters increased (Fig. 1).

Table 1 Key characteristics of the spruce forest studied

Stand formula*	Age (years)**	Stand density	Growth class**	DBH (cm)**	Height (m)**	Growing stock (m^3/ha)***
1971						
1 Layer: 9S1B, sngl: P, Ln	100	0.45	IV	24.8	17.2	210
2 Layer: 8S1B1Ln, sngl: F	70	0.10		9.0	8.0	
2020						
1 Layer: 10S, sngl: P, Ln, B	150	0.45	III	31.3	25.6	258
2 Layer: 6S2Ln2B	80	0.10		11.0	10.5	

*S—spruce, B—birch, P—pine, Ln—linden, F—fir. **Specified for the main tree species. ***Total for the entire tree stand

The first layer of larch forests has not significantly altered in composition during the past 50 years (Table 2). The number of spruce trees on the second layer of the stand increased. The growth class and stand density of the first layer have greatly increased. The growing stock of larch forests is increasing because of the increase in the diameter and height of the trees. The structural complexity of stands of larch forests has increased throughout the past 50 years (Fig. 2). With the progressive complexity of the larch stand's structure, the «peak» of small-diameter trees, as noted in 1971, over 50 years gradually passes into the medium-diameter tree category. Simultaneously, the number of large-sized trees (with a diameter more than 26 cm) is also increasing.

For the past 50 years, the growth class, stand density, and composition of pine forests have remained unchanged (Table 3). Since tree diameter and height are increasing, the growing stock of pine forests is also increasing. In parallel with the increase in large-sized pine trees, there has been a notable decline in small-diameter pine trees (Fig. 3).

The growth class and composition of fir forests have not significantly altered during the past 40 years (Table 4). The stand density has reduced greatly (by 0.3 unit). This results from the prematurely death of a great number of trees. Consequently, there was a little alteration in the average diameter, height, and growing

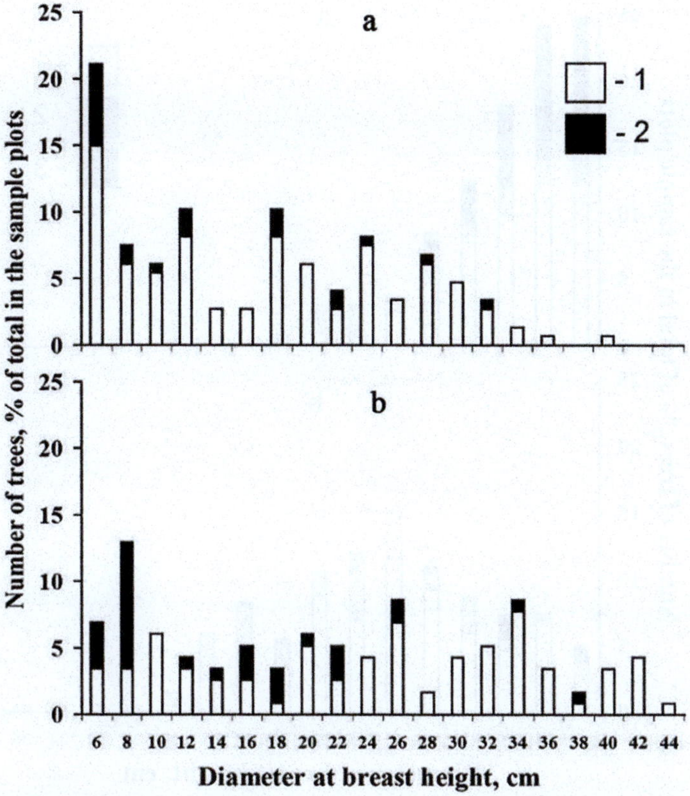

Fig. 1 The diameter-based tree distribution in spruce forests in 1971 (**a**) and 2020 (**b**) (1—Siberian spruce, 2—other tree species)

Table 2 Key characteristics of the larch forest studied

Stand formula*	Age (years)**	Stand density	Growth class**	DBH (cm)**	Height (m)**	Growing stock (m^3/ha)***
1971						
1 Layer: 5L4S1B, sngl: P	110	0.45	V	19.1	15.2	156
2 Layer: 6S3B1L + F		0.54				
2020						
1 Layer: 5L5S, sngl: F, B	150	0.61	III	23.0	24.7	254
2 layer: 8S2L + B, sngl: F		0.15				

*L—larch, S—spruce, B—birch, P—pine, F—fir. **Specified for the main tree species. ***Total for the entire tree stand

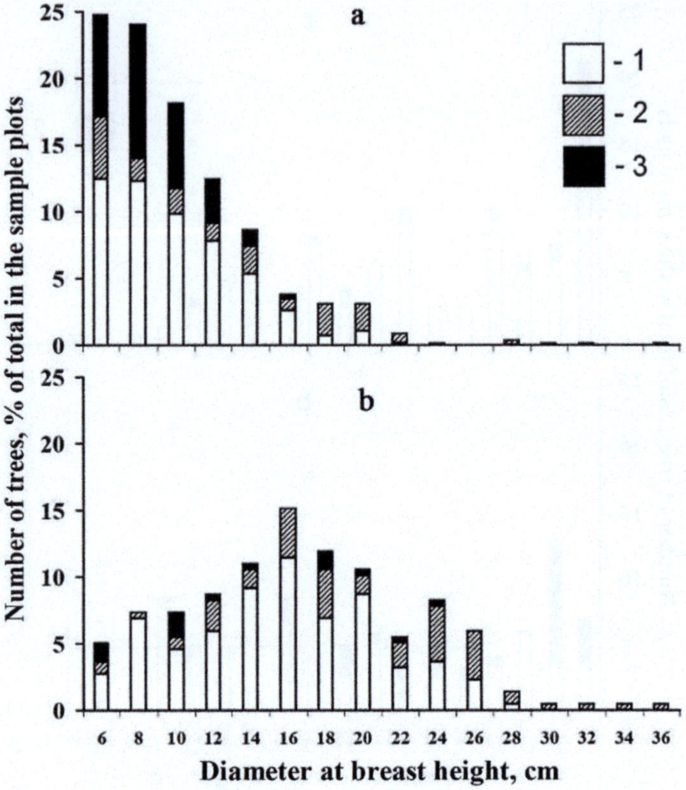

Fig. 2 The diameter-based tree distribution in larch forests in 1971 (**a**) and 2020 (**b**) (1—Siberian spruce, 2—Sukachev larch, 3—other tree species)

Table 3 Key characteristics of the pine forest studied

Stand formula*	Age (years)**	Stand density	Growth class**	DBH (cm)**	Height (m)**	Growing stock (m^3/ha)***
1971						
1 Layer: 10P + S, sngl: B	110	0.93	II	29.0	25.8	419
2 Layer: 8P1S1Ln + B	70	0.24		13.4	14.7	
2020						
1 Layer: 9P1S	150	0.80	II	30.8	26.4	470
2 Layer: 6P4S, sngl: F		0.03		13.9	13.2	

*P—pine, S—spruce, B—birch, Ln—linden, F—fir. **Specified for the main tree species. ***Total for the entire tree stand

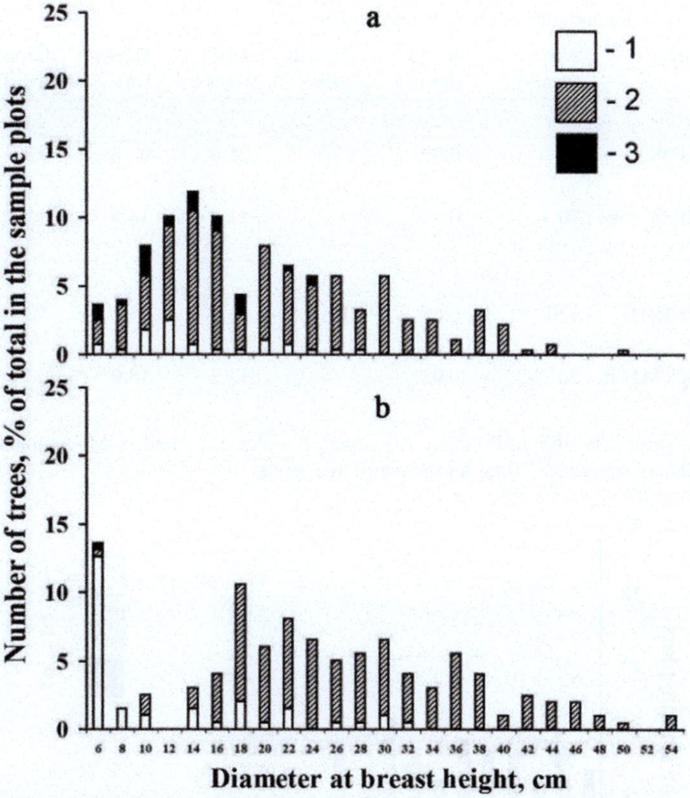

Fig. 3 The diameter-based tree distribution in pine forests in 1971 (**a**) and 2020 (**b**) (1—Siberian spruce, 2—Scots pine, 3—other tree species)

stock. Both large-sized and small-diameter trees are becoming more prevalent in the stand structure (Fig. 4).

The growth class and composition of fir-spruce forests have not significantly altered during the past 50 years (Table 5). The stand density greatly decreased because of the early demise of the trees. The growing stock has barely altered, but the average diameter and height of the trees have dramatically increased. This is also linked to the considerable early death of trees. Due mostly to the long-lived spruce trees, the structure of forests is characterised by a marked decline in small-diameter trees and a major increase in medium- and large-sized trees (Fig. 5).

Table 4 Key characteristics of the fir forest studied

Stand formula*	Age (years)**	Stand density	Growth class**	DBH (cm)**	Height (m)**	Growing stock (m^3/ha)***
1984						
1 Layer: 6F2S2B + P, sngl: A	80	0.90	I	21.0	24.8	326
2 Layer: 8F2S + B, sngl: E, Ln	70	0.15		11.4	16.5	
2020						
1 Layer: 5F4S1B, sngl: P, A	110	0.60	II	23.4	25.2	391
2 Layer: 8F1M1Ln + E, sngl: S	70	0.03		12.2	18.0	

*F—fir, S—spruce, B—birch, P—pine, A—aspen, E—elm, Ln—linden, M—maple. **Specified for the main tree species. ***Total for the entire tree stand

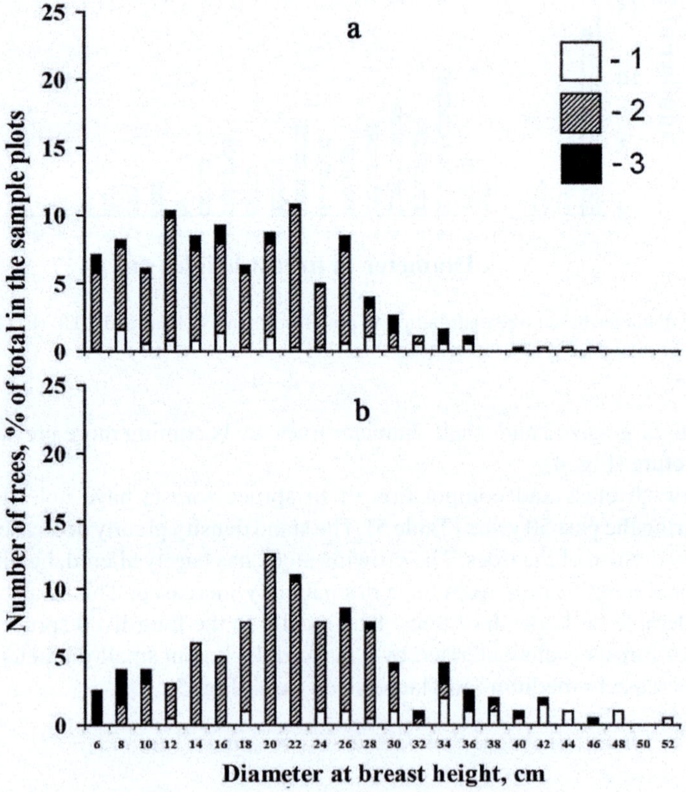

Fig. 4 The diameter-based tree distribution in fir forests in 1984 (**a**) and 2020 (**b**) (1—Siberian spruce, 2—Siberian fir, 3—other tree species)

Table 5 Key characteristics of the fir-spruce forest studied

Stand formula*	Age (years)**	Stand density	Growth class**	DBH (cm)**	Height (m)**	Growing stock (m^3/ha)***
1971						
1 Layer: 3S3F4B, sngl: P	70	0.74	I	26.3	21.8	298
2 Layer: 8F1S1Ln + E, sngl: M, A		0.28		11.2	11.9	
2020						
1 Layer: 4S4F2B, sngl: P, Ln	110	0.66	II	39.9	26.6	313
2 Layer: 7F2Ln1A, sngl: E		0.10		13.8	13.8	

*S—spruce, F—fir, B—birch, P—pine, Ln—linden, E—elm, M—maple, A—aspen. **Specified for the main tree species. ***Total for the entire tree stand

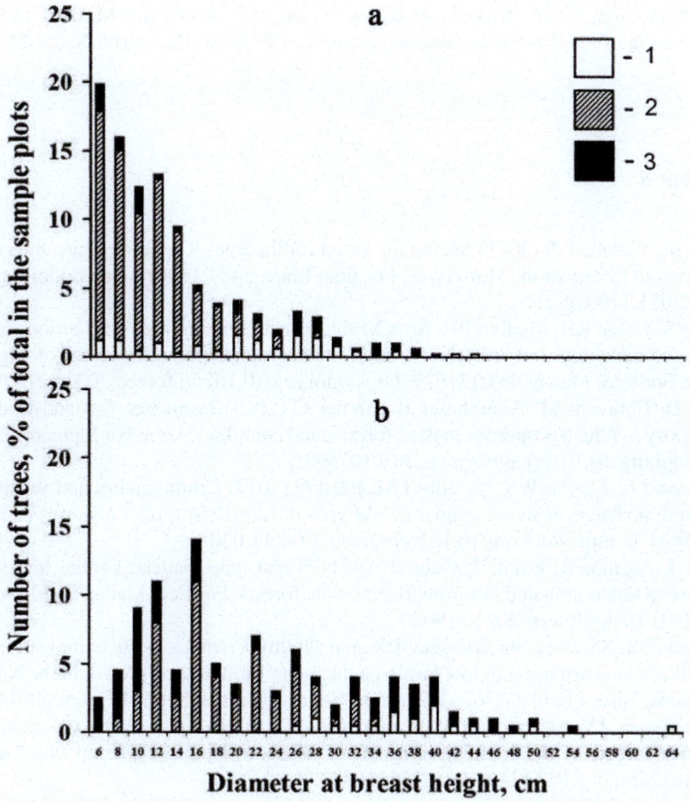

Fig. 5 The diameter-based tree distribution in fir-spruce forests in 1971 (**a**) and 2020 (**b**) (1—Siberian spruce, 2—Siberian fir, 3—other tree species)

4 Conclusions

All types of conifer forests exhibit changes in composition and forest structure, as demonstrated by a comparative examination of data covering the 50-year period of forest development (1971–2020). Due to the early mortality of a large number of trees (21–62% of the initial number of trees), there has been a considerable drop in the number of trees, whereas the percentage of large-sized trees has increased. The shift to large-sized trees and early tree death both contributed to a considerable decline in the number of small diameter (6–12 cm) trees during the study period. Tree stands in forests of spruce, fir, and fir-spruce are gradually growing because of the appearance of new generations of spruce and fir. Spruce gradually replaces pine and larch trees in pine and larch forests. The tendency of trees to conserve and self-complicate their forest structure over time is the overall pattern of the process of tree differentiation by diameter.

Acknowledgements This work is a part of the research project of the Ufa Institute of Biology, Ufa Federal Research Centre, Russian Academy of Sciences, Russia (the Ministry of Science and Higher Education of the Russian Federation, grant no. 123020700152-5 (FMRS-2023-0008)).

References

1. Maren IE, Sharma LN (2021) Seeing the wood for the trees: Carbon storage and conservation in temperate forests of the Himalayas. For Ecol Manag 487:119010. https://doi.org/10.1016/j.foreco.2021.119010
2. Khanal S, Nolan RH, Medlyn BE, Boer MM (2024) Disentangling contributions of allometry, species composition and structure to high aboveground biomass density of high-elevation forests. For Ecol Manag 554:121679. https://doi.org/10.1016/j.foreco.2023.121679
3. Seidel D, Ehbrecht M, Annighofer P, Ammer C (2019) From tree to stand-level structural complexity—Which properties make a forest stand complex? Agric For Meteorol 278:107699. https://doi.org/10.1016/j.agrformet.2019.107699
4. Burrascano S, Keeton WS, Sabatini FM, Blasi C (2013) Commonality and variability in the structural attributes of moist temperate old-growth forests: A global review. For Ecol Manag 291:458–479. https://doi.org/10.1016/j.foreco.2012.11.020
5. Boloni J, Aszalos R, Frank T, Odor P (2021) Forest type matters: Global review about the structure of oak dominated old-growth temperate forests. For Ecol Manag 500:119629. https://doi.org/10.1016/j.foreco.2021.119629
6. Kapralov DS, Shiyatov SG, Moiseev PA et al (2006). Changes in the composition, structure, and altitudinal distribution of low forests at the upper limit of their growth in the Northern Ural Mountains. Russ J Ecol 37:367–372. https://doi.org/10.1134/S1067413606060014
7. Chernen'kova TV, Morozova OV, Puzachenko MY et al (2016) Composition and structure of spruce forests of the southwestern part of Moscow region. Contemp Probl Ecol 9:820–833.https://doi.org/10.1134/S1995425516070039
8. Kulagin YuZ (1978) Ekologicheskie arealy porod-lesoobrazovatelej v rajone Ufimskogo plato (Ecological areas of the forest forming species in the area of the Ufa plateau) Russ For Sci 5:24–29

9. Martyanov NA, Batalov AA, Kulagin AYu (2002) Shirokolistvenno-hvojnye lesa Ufimskogo plato. Fitocenoticheskaya harakteristika i vozobnovlenie (Broadleaved-coniferous forests of the Ufa plateau). Ufa, Gilem
10. Zaitsev GA, Kulagin AYu, Davydychev AN (2018) The particularities of the growth of Siberian fir (*Abies sibirica* Ledeb.) in the first stages of ontogeny in conifer forests (Ufa plateau, Pre-Ural). Trees 32:511–518. https://doi.org/10.1007/s00468-017-1647-y
11. Zaitsev G, Davydychev A, Kulagin A, Giniyatullin R et al (2021) Suppressed undergrowth of Siberian spruce (*Picea obovata* Ledeb.) in early ontogeny: one-way ticket or survival strategy? Forests 12(7):851. https://doi.org/10.3390/f12070851
12. Kangas A, Maltamo M (2006) Forest inventory: methodology and applications. Springer, Dordrecht
13. Newton AC (2007) Forest ecology and conservation: a handbook of techniques. Oxford University Press, New York
14. Sukachev VN (1966) Programma i metodika biogeocenologicheskih issledovanij (Biogeocenotic research program and methodology). Nauka, Moscow
15. Kershaw JA, Ducey MJ, Beers TW, Husch B (2016) Forest mensuration, 5th edn. Wiley, Chichester
16. Anuchin NP (1982) Lesnaya taksaciya (Forest inventory). Forest industry, Moscow
17. Ross SM (2009) Introduction to probability and statistics for engineers and scientists, 4th edn. Elsevier Academic Press, San Diego

Man-Made Risks in Coal Mining, Transportation, Storage and Consumption and Prospects for Their Minimization

S. S. Timofeeva and S. S. Timofeev

Abstract Coal is one of the main sources of energy worldwide, both at present and in the foreseeable future. World proven oil reserves are sufficient for 53 years, natural gas—for 56 years at the current level of production and consumption, while for coal it is more than 130 years; alternative sources are not yet able to compete with hydrocarbon energy resources. At the same time, the production, transportation, storage and use of coal is associated with many risks—fire, environmental and other risks. Assessment of these risks, as well as the development of ways to minimize them has long been the object of research both in Russia and around the world. The article presents a brief overview of research that has been carried out by Irkutsk National Research Technical University for fifteen years at coal mining sites in the Siberian region and promising areas of joint Russian-Chinese research.

Keywords Man-made risk · Coal dust · Coal spontaneous combustion · Detection · Antipyrogen · Detector · Fire risks

1 Introduction

In the conditions of a tough political struggle for a multipolar world, sanctions being imposed on Russia, a ban on clean energy supplies brings us back to an increase in coal consumption, and consequently to the man-made and environmental risks of the coal mining industry [1].

The absolute leader in coal production in the world is China (more than 50% of coal mined in the world); Russia is one of the top ten coal-producing countries. In 2022, Russia produced 443.6 mln tons of coal and exported 210.9 mln tons of coal, including 59.52 mln tons to China and 16.7 mln tons to India [2]. Coal production in China amounted to 4.5 billion tons of coal, imports—290 million tons [3]. As all

S. S. Timofeeva (✉) · S. S. Timofeev
Irkutsk National Research Technical University, 83, Lermontova St, Irkutsk 664074, Russia
e-mail: timofeeva@istu.edu

L. V. Radionova and D. V. Ulrikh (eds.), *Advances in Ecology and Environmental Engineering*, Springer Proceedings in Earth and Environmental Sciences,
https://doi.org/10.1007/978-3-031-64423-8_12

coal-producing countries, both Russia and China encounter man-made risks at all stages of the process life cycle: mining, transportation, storage, and use (combustion and processing) of coal.

Among the specific risks of mining enterprises, which can affect the efficiency of enterprises, are:

1. External factors (economic, including: the level of coal prices, the level of demand, the level of prices for mining equipment and machinery, interest rates, exchange rates); geographical (competitiveness, remoteness from ways and buyers); social and environmental (size of resource payments, environmental policy of the state, the availability of qualified specialists, the amount of compensation payments);
2. Internal factors (mining and geological, including the period of reserves development, degree of exploration, degree of reserves depletion, the percentage of fossil losses during mining, moisture and ash content of coal, natural hazards); risks associated with work safety and the impact on the environment [4, 5].

Particular attention should be paid to the risks associated with the safety of work with a high probability of accidents and occupational diseases at work, especially when working deep underground, with less oxygen in the air, increased atmospheric pressure and high temperature, in conditions of high humidity, in the dark, with high noise and dust levels.

The impact on the environment is manifested in the disturbance of the landscape, destruction of vegetation cover, agricultural and forest lands [6–8], redistribution of groundwater flow fields and the formation of a huge amount of mine, pit, and drainage waste water, which rises to the surface, has a negative impact on the environment and is lost as a useful resource [9–11], air pollution with methane and combustion products [12], high levels of explosion and fire hazards [13], negative impact on the health of residents of mining regions and workers of coal mining enterprises [14–16] and others.

At present science knows different metrics and criteria for assessing man-made risks; the list of methods and technologies of risk assessment is constantly updated. Riskology, a science of the twenty-first century, is taking its shape, the purpose of which is to analyze, diagnose, predict, program and plan risks in different areas of activity. Any activity is potentially hazardous. A failure, material or other loss, deviation from the goal can accidentally happen. In particular, the industrial risk can be characterized by the *FAR (Fatal Accident Rate)* metric, i.e. the number of fatal accidents during 10^8 h of exposure to probable hazardous factors on the health of 1000 people. It has been established that this value is highest among the industries in the coal industry at 7.3; construction at 5.0; agriculture at 3.7; chemical industry at 1.2 [17]. The specific injury rate, the number of fatalities/injured, related to the amount of extracted raw materials, is used as a risk metric for extractive industries.

Criteria of ecological risks, including environmental risks are even more numerous; first of all they are characterized by the average life expectancy of the population in the polluted territories, pollution indices and levels, hazard ratios, carcinogenicity, loss of useful resources, fire hazard, etc. [18]. Great losses of water as a valuable resource are worth mentioning, because during coal mining the water balance is disturbed, forming large volumes of mine waters which are divided by their physical and chemical composition into acidic, mineralized and polluted by suspended particles of coal and rocks; heavy metals, oil products, oils, etc. are present in the water. The velocity of water flow through workings, as well as the depth of horizons affect the aggressiveness of mine waters. It has been established that the most aggressive acidic wastewater is present in mines developing inclined coal seams.

At present time in China 1.87 m^3 of mine water is formed per ton of mined coal and annual production of such water reaches 6.88×10^9 m^3. In Russia water inflow to mines, depending on these factors, varies in a very wide range—from 10 to 4000 m^3/h, and water availability coefficient—from 0.3 to 20 m^3/t. In terms of absolute value of water inflows mines of the Siberian region belong to the group of watered and very watered mines. The value of water inflows at the majority of mines does not remain constant during the year. Considerable increase in inflow (up to 1.5–2 times) is observed in spring period that is connected with snow melting and in period of intensive rainfall. Excess of maximum inflows over normal inflows is the greater the shallower the depth of development and the closer the connection of groundwater aquifers with surface waters. According to academician Gu Dazhao, mine water should be considered as a long-term non-traditional water resource and methods of its storage, treatment and use should be developed to exclude violation of ecological equilibrium in coal basins, in particular siltation, acidification, salinization of water bodies and streams, as well as depletion of aquifers and changes in drinking water quality.

The Department of Industrial Ecology and Life Safety at Irkutsk National Research Technical University has been conducting research work for a number of years to develop a methodology for assessing man-made risks at coal-mining and coal-consuming facilities and technologies to minimize them.

In Russia more than 60% of coal deposits are concentrated in the eastern regions in the Lena and Tunguska coal basins. The Kuznetsk and Kansk-Achinsk deposits are the leaders in terms of commercial reserves. The Kuznetsk deposit, otherwise known as Kuzbass, is the largest coal basin in Russia and the largest in the world. The largest deposits of high-grade hard coal, including coking coal, are concentrated here. It is located in Western Siberia (Kemerovo Region) and coal is extracted mainly by the shaft method. The Kansk-Achinsk coal basin is located along the Trans-Siberian Railway through the Krasnoyarsk Territory, Kemerovo and Irkutsk Regions; brown coal is mined here by open-pit method. The Irkutsk-Cheremkhovo basin stretches along the Upper Sayan from Nizhneudinsk to Lake Baikal and is divided into the Pribaikalskaya and Prisayanskaya branches. The volume of open-pit coal mining is at the level of 3.4% of the total coal production in Russia.

The objects of research were the risks of coal mining enterprises in the Irkutsk and Kemerovo Regions, including technological processes of coal mining in mines and quarries; processing at coal preparation plants, transportation and transshipment at coal terminals, combustion at thermal power plants operating on coal, use as a charge in metallurgical production.

Research is being conducted in the following directions:

- Assessment of ecological risks for environment and population caused by coal-mining enterprises [14, 19, 20];
- Development of methodology for estimating environmental load from fires at coal-mining facilities [21];
- Investigating fire hazard of coals of different deposits and causes of spontaneous combustion and emergence of endogenous fires [22, 23];
- Analysis of traumatism and occupational morbidity of personnel at coal-mining and coal-consuming enterprises [24];
- Studying dust load, disperse composition of coal dust and identifying the contribution of nano-sized particles to occupational morbidity [25–27];
- Development of methods and devices for early diagnostics of coal combustion, antipyrogenous compositions and methods of preventing fire risks [28, 29];
- Development of technologies to prevent spontaneous combustion and environmental consequences of fires [30–33];
- Development of technologies for treatment of quarry and mine waters by hybrid technologies and biotechnological methods [34, 35].

This paper summarizes the research carried out under the guidance of the author by undergraduates, graduate students and professors from the perspective of modern applied riskology, and considers innovative solutions aimed at minimizing environmental, occupational and fire risks present at coal mining and coal consumption facilities.

2 Risk Metrics for Coal Mining Technologies

As is known, the choice of coal mining technology is determined by the rock occurrence depth. Open-pit method of coal mining is mainly used when the rock occurrence depth is up to 100–150 m. Underground method of mining is used when the depth of coal seams is significant. Vertical and (or) horizontal shafts are laid to the place of coal occurrence, which form a mine.

In terms of riskology, open-pit mining is safer and provides better working conditions for coal miners than coal mines, while in terms of environmental impact, open-pit mining causes less land disturbance and has less impact on the atmosphere.

Comparison of the negative impact of open-pit and underground coal mining on the environment and workers' health was carried out according to risk metrics selected from among the most significant for the personnel, namely individual occupational risk by the number of fatal accidents, by working conditions, by emissions of coal dust into the atmosphere and the impact of atmospheric pollution.

In the work, we evaluated the risks of environmental pollution at the Cheremkhovo hard coal and Mugunsky brown coal mines and mines of Kuzbass [14, 20, 21] (Table 1).

Table 1 Environmental pollution risks in coal open pits and mines

Mining technology/ deposit	Specific volume and/or mass of emissions of coal dust and other substances, g/t	Specific volume and/or mass of emissions of suspended solids to water, g/t	Volume and/or mass of waste generation per ton of production	Exceeding of MAL for noise and at the personnel workplace, times
Underground mining, Aldarskaya mine	102.6	131.1	Enclosing rocks, kg/t 25.4	1.8
Open pit mining	256.5	129.9	Overburden rocks and mining waste, t/t	1.5
Cheremkhovo open-pit mine			29.4	
Mugunsky open-pit mine	356.3	109.3	21.2	1.2
Kasyanovskaya coal processing plant	34.0	58.2	12.1	1.4

It was necessary to assess the individual risk for the personnel of the enterprise, which represents the probability for a person to get into an undesirable situation (to be injured or killed) and social risk, which characterizes the same threats for a group of people at the same time. As it is impossible to express a group of people by one number, the social risk is expressed graphically as dependence of frequency of undesirable events in which not less than *N* people died (suffered), on this *N* number. This graphical expression of social risk is called an F/N diagram.

When assessing social risk at coal-mining facilities in Russia over a 70-year period (1943–2012), based on the analysis of statistical data collected from various sources, including MiningWiki—The Free Mining.

Encyclopedia [17, 24], it was established that group accidents in modern Russia increased significantly; the average specific injury rate in 1985–1991 was 0.53 persons/mln tons, in 1992–2007—0.57 persons/mln tons, and in 1996–2011—0.42. The maximum value of 0.99 persons/mln tons was reached in 1997 and the minimum of 0.13 persons/mln tons in 2011. Therefore, the country is currently modernizing coal mining and more and more attention is being paid to open pit mining.

The study of occupational risks of personnel engaged in technological processes of coal mining at 4 coal open pits of the Irkutsk Region: Tulun, Azeisky, Cheremkhovo and Irbeisky, as well as personnel of Kasyanovskaya coal processing plant and thermal power plant found that the risk of respiratory dust pathology is at its highest during coal transportation (Table 2).

As follows from the above data, the greatest individual occupational risk of workers at the maximum concentrations of dust in the air of the working area was revealed in the drivers of technological vehicles of Cheremkhovo open-pit mine, operator of the crushing and loading unit in the drying and heating department of

Table 2 Risk of occupational dust pathology of respiratory organs of workers engaged in mining, enrichment and use of coal

Object of research	Workplace	Shift-average dust concentration, mg/m^3	Safe work period, years	Individual risk, 10^{-3}
Cheremkhovo open-pit mine	Bulldozer operator	62.9	15.2	4.43
	EKG excavator operator	42.8	22.4	1.78
	Driver of technological vehicles	110.1	9.1	9.02
Kasyanovskaya coal processing plant	Conveyor operator in the coal receiving department	23.9	10.86	1.68
	Operator of the washing units in the main building	15.6	16.6	1.09
	Operator of the crushing and loading unit in the drying and heating department	56.4	13.3	3.97
TPP-9, Angarsk	Bulldozer operator of fuel handling shop	38.5	25.6	2.71
	Car dumper operator	29.6	12.3	2.08
	Mechanic for repair of equipment in boiler and dust preparation shops	52.6	17.3	3.70

Kasyanovskaya coal processing plant, and mechanic for repair of equipment in boiler and dust preparation shops of TPP-9. The probability of acquiring chronic diseases when working for 5–10 years ranges from 3.5 to 11.2%, depending on the level of exposure to dust-related risk factors. In case of continuous exposure to dust for 10–20 years, 8.7–26% of workers at an open-pit mine may show symptoms of chronic diseases. The maximum values of inhalation risk are observed at exposure for more than 20 years.

Thus, in terms of injuries and occupational diseases, underground coal mining is a more risky method for miners due to hazardous factors, including dust exposure;

the highest risks in open-pit mining are for drivers of line vehicles who deliver coal to warehouses, and for crushing equipment operators involved in coal preparation processes.

3 Dust Load and Occupational Pathology

Research carried out by occupational pathologists have nowadays proved that it is precisely nanoparticles in dust that have a complex of physical, chemical properties and biological effect, which often differ radically from the properties of the same substance in the form of solid phases or macroscopic dispersions. Nanosized dust particles have a number of physico-chemical features, namely, their solubility, reactive and catalytic activity significantly change, which leads to increased production of free radicals and reactive oxygen species and damage to biostructures; they have high adsorption activity, high ability to accumulate and present a high potential risk to human health [27].

It has been established that the average daily dust load in workers varies in the range from 23.9 to 110.1 mg/m^3 (Table 3).

The greatest individual occupational risk of workers at the maximum concentrations of dust in the air of the working area was revealed in the drivers of technological vehicles of Cheremkhovo open-pit mine, operator of the crushing and loading unit in the drying and heating department of Kasyanovskaya coal processing plant, and mechanic for repair of equipment in boiler and dust preparation shops of TPP-9. The probability of acquiring chronic diseases when working for 5–10 years ranges from 3.5 to 11.2%, depending on the level of exposure to dust-related risk factors. In case of continuous exposure to dust for 10–20 years, 8.7–26% of workers at an open-pit mine may show symptoms of chronic diseases. The maximum values of inhalation risk are observed at exposure for more than 20 years.

A series of studies on the dispersion composition of dust in the studied workplaces at the stages of coal mining, preparation and combustion were carried out to identify the contribution of nanosized dust fractions in the development of occupational risks for personnel and a method of isolation of coal nanoparticles was developed. Table 4 shows the fractional composition of dust formed during coal preparation; it has been established that the nanoparticle fraction reaches more than 20%.

To prevent occupational diseases of personnel engaged in mining, preparation and combustion of coal, it is necessary to pay attention to nanosized fractions of coal dust, to review standards of maximum allowable concentrations for health risk assessment and preventive measures, such as providing personal protective equipment, in particular to provide employees with innovative means and to install systems for coagulation of nanoparticles in the shops.

Table 3 Risk of occupational dust pathology of respiratory organs of workers engaged in mining, preparation and use of coal

Object of research	Workplace	Shift-average dust concentration, mg/m^3	Safe work period, years	Individual risk, 10^{-3}
Cheremkhovo open-pit mine	Bulldozer operator	62.9	15.2	4.43
	EKG excavator operator	42.8	22.4	1.78
	Driver of technological vehicles	110.1	9.1	9.02
Kasyanovskaya coal processing plant	Conveyor operator in the coal receiving department	23.9	10.86	1.68
	Operator of the washing units in the main building	15.6	16.6	1.09
	Operator of the crushing and loading unit in the drying and heating department	56.4	13.3	3.97
TPP-9, Angarsk	Bulldozer operator of fuel handling shop	38.5	25.6	2.71
	Car dumper operator	29.6	12.3	2.08
	Mechanic for repair of equipment in boiler and dust preparation shops	52.6	17.3	3.70

4 Fire Risks in Coal Mining Areas

The analysis carried out has shown that fires and explosions in mines cause the largest catastrophes in coal production [18, 21]. While fires have been the cause of 27% of all accidents in mines, the most significant accidents with the maximum number of victims have occurred and continue to occur in mines precisely because of fires (67%). Deaths and injuries at mines in 91% occurred precisely due to causes in one way or another related to spontaneous combustion of coal. The majority of

Table 4 Fractional composition of coal dust at the workplaces of Kasyanovskaya coal processing plant

Conveyor operator in the coal receiving department		Operator of the washing units	
Fraction, microns	Content, %	Fraction, microns	Content, %
0.3	10	0.3	10
0.4	20	0.4	20
0.5	30	0.5	30
0.7	40	0.6	40
0.8	50	0.7	50
0.9	60	0.8	60
1.1	70	1.0	70
1.3	80	1.2	80
1.7	90	1.6	90

accidents with the greatest loss of life is also caused by spontaneous combustion of coal. Using the example of the Kemerovo Region, we developed a methodology for a comprehensive assessment of the consequences of fires in mines and proposed to use a comprehensive index of fire hazard as a criterion, which is the sum of risks: atmospheric pollution expressed through the hazard index; integral social risk; and fire risk. Endogenous fires in the Kemerovo Region additionally release up to 250 thousand tons of hazardous substances per year, which is about 18% of the intake of pollutants from industry in the region.

5 Methods of Early Detection of Coal Spontaneous Combustion and Technology of Its Elimination

We have analyzed the existing methods for detecting spontaneous coal combustion, and found that not all of them are sufficiently effective. Like any fire, spontaneous combustion of coal is characterized by its time parameters—the earlier it is detected, the less losses it can cause before it is contained.

A detector for early diagnosis of coal spontaneous combustion has been developed and patented [28, 29]. A distinctive feature of the device is the possibility of its application in the places of the most probable source of spontaneous combustion. The proposed detector is structurally related to the detectors of aspiration type and makes it possible to ensure the safety of electronic devices located in a special chamber outside the coal mass. The detector is multi-sensory, that is, it detects a fire by several types of controlled factors from self-heating stage to self-ignition.

A method of preventing the self-ignition of coal accumulations in the mined-out space of mines has been developed and patented, which consists in periodic supply of atomized water in an air stream, as a result of which coal and rocks are moistened, and dry gas (nitrogen), which leads to a sharp intensification of evaporation of previously supplied liquid particles, accompanied by a decrease in the temperature of the coal. In order to implement the proposed method, a device and technology of alternating supply of inert composition in the mined-out space of mines has been developed. The operating principle of the device is that through it is possible to supply humidified air and dry gas alternately [32, 33].

A way of solving two existing significant environmental problems—the necessity of waste disposal and prevention of spontaneous combustion of coal—has been proposed. It has been suggested that it is possible to create an antipyrogenous composition based on finely dispersed fractions of dunite, verlite, troctolite; a number of experiments with samples of coal deposits in the Irkutsk Region have been conducted. As a result of the tests it has been established that the use of the presented samples as a mechanical film on the surface of the studied coal grades has a certain antipyrogenous effect. This allows us to draw a conclusion about the possibility of using magnesium silicate wastes as antipyrogenous compositions to prevent spontaneous combustion of coal and the feasibility of further work in this direction [29, 36].

6 Conclusion

For a number of years, INRTU has been conducting joint research with Chinese colleagues on the problems of reducing man-made risks in the coal industry, in particular on the development of hybrid technologies for wastewater treatment and creation of composite materials based on industrial waste [34, 35] and plans to expand this research by implementing joint grants to develop innovative solutions aimed at creating methods of mine water deposition in mined-out spaces, its treatment and use for domestic purposes. The most promising solutions are the creation of a complex treatment technology that combines sorption and phytoremediation technologies.

References

1. Tsivileva AE, Golubev SS (2022) Influence of sanctions on the work of coal industry enterprises. Coal 8:84–91. https://doi.org/10.18796/0041-5790-2022-8-84-91
2. Budrys Alex. Silk Road: Russian coal sails to India and China after European embargo. https://www.forbes.ru/biznes/484985-selkovyj-put-rossijskij-ugol-posle-evropejskogo-embargo-plyvet-v-indiu-i-kitaj. Accessed 20 Feb 2023
3. In 2022 China increased production of oil, gas and coal. https://1prime.ru/energy/20230117/839490475.html. Accessed 20 Feb 2023
4. Sheveleva OB, Slesarenko EV (2019) Sustainable development of a coal-mining region: technical, process and ecological aspects. Current Problems of Economy and Law 13(4):1537–1548. https://doi.org/10.21202/1993-047X.13.2019.4.1537-1548
5. Bokataya SV, Kukushkina NS (2016) Evaluation of factors affecting the financial performance of coal mining organizations in Russia. Bulletin of Professional Accountants 5:3–13. https://www.ipbr.org/assets/documents/issues/vestnik/vestnik_2016_05.pdf

6. Naumov IV (2019) Study of spatial disproportions in the processes of disturbance and reclamation of land resources in Russia. Proceedings of the Ural State Mining University 4(56):143–152. https://doi.org/10.21440/2307-2091-2019-4-143-152
7. Zenkov IV (2016) Remote sensing in solving environmental problems of forest reclamation in coal pits in Siberia. Bulletin of the Siberian State Aerospace University 17(1):36–44
8. Osintseva MA, Burova RV, Zhidkova EE (2022) Reclamation of mined-out territories of coal mines in Kuzbass. International Research Journal 9(123). https://doi.org/10.23670/IRJ.2022.123.48
9. Ermakova AYa (2013) Problems of water use in coal mining. Scientific Bulletin of MSGU 6:9–13
10. Wang Fangtian, Liang Ningning, Li Gang (2019) Damage and Failure Evolution Mechanism for Coal Pillar Dams Affected by Water Immersion in Underground Reservoirs. Geofluids (JCR:Q2Cite:2.47). https://www.webofscience.com/wos/alldb/full-record/WOS:000456590400001
11. Zhang Cun, Wang Fangtian, Bai Qingsheng (2021) Underground space utilization of coalmines in China: A review of underground water reservoir construction. Tunnelling and Underground Space Technology 107:103657. (JCR:Q1Cite:5.915)
12. Blinovskaya YaM, Mazlova EA (2019) Greenhouse gas emissions during coal mining and processing: the state of the problem and reduction technologies. Scientific Notes of RSHMU 54:145–154
13. Romanchenko SB, Naganovsky YuK, Kornev AV (2021) Innovative ways of controlling dust and explosion safety of mine workings. Notes of the Mining Institute 252:927–936. https://doi.org/10.31897/PMI.2021.6.14
14. Timofeeva SS, Luzhkov YuA (2010) Environmental and occupational risks in open-pit coal mining. Coal 1(1005):25–27
15. Piktushanskaya TE (2009) Estimation of a posteriori occupational risk of coal miners. Occupational Medicine and Industrial Ecology 1:32–37
16. Zhou LJ, Cao QG, Yu K, Wang LL, Wang HB (2018) Research on Occupational Safety, Health Management and Risk Control Technology in Coal Mines. Int J Environ Res Public Health 15(5):868. https://doi.org/10.3390/ijerph15050868
17. Khamidullina EA, Timofeeva SS, Smirnov GI (2014) Safety of coal mining in risk metrics. Safety in the Technosphere 4:3–39. https://doi.org/10.12737/5302
18. Lugovtsova NYu (2017) Assessment of fire risk in the administrative territories of the Kemerovo Region. XXI century. Technosphere Safety 2(2):68–77
19. Timofeeva SS, Smirnov GI (2016) The role of coal in the energy sector of the XXI century and risk assessment of its production and consumption. XXI century. Technosphere Safety 1(4):92–106
20. Timofeeva SS, Karpova IA (2013) Assessment of the environmental load on the atmosphere in the extraction of brown coal. Bulletin of ISTU 1(72):47–53
21. Timofeeva SS, Lugovtsova NYu (2017) Analysis of the environmental condition of the mining industry in Kuzbass. Mining Research and Information Bulletin 6:350–361
22. Timofeeva SS, Lugovtsova NYu, Yankova P, Timofeev SS (2019) Assessing the Unaccounted Environmental Pressure Caused by Endogenous Fires on the Rock Dumps of Kuzbass Overburden Rocks. IOP Conference Series Earth and Environmental Science 224(1):012057. https://doi.org/10.1088/1755-1315/224/1/012057
23. Lugovtsova NYu, Portola NL, Galsanov MV, Shevchenko NYu (2012) Endogenous fire hazard of Kuzbass mines. Bulletin of the Kuzbass State Technical University 2(90):44–47
24. Lugovtsova NY, Timofeeva SS, Statsenko SV (2019) The study of social risk in the mining enterprises of Kuzbass. Mining Journal 9. https://doi.org/10.17580/gzh.2019/09/15
25. Timofeeva SS, Murzin MV (2015) Dust load in coal mining and occupational risks. Bulletin of ISTU 5(100):68–71
26. Timofeeva SS, Murzin MA (2020) Assessing the environmental risk of mining enterprises by the integral indicator of dust emission. IOP Conference Series: Earth and Environmental Science 408(1):012067. https://doi.org/10.1088/1755-1315/408/1/012067

27. Timofeeva SS, Zueva VA (2022) Developing a technique for detecting nanoscale range particles in the air of the working area in a coal preparation plant IOP Conf. Series: Earth and Environmental Science 1061:012059. https://doi.org/10.1088/1755-1315/1061/1/012059
28. Timofeeva SS, Smirnov GI (2014) Possibilities of using fire detectors to detect the spontaneous combustion of coal in the early stages. Bulletin of ISTU 11(94):103–107
29. Timofeeva SS, Smirnov GI, Timofeev SS, Latysheva SV (2020) Specialized alam-initiating device for early detection of coal-ignition. IOP Conference Series: Earth and Environmental Science 408. https://doi.org/10.1088/1755-1315/408/1/012040
30. Timofeeva SS, Lugovtsova NYu (2017) Assessing the contribution of burning coal dumps to the radioecological situation in the Kemerovo Region. Safety in the Technosphere 6(4):31–36
31. Lugovtsova SS, Timofeeva LG, Demenkova AV, Filonov NYu (2017) Studies of total harmful gas flows on the ground surface from spontaneous mine fires in Kuzbass. Eurasian Mining 2:43–48
32. Timofeeva SS, Lugovtsova NYu (2016) The study of the complex influence of humidification and cooling on the chemical activity of coal to prevent the process of spontaneous combustion. Advances in Modern Natural Science 4:197–201
33. Lugovtsova NY, Gubanova AR (2016) Technique for Decreasing Reactivity of Coal Material to Suppress the Oxygen Absorption Process. IOP Conference Series: Materials Science and Engineering, vol. 142: Innovative Technologies in Engineering
34. Timofeeva SS, Leonova MS, Gorlenko NV, Jih-Hsing Chang, Shu-Fen Cheng (2020) Composite sorbents based on waste from the crystalline silicon production and biochar. IOP Conference Series: Earth and Environmental Science 408:012039. https://doi.org/10.1088/1755-1315/408/1/012039
35. Kuznetsova IV ,Timofeeva SS (2020) Green technologies in land recultivation for coal mining enterprises. IOP Conference Series: Earth and Environmental Science 408:012075. https://doi.org/10.1088/1755-1315/408/1/012075
36. Khudyakova LI, Voyloshnikov OV, Timofeeva SS (2014) Magnesium silicate mining wastes and technologies of their disposal: Monograph. Geo, Novosibirsk, p 177

Developing an Expert Assessment System for Green Industry Projects

A. Volosatova, T. Guseva, I. Tikhonova, and E. Averochkin

Abstract This article discusses ways to develop an expert evaluation system for projects aimed at environmental and technological transformation within the industrial sector. It analyzes both national and international green project taxonomies, underscoring that the majority of them incorporate pollution prevention and control principles based on the use of the Best Available Techniques (BAT). As part of the research, an evaluation algorithm was developed and a composite assessment criterion was refined. The article also provides the results of a comparative expert evaluation conducted on green projects within the construction materials industry across member states of the Eurasian Economic Union. In conclusion, recommendations are formulated that are aimed at fostering the balanced development of BAT and green taxonomies within the Eurasian Economic Union.

Keywords Best available techniques · Green industry projects · Green taxonomy · Expert evaluation · Eurasian economic union

1 Introduction

Sustainable development, the formation of a green economy, and the pursuit of national and international objectives are the realms that have been increasingly garnering attention across various countries. The shaping of a green economy stood out as a key priority during the Russian presidency of the Eurasian Economic Union (EAEU) in 2023. Scholars from numerous universities and research centres run research programmes devoted to the development of methodologies for working

A. Volosatova · T. Guseva · E. Averochkin (✉)
Environmental Industrial Policy Center (EIPC), 38, Stremyanny Pereulok, Moscow 115054, Russia
e-mail: averochkin@gmail.com

I. Tikhonova
Dmitry Mendeleev University of Chemical Technology of Russia, 9, Miusskaya Square, Moscow 125047, Russia

L. V. Radionova and D. V. Ulrikh (eds.), *Advances in Ecology and Environmental Engineering*, Springer Proceedings in Earth and Environmental Sciences,
https://doi.org/10.1007/978-3-031-64423-8_13

out, implementing, and assessing results of the sustainable development projects. Their efforts primarily focus on environmental regulation and the development of approaches to environmental and corporate rankings. However, no uniform expert assessment system for green projects, which is grounded in objective and transparent criteria, has been developed yet.

This article is dedicated to designing such a system for the expert assessment of projects focused on environmental and technological transformation within the industry. These projects are considered to be sustainable development tools aimed at enhancing resource and environmental efficiency, fostering a circular economy, and reducing industrial carbon intensity.

2 Designing Support Systems for Sustainable Development Projects: An Analysis of national and international Principles

Numerous strategies, policies, programs, and projects have emerged globally to achieve the Sustainable Development Goals (SDGs). Environmental projects include those aimed at climate change mitigation and adaptation, conservation of natural ecosystems and promotion of responsible production and consumption patterns [1]. In the 1980s, the term "environmental" began to be replaced by the term "green", with dedicated research groups, councils, and programs being created [2]. The ISO 14,030–3:2022 standard ("Environmental performance evaluation. Green debt instruments. Part 3: Taxonomy") published in 2022 proposes a classification of green projects, which are divided into six priority areas: (1) climate change mitigation; (2) climate change adaptation; (3) sustainable use and protection of water and marine resources; (4) transition to a circular economy, waste prevention and recycling; (5) pollution prevention and control; (6) protection and restoration of ecosystems and biodiversity [3].

According to ISO 14,030–3:2022, the key principles for creating systems to support green projects include: (1) reducing risks to the environment, natural habitats, biodiversity, human health, and welfare; (2) relying on evidence and scientific validity, incorporating research results and expert assessments; (3) assessing anticipated changes across all facets of green projects when establishing project goals and objectives.

The Russian taxonomy of green projects is largely based on these principles, with BAT-driven projects aimed at environmental and technological transformation in the industrial sector playing a great role, but the decision made in 2021 to use specific numerical indicators in the taxonomy appears irrational for two reasons. First, some of the values were drawn from BAT Reference Documents (BREFs) developed and updated in different years. Second, other values were sourced from European Union documents that lost validity in 2021, predating the approval of the Russian taxonomy.

A more logical approach, adopted in 2023, involves giving references to BREFs where relevant indicators are presented [4].

3 Developing an Expert Assessment System for Green Industry Projects

In Russia, BAT Reference Documents have been used since 2019 in the development and evaluation of draft programmes for environmental efficiency improvement and applications for environmental permits granted to large industrial manufacturers [5, 6]. The assessment process is overseen by the BAT expert community, whose activities are coordinated by the BAT Bureau [7, 8].

Projects seeking government support were, until recently, evaluated by various organizations (e.g., Skolkovo Foundation, certification bodies, etc.). As part of this study, we analysed the results of evaluating 54 projects across 12 areas of BAT application that received government support from 2018 to 2021. It was discovered that BREFs had been inaccurately chosen for 14 projects. The remaining 40 projects got soft loans totalling RUB 15.02 billion (Fig. 1).

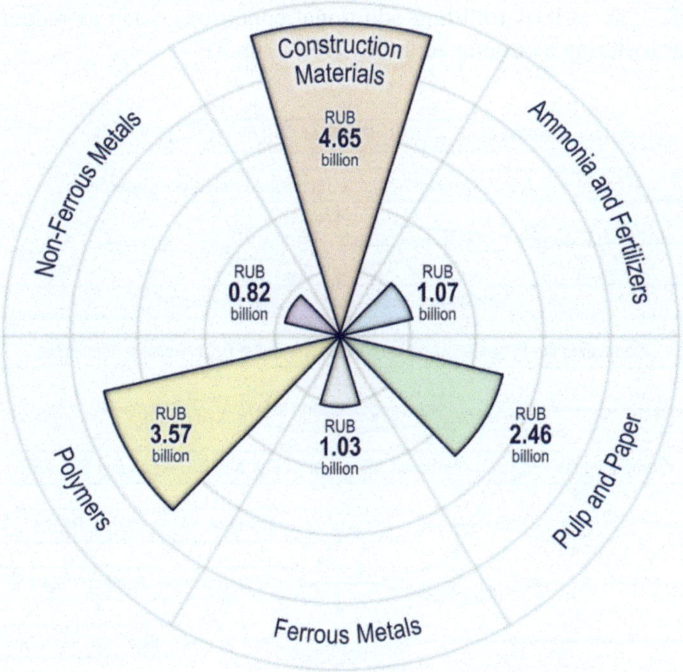

Fig. 1 Projects in BAT application areas that received government support

In several instances, support is extended to projects for establishing new enterprises whose resource and environmental efficiency fall short of BAT requirements. These results in the imprudent utilization of natural resources, increased environmental pollution, and the inability to commission newly created facilities, where compliance with BAT is a requirement established by environmental legislation. Examples of inefficient projects not meeting BAT but implemented in 2019–2022 requirements are known in such sectors as the construction materials, energy generation (large combustion plants), and chemical industry.

Information presented in BREFs is accessible to business representatives, project documentation developers, and experts, forming the foundation for enhancing the validity and transparency of decisions on supporting projects for the environmental and technological transformation of industrial facilities. It is recommended to design a uniform system for assessing green projects across all BAT application areas, drawing from the expertise of the Russian BAT expert community. The practice of expert assessment should be extended to investment projects supported within the framework of economic, industrial, and energy policy instruments [9].

The proposed expert assessment algorithm for green projects (Fig. 2) is based on the application of a composite assessment criterion, $K = K_1 \wedge K_2 \wedge K_3$.

The composite criterion K provides for the assessment of emissions targets or BAT-Associated Emission Levels (BAT-AELs) (K_1), compliance with resource efficiency indicators or BAT-Associated Environmental Performance Levels (BAT-AEPLs) (K_2), as well as fulfilling additional conditions, such as reducing carbon intensity or fostering a circular economy (K_3) (Fig. 3).

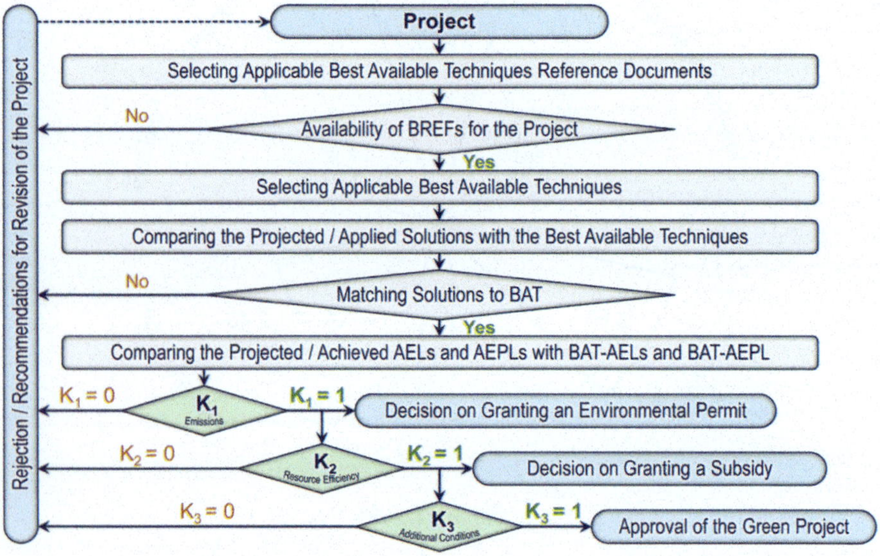

Fig. 2 BAT compliance assessment algorithm

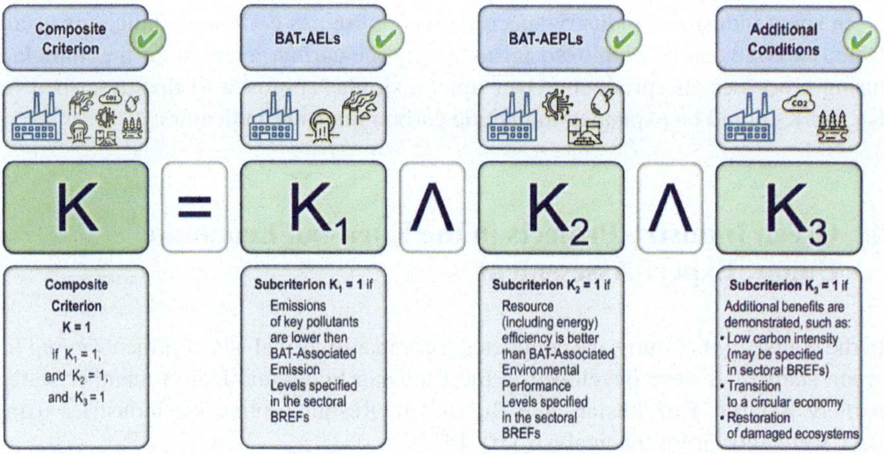

Fig. 3 Reference documents on best available techniques as sources of the assessment subcriteria

Subcriterion K_1 is a predicate with values of either 1 or 0. Achieving sectoral BAT-Associated Emission Levels is a requisite for assessing all projects. $K_1 = 1$ is a sufficient condition for making a positive decision on issuing an environmental permit.

Subcriterion K_2 is likewise a predicate, taking values of either 1 or 0. When assessing projects seeking subsidies from the federal budget and aiming to implement BAT, the second necessary and sufficient condition is achieving sectoral resource efficiency indicators (BAT-Associated Environmental Performance Levels) established in relevant BREFs. To obtain a favourable pre-financial expert assessment, both conditions $K_1 = 1$ and $K_2 = 1$ must be met.

Throughout the assessment, experts ascertain the technological and technical solutions responsible for achieving the declared (project) indicators and whether these solutions suffice to ensure compliance with established BAT requirements. Information from relevant BREFs, along with other reference documents and guidelines containing data on the best available and emerging techniques, is utilized by experts. Subcriterion K_3 is also a predicate that factors in additional project parameters (for example, reducing carbon intensity, fostering a circular economy, restoring damaged ecosystems, etc.).

Numerical values for aspects related to the circular economy or the restoration of ecosystem services have not been standardized. Therefore, experts assess whether the project plans to incorporate secondary resources into the production cycle and if the project encompasses measures such as landscape restoration or water management in depleted quarries, deciding whether the project solutions can be regarded as those providing additional environmental benefits.

In some industries, industry-specific greenhouse gas emissions indicators established in 2022 can be employed for assessing the carbon intensity of the manufacturing processes and products. Over time, a similar approach to that described for K_1 and K_2 could be expanded to include carbon intensity indicators.

4 Green Industry Projects in the Eurasian Economic Union: Expert Assessment

In the course of this study, we conducted a comparative analysis of projects vying for green status that were developed in the Eurasian Economic Union member states, namely Belarus, Kazakhstan, and Russia [9]. Resource-intensive industries using BAT were chosen for the analysis [10–14].

The rise in the production of clay blocks (characterized by a high void ratio and low density) and hollow bricks can be attributed to the diminishing use of solid brickwork in modern constructions. Also, stringent thermal resistance requirements for construction materials have been adopted universally. While past green building standards rarely touched upon the origin of construction materials, the focus has shifted to two crucial objectives: (1) ensuring the reduction of negative impact on the environment throughout the life cycle of construction materials, and (2) curbing the so-called embodied carbon, or carbon dioxide emissions accompanying the production of construction materials [15, 16].

Key environmental considerations in clay block production involve emissions of suspended solids, CO, NO_X, and SO_2 resulting from preparing brick mixtures [17, 18]. In Russia, carbon monoxide (CO), sulphur dioxide (SO_2), and nitrogen oxides (NO_X) are classified as markers, with BAT indicators established for emissions of these substances [18, 19].

The projects under analysis envision the construction of facilities with capacities ranging from 350 thousand tons to 2.5 million tons of porous clay bricks and hollow bricks per year (see Table 1). The assessment was conducted using the K criterion. Compliance with subcriterion K_1 was assessed using emission targets and compliance with subcriterion K_2 was assessed using resource efficiency indicators established in the Russian BREF 4–2023. Subcriterion K_3 covered fostering a low-carbon economy and a closed-loop production cycle.

Expert assessment results reveal similarities in technological solutions for projects A, B, and C. However, project C involves the production of hollow bricks, while projects A and B focus on porous clay blocks.

Regarding compliance with subcriterion K_1, it is noteworthy that all projects meet requirements for marker pollutant emissions. Project A, however, exhibits relatively higher CO emissions, possibly due to the addition of a significant quantity of combustible additives to the charge, such as sawdust (up to 150–160 kg per ton of product as per design documentation). Simultaneously, project A anticipates lower nitrogen oxide emissions compared to projects B and C.

Table 1 Ceramic manufacturing projects: key assessment results

Assessment criteria	Project A	Project B	Project C	BREF 4–2023
Production capacity, tons per day	2500 (porous clay blocks)	1800 (porous clay blocks)	350 (hollow bricks)	BREF 4–2023 is used for facilities with a capacity of more than 150 tons per day
K_1: Emissions				
Emissions (BAT-AELs), kg per ton of product				
CO	0.8	0.6	0.5	≤0.8
NO_X	0.3	0.5	0.5	≤0.5
SO_2	0.2	0.1	0.2	≤0.2
Ways to reduce NO_X emissions	Described in the project	Described in the project	Described in the project	Optimization of the firing process
	Described in the project	Not planned	Not planned	Application of low NO_X burners
Ways to reduce CO emissions	Described in the project	Described in the project	Described in the project	Optimization of the firing process
	Described in the project	Not planned	Not planned	Exhaust gas combustion in a furnace
K_2: Resource efficiency (BAT-AEPLs)				
Specific heat consumption, GJ/t	0.95	1.15	1.30	≤1.94 GJ per ton of end products for factories producing clay blocks; product type is not specified
Waste heat recovery	Described in the project	Described in the project	Described in the project	Waste heat recovery from a kiln system for brick drying
K_3: Fostering a circular and low-carbon economy				
Thermal conductivity, W/(m•K)	0.15–0.17	0.17–0.21	0.19–0.23	No information
Use of plant wastes (logging waste, woodworking waste, sunflower hulls, rice hulls, etc.)	Described in the project	Described in the project	Described in the project	Considered as BAT
Use of waste plastic	Not planned	Not planned	Described as a potential future solution	Not considered

Adding more combustible additives facilitates product porosity, reduces thermal conductivity (which is significant in the context of subcriterion K_3), and marginally curtails fossil fuel consumption (which is significant in the context of subcriterion K_2 and K_3, potentially reducing greenhouse gas emissions). Assuming the calorific value of sawdust to be 7 to 9 GJ per ton, the total heat released during combustion can be estimated at 1.2 GJ per ton of product, explaining the low consumption of natural gas (0.95 GJ per ton of product in project A).

It is known that heat losses from buildings, usually measured in watts per hour, are a result of the temperature difference between the inside and outside. The thermal conductivity of solid clay blocks is 0.5 to 0.6 W/(m•K), while that of the planned products (projects A, B, and C) range from 0.15 to 0.23 W/(m•K). This implies that buildings constructed with porous clay blocks will likely require less fuel for heating than those with solid bricks of the same geometry.

Comparative expert assessment results, utilizing the K criterion and BREFs, can serve as a foundation for providing recommendations on BAT compliance for the responsible sourcing purposes.

5 Developing BAT Principles and Improving the Green Project Taxonomy of the Eurasian Economic Union

Under the guidance of Academician Glazyev of the Russian Academy of Sciences, work on formulating the concept for integrating the principles of the green economy in the Eurasian Economic Union has been ongoing since 2021 [20]. In 2022, a proposition was made to incorporate the enhancement of resource efficiency and the implementation of Best Available Techniques among the core principles of green integration.

A model taxonomy of green projects for the EAEU was approved in 2022, encompassing key areas such as energy, construction, water supply and sanitation, waste management, and biodiversity protection. The document was prepared considering international experience (see for example [3, 21]). However, the general part of the relevant document only mentions the pulp and paper industry along with waste paper recycling. Sections reflecting country-specific nuances have been developed considering the environmental and climate regulations in Kazakhstan and Russia. Similar to the Russian taxonomy published in 2021, the document discussed employs specific numerical indicators of CO_2 emissions from official EU documents, which expired in 2021. It is advisable to enhance the model taxonomy by adopting the approaches outlined in ISO 14,030–3:2022 [3] and abandoning specific numerical indicators [3]. The core of the taxonomy should be grounded in principles that improve resource efficiency and introduce BAT, fostering a reduction in negative environmental impact and CO_2 emissions.

Such a decision would broaden the spectrum of green projects, aligning with the priority integration areas and the interests of EAEU member states. To improve tools for supporting green projects, it is recommended to develop methodological international standards in the realm of BAT and conduct a comparative analysis of the environmental and resource efficiency of key industries.

Considering the priority integration areas and the significant role of Russian experts in drawing up BREFs such as metallurgy, chemicals, heat and power engineering, and the production of construction materials, it is recommended forming a united BAT Expert Society [9] for the Eurasian Economic Union. Simultaneously, green projects executed through collaborative efforts by EAEU member states could undergo expert assessments based on harmonized principles.

6 Conclusion

The study has yielded the following results:

- The mechanisms ensuring the development and implementation of sustainable development projects were analyzed, proposing a classification of national and international instruments for supporting green projects covering taxonomies from Asia–Pacific countries, BRICS member states, the EAEU, the EU, and the OECD. It is demonstrated that, in relation to industrial development, all taxonomies are prepared with consideration for Best Available Techniques requirements.
- An algorithm for expert assessment of projects for environmental and technological transformation within BAT application areas was developed, and a composite assessment criterion was devised. This approach facilitates the assessment of design solutions, accounting for meeting emissions targets (K_1), resource efficiency indicators (K_2), and additional conditions (K_3) in areas such as reducing carbon intensity, fostering a circular economy, ecosystem restoration, etc.
- Based on the results of our comparative analysis of industrial development projects in EAEU member states, it is evident that expert assessment using the proposed algorithm and the K criterion allows for a judicious selection of projects that ensure high resource and environmental efficiency in BAT application areas.
- Recommendations were formulated for improving the Eurasian taxonomy and principles for selecting projects for environmental and technological transformation in EAEU member states, including (1) identifying common areas of BAT application in the context of Eurasian integration; (2) conducting a comparative analysis of resource efficiency, environmental efficiency, and carbon intensity of priority industries; (3) establishment of the Eurasian expert community in the field of BAT. These recommendations were instrumental in the preparation of the draft concept for introducing green economy principles in the EAEU.

References

1. United Nations (2015) Transforming our World: the 2030 Agenda for Sustainable Development. https://documents-dds-ny.un.org/doc/UNDOC/GEN/N15/291/89/PDF/N1529189.pdf?OpenElement. Accessed 15 Dec 2023
2. Steblyanskaya A, Bobylev S, Yan J (2023) Green Economy for Nature Conservation: New Paradigm for the Future. BRICS Journal of Economics 3(4):203–207
3. ISO 14030–3 (2022) Environmental Performance Evaluation. Green Debt Instruments. Part 3. Taxonomy
4. Guseva T, Panova S, Tikhonova I, Volosatova A, Bhimani C (2023) Resource Efficiency Enhancement as a Common Background for Green Taxonomies of BRICS Countries. Proceedings of the 23rd International Multidisciplinary Scientific GeoConference SGEM 2023 23(5.1):215–221
5. Simachev Y, Kuzyk M, Kuznetsov B, Pogrebnyak E (2014) Russia on the Path Towards a New Technology Industrial Policy: Exciting Prospects and Fatal Traps. Foresight-Russia 8(4):6–23
6. Skobelev D, Almgren R (2020) Evolution of Technology and Technology Governance. Journal of Open Innovation Technology Market and Complexity 6(2):22–38
7. Skobelev D (2019) Environmental Industrial Policy in Russia: Economic, Resource Efficiency and Environmental Aspects. Proceedings of the 19th International Multidisciplinary Scientific GeoConference SGEM 2019 19(5.3):291–298
8. Skobelev DO (2021) Building the Infrastructure for Transforming Russian Industry towards Better Resource Efficiency and Environmental Performance. Procedia Environmental Science, Engineering and Management 8(2):483–493
9. Best Available Techniques (BAT) for Preventing and Controlling Industrial Pollution. Activity 1: Policies on BAT or Similar Concepts across the World. OECD, Environment, Health and Safety, Environment Directorate (2017) https://www.oecd.org/chemicalsafety/risk-management/policies-on-best-available-techniques-or-similar-concepts-around-the-world.pdf. Accessed 15 Dec 2023
10. Ayres R et al (2011) Manufacturing: Investing in Energy and Resource Efficiency. Towards a Green Economy: Pathways to Sustainable Development and Poverty Eradication. United Nations Environment Programme, pp 245–285. https://www.researchgate.net/publication/279182413_Manufacturing_Investing_in_Energy_and_Resource_Efficiency. Accessed 15 Dec 2023
11. González I, Galán E, Miras A (2006) Fluorine, Chlorine and Sulphur Emissions from the Andalusian Ceramic Industry (Spain). Proposal for their Reduction and Estimation of Threshold Emission Values. Applied Clay Science 32(3–4):153–171
12. Mishra UC, Sarsaiya S, Gupta A (2022) A Systematic Review on the Impact of Cement Industries on the Natural Environment. Environmental Science and Pollution Research 29:18440–18451
13. Monfort E, Mezquita A, Granel R, Vaquer E, Escrig A (2010) Analysis of Energy Consumption and Carbon Dioxide Emissions in Ceramic Tile Manufacture. Castellon: Qualicer'10
14. Monroy R, Romero YA, Gelves JF (2018) Consumption of Energy in the Manufacturing of Ceramic Bricks in the Metropolitan Area of Cúcuta, Colombia. IOP Conf. Series: Journal of Physics. Conf. Series 1126:012014. https://doi.org/10.1088/1742-6596/1126/1/012014
15. Sbahieh S, Serdar MZ, Al-Ghamdi SG (2023) Decarbonization Strategies of Building Materials Used in the Construction Industry. Materials Today Proceedings. https://doi.org/10.1016/j.matpr.2023.08.346
16. European Commission (2002) Sustainable Construction Final Report. http://europa.eu.int/comm/enterprise/construction/suscon/finrepsus/susfin.htm. Accessed 15 Dec 2023
17. European Commission (2023) Best Available Techniques (BAT) Reference Document for the Ceramic Manufacturing Industry. Joint Research Centre. Institute for Prospective Technological Studies. Formal draft
18. BREF (ITS) 4 (2023) Information and Technical Reference Document on Best Available Techniques for the Ceramic Manufacturing Industry. Rosstandard, 2023 (in Russian)

19. Zakharov A, Bezmenov A, Goldobina V (2015) Resource Efficiency of Ceramic Production as a Modern Trend. Gaudeamus Igitur 1:17–21 (in Russian)
20. Glazyev S (2021) How to Build a Great Eurasian Partnership? Eurasian Integration: Economy, Law, Policy 15(3):9–14 (in Russian)
21. Marcos S, Castrillo Lara MJ (2021) Sustainable Finance in Europe: The EU Taxonomy and Green Bond Standard. In: Handbook of Research on Global Aspects of Sustainable Finance in Times of Crises. IGI Global, Berlin, pp 114–130

The Possibility of Assessing the Pinery State Under Technogenic Pollution by NIR Spectroscopy and Soft Independent Modelling of Class Analogy

A. Khokh

Abstract This paper presents the results of the comparative analysis of the near-infrared spectra of pine needles of the current year, as well as pine needles of 1–3 years of life from territories experiencing different degrees of manmade load. Principal component analysis and soft independent modelling of class analogy were applied to evaluate spectral data. The potential possibility of using the NIR spectroscopy method to assess the impact of technogenic pollution on the stability of the pinery development has been proved by experiments. To differentiate the studied samples of needles, classification SIMCA models were developed, while the maximum classification accuracy was revealed for needles of the 2nd year of life: it was 93.11 ± 2.22% for the training sample, 92.27 ± 1.72% for the test sample.

Keywords Scotch pine · NIR spectroscopy · Needles · SIMCA · Technogenic pollution · Urban environment

1 Introduction

Environmental issues are relevant for the entire Republic of Belarus and for each of its regions, especially large industrial centres.

At present, Scotch pine (*Pinus silvestris* L.) plays an important role in improving the urban environment ecological situation. At the same time, it is in cities that it is constantly affected by a complex of unfavourable environmental and anthropogenic factors, the result of which is impaired growth and development phenorhythms, improper formation and functioning of the assimilation organs, a decrease in productivity (growth) and regulatory environmental functions [1–4]. Under such circumstances, pine plantations cannot act as ecologically full-fledged ecosystems, and ecosystem mechanisms for the regulation of sustainability cannot show their effect.

A. Khokh (✉)
Scientific and Practical Center of The State Forensic Examination Committee of The Republic of Belarus, 25, Philimonova Str, Minsk 220114, Republic of Belarus
e-mail: 1ann1hoh@gmail.com

L. V. Radionova and D. V. Ulrikh (eds.), *Advances in Ecology and Environmental Engineering*, Springer Proceedings in Earth and Environmental Sciences,
https://doi.org/10.1007/978-3-031-64423-8_14

The high level of technogenic load in the city implies the development and implementation of objective pollution control methods to assess the current state of pine plantations and trends in the situation development in the future. To enhance the reliability of environmental assessments and forecasts, it is necessary to increase the number of research methods and procedures and the studied parameters.

The purpose of this paper is to evaluate the possibility of studying the resistance of pine plantations to the combined impact of anthropogenic factors by spectroscopy in the near infrared region (hereinafter referred to as NIR spectroscopy).

In recent years, NIR spectroscopy has been rapidly developing finding its practical application, including for studying plant objects, since it is non-destructive, does not require expensive consumables, and is characterized by a high analysis rate [5–8]. Approximately the NIR region is located between 13,300 and 4000 cm^{-1} (from 740–780 to 2500 nm) [9]. While the spectra obtained in the mid-IR region register mainly atomic vibrations in the individual chemical bonds of most molecules, the corresponding NIR spectra show the so-called overtones and combination bands.

2 Methods

The study object was the needles of the current year, as well as needles of 1–3 years of life, selected at the end of August 2022 in the territory of Minsk, as well as in the background (control) territory of the State Nature Protection Institution "Berezinsky biosphere reserve". The study included 7200 pairs of needles from 360 trees.

Needle samples were scanned on a desktop scanner Epson Perfection V19 («Epson», Indonesia). Then, using the scanned images, the length of the left and right needles in a pair was calculated using the MCview program (LOMO-Microsystems, RF) with an accuracy of 0.01 mm.

The fluctuating asymmetry index (FAI) was calculated by the following formula [10]:

$$\mathrm{FAI} = \frac{2(L_1 - L_2)}{(L_1 + L_2)}$$

where L_1 is the length of the left needle in a pair, L_2 is the length of the right needle in a pair.

Then the needles were dried at 30–35 °C to achieve the residual moisture content—NMT 7%. Next, the samples from one tree (5 pairs) were crushed separately for each year of life in an agate mortar, then the powder was placed in a Petri dish made of optical transparent glass and NIR diffuse reflection spectra were recorded using a portable NIR spectrometer MicroNIR OnSite №1–00247 with a diode-matrix detector (VIAVI, USA) in the range of 10,526–6060 cm^{-1} with a resolution of 2 cm^{-1} after averaging the accumulated spectrograms containing 64 scans. The spectra were recorded using the MicroNIR™ Pro 1700 software package (version 2.5.1) (VIAVI, USA).

All subsequent computational experiments were carried out on the second-order derivatives of the NIR spectra using the Unscrambler X (version 10.4.1) (CAMO, USA; the license key for the program—MKCNDKKAKBNRZSMNTSNSKFBR-MMZRKK) software package.

Classification procedures for the NIR spectra were carried out using the principal component analysis (hereinafter referred to as PCA) [11]. Classification models were built using the Soft Independent Modelling of Class Analogy (SIMCA) [12].

Both in the PCA and in the construction of classification models, the algorithm of non-linear iterative partial alternating least squares (NIPALS) was used [13].

The models were tested by the cross-validation method, analysis of the Hotelling statistics plots and the explained variance plots [14].

3 Results

Based on the calculated biometric indicators, the needle samples were grouped into 6 samples corresponding to the fluctuating asymmetry index score, which made it possible to analyze the impact of the manmade load from the normal value to the critical.

As can be seen in Table 1, as habitat conditions deteriorate, the length of needles decreases by an average of 1.2 times, which is confirmed by the high significance ($p < 0.0001$) of one-way analysis of variance (ANOVA).

At the same time, no statistically significant differences have been revealed between the needles of different years of life for samples with FAI scores of 1–3 and 5. For samples with an FAI score of 4, the F-test was 24.6, $p < 0.0001$, with an FAI score of 6, the F-test was 3.03, $p = 0.028$.

At the first stage, the obtained NIR spectra of needles were analyzed by PCA. The samples were preliminarily divided into groups according to the FAI score; the age of the needles was not taken into account. It has been established that the first principal component (PC-1) explains 80% of the variance, it reflects the direction of the greatest change in the data. The second and subsequent components mainly contain noise [16], and therefore may not be considered further (the second main component explains 18% of the original variance, the third—2%). As can be seen in Fig. 1, the points reflecting the studied needle samples were combined into 3 classes. The first class includes samples with an FAI score of 1, which corresponds to a state of relative ecological well-being; the second class includes the samples with FAI scores of 2, 3, and 4, which corresponds to the initial phase of anthropogenic transformation and a highly nonequilibrium state; the third class includes the samples with FAI scores of 5 and 6, which corresponds to the state of anthropogenic degression (Fig. 1).

Next, the spectral data obtained for 6 samples collections (FAI score = 1–6), separately for the needles of the current year, as well as for the needles of 1–3 years of life, were randomly divided five times into training (calibration sets, 2/3 spectra) and test (validation sets, 1/3 of the spectra) samples.

Table 1 Average needle lengths and developmental stability

The sampling = FAI score [15]		The needles of the current year				The needles of the first year of life			
		L_1, mm	L_2, mm	$L_{everage}$, mm	FAI	L_1, mm	L_2, mm	$L_{everage}$, mm	FAI
1*	M_x	56.7	56.4	56.6	0.0025	57.0	56.8	56.9	0.0019
	± *SD*	7.6			0.0003	8.3			0.0002
2	M_x	53.6	53.3	53.5	0.0036	51.7	51.4	51.6	0.0032
	± *SD*	9.7			0.0007	8.1			0.0005
3	M_x	47.4	47.0	47.2	0.0046	49.9	49.4	49.6	0.0045
	± *SD*	8.2			0.0009	8.4			0.0005
4	M_x	48.9	48.3	48.6	0.0052	49.1	48.5	48.8	0.0057
	± *SD*	7.6			0.0004	8.6			0.0005
5	M_x	49.4	48.8	49.1	0.0068	49.4	48.7	49.1	0.007
	± *SD*	9.4			0.001	10.4			0.0005
6	M_x	45.5	44.9	45.2	0.0092	47.9	47.2	47.6	0.0092
	± *SD*	8.6			0.0009	9.2			0.0006
		The needles of the second year of life				The needles of the third year of life			
1	M_x	57.8	57.6	57.7	0.0018	56.2	56.0	56.1	0.0021
	± *SD*	7.9			0.0003	9.3			0.0003
2	M_x	53.0	52.7	52.8	0.0031	51.8	51.5	51.6	0.0034
	± *SD*	8.8			0.0003	9.3			0.0004
3	M_x	49.3	48.8	49.0	0.0045	50.2	49.8	50.0	0.0043
	± *SD*	8.8			0.0005	9.2			0.0003
4	M_x	49.6	49.0	49.3	0.0059	43.7	43.2	43.5	0.0058
	± *SD*	8.4			0.0006	10.2			0.001
5	M_x	48.7	48.0	48.3	0.0071	48.9	48.2	48.6	0.0064
	± *SD*	10.1			0.0009	7.6			0.0004
6	M_x	47.0	46.1	46.5	0.0092	47.1	46.5	46.8	0.0094
	± *SD*	8.9			0.001	11.2			0.0008

Note M_x is the arithmetic mean, SD is the standard deviation; *Assessment of the anthropogenic factor influence: 1—normal, 2—weak, 3—moderate, 4—high, 5—very high, 6—critical

12 abnormal samples were excluded from both sets; those whose primary properties clearly did not correspond to the secondary ones and fell outside the boundaries of the decision-making area. To determine the abnormal samples, the primary properties of all samples of the calibration sets (spectra) were evaluated based on the calculations of the Mahalanobis distance [17].

On the basis of training samples, classification SIMCA models were constructed using five-block cross-validation. Each sample (target class) was modeled by the PCA method separately, independently of the others, and checked for possible outliers. The calculation of the distances between the studied samples in relative units ranged

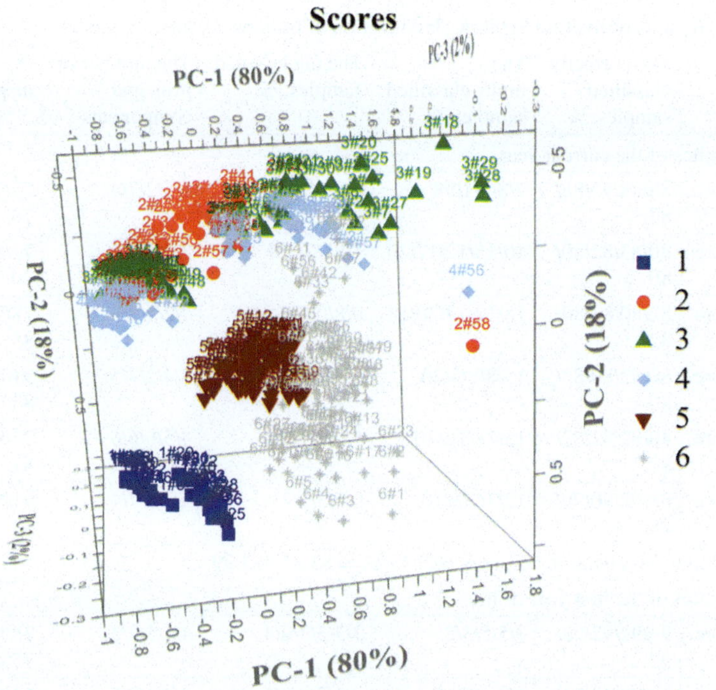

Fig. 1 Score matrix for the studied samples in 3d projection

from 7.96 to 99.7, which is considered suitable for a reliable distinction between the study items [12].

Table 2 presents the SIMCA classification results of training samples of needles of different years of life.

Table 2 show that the average classification accuracy of the training samples ranged from 83.83% for the needles of the current year to 93.11% for the needles of the second year of life. Moreover, the optimal number of principal components and % dispersion for PC-1 are shown. Thus, the minimum number of principal components was 2, the maximum—7. Therefore, the dispersion of PC-1 ranged from 69 to 92%, respectively.

At the next stage, the validation of the constructed model was carried out. For this purpose, validation sets were used that contained samples not involved in the model construction.

Table 3 presents the classification results obtained for the test samples.

Table 3 shows that the average classification accuracy of the test samples ranged from 81.72% for the needles of the current year to 92.27% for the needles of the second year of life.

Table 2 Results of fivefold SIMCA classification of training samples of needles

	The correctly classified samples, %	The multi-classified samples, %	The unclassified samples, %	The number of principal components	PC-1 dispersion, %
The needles of the current year					
FAI score = 1	88/88/92/88/83	7/8/5/10/8	5/3/3/2/8	6/5/2/7/4	75/75/78/78/79
FAI score = 2	80/83/82/80/80	10/15/12/17/20	10/2/7/3/0	3/6/3/5/5	72/89/77/91/70
FAI score = 3	82/80/83/80/82	18/12/17/20/18	0/8/0/0/0	7/2/4/2/3	78/78/90/70/85
FAI score = 4	80/83/80/83/80	15/8/8/5/20	5/8/12/12/0	6/4/2/7/5	91/89/70/76/83
FAI score = 5	80/82/82/83/82	15/18/12/7/18	5/0/7/10/0	4/5/6/6/6	76/83/72/86/81
FAI score = 6	88/92/90/90/88	12/3/7/8/10	0/5/3/2/2	4/3/6/7/4	91/89/70/76/83
The average classification accuracy ± SD (%) = 83.83 ± 3.99					
The needles of the first year of life					
FAI score = 1	93/92/92/93/92	3/5/5/3/8	3/3/3/3/0/4	4/4/5/2/2	78/85/86/75/75
FAI score = 2	88/90/90/83/92	5/7/10/8/5	7/3/0/8/3/5	5/2/3/3/5	71/82/75/76/86
FAI score = 3	85/83/80/83/88	10/7/15/12/8	5/10/5/5/3/6	6/6/5/6/2	77/88/73/81/76
FAI score = 4	80/83/85/83/80	10/8/10/12/13	10/8/5/5/7/7	7/2/7/3/7	69/80/78/88/73
FAI score = 5	88/87/82/83/82	8/10/13/10/8	3/3/5/7/10/6	6/6/7/4/6	77/78/87/74/91
FAI score = 6	92/92/90/90/92	5/8/7/8/5	3/0/3/2/3/7	7/7/3/3/5	72/90/86/70/87
The average classification accuracy ± SD (%) = 87.11 ± 4.42					
The needles of the second year of life					
FAI score = 1	93/95/93/93/95	7/5/5/5/5	0/0/2/2/0	5/7/2/3/2	80/70/91/85/90
FAI score = 2	95/90/90/90/92	7/5/5/5/5	0/3/5/0/0	3/6/5/4/5	81/84/74/77
FAI score = 3	90/92/92/95/90	10/3/8/5/10	0/5/0/0/0	4/2/6/2/3	90/88/73/76/89
FAI score = 4	95/92/95/93/92	3/8/5/7/3/2	2/0/0/0/5	3/4/4/7/6	76/79/90/91/74
FAI score = 5	95/90/90/93/95	5/8/8/5/5	0/2/2/2/0	7/2/4/7/3	85/91/74/85/84

(continued)

Table 2 (continued)

	The correctly classified samples, %	The multi-classified samples, %	The unclassified samples, %	The number of principal components	PC-1 dispersion, %
FAI score = 6	97/95/95/97/95	2/3/5/3/5	0/2/2/2/0	6/4/7/4/3	87/73/74/84/77
The average classification accuracy ± SD (%) = 93.11 ± 2.22					
The needles of the third year of life					
FAI score = 1	90/88/90/93/92	8/5/7/3/5	2/7/3/3/3	4/7/7/5/2	87/89/89/87/83
FAI score = 2	88/90/85/85/85	7/10/10/15/10	5/0/5/0/5	2/5/4/7/4	73/73/89/85/76
FAI score = 3	90/88/88/85/90	10/12/8/8/7	0/0/3/7/3	3/4/3/7/6	80/84/75/82/72
FAI score = 4	88/88/88/92/88	8/7/8/5/8	3/5/3/3/3	6/7/7/6/7	70/84/73/84/86
FAI score = 5	88/90/90/92/95	12/10/5/8/5	0/0/5/0/0	3/7/4/5/3	77/72/86/71/70
FAI score = 6	90/92/90/90/93	5/8/7/3/7	5/0/3/7/0	7/7/5/3/4	74/92/87/87/88
The average classification accuracy ± SD (%) = 89.44 ± 2.45					

4 Discussion

Environmental pollution has become global and chronic. Under these conditions, the possibility of monitoring the state of green spaces by simple and fairly reliable methods and tools is of great importance.

As the results of the study showed, one of these tools is NIR spectroscopy. This method has a set of advantages: ease of use, sufficient rapidity (one spectrum is taken on average in 10–20 s), minimal sample preparation, no need for chemical reagents, and the possibility of non-destructive analysis. But despite the apparent method simplicity, the analysis of natural objects, including needles, involves taking into account a number of factors, among which, many are driven by the structural features of the object of analysis and the method of obtaining NIR spectra. Moreover, the calculated model is strongly influenced by factors associated with the creation of calibration and validation sets. Thus, the presence of abnormal samples in any of the sets directly affects the values of the statistical characteristics calculated when checking the model, and can mislead the researcher. Also, the uneven distribution of samples in the range of the analyzed properties changes in the calibration set results in the fact that the model will show the required accuracy of the results in that part of the range where there were more samples and will be unrepresentative in the rest of the range. An uneven distribution in the validation set can contribute to unrepresentative validation of the model, which will also further affect the analysis

Table 3 Results of fivefold SIMCA classification of training test of needles

	The correctly classified samples, %	The multi-classified samples, %	The unclassified samples, %
The needles of the current year			
FAI score = 1	82/88/82/82/83	10/8/12/13/7	8/4/6/5/10
FAI score = 2	80/80/82/80/80	15/17/10/12/17	5/3/8/8/3
FAI score = 3	82/80/83/80/82	12/12/5/13/12	6/8/12/7/6
FAI score = 4	80/83/80/83/80	12/12/13/13/20	8/5/7/4/0
FAI score = 5	80/82/82/83/82	17/12/18/17/18	3/6/0/0/0
FAI score = 6	88/82/82/80/80	7/12/13/10/15	5/6/5/10/5
The average classification accuracy ± SD (%) = 81.72 ± 2.17			
The needles of the first year of life			
FAI score = 1	87/87/85/85/85	7/13/15/15/15	6/0/0/0/0
FAI score = 2	82/82/87/83/82	10/18/13/15/18	8/0/0/2/0
FAI score = 3	82/83/80/83/82	12/17/20/8/10	6/0/0/9/8
FAI score = 4	80/82/85/83/82	13/18/15/17/18	7/0/0/0/0
FAI score = 5	88/82/82/83/82	12/7/18/17/12	0/11/0/0/6
FAI score = 6	87/87/85/87/87	7/5/15/5/13	6/8/0/8/0
The average classification accuracy ± SD (%) = 83.78 ± 2.35			
The needles of the second year of life			
FAI score = 1	90/95/92/93/90	10/5/8/7/10	0/0/0/0/0
FAI score = 2	92/90/90/90/92	8/10/10/7/8	0/0/0/3/0
FAI score = 3	93/92/92/92/92	7/8/8/8/8	0/0/0/0/0
FAI score = 4	95/92/90/93/92	3/5/3/7/5	2/3/7/0/3
FAI score = 5	95/92/92/93/92	5/2/8/7/8	0/6/0/0/0
FAI score = 6	92/95/95/95/93	8/3/2/5/7	0/2/3/0/0
The average classification accuracy ± SD (%) = 92.27 ± 1.72			
The needles of the third year of life			
FAI score = 1	90/88/90/93/88	5/12/10/7/12	5/0/0/0/0
FAI score = 2	87/90/85/87/85	13/10/15/13/15	0/0/0/0/0
FAI score = 3	87/87/87/85/90	5/7/7/7/10	8/6/6/8/0
FAI score = 4	90/90/88/87/88	10/10/12/5/12	0/0/0/8/0
FAI score = 5	87/90/90/90/95	13/2/2/10/5	0/8/8/0/0
FAI score = 6	88/88/90/90/88	12/7/10/10/12	0/8/8/0/0
The average classification accuracy ± SD (%) = 88.61 ± 2.28			

accuracy. However, modern software used in the calculation of models makes it possible to eliminate such factors at the stage of formation of a calibration and validation set of samples, moreover, it does this automatically [18–20].

In the study, the results obtained on test samples are similar to those obtained on training ones, which indicates the stability of the developed classification models and the reliability of the obtained ones. At the same time, it should be noted that the needles of the second year of life are characterized by the maximum accuracy of classification and the minimum coefficient of variation, which indicates a small spread relative to the arithmetic mean.

5 Conclusion

Thus, the paper proves that NIR spectroscopy and chemometric modelling can help to build multidimensional calibration models that allow differentiation of Scots pine needles from areas experiencing varying degrees of manmade load. Soft independent modelling of class analogy made it possible to develop models with a sufficiently high classification accuracy (in all cases > 80%). On the basis of the experiments carried out, needles of the second year of life are recommended as an object of study for assessing the state of Scotch pine plantations under conditions of technogenic pollution. In this case, the average accuracy of the developed SIMCA classification model equalled 93.11 ± 2.22 (%) for the training sample and 92.27 ± 1.72 (%) for the test sample.

The results obtained in the work can be used in the implementation of regional environmental programs at various levels, local monitoring of pine plantations and the development of recommendations for taking measures to reduce the damage caused by pollution.

References

1. Sut-Lohmann Magdalena et al (2020) Accumulation of airborne potentially toxic elements in Pinus sylvestris L. bark collected in three Central European medium-sized cities. Ecotoxicology and Environmental Safety 200:110758. https://doi.org/10.1016/j.ecoenv.2020.110758
2. Shiretorova VG et al (2023) The influence of technogenic pollution on the Pinus sylvestris pollen chemical composition. E3S Web of Conferences. EDP Sciences 407:01012. https://doi.org/10.1016/j.ufug.2022.127819
3. Veselkin DV (2023) Urbanization increases the range, but not the depth, of forest edge influences on Pinus sylvestris bark pH. Urban Forestry & Urban Greening 79:127819. https://doi.org/10.1016/j.ufug.2022.127819
4. Tatarintsev AI et al (2022) The Pathological Status of Pinus sylvestris L. Understory Affected by Anthropogenic Air Pollution Stress (Case Study of Forests near Krasnoyarsk). Land 11(10):1625. https://doi.org/10.3390/land11101625

5. Prananto JA et al (2020) Near infrared (NIR) spectroscopy as a rapid and cost-effective method for nutrient analysis of plant leaf tissues. Advances in Agronomy 164:1–49. https://doi.org/10.1016/bs.agron.2020.06.001
6. Pan X et al (2022) Identification of softwood species using convolutional neural networks and raw near-infrared spectroscopy. Wood Material Science & Engineering, pp 1–11. https://doi.org/10.1080/17480272.2022.2130822
7. Khokh AN, Zvyagintsev VB (2022) Identifying Timber Species by Diffuse Reflection Spectra in the Near-Infrared Region Using a Linear Discriminant Analysis. Theory and Practice of Forensic Science 17(1):50–57. https://doi.org/10.30764/1819-2785-2022-1-50-57
8. Munyendo L et al (2023) Novel method for the detection of adulterants in coffee and the determination of a coffee's geographical origin using near infrared spectroscopy complemented by an autoencoder. International Journal of Food Science & Technology 58(3):1284–1298. https://doi.org/10.1111/ijfs.16283
9. Xiaobo Z et al (2010) Variables selection methods in near-infrared spectroscopy. Analytica chimica acta 667(1-2):14–32. https://doi.org/10.1016/j.aca.2010.03.048
10. Palmer AR (1994). Fluctuating asymmetry analyses: a primer. In: Developmental instability: its origins and evolutionary implications: proceedings of the international conference on developmental instability: its origins and evolutionary implications, Tempe, Arizona, 14–15 June 1993 Springer Netherlands, pp 335–364
11. Hasan BMS, Abdulazeez AM (2021) A review of principal component analysis algorithm for dimensionality reduction. Journal of Soft Computing and Data Mining 2(1):20–30. https://publisher.uthm.edu.my/ojs/index.php/jscdm/article/view/8032
12. Bächle H et al (2012) Classification of thermally modified wood by FT-NIR spectroscopy and SIMCA. Wood Science and Technology 46:1181–1192. https://doi.org/10.1007/s00226-012-0481-z
13. Ochoa-Muñoz AF et al (2019) Missing data in multiple correspondence analysis under the available data principle of the NIPALS algorithm. Dyna 86(211):249–257. https://doi.org/10.1007/s00226-012-0481-z
14. Granstedt JL et al (2022) Learned Hotelling observers for use with multi-modal data. Medical Imaging 2022: Image Perception, Observer Performance, and Technology Assessment. SPIE 12035:262–268. https://doi.org/10.15446/dyna.v86n211.80261
15. Skripalshikova LN, Stasova VV (2014). Bioindicative Parameters of Stand Growth Stability in the Disturbed Landscapes of Krasnoyarsk and Achinsk Forest-Steppe Regions. Siberian Journal of Forest Science 2:62–72
16. Kherif F, Latypova A (2020) Principal component analysis. Machine Learning. Academic Press, pp 209–225. https://doi.org/10.1016/B978-0-12-815739-8.00012-2
17. Brereton RG et al (2016) Re-evaluating the role of the Mahalanobis distance measure. Journal of Chemometrics 30(4):134–143. https://doi.org/10.1002/cem.2779
18. Urias RWP et al (2015) IMMAN: free software for information theory-based chemometric analysis. Molecular diversity 19:305–319. https://doi.org/10.1007/s11030-014-9565-z
19. Pierce KM et al (2012). Review of chemometric analysis techniques for comprehensive two dimensional separations data. Journal of Chromatography A 1255:3–11. https://doi.org/10.1016/j.chroma.2012.05.050
20. Chapman J et al (2020) Combining chemometrics and sensors: Toward new applications in monitoring and environmental analysis. Chemical Reviews 120(13):6048–6069. https://doi.org/10.1021/acs.chemrev.9b00616

Practical Application of Groundwater Regime Monitoring at Middle Ural Deposits the Practice of Application of the Results of Regulatory Observations for Groundwater at the Fields of the Middle Urals

V. A. Pochechun and A. I. Semyachkov

Abstract This paper presents and justifies one of the hydrogeoecological research methods, namely, the study of groundwater regime to accurately forecast hydrogeological and geoecological conditions of deposit mining and develop highly efficient actions to provide the safety of deposit operation, prevent potential adverse consequences, and minimize the environmental and economic damages during the post-mining period. The Urals are famous for the deposits of iron, copper, nickel, zinc, manganese, and chromite ores, as well as gold, platinum, gemstones, asbestos, bauxites, coal, and many other resources. Hydrogeological conditions are one of the key factors that can complicate the mining of these deposits. Groundwater flows into the mines, complicates headwork and resource extraction, and requires the construction of mine and quarry water removal facilities and drainage systems, which affects the economics of mining company operations. Groundwater influx may have catastrophic results, often leading to human casualties. All of the above calls for the study of the hydrogeological conditions at mineral resource deposits at all of the lifecycle stages from exploration to mining company shutdown.

Keywords Deposit · Groundwater · Regime monitoring · Monitoring network · Chemical pollution · Hydrogeological conditions · Water inflow · Drainage activities

V. A. Pochechun (✉) · A. I. Semyachkov
Institute of Economics of the Ural Branch of the Russian Academy of Sciences, 29, Moskovskaya Street, Yekaterinburg 620014, Russia
e-mail: viktoriyapochechun@mail.ru

L. V. Radionova and D. V. Ulrikh (eds.), *Advances in Ecology and Environmental Engineering*, Springer Proceedings in Earth and Environmental Sciences,
https://doi.org/10.1007/978-3-031-64423-8_15

1 Introduction

The groundwater parameters during the construction and operation of deposits may both increase or decrease. The study of groundwater regime stands for the study of the patterns of water influx and level fluctuations over time due to natural or man-made factors. This problem can be solved relative to other parameters like the chemical composition, temperature, and gas composition of drainage water, as well as other parameters [1–4].

Regime monitoring can be used at different deposit development stages because it can provide some results even during the first years of operation, and their reliability only increases over time. In other words, it is never too late to launch regime monitoring but the quality of the data obtained will be higher for longer periods [5–8].

The majority of solid mineral deposits in the Urals feature simple hydrogeological conditions. An open-cut drainage system (with a sump) is the most rational and sufficiently efficient drainage method in this scenario. The regime monitoring at such deposits is normally reduced to the regular volumetric assessment of the quarry removal system performed once a year or sometimes once a season [9] and monitoring the quality of groundwater at the same intervals.

2 Methods

This work uses hydrogeological (deposit groundwater regime and balance monitoring), geochemical, and geoecological (groundwater quality and pollution monitoring) research methods, as well as observation and statistical and cartographic data processing methods. The research is based on a large amount of accumulated data and signature fieldwork [10–13].

3 Results

The practical research on one of the deposits, namely, the Volkovsky copper-iron-vanadium mine (Lavrovo-Nikolaevsky mine), showed that measuring the quarry water influx during operation is mostly formal unless there are conditions that complicate mining work, such as the impossibility of using blast-hole drilling due to the high local water content of specific quarry banks, reduced quarry water removal quality, etc. To solve these problems, at some point during the operation of the Volkovsky deposit, a network of regime monitoring holes was drilled within the mining claim, mainly to assess the impact of the copper heap leaching unit (HLU) on the quality of groundwater. As part of the research, a regime network of 26 holes was created using the normal profile principle directed from the quarry borders to the groundwater feed and discharge borders. Its purpose, among other things, was to outline the borders of the depression pit formed due to the long-term operation of the drainage system [14–16].

In terms of representation, the monitoring network of the Volkovsky deposit was one of the few positive examples of regime activity arrangement at this scale and detail. These activities helped solve a number of important problems: for instance, it helped define the borders of the depression pit, calculate the actual volume of quarry water influx, identify its key sources while also assessing the proportion of resources obtained from the Chernushinskoye swamp and exclude the Laya river, whose valley is about 1 km away from the spoil bank of the Lavrovo-Nikolaevsky quarry from the list of sources, using modern graphic PC software.

The groundwater level in the monitoring holes was measured for two years at least once a month. The processing of these data with graphic methods led us to the following conclusions:

- the observed groundwater level drop rate within the depression pit formed in the range of the open-cut drainage of the Lavrovo-Nikolaevsky quarry was 0.5–0.7 m a year (Fig. 1);
- the outline of the depression pit follows the outline of the quarry. On the spoil (eastern) bank, the depression pit outline changes insignificantly, and it mostly expands to the west, while the northern and southern borders of the pit correspond with the water-parting lines. Following the hydroisobath map, we identified four water-conducting zones, with background water transmissibility values in practically impervious blocks below 24 m^2/day, while water-conducting zones feature water transmissibility values between 50 and 100 m^2/day;
- the average absolute groundwater level on the western bank of the quarry is 10 m lower than on the eastern one, which is attributed to its greater contribution to the formation of the total water influx, even though the main constant-head border, the Laya River, lies to the east of the quarry. The proportion of infiltration recharge in the formation of the total quarry water influx is not important, which can be confirmed by the stability of depression pit borders compared with winter and summer conditions—only in summer the groundwater level rises by one meter on average;
- near the eastern bank, the quarry completely uncovered an exogenic conjugated fracture area. The groundwater movement in this area mainly occurs within the shallow (up to 1–2 m) drilling-induced fracture area created by the drilling and blasting in the quarry. This results in a very insignificant amount of drainage water on this side, and the main development of the depression pit on the western side, while the resources attracted to the formation of the total water influx are mainly taken from the Chernovskoye swamp. The slope of the depression pit is greater on the eastern bank.
- On low quarry horizons, the positioning of groundwater is determined by the shape of the banks. The largest high drilling-induced fracture area on the bank probably acts as water drainage.

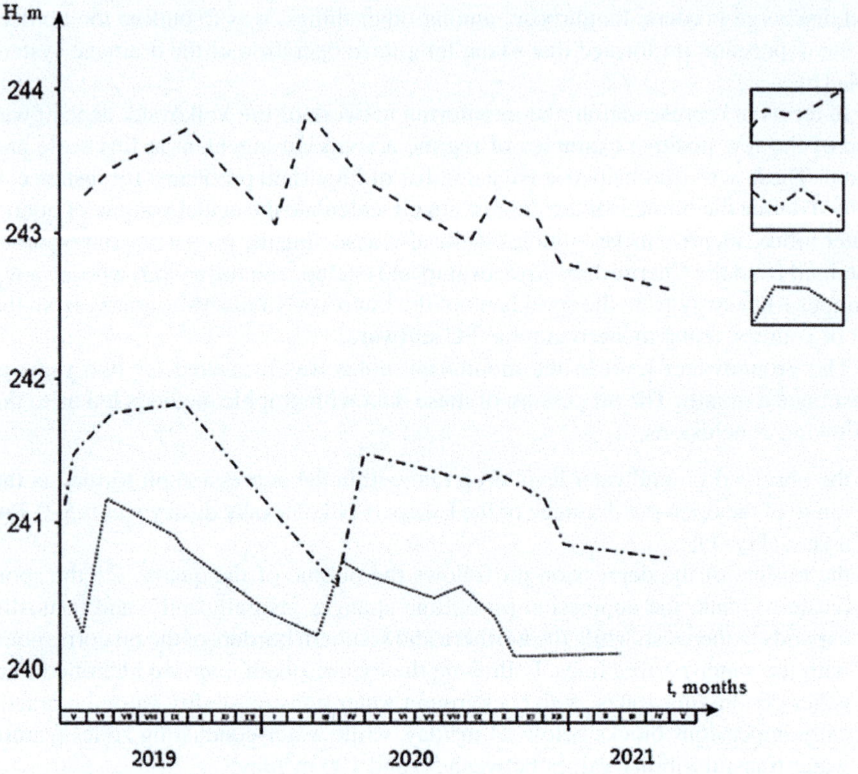

Fig. 1 The results of regime monitoring of groundwater level changes in the following wells: 1—well 7; 2—well 4; 3—well 6

- The formation of a drilling-induced fracture area is primarily connected to blasting operations.
- Near the western bank, there are no depression pit shape anomalies, which may be attributed to the higher filtering properties of the water-bearing rocks.

The research data for the regime monitoring of groundwater quality in the wells of the Volkovsky deposit from 2019 to 2021 show that the chemical composition of groundwater within the Lavrovo-Nikolaevsky area changed from hydrocarbonate/calcium to sulfate/sodium compared to the deposit mining start date (1961). We also observed a steady increase in the concentrations of pollutants: the content of sulfates over this period increased from 30 to 850 mg/dm^3, the pH value increased from 5.8 to 7.8, the content of copper—from 0.02 to 2.97 mg/dm^3, zinc—from 0.1 to 1.63 mg/dm^3, total iron—from 0.1 to 1.67 mg/dm^3. The chemical composition of the exploration wells located to the north of the quarry in an area undisturbed by the mining operations retained a hydrocarbonate/calcium character with a mineralization of 0.2–0.4 g/dm^3. No natural factors behind the changing of the chemical composition of groundwater were found. However, when we compared the concentration values of anions, cations, and microelements in the groundwater and the anion-cation and microcomponent composition of the Laya River, we concluded that the chemical composition of groundwater within the deposit did not depend on the chemical composition of the river. Thus, there is no hydraulic link between them, i.e. the overburden is sufficiently water-resistant. This means that if the water of the Laya River becomes contaminated with metals and other pollutants, the groundwater will not be contaminated [17–20].

The spread of pollution in the groundwater of the deposit is limited to the depression pit shown in the hydroisobath map (Figs. 2 and 3). It covers an area of about 6 km^2.

The regime monitoring conducted at the Volkovsky mine in the late 1980s—early 1990s provided quite valuable data that were used for the calculation of the total water influx to the prospective quarry in the area using the analogy method.

The hydrogeological conditions of the Goroblagodatskoye iron ore deposit located in the Tagilo-Kushvinky district of Sverdlovsk Oblast, like the Volkovsky deposit, were classified as complex. However, the complexity of the hydrogeological conditions at this deposit is mainly attributed to the specifics of its formation rather than

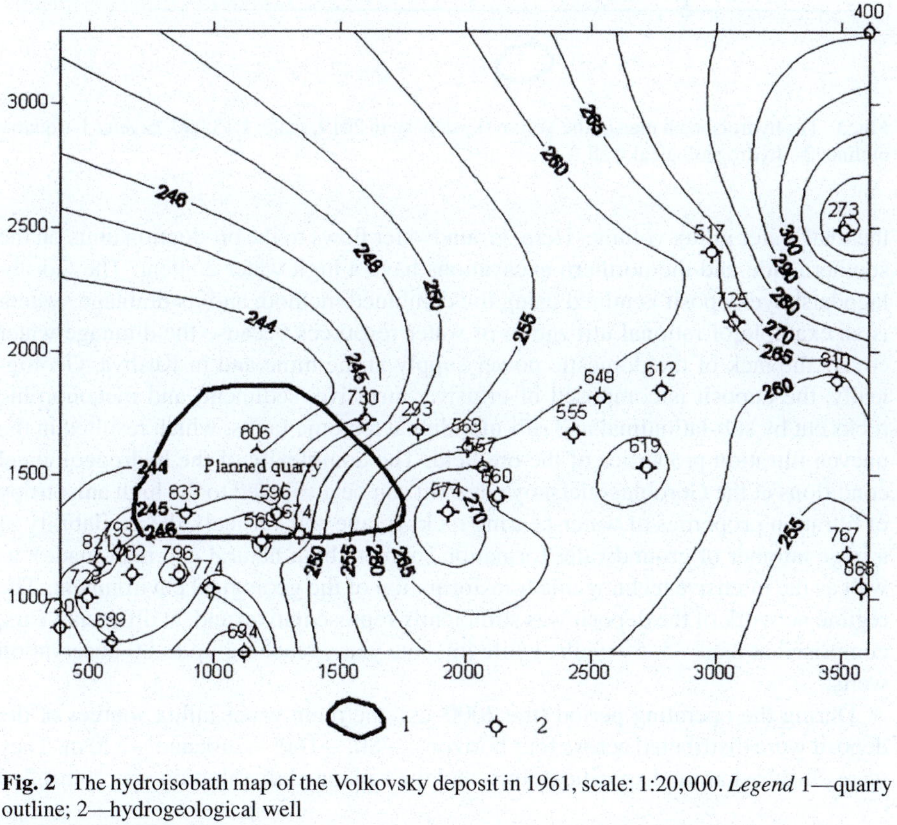

Fig. 2 The hydroisobath map of the Volkovsky deposit in 1961, scale: 1:20,000. *Legend* 1—quarry outline; 2—hydrogeological well

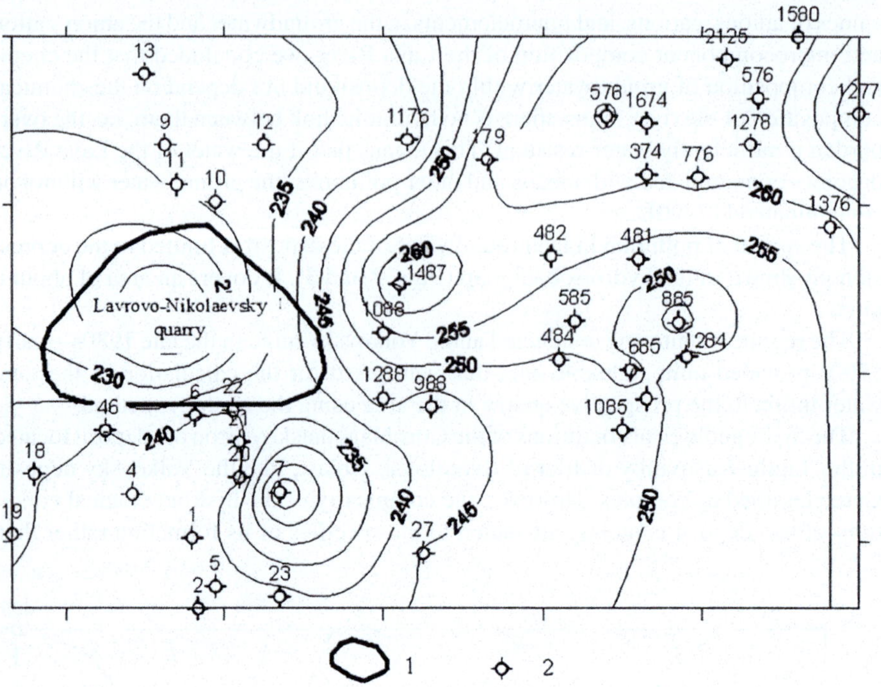

Fig. 3 The hydroisobath map of the Volkovsky deposit in 2019, scale: 1:15,000. *Legend* 1—quarry outline; 2—hydrogeological well

the total water influx volume. Here, groundwater flows to the production units on the southern side and the northern excavations have a high water content. The Goroblagodatskoye deposit is mined using the combined method, and its drainage system is an example of rational utilization of water resources because the drainage water covers the lack of the domestic power supply at the mine and in Kushva. Geologically, the deposit is composed of effusive, intrusive, sediment, and metamorphic rocks cut by sub-latitudinal and sub-meridional tectonic faults, which resulted in the uneven filtration properties of the ore field. The complexity of the hydrogeological conditions at the Goroblagodatskoye deposit can be attributed to the high anisotropy of filtration properties of water-bearing rocks (plane and vertical), the availability of a large number of groundwater formation sources (both natural and man-made), as well as the intensive technogenic transformation of the geological environment. The regime network of the deposit was sufficiently representative, and, at different years, comprised at least 35–55 wells, while in some years reaching up to 80 observation wells.

During the operating period (the 2000–es), the main water influx sources of the deposit were distributed across four horizons: −80; −160; −240; and −320 m. They were clustered around specific points, and their planned distribution was controlled by the tectonic structure of the ore field combined with the lithological composition of the water-bearing rocks.

The dependencies between the water influx to the mine excavations of the Goroblagodatsky deposit and the amount of precipitation in the ore field were analyzed using the pair correlation method. We researched the amount of drained fluid precipitation (excluding evaporation) and the moisture content in the slow cover before the thaw (solid part). The results are summarized in Table 1.

The performed calculations show that the water influxes of the Goroblagodatsky mine are affected by natural and man-made regime-shaping factors. The natural factors primarily include the groundwater formation conditions and uneven filtration properties of the mass, and the man-made ones include the mining operation conditions and the drainage system.

During the underground mining of the deposit, the drainage is based around the mining horizon that receives the main water influx. Ever since the commissioning of the Goroblagodatsky mine, the water influx has been growing and its sources clustered and distributed according to the depth of the mine.

The analysis of correlations between the water influx and precipitation shows that the water influx generated at upper horizons (until the depth of 60 m) is closely linked

Table 1 Yearly average influx to the mine excavations at the Goroblagodatskoye deposit and precipitation data

Year	Yearly precipitation, mm	Precipitation excluding evaporation, mm	Yearly average water influx to the Central quarry, m^3/h	Yearly average water influx to the Southern mine, m^3/h	Yearly average water influx to the entire mine, m^3/h
2005	387.5	162.2	133.3	236.8	370.6
2006	570	207.2	161.7	247.4	404.1
2007	503.3	226.8	185.4	280.4	472.5
2008	670.1	322.4	219.6	254.6	474.4
2009	446.3	–	–	–	492
2010	453.5	188.5	172	260.5	435.2
2011	601.4	297.1	183	243.2	431.2
2012	653.4	325.5	182.3	229.1	378.1
2013	542.1	242.8	145.8	300.2	445.9
2014	640.7	313.3	137.6	357.6	495.2
2015	547.9	279.8	141.9	423.1	565
2016	505.5	185.4	131.6	372.8	504.4
2017	443.7	157.5	128	360.5	407.6
2018	423.2	158	149.3	376.6	–
2019	512.5	174.3	84.2	–	536.7
2020	544	304.8	–	–	609.8

to precipitation while for lower horizons this link becomes weak and practically disappears after −120 m.

The analysis of the hydrogeological situation at the Goroblagodatsky deposit where the natural groundwater regime was disturbed by mining helped us provide a quantitative assessment of the key structural water balance components that indicate that man-made factors are prevalent in the formation of the total water influx of the deposit as a whole (see Table 2).

Another example of applying the regime monitoring data in practice is the study of the hydraulic link between surface water and groundwater [21]. The majority of rivers in the Urals are small. The irregularity of their valleys is so great that leaving them unaccounted may result in significant inaccuracies of hydrogeological forecasts including the assessment of mine influx amounts. With a monitoring profile that consists of 2–3 wells and is orthogonal to the nearest river valley, we can use groundwater level and river level monitoring data from the same time to calculate the filtration resistance of the course (a river irregularity parameters) using the level proportion method. The irregularity parameter of the Laya River (Volkovsky mine) calculated using this method was 500 m.

The reviewed groundwater regime monitoring method for complex hydrogeological research can also be used for the planning of efficient drainage activities at deposits.

The considered method of routine observations of groundwater, as part of complex hydrogeological studies, can also be used when planning a system of effective drainage measures in fields.

It is convenient to divide the issues of deposit drainage into strategic and tactical. Strategic issues include justification for the drainage system and method.

The specific flow rate of wells is determined mainly by the filtration properties of microheterogeneity elements. The effectiveness of drainage devices is determined

Table 2 The calculation results for the total water influx balance components at the Goroblagodatsky deposit

Balance component	Amount of water, m^3/h	Element proportion in the entire balance, %
Total water influx	500	100
Reserve volume (within the depression pit)	30	6
Attracted resource (central and Saldinskoe sludge dumps)	300	60
Constant-head borders	30	6
Verkhne-Turinsky reservoir	5	2
Kushva river	3	1
Kushvinsky reservoir	2	0
Infiltration feed	100	20
Quarry and caving area	35	7

primarily by their water abundance. Therefore, the installation of drainage devices must be carried out in zones of maximum specific flow, which are essentially elements of microheterogeneity that have high water conductivity coefficients, determined by trial experimental filtration work. Due to the fact that the permeability of microheterogeneity elements has a chaotic distribution within the genetic types of rocks, only discrete installation of drainage devices is possible in the fields of the Urals.

Vertical hydrodynamic variability of filtration properties determines the depth of drainage measures. When ore bodies occur near the surface with mining by collapse systems, the collapse zone will completely open the aquifer and will be the main basis for groundwater drainage. The underground method of drainage at exposure horizons is effective only within the aquifer, that is, in effusive-pyroclastic and sedimentary rocks to a depth of 300 m, and in intrusive rocks to a depth of 50 m. Laying drainage devices below these depths is ineffective, since they will located in a relatively aquifer. Due to the development of mining operations in plan and depth, drainage devices located in the aquifer can be undermined, which will not lead to a decrease in the flow of groundwater to the mining blocks, but, on the contrary, to an increase. Thus, in these conditions, it is necessary either to constantly move drainage devices at the depth of the base of the aquifer from the ore zone, ahead of the mining front, or to create a drainage system in conjunction with a longer life of the deposits (preferably at the end of mining). At the initial stage of development, the drainage system can be represented by surface water-reducing wells with a depth of 300 m in sedimentary and effusive-pyroclastic rocks and 50 m in intrusive rocks. They can then be converted into through filters as mining progresses. Depending on the specific conditions, it is possible to use underground drainage wells and units, timing them to the base of the aquifer at some distance from the ore-skarn zone so that they are not undermined and do not fall into the collapse zone.

In any case, drainage devices must be located in the most water-abundant zones associated with the power source and unloading of groundwater in mine workings. Only in this case can proper drying efficiency be achieved. The search for such zones is the tactic of drainage measures in the fields of the Urals. Their determination is proposed to be carried out in two stages.

In the first stage, the search is carried out on a computational model constructed as a result of schematization and confirmed by epignosis modeling, reflecting the planned filtration macroheterogeneity of rocks and the balance of groundwater in the upper hydrodynamic zone. By specifying actual or predicted groundwater levels (type 1 boundary conditions) in discretization blocks where mine workings are located or proposed, it is possible to estimate the change in water inflow to them depending on water withdrawal from other blocks. As a result of calculating various options, an optimal solution is established—a block of rocks is determined, the extraction of groundwater from which at a given flow rate will most effectively reduce the water inflow to the mine workings.

At the second stage, it is necessary to take into account the planned filtration microheterogeneity of rocks, that is, to determine within the discretization block models of zones with increased filtration properties, caused by elements of microheterogeneity. The drainage devices installed in them must have high specific flow rates. The search for such zones must be carried out by control and exploration drilling. The placement of exploratory wells can be controlled by geological, structural and hydrogeological analysis.

Based on geological and structural analysis, it is possible to determine the position of zones of tectonic disturbances in which increased filtration properties of rocks are expected. However, an analysis of the relationship between the water conductivity of genetic types of rocks and tectonic disturbances in them showed that there is no significant difference between them. This means that the feature in question cannot be used as a search criterion.

Hydrogeological analysis includes the analysis of hydroisohypsum maps, on which zones of increased permeability are displayed as areas with a reduced pressure gradient. The effectiveness of this feature is determined by the detail of the regime network.

4 Conclusion

Due to the high cost of drainage measures, their feasibility study should be based on convincing geofiltration calculations, combining analytical assessments and mathematical modeling. The key link in such geofiltration tasks is considered to be a reliable schematization of natural conditions. This schematization can be carried out on the basis of routine observations of groundwater, which is one of the main methods in hydrogeological and geoecological studies in the fields of the Urals.

References

1. Kolesnikova LA, Kovalchuk TV (2021) Problemsand prospects of environmental safety of mining regions. Mining Informational and Analytical Bulletin 2–1:275–286. https://doi.org/10.25018/0236-1493-2021-21-0-275-286
2. Ignatenko YuV (2018) Assessing the quality of groundwater near the sludge dump of PNTZ OJSC. The Ural Mining School to the Regions, a collection of papers from the international research and practice conference, Yekaterinburg, April 16–17, 2018, Ural State Mining University, pp 588–589
3. Dalatkazin TSh, Kayumova AN (2020) Studying the hydrogeological conditions to ensure the safety of Sokolovsky iron ore deposit. News of the Higher Institutions. Mining Journal 1:133–141
4. Karabin MM (2021) Assessing the impact of hydrogeological processes and stability underworking for a coal quarry mass. News of the Higher Institutions. Mining Journal 1:36–44
5. Ignatyeva M, Yurak V, Dushin A (2022) Valuating natural resources and ecosystem services: systematic review of methods in use. Sustainability (Switzerland) 14 (3):1901.https://doi.org/10.3390/su14031901
6. Ignatyeva, MN, Yurak VV, Dushin AV, Polyanskaya IG (2021) Assessing challenges and threats for balanced subsoil use Environment. Development and Sustainability 23(12):17904–17922.https://doi.org/10.1007/s10668-021-01420-1

7. Valiev NG, Golik VI, Propp VD, Bolgova AI, Ovsyannikov MS (2022) Regularities of environmental protection process management. Mining Informational and Analytical Bulletin 11–1:40–50. https://doi.org/10.25018/0236_1493_2022_111_0_40
8. Abaturova IV, Savintsev IA, Storozhenko LA et al (2020) Stages of studying the engineering and geological conditions at mineral resource deposits from exploration to mining. (Geoecology, rational environmental management). News of the Higher Institutions. Mining Journal 7:83–91
9. Ekimova OA, Parfenova LP, Bichukina IA (2022) Assessment of groundwater inflows into the opencast mine during the development of a copper-iron-vanadium deposit in the Middle Urals. Mining Informational and Analytical Bulletin 5–1:70–81. https://doi.org/10.25018/02361493_2022_51_0_70
10. Abaturova IV, Bykova AV (2020) Methods of monitoring exogenetic processes in natural-technical systems. Engineering and Mining Geophysics.https://doi.org/10.3997/2214-4609.202051086
11. Abaturova IV, Storozhenko LA, Pisetsky VB, Savintsev IA (2020) Use of geological and structural analysis in evaluating engineering and geological conditions of mineral deposits. Engineering and Mining Geophysics, 2020. https://doi.org/10.3997/2214-4609.202051096
12. Kozin VZ, Komlev AS (2022) Sampling effects related to the asymmetry of distributions of mass fractions of components in point. Mining Informational and Analytical Bulletin 11–1:107–118. https://doi.org/10.25018/0236_1493_2022_111_0_107
13. Kozin VZ, Komlev AS (2021) Calculation of fundamental sampling error. Mining Informational and Analytical Bulletin 1–11:265–275. https://doi.org/10.25018/0236_1493_2021_111_0_265
14. Tagiltsev SN, Tagiltsev VS, Surganov SV, Kurichenko AA (2021) The main complicating factors in conducting and processing filtration testing of single wells. 17th Conference and Exhibition Engineering and Mining Geophysics, 2021. https://doi.org/10.3997/2214-4609.202152063
15. Tagiltsev SN, Tagiltsev VS, Surganov SV, Kurichenko AA (2021) Use of hydrodynamic characteristics of the underground water flow for processing single pumping data. 17th Conference and Exhibition Engineering and Mining Geophysics, 2021. https://doi.org/10.3997/2214-4609.202152023
16. Tagilcev SN, Cherednichenko AV, Melnik VV (2020) Aggregation methods hydrogeomechanical, electrical and dowsing to select areas of hydrogeological wells. Mining Informational and Analytical Bulletin. 3–1:236–246. https://doi.org/10.25018/0236-1493-2020-31-0-236-246
17. Pochechun VA, Semyachkov AI, Kurbanov IKO (2021) Surfaceand groundwater geoecology in the Middle Urals: a case-study of Polevskoy and Kachkanar mining districts. Sverdlovsk region Mining Informational and Analytical Bulletin 1–11:90–97. https://doi.org/10.25018/0236_1493_2021_111_0_90
18. Boltyrov VB, Storozhenko LA, Sapsay MA (2021) Cumulative ecological impact in the territory of long-term disposal of old mining waste. Mining Informational and Analytical Bulletin 5:202–217. https://doi.org/10.25018/0236_1493_2021_52_0_202
19. Storozhenko LA, Petrova IG, Klokova YV, Lymar IO (2020) Impact of open-pit mining on the geological environment. Engineering and Mining Geophysics. https://doi.org/10.3997/2214-4609.202051132
20. Makarov AB, Talalay AG, Guman OM, Khasanova GG (2022) Technogenic deposits and the specifics of their impact on natural environments. News of the Higher Institutions. Mining Journal 3:120–129
21. Ribnikova L, Navolokina V (2020) Analysis and Feasibility of Measures to Minimize the Impact of Acid Mine Waters Discharged by Abandoned Copper-Sulphide Mines on Hydrosphere of the Tagil River. E3 S Web of Conferences 177:04009. https://doi.org/10.1051/e3sconf/202017704009

The Behavior of Oil Pollution Composition in Permafrost Soils Affected by Diesel Spill

Yu. S. Glyaznetsova, I. N. Zueva, O. N. Chalaya, S. Kh . Lifshits, and O. S. Lvova

Abstract The article presents an experimental study of oil pollution composition during biological treatment of permafrost soils contaminated with diesel fuel. The experiment was conducted at an oil depot with the use of a biological product based on indigenous hydrocarbon-oxidizing microorganisms isolated from permafrost soils. To characterize the composition of oil pollution and its changes, and to assess the transformation of petroleum hydrocarbons, informative analytical indicators obtained by physical and analytical methods were used. It was revealed that the change in the initial composition of oil pollution depends on biochemical processes of hydrocarbons destruction along with the processes of oil hydrocarbons evaporation. After 2.5-month bioremediation, involving the incubation of the biological product, the composition of pollution changed: the share of resinous-asphaltene components increased, whereas the concentration of hydrocarbons and residual oil decreased. The share of oxygen-containing groups and bonds in the chemical structure increased significantly. Redistribution within both homologous series and between different series of homologues was established in the composition of alkane hydrocarbons. The obtained results indicate changes in the pollution composition during soil purification due to biodegradation of petroleum hydrocarbons.

Keywords Bioremediation · Hydrocarbons · Oil pollution · Permafrost soils · Asphaltenes · Resins components

1 Introduction

The Republic of Sakha (Yakutia) has large oil and gas fields. The Republic's Energy Strategy until 2030 includes the creation of the oil refinery industry [1].

Yu. S. Glyaznetsova (✉) · I. N. Zueva · O. N. Chalaya · S. K. . Lifshits · O. S. Lvova
Federal Research Center "Yakutsk Science Center SB RAS", Institute of Oil and Gas Problems, Siberian Branch of the Russian Academy of Sciences, Petrovsky St, Yakutsk 677980, Russia
e-mail: gchlab@ipng.ysn.ru; glyaz1408@mail.ru

L. V. Radionova and D. V. Ulrikh (eds.), *Advances in Ecology and Environmental Engineering*, Springer Proceedings in Earth and Environmental Sciences,
https://doi.org/10.1007/978-3-031-64423-8_16

Balanced and rational use of natural resources is one of the priorities of the Russian Federation State Energy Policy [2]. Today, to sustain the republic's population, oil products are imported from other regions. Due to the intensive development of the oil and gas industry, the prospect of hydrocarbon production on the Arctic shelf, and the exploitation of existing oil supply facilities, human impact on the environment is increasing rapidly.

According to the Ministry of Ecology, Nature Management and Forestry, the number of oil and oil products (OP) spills has exceeded 316 cases over the past 12 years, including those in the Arctic zone [3]. Most spills are caused by road traffic accidents. In addition, spills occur due to the technical condition of tanks, during pumping of oil products, as well as due to natural and man-made disasters (floods, fires, etc.).

The consequences of oil spills are difficult to assess since oil pollution disrupts natural processes and changes living conditions for all organisms in the habitat. In permafrost soils, the oil pollution degradation processes take much longer than in regions with milder climates. Permafrost soils have a low capacity for self-purification. Without applying some measures to remediate the soil in a northern climate, oil degradation can last for over 50 years [4–6]. A low thickness organogenic horizon that forms in permafrost soils means that microbiological processes are slowed down, and the content of basic biogenic elements is low. The presence of permafrost complicates the transformation of petroleum hydrocarbons (HCs); the permafrost layer serves as a geochemical barrier where oil components accumulate and the direction of oil migration shifts from vertical to horizontal.

Tackling the remediation of oil-contaminated soil requires an integrated approach. Improved environmental monitoring of disturbed lands to judge whether the treatment has been effective is as important as the development of remediation methods themselves. These tasks become especially relevant due to the urgent need to create effective methods of remediating oil-contaminated soils in harsh northern climates.

The remediation of oil-contaminated soils involves physical, chemical, and biological methods of treatment, as well as their combinations [6–8]. All methods are divided into artificial and those intensifying natural processes of self-remediation. Different purification methods are applied depending on the timeframe and degree of soil contamination; and therefore, identifying a differentiated pollution pattern is of utmost practical significance for selecting the most effective cleaning methods, as well as for minimizing the cost of their implementation.

An analysis and systematization of existing methods of cleaning up oil-contaminated soils, both according to exciting literature data and the results of our own research, showed that the biological method of purification proved to be the most effective and environmentally friendly solution in permafrost conditions. This method utilizes the application of biopreparations based on hydrocarbon-oxidizing microorganisms (HCOM) by sowing perennial plant varieties, and proved to be effective [6, 8–15].

The purpose of the research was to evaluate the effectiveness of the developed technology for bioremediation of oil-contaminated permafrost soils.

2 Materials and Methods

An experiment on biological remediation of soils contaminated by a diesel spill was conducted at an oil depot located in central Yakutia (N 58.970185 E 126.233816). To reach the aim, 5 experimental sites were allotted. According to [16], the study area soils are characterized by low hydrocarbon pollution resistance.

The experimental sites were treated with a biological product developed at the Institute based on indigenous HCOM isolated from permafrost soils. The advantage of this biological product is the ability of the microorganisms to develop at low positive temperatures (above +4 °C) and under conditions of insufficient aeration and humidity [8].

We studied soil samples taken from the experimental sites before and after the biological treatment. The process of oil bioremediation was carried out between June and August. The exposure time of the biological product was 2.5 months.

The concentration of the OP was determined by the yield of bitumoids, extracted by chlorophorm. The complex of analytical studies included the following: studying the structural-group composition of soil extracts by IR-Fourier spectroscopy on Protege 460 spectrometer (Nicolet, USA) within the range of wave numbers 500–4000 cm^{-1}, certified in accordance with ISO 9000 and 9002 standards by the Gosstandard of Russia. The spectra were interpreted using the atlas of IR-spectra and tables of wave numbers. Liquid adsorption chromatography was used for the fractional division of soil extracts into hydrocarbon, resinous, and asphaltene resin components.

The analysis of alkanes was carried out using an Agilent 6890 chromatography-mass spectrometer (USA), which has an interface with a high-performance mass selective detector Agilent 5973N, with a quartz capillary column 30 m × 0.25 mm and an HP-5MS impregnated phase; the carrier gas was helium, the flow rate was 1 mL/min. The temperature was changed from 100 to 300 °C at a rate of 6 °C/min, at an evaporation temperature of 320 °C, the ionizing voltage of the source was 70 eV. The identification of compounds was based on the retention time with comparison of the obtained mass fragmentograms with the spectra available in the system library. More detailed research methods are given in the work [10, 17].

The measurements were taken three times. Figures and tables present the data in arithmetic mean values. The results obtained in comparison with the control data are statistically reliable.

3 Results

The initial content of OP in the soils ranged from 24533 to 67528 mg/kg (Table 1). According to the classification [18], the level of contamination was graded as high and very high. In 2.5 months after the bioremediation, the residual content of OP in the soils of the experimental sites decreased significantly (Table 1).

Table 1 Geochemical properties of soils before and after bioremediation

Experimental site no.	Bioremediation stage	OP content (mg/kg)	Group component composition of soil extracts (%)		
			HCs	Resins	Asphaltenes
1	Initial	67528	89.5	7.4	3.1
	After 2.5 months	3804	6.8	58.9	34.3
2	Initial	67043	88.8	8.4	2.8
	After 2.5 months	3812	8.3	52.7	39.0
3	Initial	70495	92.5	5.6	1.9
	After 2.5 months	5979	69.4	21.1	9.5
4	Initial	32148	79.3	11.8	8.9
	After 2.5 months	15485	72.5	18.6	8.9
5	Initial	24533	87.0	9.8	3.2
	After 2.5 months	13677	70.6	20.3	9.3

The decreased OP concentration was accompanied by a change in the structure-group composition of the soil extracts. Figure 1 shows IR spectra of the extracts of soil samples from the experimental sites No. 1 and No. 5 with maximum and minimum initial OP content. The configuration of the IR-spectra of the initial samples (before bioremediation) was determined exclusively by the hydrocarbon components: predominance of methylene, methyl groups, and aromatic cycles—a set of intense absorption bands (a.b.) in the range of 650–1000 cm^{-1}, 1380, 1460 and 1600 cm^{-1}. Such composition is indicative of OP and oil contaminated soils.

In 2.5 months after the biological treatment, the type of IR-spectra changed. The spectrum of site No. 1 sample demonstrated intense a.b. at 1700–1740 and 1170

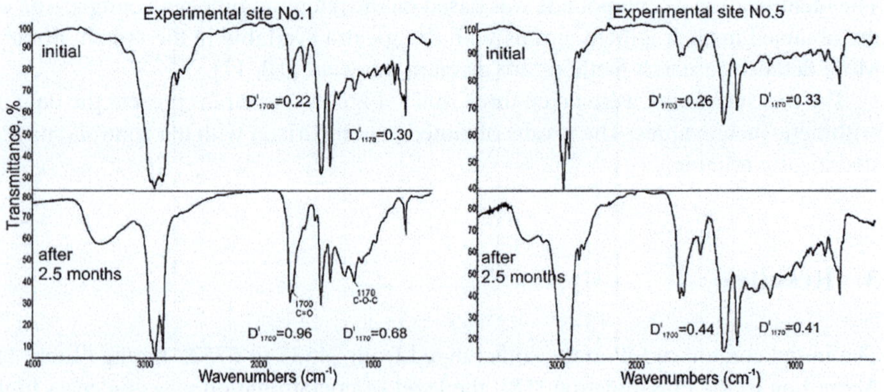

Fig. 1 IR-spectra of soil extracts samples from experimental sites No. 1 and No. 5 before and after bioremediation

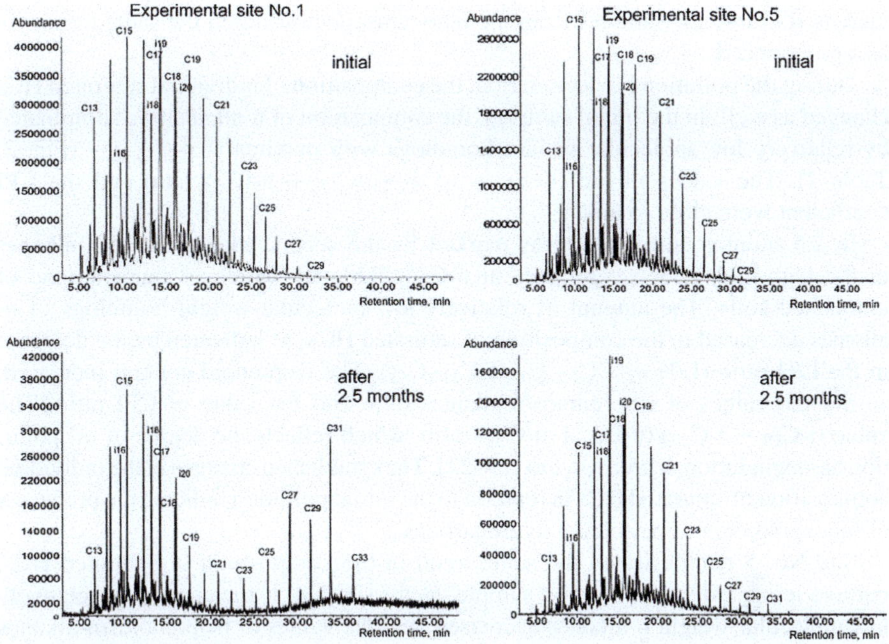

Fig. 2 Mass chromatograms (m/z of 57) for saturated HCs in soil extracts of the samples from experimental sites No. 1 and No. 5 before and after bioremediation. C_{13} … C_{31}—n-alkanes; i_{16}… i_{20}—isoprenoids

cm^{-1}, which indicates the processes of biochemical oxidation of petroleum hydrocarbons. At the same time, according to GC/MS data, a small amount of relatively low molecular weight alkane hydrocarbons was detected in the composition of saturated HCs (Fig. 2).

Changes in the composition of soil extracts were also observed in other experimental sites, but to a lesser extent in comparison with site No. 1. Traces of residual oil contamination were detected in the chemical composition of site No. 5 soils, which was indicated by the presence of a.b. of aromatic HCs in the area of 1600 cm^{-1}. However, the IR spectrum of this sample also contained intensive a.b. of oxygen-containing groups and bonds (Fig. 1), which indicated the biodegradation of HCs due to the activity of HCOM introduced with the biopreparation. It was also indicated by increased relative absorption coefficients of carbonyl groups ($D^{I}{}_{1700}$) and ether bonds ($D^{I}{}_{1700}$). The ratios were calculated as the ratio of the optical densities of a.b. 1700 and 1170 cm^{-1} to the optical densities of a.b. of the methyl and methylene groups at 1460 cm^{-1} (Fig. 1). For site No. 1, these coefficients increased four-fold for carbonyl groups and increased doubled for ether bonds. For site No. 5, the changes in the coefficients were manifested to a lesser extent.

In the group component composition, the content of HCs significantly decreased in soils of sites No. 1 and No. 2, while the content of resins and asphaltenes increased (Table 1). Samples from the other sites demonstrated the same trend, yet less pronounced.

During the pollution biodegradation, the composition of individual saturated HCs changed as well. In the initial samples, the composition of n-alkanes was dominated by relatively low molecular weight homologs with maxima at n-$C_{15,16,17}$ (Fig. 2; Table 2). The values of the (i-C_{19} + i-C_{20})/(n-C_{17} + n-C_{18}) ratio and the CPI coefficient were close to unity.

In 2.5 months, site No. 1 was marked by the second maximum of n-alkanes in the high-molecular-weight area at n-C_{31} (Table 2), which is characteristic of untouched soils. The amount of relatively low-molecular-weight homologs of n-alkanes decreased in the composition of saturated HCs, as indicated by the decrease in the L/H ratio (L/H = ΣC_{12}–C_{20}/ΣC_{21}–C_{33}). The isoprenoid content increased, so did the values of isoprenoids/n-alkanes ratio and the value of CPI ratio. The value (i-C_{19} + i-C_{20})/(n-C_{17} + n-C_{18}) ratio, which reflects the degree of oil pollution biodegradation, increased, too [19–22]. The established changes in the individual composition of saturated HCs in relation to the initial pollution indicate the processes of biodegradation of petroleum hydrocarbons.

Site No. 5 demonstrated the same trend in the composition of saturated HCs, but to a lesser extent. In this soil sample, as for site No. 1, the amount of relatively low molecular weight n-alkanes decreased, and the values of isoprenoids/n-alkanes and (i-C_{19} + i-C_{20})/(n-C_{17} + n-C_{18}) increased. In contrast to the soils of site No. 1, the maximum of n-alkanes remained in the low molecular weight area at n-$C_{17,18}$. The value of the CPI coefficient hardly showed any changes and was close to unity (Table 2). It is probable that soils of this site were less affected by biodegradation processes.

Table 2 Trends in the individual composition of saturated HCs

Parameters	Experiment sites			
Bioremediation stage	No. 1		No. 5	
	Initial	After 2.5 months	Initial	After 2.5 months
L/H	3.53	1.00	3.74	2.37
Maximum n-alkanes	n-$C_{15,16,17}$	n-C_{15}, n-C_{31}	n-$C_{15,16}$	n-$C_{17,18}$
CPI	0.98	1.90	0.96	1.00
Isoprenoids/n-alkanes	0.33	0.51	0.36	0.52
(i-C_{19} + i-C_{20})/(n-C_{17} + n-C_{18})	1.02	1.50	0.95	1.33

4 Discussion

The duration of bioremediation process depends on many factors, such as the soil and climatic conditions and the nature of pollution [5, 23]. According to the authors [6, 16, 24, 25], the approximate period of biological purification in different natural areas can take between 2 and 5 years or more. To achieve the desired effect of soil remediation from oil pollution, it is necessary to take into account the peculiarities of petroleum hydrocarbons transformation in the soil.

When analyzing the results of work performed, it is possible to overestimate the effectiveness of biopreparations in case of overlooking the loss of petroleum hydrocarbons due to evaporation of light fractions. Up to 75% of oil from its initial amount is lost due to evaporation [6, 26, 27]. Using a complex of analytical methods makes it possible to identify directional changes in the composition of petroleum hydrocarbons, indicating the processes of degradation of oil pollution in soil. They include a decrease in the concentration of oil with a simultaneous decrease in the proportion of hydrocarbons; an increased share of resins and asphaltenes; redistribution of hydrocarbons in the composition of alkanes and arenas; changes in the chemical structure of extracts; and differences in the distribution of stable carbon isotopes ($^{13}C/^{12}C$) of oil and soil organic matter [8, 9, 12, 14, 28–31].

The conducted bioremediation experiment established changes in the chemical composition of soils towards reduced content of petroleum hydrocarbons, as well as in the nature of their distribution, and an increased share of oxygen-containing compounds, all signaling the ongoing processes of oil pollution biodegradation. However, some areas of the oil depot remain under the residual impact of the oil pollution, e.g. experimental sites No. 4 and No. 5. Thus, it is necessary to conduct a re-treatment of these soils. Some studies have shown that acceleration of diesel biodegradation in Arctic soils can be achieved by adding nutrients to the soil [31, 32].

Uneven distribution of pollution in the soil profile can result in large difference in the values of the initial OP content and the different nature of changes in the composition of petroleum hydrocarbons. The nature of the OP spread depends on the distance from the source of pollution and the terrain specifics.

A significant decrease in the residual content of OP was found at the experimental sites where bioremediation work was carried out, which indicates the effectiveness of the measures taken. The highest efficiency of biological treatment of soils was established on experimental sites No. 1, 2, and 3. The given data indicate a significant degradation of oil pollution and testify to the high efficiency of the conducted works on cleansing contaminated soils with the developed biopreparation.

5 Conclusions

The complex of physico-chemical analysis methods used in the study showed their high informative value in studying changes in the composition of the initial oil pollution on the territory of the oil depot.

It was established that changes in the OP content and the composition of oil pollution in permafrost soils occur due to the evaporation of light HCs and biodegradation of HCs, caused by the activity of the introduced indigenous HCOM.

The achieved directional changes in the composition of saturated HCs, the redistribution of HCs both within homologous series and between different series of homologues, as well as changes in the chemical structure of soil extracts against the background of decreased residual content of OP indicate the biodegradation of oil pollution. The experimental results showed the effectiveness of bioremediation of oil-contaminated permafrost soils using the developed biopreparation.

The conducted study and the analysis of literature data showed that biological treatment methods based on the use of microbial preparations proved to be the most environmentally friendly solution that excludes negative consequences for the natural environment.

Considering the rapid development of the oil and gas complexes in northeast Russia, the obtained results can be used when developing technologies for the remediation of disturbed lands.

Acknowledgements The results of the study were obtained under the Governmental Order of the Ministry of Science and Higher Education of the Russian Federation (122011200369-1) with the use of scientific equipment of the Center of Collective Use, Federal Research Center, Yakutsk Scientific Center, Siberian Branch, Russian Academy of Sciences.

References

1. Petrov NA, Efimov VM, Alekseev NN et al (2012) State and main tasks of innovative development of the oil and gas complex of the Republic of Sakha (Yakutia). Science and education 1:35–39
2. Government of the Russian Federation (2020) Order on approval of the Energy Strategy of the Russian Federation for the period up to 2035, June 9, 2020, No 1523-r, Moscow, Russia
3. Government of the Republic of Sakha (Yakutia). Ministry of Ecology, Nature Management and Forestry of the Republic of Sakha (Yakutia) (2023). State Report on the State and Environmental Protection of the Republic of Sakha (Yakutia) in 2022. Yakutsk. https://minpriroda.sakha.gov.ru/doki/doklady-o-sostojanii-okruzhajuschej-sredy. Accessed 15 Jan 2023
4. Camenzuli D, Freidman BL (2015) On-site and in situ remediation technologies applicable to petroleum hydrocarbon contaminated sites in the Antarctic and Arctic. Polar Research 34:24492. https://doi.org/10.3402/polar.v34.24492
5. Filler DM, Snape I, Barnes DL (2008) Bioremediation of petroleum hydrocarbons in cold regions. Cambridge University Press. https://doi.org/10.1017/CBO9780511535956
6. Oborin AA, Khmurchik VT, Ilarionov SA et al (2008) Oil-contaminated biocenosis. Perm, Perm State Technical University, Ural Branch of the Russian Academy of Sciences

7. Fingas M (2010) Oil spill science and technology: prevention, response, and clean up. USA. Gulf Professional Publishing
8. Shihranov OG, Glyaznetsova YuS, Erofeevskaya et al (2015) Bioremediation methods of oil-contaminated soil for the climatic conditions of the far north and the assessment of their effectiveness. Science and Technologies: Oil and Oil Products Pipeline Transportation 17(1):90–97
9. Borah D, Yadav RNS (2017) Bioremediation of petroleum-based contaminants with biosurfactant produced by a newly isolated petroleum oil degrading bacterial strain. Egyptian Journal of Petroleum 26(1):181–188. https://doi.org/10.1016/j.ejpe.2016.02.005
10. Glyaznetsova YuS, Zueva IN, Chalaya ON (2020) Experience in the remediation of oil polluted soils of the Arctic zone of Yakutia. IOP Conf. Series: Earth and Environmental Science 459. https://doi.org/10.1088/1755-1315/459/5/052010
11. Jane PK, Gupta VK, Gaur RK et al (2011) Bioremediation of petroleum oil contaminated soil and water. Researcher J Environ Toxicol 5(1):1–26. https://doi.org/10.3923/rjet.2011.1.26
12. Jovančićević B, Antić M, Pavlović I et al (2008) Transformation of petroleum saturated hydrocarbons during soil bioremediation experiments. Water Air Soil Pollut 190(1–4):299–307. https://doi.org/10.1007/s11270-007-9601-z
13. Lifshits SKh, Glyaznetsova YuS, Chalaya ON, Zueva IN (2017) Increase in remediation processes of oil-contaminated soils. Remediation J 28(1):97–104.https://doi.org/10.1002/rem.21546
14. Rogozina EA (2010) Geochemical changes in oil composition for biodegradation. Prospect and protection of mineral resources 4:63–68
15. Stallwood B, Shears J, Williams PA et al (2005) Low temperature bioremediation of oil-contaminated soil using biostimulation and bioaugmentation with a *Pseudomonas sp.* from maritime Antarctica. J Appl Microbiol 99:794–802. https://doi.org/10.1111/j.1365-2672.2005.02678.x
16. Gennadiev AN, Pikovskii YuI (2007) The maps of soil tolerance toward pollution with oil products and polycyclic aromatic hydrocarbons: methodological aspects. Eurasian Soil Science 40(1):70–81. https://doi.org/10.1134/S1064229307010115
17. Kashirtsev VA (2003). Organic geochemistry of the naphtides of the East of Siberian platform. Yakutsk, Publishing house of the Siberian Branch of the Russian Academy of Sciences
18. Gol'dberg VM, Zverev VP, Arbuzov AI et al (2001) Technogenic pollution of natural waters by hydrocarbons and its ecological consequences. Moscow
19. Brodskii ES, Butkova OL, Shelepchikov AA et al (2010) Identification of sources of oil spills in soils and sediments by gas chromatography/mass spectrometry. J Analytical Chemistry 65(14):1524–1531. https://doi.org/10.1134/S1061934810140133
20. Philp RP (1994) High-temperature gas chromatography for the analysis of fossil fuels: a review. J High Resolution Chromatography 17(6):398–406
21. Ibatullina IZ, Yakovlev AS, Semenova TA (2012) Specific features of oil biodegradation in meadow-chestnut soils of the Stavropol region (model experiment). Eurasian Soil Science 45(3):329–336. https://doi.org/10.1134/S1064229312030040
22. Cheng D, Dou L, Chen Q et al (2022) Geochemical characteristics and origins of biodegraded oils in the Bongor Basin (Chad) and their implications for petroleum exploration. Energy Exploration & Exploitation 40(2):682–700. https://doi.org/10.1177/01445987211069582
23. Yang SZ, Jin HJ, Wei Z et al (2009) Bioremediation of oil spills in cold environments: a review. Pedosphere 19:371–381. https://doi.org/10.1016/S1002-0160(09)60128-4
24. Margesin R, Schinner F (2001). Biodegradation and bioremediation of hydrocarbons in extreme environments. Appl Microbiol Biotechnol 56:650–663. https://doi.org/10.1007/s002530100701
25. Yuniati MD (2018) Bioremediation of petroleum-contaminated soil: A Review. IOP Conf. Series: Earth and Environmental Science 118:012063. https://doi.org/10.1088/1755-1315/118/1/012063
26. Fingas MF (1995). A Literature Review of the Physics and Predictive Modelling of Oil Spill Evaporation. J Hazardous Materials 42:157–175

27. Teng J, Yasufuku N, Liu Q et al (2014). Experimental evaluation and parameterization of evaporation from soil surface. Nat Hazards 73:1405–1418. https://doi.org/10.1007/s11069-014-1138-z
28. Drugov YuS, Rodin AA (2007) Environmental analyses for oil spills and oil products. Binom, Moscow
29. Wang Z, Fingas M, Page DS (1999) Oil spill identification. J Chromatography A 843:369–411. https://doi.org/10.1016/S0021-9673(99)00120-X
30. Zyakun AM, Baskunov BP, Zakharchenko VN et al (2014) Bioremediation of oil-polluted soils: using the [^{13}C]/[^{12}C] ratio to characterize microbial products of oil hydrocarbon biodegradation. Appl Biochemistry and Microbiology 50(5):483–491. https://doi.org/10.1134/S00036838 14040152
31. Kauppi S, Sinkkonen A, Romantschuk M (2011) Enhancing bioremediation of diesel-fuel-contaminated soil in a boreal climate: comparison of biostimulation and bioaugmentation. International Biodeterioration & Biodegradation 65:359–368. https://doi.org/10.1016/j.ibiod.2010.10.011
32. Thomassin-Lacroix EJM, Eriksson M, Reimer KJ et al (2002) Biostimulation and bioaugmentation for on-site treatment of weathered diesel fuel in Arctic soil. Appl Microbiol Biotechnol 59:551–556. https://doi.org/10.1007/s00253-002-1038-0

Research and Analysis of Energy Dissipation in Spiral Channel Shaft

Hao Liu and S. V. Fedorov

Abstract This research focuses on a spiral channel shaft as an example of an efficient flow energy dissipation setup for water disposal sys-tems. The energy dissipation mechanism in the shaft combines several features like the use of friction losses due to water flowing over the spiral channel and the drop of the flow into the water cushion in the tail race. The hydrodynamic properties of the flow are analyzed with computer modeling. The numerical model-ing of the two-phase gas and fluid flow in the spiral channel shaft used the Euler-Euler multiphase flow model and the k-ε turbulence model in the ANSYS FLUENT software. To assess the modeling quality, the authors developed a scaled laboratory setup of a drop shaft. The model and laboratory experiments confirm the efficiency of energy dissipation.

Keywords ANSYS FLUENT · Water disposal system · Drop shaft · Energy dissipation · Spiral channel

1 Introduction

Various drop shaft structures (cascade drops, standing pipe shafts, ogee weirs, and races) are used in water disposal system design to connect collectors at different depths [1, 2]. The flow motion is typically associated with high velocities and dynamic loads on the structural elements of the tail race. Thus, the structures used shall efficiently dissipate energy while transporting the flow over the shaft [3]. For instance, cascade shafts dissipate energy by letting the flow drop from one step to another. In spiral channel shafts, the flow swirls and presses itself against the wall due to the centrifugal force. The energy is dissipated evenly due to the friction in the channel along the entire structure height.

H. Liu (✉) · S. V. Fedorov
St. Petersburg State University of Architecture and Civil Engineering, 4, 2nd Krasnoarmeiskaya Str., St. Petersburg 190005, Russia
e-mail: a304329703@gmail.com

L. V. Radionova and D. V. Ulrikh (eds.), *Advances in Ecology and Environmental Engineering*, Springer Proceedings in Earth and Environmental Sciences,
https://doi.org/10.1007/978-3-031-64423-8_17

The spiral channel shaft can be used in water disposal systems with a large wastewater flow and significant depth of sewage collector [4]. The movement of wastewater is a complex two-phase turbulent movement. The water swirls in the channel and drags along the air and gases present under the roof of the structure. The water cushion in the dissipation basin slows down the swirling flow and drains it to the collector. This results in the significant anisotropy of local velocity and pressure in the flow [5, 6].

To optimize the spiral channel shaft design and implement specific water disposal projects, it is necessary to assess the efficiency of the structure over an extensive range of flow and drop height values. This requires a large amount of modeling experiments. In this research, we developed a drop shaft model in the ANSYS FLUENT software and tested the modeling results using a laboratory setup.

2 Research Relevance

The majority of drop shafts feature high-velocity flow dropping that causes quick and significant wear of reinforced concrete structures. The process is aggravated by the degassing of the incident flow causing the emission of hydrogen sulfide and increased gas corrosion. Thus, it is important to identify the drop shaft design that can reduce the negative impacts of the flow to increase the service life and reliability of the structure and the entire water disposal system. Currently, the number of research works on spiral channel shaft designs is low. Further research into their functional parameters shall help increase the use of these designs in the water disposal systems of large cities.

Hydraulically, the drop shaft can be divided into the head race and the tail race. In the head race, the inlet carries the wastewater to the shaft. Then the water is carried down the shaft with chutes, structural elements, pipelines, and other facilities. After that, the flow ends up in the dissipation basin of the tail race [7]. Note that the spiral channel flow is quite complex, and its parameters depend on multiple factors, including shaft dimensions, channel slope, and wastewater flow rate. Studying the patterns in the drop shaft flow and obtaining dependencies for spiral channel and dissipation basin modeling in this case requires using the Navier–Stokes equation system. The mathematical model is used because it is impossible to perform research in the field, and it can be used to analyze the hydrodynamic parameters of drop shafts of different sizes.

The mathematical model was developed in the ANSYS FLYENT finite-element analysis package. Currently, this approach is widely used in research and engineering because it helps analyze a phenomenon at the lowest cost. Numerical modeling methods easily provide a velocity and pressure profile for the wastewater flow, as well as the volume ratio of the water and air and gas mix, which are hard to obtain from hydraulic models or field tests [8, 9]. Note that the obtained modeling results have to undergo quality and quantity verification. For this purpose, we can use field observations, known calculation dependencies, and laboratory experiment results.

To test the model of the shaft proposed in this paper, we developed a geometrically accurate laboratory setup.

Comparing the model and laboratory flow parameters shall help adjust the model parameters and make sure the calculated results are the same as the real flow parameters in the spiral channel shaft. Further on, we are planning to perform series modeling for various shaft dimensions, spiral channel parameters, and water flow rates.

The modeling and laboratory experiment results shall help establish a calculation and design procedure for spiral channel shafts and improve the understanding of energy dissipation of the water flow transported over them.

Of the publications available one shall note papers [10, 11] that confirm the feasibility of using spiral channel shafts and the importance of experimental analysis of hydraulic parameters for the structure optimization.

Developing this area of knowledge is important for the engineering practices of Russia and China. Studying the functional parameters of spiral channel shafts shall have a positive effect on their usage in water disposal systems.

3 Equations

3.1 Mathematical Model

The two-phase model of water and air flow in a drop shaft can be based on equations describing the movement of each phase separately. This requires the following assumptions: (1) Both phases are viewed as continuous media that mix with one another; (2) To describe how the movement of each phase changes over time, we use average-volume mass and momentum equations; (3) The pressure and velocity in the process in question are low, thus, both phases can be deemed non-compressed, and the densities of the gas and the fluid can be deemed constant; (4) The model excludes the transfer of weight and heat between the fluid and gas phases.

Review the key equations used in modeling.

The mass and momentum conservation equations are written as follows:

$$\frac{\partial a_i}{\partial t} + \nabla g\left(a_i, \overset{>}{ui}\right) = 0 \tag{1}$$

$$\frac{\partial a_i \rho_i u_i}{\partial t} + \nabla g\left(a_i\, \rho_i\, u_i\, a_{ij}\right) = -\, a_i\, g \nabla\, p_i + \nabla g(a_i\, \tau_i) + F_i + a_i\, \rho_i\, g \tag{2}$$

where *i* is the number of the phase for the fluid phase (water) or gaseous phase (air); *j* is the spatial coordinate (x, y, z); *t* is the time in seconds; *a* is the volume ratio of the phase; *u* is the local flow velocity in m/s; ρ is the density of the medium in kg/m^3; *p* is the pressure in Pa; τ is the viscous stress tensor in Pa; F_i is the external force that affects the flow of the medium in N/m^3; *g* is the gravity acceleration in m/s. In this case, we only accounted for the gravity force $F_i = F$ affecting the two-phase flow.

The gravity force expression looks as follows.

$$F = -0.75\, a_g (1 - a_g)\, \rho_l \frac{C_D}{d_g} |u_g - u_l| (u_g - u_l) \tag{3}$$

where g is the gaseous phase, l is the liquid phase; d_g is the bubble diameter in meters; C_D is the friction coefficient determined using the Schiller-Naumann friction model depending on the movement mode:

$$C_D = \begin{cases} \frac{24(1+0.15\mathrm{Re}^{0.687})}{\mathrm{Re}} & \mathrm{Re} \le 1000 \\ 0.44 & \mathrm{Re} \le 1000 \end{cases} \tag{4}$$

where Re is the Reynolds number for the primary (dragging) phase g relative to compared to the secondary (dragged) phase i can be calculated as follows:

$$\mathrm{Re} = \frac{\rho_g |u_l - u_g| d_g}{\mu_g} \tag{5}$$

where μ_g is the dynamic viscosity of the primary (dragging) phase in Pa s.

The vortex formation in the water phase is addressed with the standard k-ε model. The k-ε model consists of two equations: the turbulent kinetic power transfer equation (k) and the dissipation equation (ε).

Equations k and ε look as follows:

$$\frac{\partial}{\partial t}(\rho \varepsilon) + \frac{\partial}{\partial x_i}(\rho k u_i) = \frac{\partial}{\partial x_i}\left[\left(\mu + \frac{\mu_i}{\sigma_k}\right)\frac{\partial k}{\partial x_j}\right] + G_k + G_b - \rho\varepsilon \tag{6}$$

$$\frac{\partial}{\partial t}(\rho \varepsilon) + \frac{\partial}{\partial x_i}(\rho \varepsilon u_i) = \frac{\partial}{\partial x_i}\left[\left(\mu + \frac{\mu_i}{\sigma_\varepsilon}\right)\frac{\partial k}{\partial x_j}\right] + G_{1\varepsilon}\frac{\varepsilon}{k}(G_k + G_{3\varepsilon} G_b) - G_{3\varepsilon}\,\rho\frac{\varepsilon^2}{k} \tag{7}$$

where G_k is the turbulent kinetic energy generated from the average velocity gradients, m^2/s^2; G_b is the kinetic energy of the buoyant force in m^2/s^2; σ_k and σ_ε are the variable factors with assumed values $\sigma_k = 1.0$ and $\sigma_\varepsilon = 1.3$; μ is the dynamic viscosity; ρ is the average density of the two-phase flow, kg/m^3; $C_{1\varepsilon}$, $C_{2\varepsilon}$, $C_{3\varepsilon}$, and C_μ are the experimental constants with assumed values $C_{1\varepsilon} = 1.44$, $C_{2\varepsilon} = 1{,}92$, $C_{3\varepsilon} = 1.2$, $C_\mu = 0.09$; μ_t is the turbulent viscosity in Pa s that can be calculated using the following equation [12, 13]:

$$\mu_t = \rho\, C_\mu \frac{k^2}{\varepsilon} \tag{8}$$

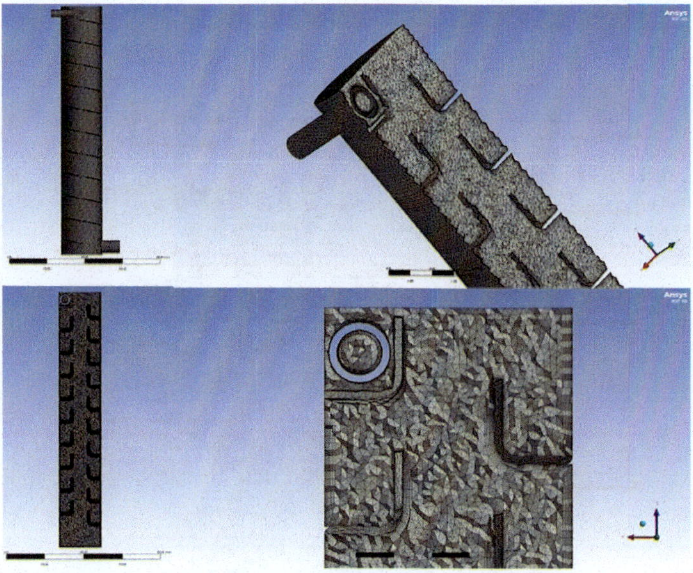

Fig. 1 Computational grid

3.2 Numerical Modeling and Model Parameter Adjustment

For tests, we developed a scaled model of a real shaft with a drop height H = 60 m, shaft diameter D = 10 m, and water flow rate Q = 6.94 m^3/s. Using the Froude number, we scaled the structure to 1:60. As a result, we obtained the following model dimensions: H_M = 1 m, D_M = 0.17 m, Q_M = 0.20 m^3/h. The inlet diameter was d_p = 0.04 m. Channel dimensions: width b_l = 0.036 m, wall height h_l = 0.02 m, channel slope i_l = 0.215. During the modeling, we solve a non-stationary problem in ANSYS FLUENT. The total flow passage time through the structure was assumed T = 200 s. To improve problem convergence, the time increment was set at Δt = 0.1 s. The computational domain of the model stipulated the presence of two media, water and air, with properties standard for 25 °C.

To improve the modeling results, we focused on the quality of the computational grid. Figure 1 shows the generated grid. Since the cavity of the shaft has a complex shape, we used an unstructured tetrahedral grid with a cell size of 0.005 m.

3.3 Laboratory Setup

The dimensions of the laboratory setup were the same as those used in modeling (Fig. 2). The drop shaft body (5) was made of transparent plexiglass and installed in a metal frame (7). The body had two pipes with flange connections for the inlet (3)

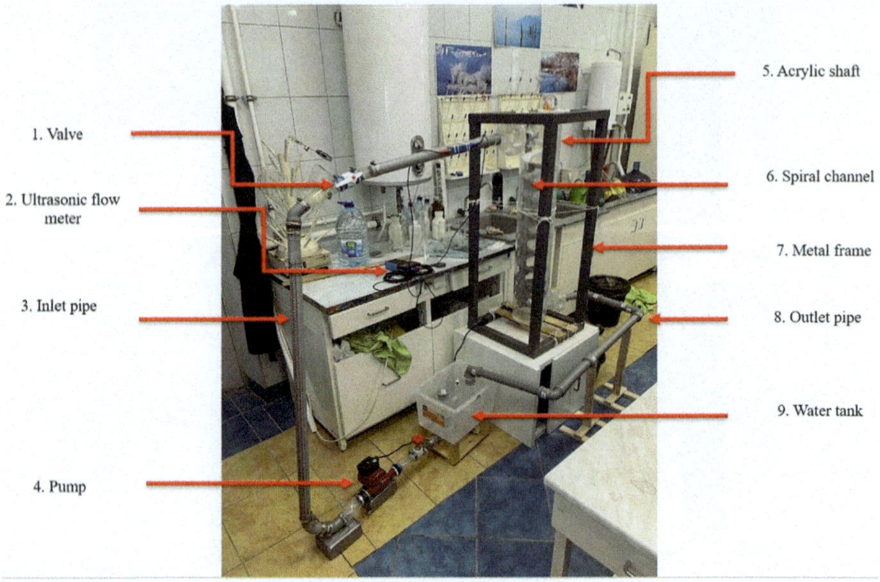

Fig. 2 The laboratory setup of the spiral channel shaft

and outlet (8) pipelines. We used 3D printing to produce the spiral channel. The 3D printing was used to produce single spiral segments that were further put together inside the body (6).

To induce the water circulation in the setup, we installed a water tank (9) from which the water was supplied to the head race of the shaft using the pump (4). Having passed through the shaft, the water returned to the tank (4). To control the circulation water flow, we used the valve (1) on the pressure pipe. To measure the flow rate and velocity of the circulating water, we used a portable ultrasound flow meter (2).

4 Results and Analysis

4.1 Discussion of Modeling Results

Figure 3 shows the volume ratio profile of shaft water in the computational domain for different periods. The calculation shows that initially, water enters the first segment of the spiral channel. Then the water drains down into other channel segments due to the gravity force. Swirling induces a centrifugal force that presses the flow against the channel wall. Then the flow reaches and fills the dissipation basin, and then leaves the computational domain over the outlet pipe. The modeling results are consistent with our expectations.

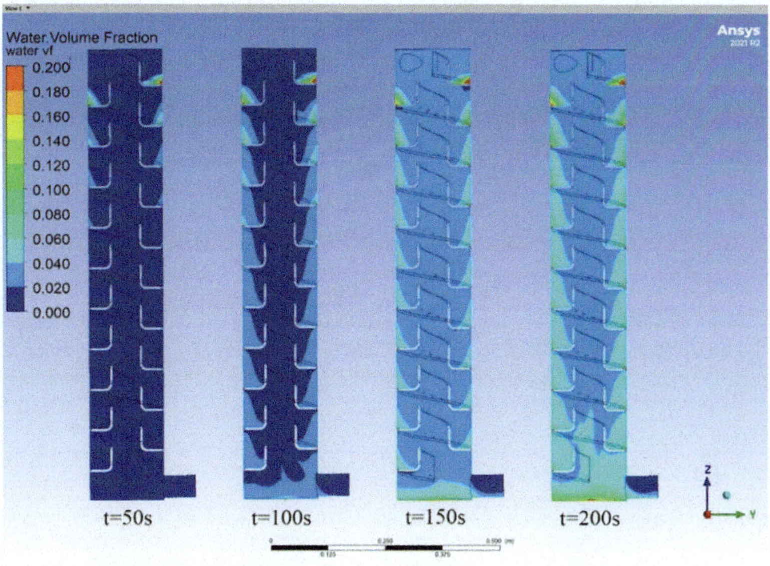

Fig. 3 The volume ratio of water

To assess the flow rate in the spiral channel, we selected three cross-sections: 1—in the first spiral channel segment along the flow, 2—in the middle of the shaft, and 3—in the last segment before the dissipation basin. Figure 4 shows the average water flow rate dependencies for the selected cross-sections.

When the movement stabilizes, the average flow depth in Cross-Section 1 is 0.855 m, in Cross-Section 2—0.415 m, and in Cross-Section 3—0.055 m.

4.2 Laboratory Setup Results

The flow rate of the water supplied to the setup was $Q = 0.196$ m^3/h. The velocity of water in the inlet pipe was $V_M = 0.05$ m/s. The movement in the setup stabilized over a period of T = 12 s after it started.

Figure 5 shows the flow of water in the spiral channel moving due to gravity. Similar to model calculations, swirling generated a centrifugal force that pressed the flow to the outer shaft wall. As the flow moves along the channel, it accelerates, and the depth of the water is reduced to 0.5 mm at the spiral channel outlet. The flow enters the dissipation basin at a tangent to its outer wall. The presence of the water cushion efficiently dissipates the energy.

To assess the velocities in the same cross-sections as in the model, we measured the wetted perimeter χ of the channel and the average water layer depth b. Table 1 shows the changes in the hydraulic parameters and the average flow velocity.

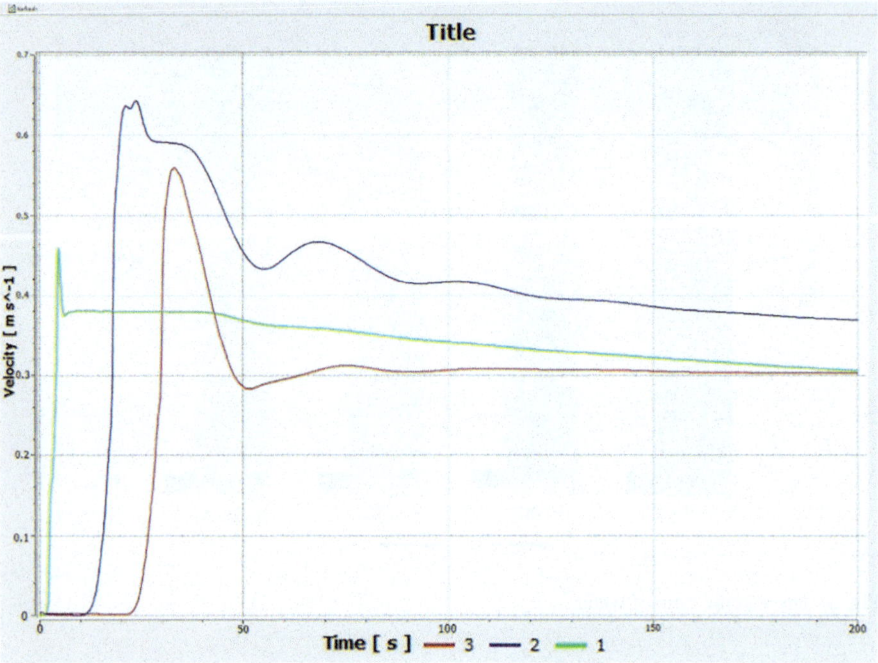

Fig. 4 Water flow rate dependencies in the selected cross-sections

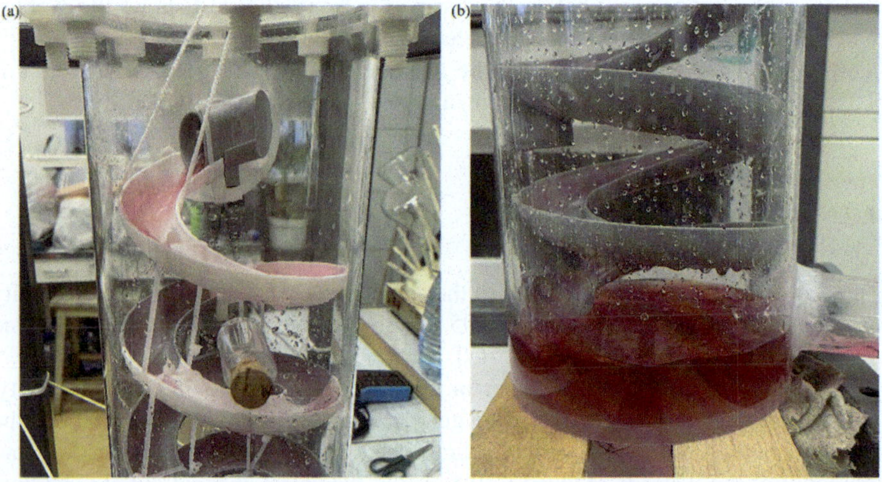

Fig. 5 Flow motion in the spiral channel: **a** head race, water flow inlet; **b** tail race, water cushion formation in the dissipation basin

Table 1 The changes in the hydraulic parameters and the average flow velocity

Cross-section	b (m)	χ (mm)	ω (m^2)	V (m/s)
1	0.007	0.035	0.000245	0.22
2	0.005	0.037	0.000185	0.29
3	0.004	0.037	0.000148	0.36

The flow velocity at the channel outlet is 1.6 times higher than the inlet velocity. When the water cushion floods the spiral channel, the velocity gradually drops to 0.05 m/s without significant fluid vortex formation in the dissipation basin.

The obtained hydraulic parameters of the flow confirm the model calculation results. The velocity values in the selected cross-sections differ from the modeled ones by 3.3%. The water flow depth differs from the model value by 2.1%.

Thus, we can claim that the developed model yields the same quality and quantity data as the experiments, and can be used for the further study of shaft operation patterns taking into account the changing of its geometry.

The spiral channel shaft allows for more efficient dissipation of the flow energy in the tail race compared to other drop designs. The transported water moved along the spiral channel in one solid flow. Observations showed the absence of breaks, splits, or intensive vortex formation. No significant turbulence or hydraulic jumps have been observed in the dissipation basin.

These results justify the further application of this structure in engineering practices and further research in this area.

5 Conclusion

This article reviewed the potential usage of a spiral channel drop shaft in water disposal systems. To analyze the operating patterns and obtain calculation dependencies, we suggested a mathematical model based on the Navier–Stokes equation system solved using the finite-element analysis in the ANSYS FLUENT software. The quality and quantity parameters of the model were verified through laboratory experimentation with a scaled drop shaft setup. The difference between the model and experimental velocities remained between 3.3 and 20%, while the difference in the channel fill rate values was between 2.1 and 9.8%. The results obtained justify the use of this mathematical model for further research on drop shafts with different dimensions.

The results of laboratory setup experiments show that it can efficiently dissipate the flow energy in a spiral channel drop shaft.

References

1. He Zhenjun, Wang Bin, Yang Yu et al (2017) Research and application progress of shafts in municipal drainage systems. J. China Water Supply and Drainage 33(10):49–53
2. Men Xuan, Li Dong, Zhang Jie (2015) Construction status of deep tunnel drainage systems at home and abroad and its enlightenment. J. Hebei Industrial Science and Technology 32(05):438–442
3. Ni Hangen (2000) Efficient Energy Dissipation Engineering. Dalian University of Technology Press, October 2000
4. Dong Xinglin, Guo Jun, Xiao Bailin et al (2000) Design research on whirlpool shaft type spillway tunnel with high water head and large discharge. Journal of Hydraulic Engineering 11:27–33
5. Dong Xinglin, Yu Minmin, Wu Zengmou et al (2002) Research and application of diversion tunnel reconstruction of cyclonic spillway tunnel. J Hydropower 2:27–29
6. Chen Yan (2016) Analysis and reference on comprehensive rainwater management in Japanese cities. J China Water Supply and Drainage 32(10):42–47
7. Dong Yulian, Liu Yakun, Sun Hongliang, Zhou Le (2013) Research on composite energy dissipation of multi-stage water drop and underflow in curved spillway. Journal of Water Conservancy and Construction Engineering 11(05):1–4
8. Chanson H (1999) Energy dissipation and drop structures in ancient times: the Roman dropshafts. Water 99 Joint Congress, pp 987–992
9. Chanson H (2004) Hydraulics of rectangular dropshafts. Journal of Irrigation and Drainage Engineering 130(6):523–529
10. Chanson H (2004) Understanding air-water mass transfer in rectangular dropshafts. Journal of Environmental Engineering and Science 3(5):319–330
11. Khatsuria RM (2004) Hydraulics of spillways and energy dissipators. Marcel Dekker Inc., NewYork, USA, p 680
12. Li Er, Zeng Xiangying, Zou Huijun et al (2023) Selection of inflow shafts for Wuhan Dadong Lake Deep Tunnel based on numerical simulation. China Water Supply and Drainage 39(15):108–114. https://doi.org/10.19853/j.zgjsps.1000-4602.2023.15.016
13. Liu Changyong (2022) Numerical simulation of shaft energy dissipation and spillway tunnel based on CFD. Water Conservancy Science and Cold Area Engineering 5(08):10–13

Photocatalytic Degradation of Methylene Blue Dye Using H_2O_2 on Ceramic Membrane Supported by Vanadium-Containing Layer

Hein Myat Lwin and O. V. Yarovaya

Abstract This article focuses on the photocatalytic degradation of methylene blue dye over ceramic membrane based on α-Al_2O_3 with a vanadium-containing layer, using hydrogen peroxide as an oxidizer. The V_2O_5 catalytic layer was applied to the external surface of the ceramic membrane using the sol–gel method. The applied layer was subjected to thermal treatment at 700 °C, and their structure and morphology were analyzed. The supported vanadium layer was characterized using XRD, ASAP, and oxidimetric titration methods. The process of catalytic degradation of methylene blue was studied using different masses of the V_2O_5 catalyst applied to the membrane surfaces, and the reaction rate constant was calculated. The results show that the mass of the catalyst applied has a positive effect on the catalytic degradation of methylene blue.

Keywords Ceramic membranes · Vanadium-containing layer · Photocatalytic degradation · Methylene blue

1 Introduction

Methylene blue is widely used in microscopy, histology, and microbiology for its cell and tissue staining capabilities. It also has industrial applications in textile dyeing, printing, and papermaking. However, the release of methylene blue into the environment through industrial effluent has become a significant concern due to its negative impacts on aquatic organisms and human health. Consequently, there is an urgent need for environmental control measures [1, 2].

Traditional treatment methods, such as adsorption and coagulation, often fail to effectively remove methylene blue from water solutions. Therefore, there is a pressing requirement for the development of efficient and sustainable approaches to degrade

H. M. Lwin (✉) · O. V. Yarovaya
D. I. Mendeleev University of Chemical Technology of Russia, 9, Miusskaya Square St, Moscow 125047, Russia
e-mail: heinmyatlwin2468@gmail.com

L. V. Radionova and D. V. Ulrikh (eds.), *Advances in Ecology and Environmental Engineering*, Springer Proceedings in Earth and Environmental Sciences,
https://doi.org/10.1007/978-3-031-64423-8_18

methylene blue contaminants. Catalytic degradation is considered one of the most promising methods due to its high efficiency, cost-effectiveness, and minimal environmental impact [2]. Supported catalysts are commonly used to enhance catalytic decomposition efficiency by acting as adsorbents for heavy metal removal. These catalysts provide a large surface area and protect active particles, preventing aggregation and deactivation [3]. To further improve decomposition efficiency, oxidizing agents, such as hydrogen peroxide and ozone, are employed. These agents react with the catalyst surface, producing highly reactive oxygen species that efficiently oxidize organic pollutants, breaking them down into simpler and less harmful compounds. Oxidizing agents also assist in regenerating active sites on the catalyst, thereby improving overall efficiency and extending the catalyst's lifespan [4].

The concept of catalyst support refers to the material or structure that provides a surface area for catalytic reactions. Various support materials, including activated carbon, zeolites, metal oxides, and carbon nanotubes, are used. Ceramic membranes have emerged as promising support materials due to their unique properties. Ceramic membranes consist of inorganic materials, typically oxides, with a highly porous structure, exceptional temperature stability, chemical stability, and good mechanical strength [5].

Vanadium pentoxide (V_2O_5), one of the metal oxides, is commonly n-type with a band gap energy of 2.80 eV. It has been widely utilized in various applications such as sensors, actuators, lithium-ion electrode materials, catalysts, and UV-driven reversible switches [6]. In recent studies, researchers have successfully synthesized V_2O_5 crystallite as a photocatalyst. In their work, they demonstrated that hollow V_2O_5 microspheres exhibited remarkable photoactivity in the degradation of rhodamine B under UV light exposure [7]. In another study [8, 9], V_2O_5/TiO_2 and Graphene-V_2O_5 nanocomposites were prepared and showed photocatalytic activities in methylene blue degradation. Therefore, the preparation of a ceramic membrane with a supported layer of V_2O_5 become a reliable solution for the catalytic degradation of methylene blue.

2 Experimental Procedures

2.1 Materials

The following reagents were used in the work: vanadium (V) oxide, "analytical grade"; hydrogen peroxide H_2O_2, "special purity grade"; sulfuric acid H_2SO_4, "chemically pure grade"; sodium hydroxide NaOH, "analytical grade", phenylanthranilic acid (indicator), "analytical grade" and methylene blue (blue) "analytical grade". The carriers used were porous tubular ceramic membranes based on α-Al_2O_3 (85%) with an internal diameter of 6.5 mm, a length of 50 mm with a wall thickness of 1.5 mm, obtained using technology developed at the Department of Chemical

Table 1 Main characteristics of porous tubular ceramic membranes based on α-Al_2O_3

No.	Characteristics	Value
1	Diameter, mm	6.5
2	Wall thickness, mm	1.5
3	Pore radius, μm	0.8
4	Open porosity, %	45 ± 3
5 6	Most probable particle size, μm	15
	Specific surface area, m^2/g	0.3
7	Electro kinetic potential of particles (pH 7), mV	+10 ± 5

Technology of Ceramics and Refractories of the MUCTR University. Their detailed characteristics are given in Table 1.

2.2 *Preparation of Membranes with Catalytically Active Layer of V_2O_5*

Sol–gel method was selected for preparing membrane with catalytic supported layer of V_2O_5. To obtain a V_2O_5 sol (0.5 mass percent), the thermolysis of V_2O_5 powders in hydrogen peroxide solutions was carried out. A solution of hydrogen peroxide (H_2O_2) was added to a round-bottom flask containing V_2O_5 powder 0.5 g in the mole ratio [1]:[30] and stirred until a homogeneous yellow mass was formed. Then the required amount of water was added and heated to boil. After boiling for 5 min, the color changes from yellow to orange and after 3 min turns into dark red. Basic colloidal chemical properties of the obtained $V_2O_5.nH_2O$ sol are presented in Table 2.

The supported layer was obtained from synthesized $V_2O_5.nH_2O$ sols (0.5 wt. %) by filtration the dispersion medium through the outer layer of ceramic membrane using a laboratory vacuum pump. Schematically, the process of producing membranes with a supported layer was as follow: First, the tubular membrane was closed at one end with a rubber cap and the other end of the membrane tube was left open, then the open end of the tube was connected to a plastic hose. Next, one

Table 2 Basic colloidal chemical properties of $V_2O_5.nH_2O$ sol

No.	Properties	Value
1	Concentration, %wt	0.5
2	Electrophoretic mobility, m^2/Vs	-1.38×10^{-8}
3	Electric potential, mV	-19 ± 1
4	Ionic strength, mol/m^3	3.178×10^{-2}
5	pH	2.4 ± 0.2

end of the tubular substrate was sealed, and a vacuum pump was connected to the plastic hose through a Bunsen flask. The substrate was immersed vertically in the sol and a fixed volume of sol was passed through the substrate under the action of a vacuum pump. In this case, partial fixation of particles occurs on the surface of aluminum oxide grains forming the layer on the outer layer of ceramic membrane. After drying the applied layer, the sample was heat treated at 700 °C for 60 min, which is the most suitable thermal temperature that preserves the layers thermally and chemically better than other treatment temperatures [10]. During the heat treatment process, sample was kept at 200 °C for 30 min to remove water adsorbed on the surface. According to procedure above, the outer surface of ceramic membrane at different length was coated with V_2O_5 layer with the same theoretically calculated thickness layers of 7 μm as this value of thickness is optical for obtain the homogeneous supported layer. The theoretical mass of particles and the thickness of the resulting layers were calculated in according to literature [11].

2.3 *Characterization*

Morphology of the supported layer was analyzed by Scanning Electron Microscope (SEM) JEOL 1610LV (JEOL, Japan). The crystallographic structure of the samples was established using powder X-ray diffraction analysis (XRD, with Cu-Ka radiation). Absorption spectra of aqueous solutions methylene blue were recorded on a brand spectrophotometer LEKI SS2110UV in plastic or quartz cuvettes with an absorbing layer thickness of 10 mm in the wavelength range 500–800 nm. Surface properties and pore structure of the supported V_2O_5 nanostructures were studied by ASAP 2020 MP (Micrometrics Instrument Corp., USA).

2.4 *Determination of the Mass of the Supported Layer*

To determine the mass of the layer supported on the surface of the ceramic membrane, oxidimetric titration of VO_4^- ions was used. To do this, the membrane was immersed in 50 ml of a 1 M NaOH solution; after the supported layer dissolved, the membrane was removed from the solution and determined vanadium (V), by titrating using Mohr's salt $(NH_4)_2Fe(SO_4)_2$. Previously, 5 M sulfuric acid was added to the test solution in a volume ratio of 1:1. Phenylanthranilic acid was used as an indicator. The equivalence point was determined when the solution transitions from violet to green [12]. In addition, to wash the supported layer obtained after heat treatment, membranes were boiled in NaOH using a refrigerator. The reaction that occurs during the reaction is shown in the following Eq. 1 [12].

$$2H_3VO_4 + 2(NH_4)_2Fe(SO_4)_2 + 3H_2SO_4 \rightarrow V_2O_2(SO_4)_2$$

$$+ Fe_2(SO_4)_3 + 2(NH_4)_2SO_4 + 6H_2O \quad (1)$$

The vanadium (V) concentration and V_2O_5 content are calculated using formulas (3) and (4).

$$V_{average} = \frac{V_1 + V_2 + V_3}{3} \quad (2)$$

$$Vanadium(V)concentration, mole/liter = \frac{V_{average} * N}{V_{analysis} * 2} \quad (3)$$

$$V_2O_5, g/liter = \frac{V_{average} * N * 91}{V_{analysis}} \quad (4)$$

where: $V_{average}$—The average volumes of Mohr's salt used for titration, ml; V_1, V_2, V_3—Volumes of Mohr's salt used for titration of each of the three samples, ml; N—Normality of Mohr's salt solution, mole/liter; $V_{analysis}$—Volume of the analyzed solution (10 ml); 91—Equivalent mass of vanadium oxide (1/2 V_2O_5).

2.5 *Catalytic Degradation of Methylene Blue*

The degradation of methylene blue was carried out in a batch reactor. The membranes were immersed in a dye solution; hydrogen peroxide was added and stirred until the solution depreciated. The initial concentration of methylene blue ($C_{16}H_{18}ClN_3S$) is—0.018 g/liter (0.0075 M) and the mole ratio of dye to hydrogen peroxide is 1:15. UV lamp (8-Watt, 200 Volt, length 30 cm) is used as the source of UV light. A simple of the solution was taken every 30 min and the absorption spectrum was analyzed on a spectrophotometer. The degree of decolorization was calculated at maximum wavelength of 665 nm using the Eq. 5.

$$D = \frac{(A_0 - A_t)}{A_0} * 100 \quad (5)$$

where: A_o = Optical density at time $t = 0$ min, A_t = Optical density at time t min.

3 Results and Discussion

3.1 Characterization of V_2O_5 Supported Layer

To confirm the chemical composition of the supported layer on the ceramic membrane, the sol was dried at room temperature on a glass plate to form a xerogel. Thermal treatment of xerogel was also carried out under the same thermal procedure conditions for the membrane with supported layer. Their crystallinity was evident from the X-ray diffraction pattern shown in Fig. 1. It has been proven that by X-ray phase analysis that the deposited xerogel layer before thermal treatment is $V_2O_5.1.6H_2O$ (Fig. 1a). Its diffraction peaks in Fig. 1a were in good agreement with the standard X-ray diffraction pattern of CuO (JCPDS card no. 40–1296). The peaks observed at $2\theta = 22.7°$, $25.54°$, $30.49°$, $38.50°$, $47.07°$ and $50.55°$were assigned to Bragg reflections (003), (001), (004), (005), (006), (007) respectively. At a heat treatment temperature of 700 °C, the oxidation state of V_2O_5 does not change according to the V–O phase diagram [13]. Its diffraction peaks in Fig. 1b were in good agreement with the standard XRD pattern of CuO (JCPDS PDF card no. 41 1426). Peaks observed at $2\theta = 15.349°$; 20.26; 21.72; 26.172; 31.0; 32.364; 33.294; 34.28; 42.0; 45.45; 47.322; 47.84; 52.51; 58.0; 61.07, 62.07; 75.97 and 74.5° were attributed to the Bragg reflections (200), (001), (101), (110), (301), (011), (111), (310), (102), (411), (600), (302), (402), (221), (321), (710), (621), (313) respectively.

To assess the morphology of the deposited layer, the substrate surfaces were analyzed using SEM micrographs and the results are shown in Fig. 2. Before heat treatment, it can be seen that the $V_2O_5.1.6H_2O$ layers are represented by overlapping nanosheets with a thickness of about 25 nm without infiltration of the layer into the pores of the membrane (Fig. 2a, b). At the thermal temperature of 700 ºC, the layer penetrates the pores of the membrane, and it is found that the two main crystal structures of the vanadium compound are represented by nanorods with the size of

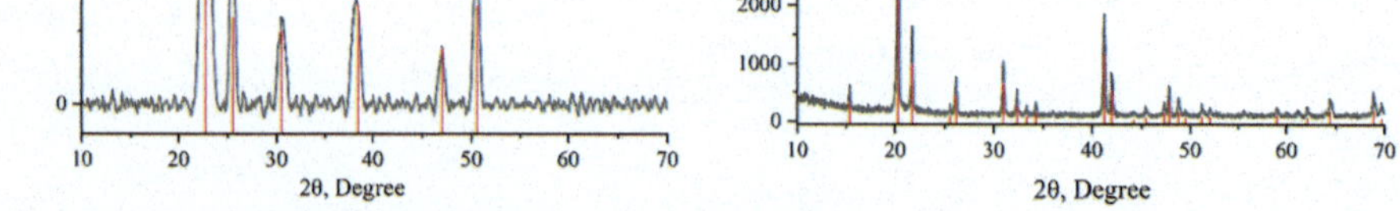

Fig. 1 X-ray diffraction patterns of **a** $V_2O_5.1.6H_2O$ xerogel obtained after drying the sol at room temperature; **b** α-V_2O_5 powder obtained after heat treatment at 700 °C for one hour

0.5 μm * 50 nm and its aggregates, and micro plates with the size of 0.5 μm (Fig. 2d, e). Knowing from X-ray diffraction analysis data and from the V–O phase diagram [13] that α-V_2O_5 is a thermodynamically stable phase, the change in morphology at 700 ºC could be related to the melting point since its melting point of V_2O_5 is at a temperature of about 670 ºC [14] and may also be related change to $AlVO_4$ because according to the V_2O_5-Al_2O_3 system, $AlVO_4$ can arise in the reaction of V_2O_5 and Al_2O_3 [15]. To determine the mass of the applied layer after heat treatment, an analysis was carried out according to Sect. 2.4, and the results of the mass analysis according to the length of the membrane that was supported by vanadium layer are shown in Table 3.

Another important factor is the specific surface area of the resulting supported layers. The N_2 gas adsorption/ desorption isotherm of $V_2O_5.1.6H_2O$ xerogel before thermal treatment and isotherm of V_2O_5 nanoparticles after thermal treatment at 700 °C were also studied. These isotherms (Fig. 3a and b) showed a type IV curve with a hysteresis loop of H3-type for both V_2O_5 nanostructures based on the IUPAC classification. The hysteresis loop above $P/P0 = 0.4$ suggests the presence of mesoporous structures in the pod shaped V_2O_5 nanostructures. As displayed in the BHJ pore size distribution curve (Fig. 4a and b), most of the pores had size of between 10 and 50 nm with a pore size distribution range of ~ 2–100 nm. BET surface area of the $V_2O_5.1.6H_2O$ xerogel is 0.6861 m^2g^{-1} and V_2O_5 nanoparticles (after thermal treatment at 700 °C) is 0.4237 m^2g^{-1}, respectively. Showed that although specific surface area and total pore volume of nanoparticles after thermal treatment is decreased lower average pore diameter.

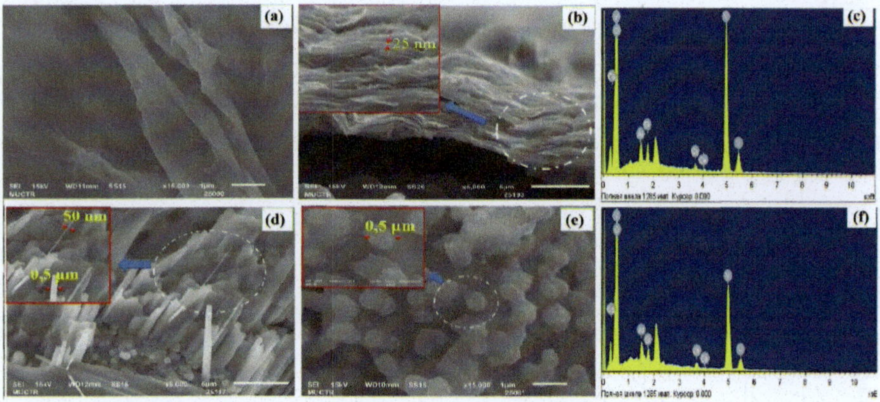

Fig. 2 SEM micrographs of the **a** surface, **b** surface cleavage and **c** EDX spectrum of the supported layer without thermal treatment; and **d, e** micrographs of the surface and **f** EDX spectrum of the supported layer after thermal treatment at 700 °C

Table 3 Effect of catalyst mass loaded on change in methylene blue conversion and reaction rate constant during the catalytic reaction

No.	Catalyst	Length of membrane, cm	Total mass of the layer in terms of V_2O_5, mg	Conversion after 3 h, %	Rate constant	
					$min^{-1} \times 10^3$	min^{-1}/ mg
1	$1V + H_2O_2$	4.1	19.1	41.75	3.2	0.16 ± 0.01
2	$2V + H_2O_2$	8.2	38.2	66.58	6.2	
3	$3V + H_2O_2$	12.3	57.3	80.72	8.8	
4	$1V + UV + H_2O_2$	4.1	19.1	81.82	9.6	0.5
5	$2V + UV + H_2O_2$	8.2	38.2	86.40	10.7	0.281
6	$3V + UV + H_2O_2$	12.3	57.3	90.29	12.3	0.215
7	H_2O_2	–		21.99	1.4	

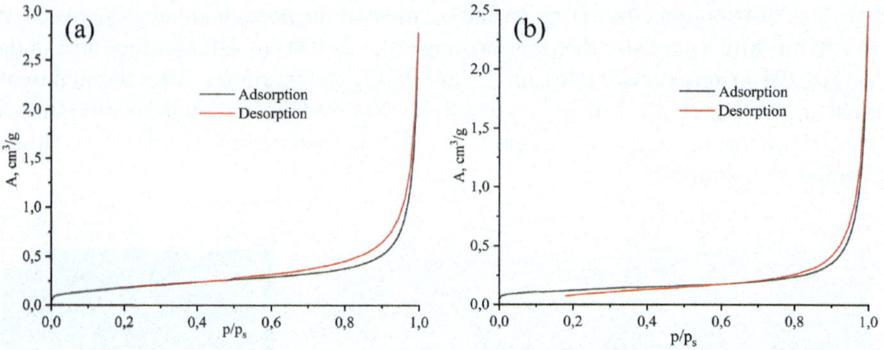

Fig. 3 N_2 adsorption–desorption isotherm of **a** $V_2O_5.1.6H_2O$ xerogel and **b** V_2O_5 powder (after thermal treatment at 700 °C)

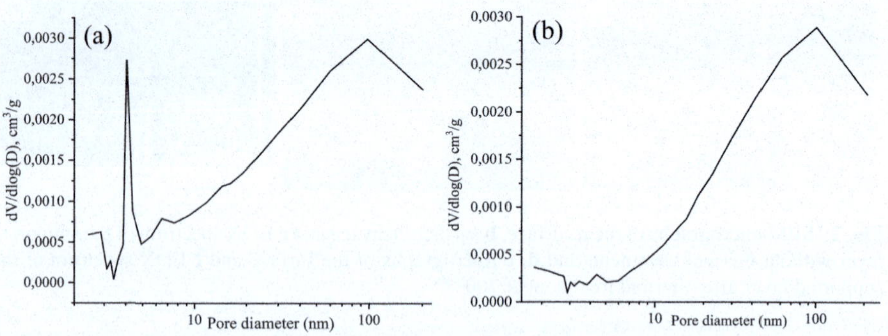

Fig. 4 BJH pore size distribution for **a** $V_2O_5.1.6H_2O$ xerogel and **b** V_2O_5 powder (after heat treatment at 700 °C)

3.2 Catalytic Degradation of Methylene Blue

Photocatalytic degradation of methylene blue is carried out at different masses of loaded catalyst (V_1-V_3) using the oxidizing agent H_2O_2. Also carried out the catalytic decomposition of methylene blue without UV radiation. The reaction conditions are shown in Sect. 2.5. As shown in Fig. 5a, the absorbance of methylene blue decreases with time, and the maximum adsorption wavelength of methylene blue is observed at 665 nm. So, to compare the effect of mass loading and determine the time dependence of optical density D, the value of D at 665 nm was chosen. According to Figs. 5b and 6a, it is observed that increasing the mass of the catalyst has a positive effect on the reaction rate, and with the help of UV irradiation, the reaction is clearly increased.

Based on Fig. 5b, the degree of discolorization was calculated using Eq. 5, and a graph of the degree of discolorization versus time was plotted as shown in Fig. 6a. The reaction kinetic was calculated using the $\ln A_t/A_0$—time graph. The conversion of methylene blue after reaction for 3 h and the kinetic rate of the reaction results

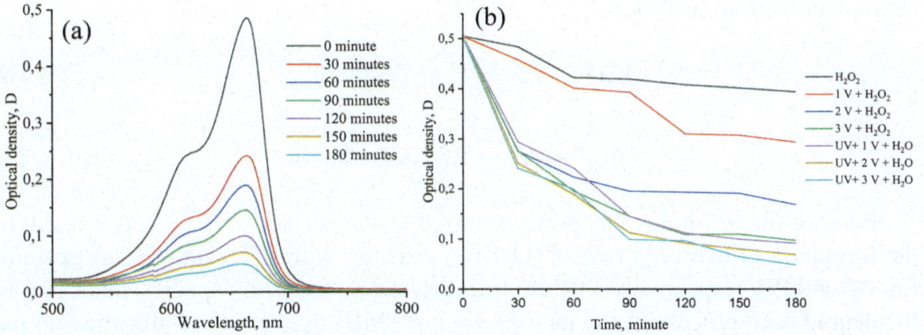

Fig. 5 **a** Absorption spectra of methylene blue as a function of time during catalytic degradation (3 V + UV light + H_2O_2) and **b** Comparison of the dependence of the optical density of methylene blue as a function of time at different catalyst loading ($\lambda = 665$ nm)

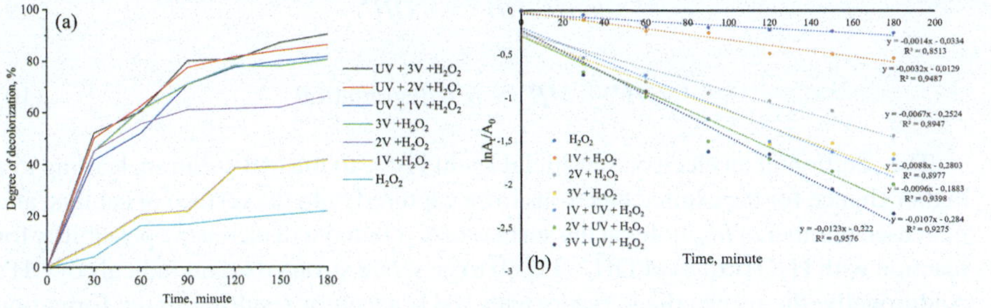

Fig. 6 **a** Comparison of the dependence of the degree of decolorization of methylene blue as a function of time at different catalyst loading ($\lambda = 665$ nm) and **b** Dependence of the logarithm of the optical density of the dye on time during catalytic degradation process

calculated from the first order rate equation are shown in Table 3. Vanadium 3 V loaded membranes removed 97% of methylene blue in 3 h of photocatalytic reaction.

4 Reaction Mechanism

4.1 Photocatalytic Degradation Mechanism ($UV + H_2O_2 + V_2O_5$)

When a photon energy is larger than the band gap ($h_v > E_g$), the light is absorbed by V_2O_5, and the conduction band electrons (e_{CB}^-)and valence band holes (h_{VB}^+)are produced because of electron excitation [16]. According to literature, [17–19] mechanism for photocatalytic degradation reaction can be concluded as following. First, a number of the generated e_{CB}^- and h_{VB}^+ pairs are engaged in the volume recombination process and some others are transferred to the surface of catalyst to contribute in the chemical reactions as follow:

$$V_2O_5 + h_v \rightarrow V_2O_5(e_{CB}^- + h_{VB}^+) \tag{6}$$

$$h_{VB}^+ + MB \rightarrow MB\,\text{degradation} \tag{7}$$

Some of the h_{VB}^+ holes that are transferred to the surface of the catalyst result in the formation of hydroxyl radical ($OH^•$) by reacting with H_2O and OH^- as given in Eq. (8) and (9), respectively. $OH^•$ radical, as a strong oxidizing species, takes part in the degradation process of the methylene blue (MB) dye, which is absorbed on the surface of catalysts, based on the reaction of Eq. (10).

$$h_{VB}^+ + H_2O \rightarrow H^+ + OH^• \tag{8}$$

$$h_{VB}^+ + OH^- \rightarrow OH^• \tag{9}$$

$$OH^• + MB \rightarrow MB\,\text{degradation} \tag{10}$$

The addition of an electron sacrificial agent H_2O_2 to the MB solution, leading to a higher chance for the excited electrons to be captured onto the surface of catalyst and increased number of h_{VB}^+ holes at the surface of V_2O_5 and will increase a possibility for reaction with H_2O (Eq. 8) or OH^- (Eq. 9) to generate strong oxidant radicals of $OH^•$. Additionally, the received e_{CB}^- reacts with H_2O_2 acceptor resulting in the formation of $OH^•$ oxidant, which improves the photocatalytic activity of the V_2O_5 (Eq. 11). Moreover, H_2O_2 can directly produce hydroxyl radicals by absorbing a photon energy according to (Eq. 12). Moreover, by reacting with $OH^•$ radicals (Eq. (13)) and h_{VB}^+

holes (Eq. (14)) to produce $OOH^{\bullet}$ radicals.

$$e^-_{CB} + H_2O_2 \rightarrow OH^{\bullet} + OH^- \quad (11)$$

$$H_2O_2 + hv \rightarrow 2OH^{\bullet} \quad (12)$$

$$H_2O_2 + OH^{\bullet} \rightarrow H_2O + OOH^{\bullet} \quad (13)$$

$$H_2O_2 + hv^+_{VB} \rightarrow H^{\bullet} + OOH^{\bullet} \quad (14)$$

Moreover, the oxygen molecules (O_2) adsorbed onto the surface of catalyst, play a crucial role in the improvement of the photocatalytic efficiency by generation of superoxide ($O_2^{-\bullet}$) radicals (see Eq. (15)). Consequently, ($O_2^{-\bullet}$) species react with hydrogen ions (H^+) from water splitting, based on Eq. (16), and an available electron sacrificial agent such as H_2O_2, based on Eq. (17), to form $OOH^{\bullet}$ and $OH^{\bullet}$ radicals.

$$e^-_{CB} + O_2 \rightarrow O_2^{-\bullet} \quad (15)$$

$$O_2^{-\bullet} + H^+ \rightarrow OOH^{\bullet} \quad (16)$$

$$O_2^{-\bullet} + H_2O_2 \rightarrow OH^{\bullet} + OH^- + O_2 \quad (17)$$

Finally, these strong oxidizing radicals are essential for degradation of the MB dye because radicals can convert organic compounds into minerals. These organic compounds are degraded by losing their electrons and eventually releasing CO_2 and H_2O based on Eq. (18):

$$MB + (OOH^{\bullet}/O_2^{-\bullet}/OH^{\bullet}) \rightarrow H_2O + CO_2 \text{ (degradation)} \quad (18)$$

4.2 *Catalytic Degradation Mechanism ($V_2O_5 + H_2O_2$)*

The reaction kinetic between methylene blue, H_2O_2 and V_2O_5/Al_2O_3 catalyst was reviewed according to literature [20, 21]. H_2O_2 molecules interact with the V_2O_5 surface, which activates the peroxide by breaking the O–O bond. This leads to the generation of highly reactive hydroxyl radicals ($^{\bullet}OH$) on the catalyst surface. First, The H_2O_2 was first dissociated into hydrogen ions (H^+) and hydroperoxide (OH^-) which is the conjugate base. The OH^- can interact with a non-dissociated H_2O_2 molecule, to release oxygen, water and hydroxyl radicals ($OH^{\bullet}$). The hydroxyl radicals play a crucial role in the degradation of methylene blue. The positively charged

MB molecules can adsorb on the surface of V_2O_5 since the catalyst surface is negatively charged. The generated hydroxyl radicals attack the adsorbed methylene blue molecules, which leads to the cleavage of the dye molecules into smaller fragments.

$$H_2O_2 \leftrightarrow H^+ + OH^- \tag{19}$$

$$H_2O_2 + OH^- \rightarrow H_2O + O_2 + OH^\bullet \tag{20}$$

$$H_2O_2 + OH^\bullet \rightarrow H_2O + OOH^\bullet \tag{21}$$

$$MB + (OH^\bullet/OOH^\bullet) \rightarrow \text{degraded products} \tag{22}$$

5 Conclusion

In conclusion, the photocatalytic degradation of methylene blue on a ceramic membrane supported with a vanadium-containing layer was successfully carried out and it was found that the mass of the catalyst loaded has a positive effect on the reaction rate. In addition, noticed that a simple catalytic degradation process without UV radiation has lower efficiency than the photocatalytic degradation process.

References

1. Idrees K, Khalid S (2022) Review on methylene blue: its properties, uses, toxicity and photo degradation. J Water 14:1–30. https://doi.org/10.3390/w14020242
2. Peter O, Timothy OA (2022) Methylene blue dye: toxicity and potential elimination technology from wastewater. Results in Engineering 16:1–17. https://doi.org/10.1016/j.rineng.2022.100678
3. Bailie JE, Hutchings GJ (2001) Support catalysts. Encyclopedia of Materials: Science and Technology 2:8986–8990. https://doi.org/10.1016/B0-08-043152-6/01620-X
4. William HG, Joon WK (2008) The chemistry of water treatment processes involving ozone, hydrogen peroxide and ultraviolet radiation. Ozone: Science and Engineering 9:335–352. https://doi.org/10.1080/01919518708552148
5. Sumayyah MA Othman (2019) Catalysts of the preparation and industrial importance of catalysis and catalyst deactivation. International Journal of Advanced Chemistry Research 1:23–27
6. Thi Dieu HN, Vo Khoung D (2021) Electronic and optical excitation properties of vanadium pentoxide V_2O_5. Computational Materials Science 198:1–8. https://doi.org/10.1016/j.commatsci.2021.1106
7. Hai-Long F, Hui-Jing Z (2008) Synthesis of hollow V_2O_5 microspheres and application to photocatalysis. Solid-state science 10:1276–1284. https://doi.org/10.1016/j.solidstatesciences.2007.12.026

8. Yang XH, Fu HT (2006) Synthesis of V_2O_5@TiO_2 core–shell hybrid composites for sunlight degradation of methylene blue. RSC Adv 6:34103–34109. https://doi.org/10.1039/C6RA03376H
9. Mahalingam S, Ali A (2015) Enhanced photocatalytic performance of Graphene -V_2O_5 nanocomposite in the degradation of methylene blue dye under direct sunlight. ACS Appl. Mater. Interfaces 7:1405–14911. https://doi.org/10.1021/acsami.5b02715
10. Khein ML, Yarovaya OV (2023) Analysis of the effect of pH on the chemical stability of supported catalysts based on V_2O_5. Scientific measurement, pp 128–130. https://doi.org/10.46916/04072023-4-978-5-00215-045-8 (in Rus.)
11. Khein ML, Yarovaya OV (2022) Development of a method for producing V_2O_5 dispersions to obtain deposited layers on the outer surface of Al_2O_3-based tubular membranes. Success in chemistry and chemical engineering. tech. 13:132–134. (in Rus.)
12. Bakhvalov AV (2004) Adaptation of the method for determining vanadium (V) compounds for the analysis of alkaline electrolyte in electrolysis plants. Problems of modern science and education 6:17–21 (in Rus.)
13. Zhanmin C, Shuai L, Wei X (2015) Critical evaluation and thermodynamic optimization of the V–O system / Computer Coupling of Phase Diagrams and Thermochemistry. Computer Coupling of Phase Diagrams and Thermochemistry. 51:241–251
14. Govindarajan D, Gopalakrishnan R (2019) Supercapacitor behavior and characterization of RGO anchored V_2O_5 nanorods. Journal of Materials Science: Materials in Electronics 30:16124–16155
15. Xin Z, Zhenhua X, Limin H (2016) Hot corrosion behavior of $LaTi_2Al_9O_{19}$ ceramic exposed to vanadium oxide at temperatures of 700–950°C in air. Corrosion Science 104:310–318. https://doi.org/10.1016/j.corsci.2015.12.024
16. Boshra GS, Ali KZ (2020) Doped ZnO nanostructures with selected elements - structural, morphology and optical properties: A review. Creamic International. 45:5507–5520. https://doi.org/10.1016/j.ceramint.2019.11.051
17. Ahad M, Mohammad Khoshghadam-P (2020) Synergetic photocatalytic effect of high purity ZnO pod shaped nanostructures with H_2O_2 on methylene blue dye degradation. Journal of Alloys and Compounds 845:1–36. https://doi.org/10.1016/j.jallcom.2020.156333
18. Muhammad MS, Naveed AS (2020) Preparation and characterization of Vanadium pentoxide (V_2O_5) for photocatalytic degradation of monoazo and diazo dyes. Surfaces and Interfaces 19:1–9. https://doi.org/10.1016/j.surfin.2020.100502
19. Hanadi AM, Seroor AK, Mohammed IB (2021) Photolysis of methylene blue dye using an advanced oxidation process (UV/H_2O_2). Materials Science and Engineering 25:59–67. https://doi.org/10.31272/jeasd.25.1.5
20. Abdelhamid M El-Sawy, Ali H Gemeay (2021) Catalytic degradation of methylene blue in aqueous solution by H_2O_2 and SiO_2-NH_2-Cu(II)@SiO_2 nanoparticles as catalyst. Journal of Molecular Liquids 341:1–15. https://doi.org/10.1016/j.molliq.2021.117422
21. Lukasz W, Maria Z (2017) Insight into pathways of methylene blue degradation with H_2O_2 over mono and bimetallic Nb, Zn oxides. Applied Catalysis B: Environmental 224:634–647. https://doi.org/10.1016/j.apcatb.2017.11.008

Efficiency of Using Granular Activated Carbon Obtained from Tamarind Fruit Processing Waste for the Purification of Multi Components Impurities in Industrial Draining Wastewater

Saw Win Myint, Zaw Ye Naing, A. V. Nistratov, and V. N. Klushin

Abstract Activated carbon is widely used in water treatment and purification due to its ability to remove water pollutants in this study, the researchers focused on the production of granular activated carbon from tamarind fruit peel. The carbon was prepared by steam activation, and its structural-adsorption and absorption properties were examined. The information related to the production of granular activated carbon from tamarind fruit peel and its structural-adsorption and absorption properties are reflected. The results of a comparative assessment of the efficiency of purification from organic impurities according to the criterion of total organic carbon of multicomponent wastewater from the territory of the coke oven and by-product production of "Moskok" under dynamic conditions using the above-mentioned activated carbon and commercial adsorbent of Russian production of the BAC brand are presented and discussed. In conclusion, these abstract highlights the production of granular activated carbon from tamarind fruit peel and its application in the purification of wastewater. The comparative assessment presented in the study provides insights into the efficiency of different adsorbents in removing organic impurities from multicomponent wastewater. Furthermore, the potential of activated carbon derived from tamarind fruit peel for the removal of specific contaminants has been explored in other studies.

Keywords Tamarind fruit · Fruit shell · Thermal processing · Steam activation · Activated carbon · Organic impurities · Dynamic sorption · Wastewater treatment

S. W. Myint (✉) · Z. Y. Naing · A. V. Nistratov · V. N. Klushin
Mendeleev University of Chemical Technology of Russia, 9, Miusskaya Sq, Moscow 125047, Russia
e-mail: sawwinmyint86@gmail.com

L. V. Radionova and D. V. Ulrikh (eds.), *Advances in Ecology and Environmental Engineering*, Springer Proceedings in Earth and Environmental Sciences,
https://doi.org/10.1007/978-3-031-64423-8_19

1 Introduction

In recent decades, the deep removal of hazardous pollutants from wastewater and natural water using activated carbons obtained from plant waste has received increasing attention as an important area in solving numerous environmental protection problems. In present-day Myanmar, formerly known as Burma, a country in the tropical belt of Southeast Asia with a diverse and rich natural resource, whose economy is predominantly agricultural, the government is implementing an ensemble of initiatives aimed at accelerating its industrial and industrial development [1].

Today, Myanmar has hundreds of fast-growing industrial zones, most of which are made up of small-scale enterprises. These small enterprises lack the capital to invest in equipment and facilities to deal with waste and pollution of the biosphere caused by their production activities. Under the current conditions, government planning is struggling to keep pace with industrial growth [2]. In particular, this results in the lack of a clear plan for the treatment of solid and liquid wastes and difficulties in ensuring the implementation and enforcement of regulations concerning, for example, appropriate landfills, as well as pollution control laws that limit the negative effects of industrial growth [3]. At the same time, according to a study by the Green Homeland Development Association [4], the level of pollutants in wastewater has increased significantly in the last two years in the industrial zones of Hlaing Taryar and Shwe Pyi Tar. Pollution of the country's natural waters is caused in part by mining (as a result of, inter alia, drilling and oil and gas production), mining, agriculture and many enterprises related to the processing of its crops, which leads to the introduction of hazardous substances in the form of a variety of organic and inorganic compounds, including pesticides, heavy metals, radionuclides and other pollutants [5].

Among the means of their deep extraction from aqueous solutions, the most significant role belongs to activated coals, the production of which is absent in Myanmar [4] although the country has a significant amount of large-tonnage plant waste, the ensemble of which, judging by the literature data [6], can potentially be a high-quality raw material for the production of these adsorbents, which is very important for the national economy due to the high cost of such products on world markets and the obvious possibility of selling its own production with subsequent use, along with other methods, for the necessary processing of various adsorbents [7]. Tamarind (Tamarindus Indica L.) is widely cultivated in India, Thailand, Indonesia, Myanmar and the Philippines [8]. Its leguminous fruits, used mainly for food purposes, consist of pulp, covered with a hard shell of seeds and peel. Seeds make up 30–40% of the fruits with a large amount of agri-biproducts [8]. The peel of tamarind fruits used is usually classified as food production waste, and from the seeds, judging by the literature [9], it is possible to obtain activated carbons suitable for purifying industrial emissions and discharges of organic compounds, including dyes, and heavy metals. Along with this, the use of non-renewable but traditionally used natural resources in the form of fossil coals, wood, cokes, semi-cokes and other carbon-containing materials to produce commercial activated carbon is considered expensive. Therefore, many researchers obtain alternative adsorbents, which are cheaper and relatively effective compared to commercial ones, from agricultural waste and various food industries, and their costs when extracting pollutants that are in low concentrations in the treated environments due to the laws of sorption equilibrium and the quality of the absorbers used is improved. In particular, the authors obtained activated carbon from the peel of tamarind fruits with a sufficiently high absorption capacity for organic pollutants of multicomponent effluents [10]. As a representative of these, a sample of storm water from the territory of the coke-chemical production of "Moskok" JSC was used. The task of deep cleaning them remains relevant for the enterprise until recently, and the unstable composition is characterized by the data in Table 1 according to [11].

In this regard, the authors assessed the feasibility of transforming the waste of food production in Myanmar, represented by tamarind fruit shell (TFS), into active coals, with the involvement of the most acceptable pyrolysis operations in this country and activation of the resulting carburized residue with water vapor. The choice of this raw material is associated with publications, according to which active carbons with good absorption properties are obtained from tamarind tree care waste (cut branches, fallen leaves, stem wood) and the seeds of their fruits. In the paper [12], on the basis of the data of thermographic studies of a sample of tamarind fruits and its components, the rational limits of thermal effect on CBT during pyrolysis were established, the expedient conditions for the implementation of the process itself were characterized in [13], and such activations by water vapor of the obtained carbonisate were described in [14]. A number of applied properties of carbon adsorbents obtained on the basis of CBT have been described in [15].

Below are the conditions of phase contact and the results of comparative studies of the adsorption dynamics of the sum of organic pollutants from a sample of multicomponent effluents from the territory of the coke oven and by-product production of Moskov JSC with the specified activated carbon and those of the Russian production of the BAC brand, which also have a plant nature and are oriented according to [16] for wastewater treatment.

Table 1 Technical indicators of activation process for carbonizate tamarind fruit shell

No.	t, (min)	I, (mg/g)	MB, (mg/g)	V_{Σ}, (cm^3/g)	Vs, (cm^3/g)			Yield, (%)
					C_6H_6	CCL_4	H_2O	
1	0	450.00	332.00	1.0	0.260	0.238	0.284	42.80
2	30	762.50	330.00	1.0	0.250	0.221	0.20	41.9
3	60	762.00	321.12	1.1	0.220	0.200	0.154	39.05

2 Experimental Part

Adsorption processes for the removal of organic pollutants, dyes and heavy metals are widely known. The choice of adsorbent is the basis for the successful development of the adsorption process, its efficiency and operation. A viable adsorbent must have high adsorption interaction with target pollutants and the ability to remove them [17]. These adsorbents may be synthetic or derived from natural sources. Some of the common types of adsorbents include organic metal frameworks, graphene-based materials, activated carbon (AC), zeolites, clay minerals, biomass, carbon nanotubes, polymeric materials, etc. [18]. The potential purpose of active carbon is determined by the raw material, type of technology and production conditions, and another important advantage is carbon-containing raw materials [19, 20]. The use of carbon adsorbents such as activated carbon obtained from biomass waste inspires optimism due to their low cost, high efficiency, ease of synthesis, environmental safety, sustainability and renewability [21–23].

The care of tamarind trees and the processing of their crops accompanies the formation of a number of types of waste, represented mainly by fallen and cut branches, foliage, stem wood (trimmings, chips, sawdust), spoiled fruits, pods and seeds. According to the available literature, such plant wastes can serve as secondary raw materials for the production of activated carbons. In the study, the daily runoff of daily freshness of output No. 1 of the above-mentioned enterprise, characterized by a value of 85 g/m^3 of total organic carbon (TOC), was used. This runoff is variable over time in terms of volume, presence and content of impurities, and its fairly detailed characteristics are presented in [24]. The phase contact was carried out in a vertically mounted glass tubular reactor with an internal diameter of 12.3 mm, equipped with a pressure vessel, a partition in the form of a thin stainless-steel mesh rolled into a cylinder and a stopcock. In order to minimize the wall effect [25], the activated carbons for testing were prepared in the form of fractions with a grain diameter of 0.4–0.6 mm, placing them in the reactor in the form of layers of different heights. The experiments were carried out at room temperature (20–23 °C) and a flow rate of 0.0024 m^3/h by periodic sampling of filtrate to determine the TOC in them by the chromatographic method using a gas chromatograph of the "Tsvet" 500M brand. Until recently, the task of their deep cleaning has remained relevant for the enterprise, and the unstable composition is characterized by the data in Table 1 according to [26].

3 Result and Discussion

To be obtained for optimal parameters for final products of tamarind fruit shell, carried out the rational processing of pyrolysis and activation. The experimental scales of pyrolysis are taking in the stainless-steel reactors, which is controlling with the voltage regulators for desire heat energy. By varying the temperature from 150 to 550 °C, obtaining carbonizate (biochar) is analyzing with methyl blue and iodine for determination of abilities of absorption and then adsorption with C_6H_6, CCL_4 and H_2O is knowing for meso and micro pore. The results of experimental analysis are showing in table forms. In Table 1, showing results are taking out at different thermal exposures of 0, 30, 60 min, at an activation temperature of 750 °C with an intensity of 5 °C/min and a steam consumption of 5 g/g product.

As showing in Table 1, there is one suitable condition for further activation process at this temperature. It is the non-thermal holding time because of its percentage of yield final product.

In the Table 2, the results are obtaining by varying the intensities of thermal changing at different intensities of 5, 10, 15 °C/min without isothermal support and a steam flow rate of 5 g/g product with a final temperature of 750 °C and observed that the porosities (micro, meso and macro) especially in micropore of obtaining activated carbon is different almost 2 times when the adsorption of benzene, which is taking out at the room temperature, for 5 and 10°C/min. For 15°C/min, consider about lack of micropore and enrichment of meso. After this stage of analysis, there are choosing the optimum condition for activation.

Optimization of pyrolysis and combined-cycle activation conditions was carried out in [27], and the technical indicators of the obtained carbonizate (biochar) and activated carbon are shown in the form of Table 2. The conditions of the processing processes are the heating intensity of pyrolysis and activation is 10 °C/min, the final holding time of pyrolysis is 0 min, and the activation time is 60 min. One the important factor of obtaining AC is as mentioned as above is yield of products and 53% for pyrolysis and 66% for activated carbon from raw material is acceptable. The pore structure is developed by steam activation almost 3 times and abilities of removing of iron and methyl blue is also improved about 2 and 3, 5 times respectively. The properties of the using activated carbons and its carbonizate for dynamic sorption is representing in the Table 3.

The technical parameters of using wastewater from "Moskok" is showing in Table 4 and the data is collecting from work [28].

Table 2 Technical indicators of activation of tamarind carbonate at different thermal intensities

No.	Thermal intensities (°C/min)	I, (mg/g)	MB (mg/g)	VΣ (cm^3/g)	Vs (cm^3/g)			Yield, (%)
					C_6H_6	CCL_4	H_2O	
1	5	558.80	340.74	1.10	0.26	0.238	0.284	36.80
2	10	450.00	332.00	1.0	0.131	0.20	0.202	46.90
3	15	431.80	313.30	1.0	0.167	0.094	0.189	49.05

Table 3 Technical indicators of carbonizate (K) and activated carbon (AC) tamarind shell

№.	Index	T, °C	I, mg/g	MB, mg/g	V_Σ, cm^3/g	Vs, (cm^3/g)			Yield, (%)
						C_6H_6	CCL_4	H_2O	
1	Products								
2	Carbonizate	350	184	266	0.60	0.044	0.075	0.161	53
3	Activated carbon	700	711	366	0.90	0.118	0.023	0.17	66

Table 4 Characteristics of wastewater from outlet no. 1 of the Moscow coke and gas plant (the numerator and denominator are data of adjacent years)

№.	Index	MPC*, mg/cm^3	Number of analyses	Higher MPC	Content, mg/cm^3		
					Average	Min.	Max.
1	pH	6.5–8.5	21/24	0/4	7.5/7.3	7.0/2.0	8.1/10.5
2	Phenols	0.001	21/24	21/24	0.09/0.15	0.004/ 0.02	0.29/ 1.69
3	Petroleum products	0.05	21/24	21/24	1.49/1.94	0.28/ 0.19	7.95/ 5.10
4	Dry remaining	1000	21/24	18/16	1853/1347	551/447	3670/ 5228
5	Cyanides	0.05	21/24	1/7	0.01/0.05	0.00/ 0.00	0.08/ 0.23
6	Total iron	0.10	21/24	21/24	2.67/4.33	0.62/ 0.13	7.69/ 13.53
7	Chlorides	300	21/24	15/7	516/301	73/42	1644/ 1650
8	Transparency	не < 13	21/24	2/3	18/18	8/3	25/28
9	Suspended solids	10.75	21/24	21/22	43.0/89.6	11.0/2.3	262.0/ 526.9
10	Ammonium nitrogen	0.40	21/24	21/22	6.4/4.56	1.9/0.3	24.3/ 11.2
11	Sulfates	100	21/24	21/21	406/222	108/31	941/453

*Maximum permissible concentration (MPC) for the water of fishery reservoirs

4 Results and Discussion

Studying the dynamic sorption system is interested in modelling the industrial scales sorption system for water or gas purification. The output curves of the studied adsorption dynamics processes are shown Figs. 1 and 2.

The obtained output curves make it possible to determine important indicators of the effectiveness of adsorbents: the time of protective action of the grain layer (τ_{TH}) and phase equilibrium time (τ_{eq}). The first of them naturally increases with the height of the layer L, and the values of the τ_{TH} at L = const are close to both active

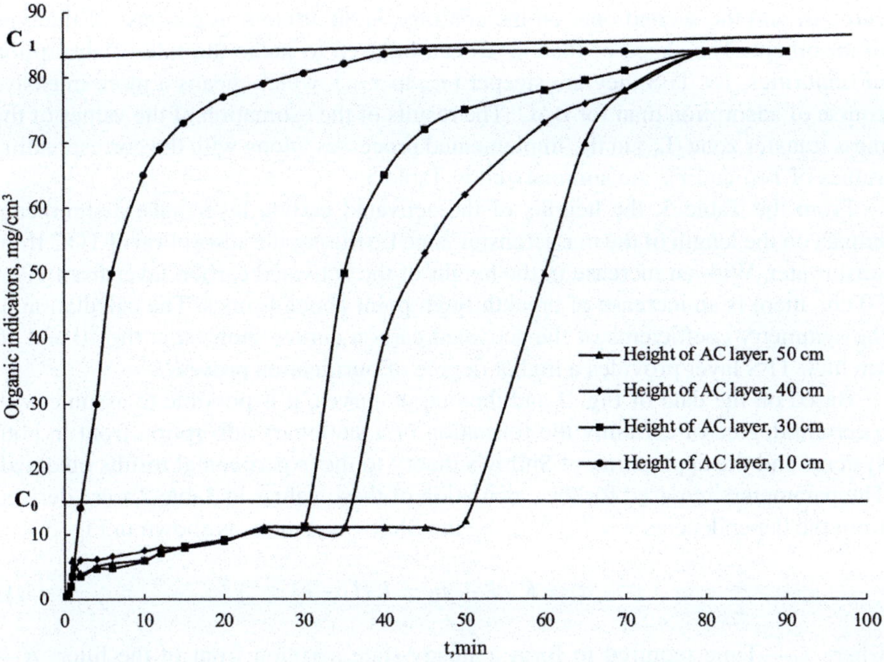

Fig. 1 Output adsorption dynamics curves using active carbons based on TFS. $\mathbf{C_0}$—initial concentration, $\mathbf{C_1}$—final concentration

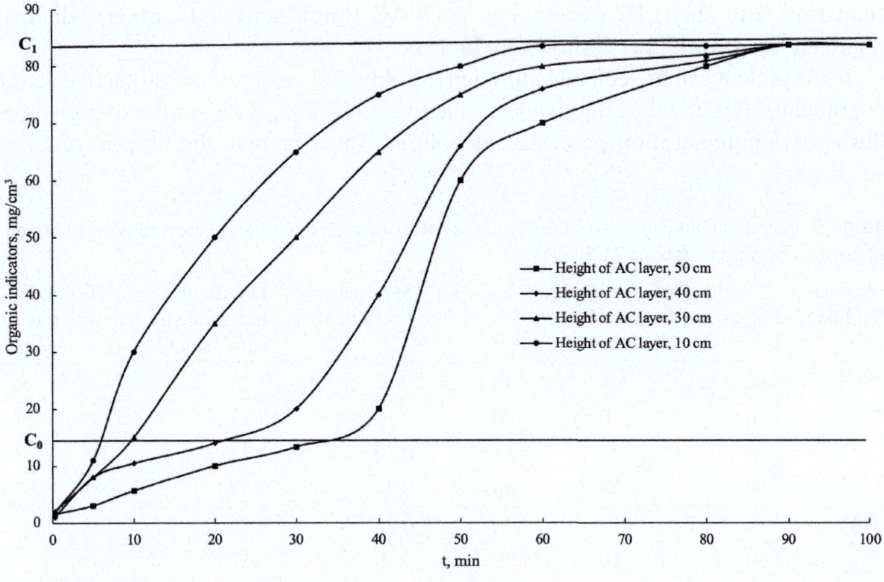

Fig. 2 Output adsorption dynamics curves using active carbons based on BAC. $\mathbf{C_0}$—initial concentration, $\mathbf{C_1}$—final concentration

carbons, and the second one, on the contrary, is 80–90 min, regardless of the type of adsorbent. The shape of the curves can be used to judge the rate of absorption of impurities; for TFS they are steeper (τ_{eq} to τ_{TH}), which means a more intensive course of adsorption than for BAC. The results of the estimation of the values of the mass transfer zone (L_o) in the implemented processes, along with the corresponding values of τ_{TH} and τ_o, are summarized in Table 5.

From the Table 5, the heights of the activated carbon layer have a significant impact on the length of the mass transfer zone Lo during the adsorption of TOC from wastewater. With an increase in the height of the activated carbon layer from 10 to 50 cm, there is an increase of breakthrough point about 4 times. The calculation of the symmetry coefficients of the activated carbon curves shows that the 40 and 50 cm high TFS layer provides a higher degree of purification process.

Based on the data of Fig. 2 and the output curves, it is possible to assume with a certain degree of certainty the formation of a stationary adsorption front in both systems and the application of Shilov's theory to the experimental results obtained. The parameters required for the calculation of these values in Table 2 were derived from the dependencies $\tau = f(L)$, using the Shilov equation, as shown in Fig. 3.

$$\tau = K \cdot L - \tau_0 = K(L - h) \tag{1}$$

where τ_0—Time required to form a steady-state sorption front in the filter; K—coefficient of protective action; h—value that characterizes the unused (at the time of the slip) height of the adsorbent layer, L_0—Length of mass transfer zone.

Comparation of the dynamic characteristics of using activated carbon TFS (tamarind fruit shell) by experiment and BAC (birch activated carbon) which is obtained from work [29] is shown in Table 6.

In the table it can be seen that although the obtained values of the protective action coefficients measure the effectiveness of activated carbon (K) is similar to each other during dynamic sorption processes of both activated carbon, the time τ_0 required

Table 5 Technical indicators of obtaining curves of adsorption dynamics of the removing of organic impurities by active carbons TFS and BAC

Activated carbon	Height of AC layer, cm	τ_{TH}, min	T_{eq}, min	Symmetrical coefficient, ϕ	Length of mass transfer zone L_o, cm	TOC content in water at $\tau_{TH.}$, mg/cm^3
TFS	10	2	40	0.17	1.99	14.0
	30	30	80	0.19	2.42	12.0
	40	35	80	0.30	3.45	10.0
	50	53	80	0.34	7.27	9.0
BAC	10	5	90	0.23	2.25	11.0
	30	10	90	0.19	2.82	8.0
	40	23	90	0.25	5.14	8.0
	50	35	90	0.25	8.28	5.0

Fig. 3 Experimental dependencies of the protective time of activated carbon layers based on TFS and BAC on their height

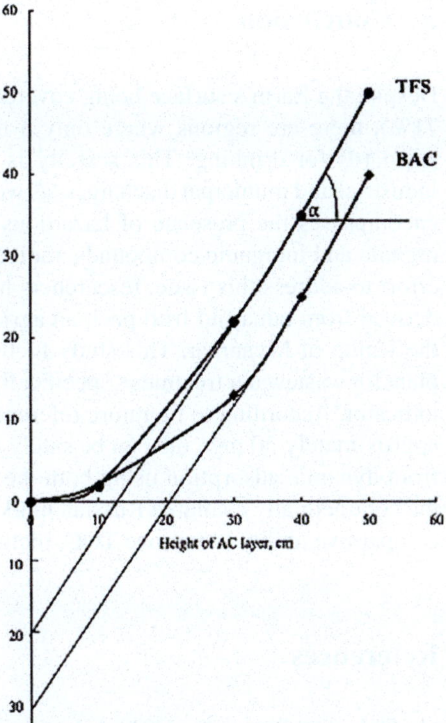

Table 6 Dynamic characteristics of activated carbons TFS and BAC

Type of AC	Coefficient protective action, K	h, cm	τ_{o}, min	L_o, cm
TFS	0.80	12	20	28
BAC	0.85	20	30	32

for the formation of a stationary sorption frontal in the column of activated carbon for tamarind fruit shell is more quickly than birch AC, because the unused value (h) is shorter for activated carbon from TFS, and although the values of the length of the mass transfer zone Lo is higher for BAC, which means most of the layer of birch wood base activated carbon is working nearly 64%. From the comparisons of obtaining results of dynamic sorption using activated carbon base on Myanmar' raw material are qualified for purification of organic impurities from wastewater.

5 Conclusion

Despite the Earth's surface being covered by a significant amount of water (around 71%), there are regions where only a mere 1% of the water meets international standards for drinking. This scarcity is largely due to water pollution caused by industrial and municipal discharges, as well as agricultural activities. Water pollution encompasses the presence of hazardous substances like heavy metals, pesticides, organic and inorganic compounds, radioactive elements, and other pollutants. In an effort to address this issue, researchers have explored the use of activated charcoal derived from tamarind fruit peel, an agricultural waste abundant in the Republic of the Union of Myanmar. This study focuses on its successful application at a coke plant for wastewater treatment, specifically targeting the Total Organic Carbon (TOC) indicator. According to literature references [30], treated water with a TOC level of approximately 30 mg/cm3 can be safely discharged into water bodies. The findings from dynamic adsorption using both the tamarind fruit peel activated charcoal and the commercially available Russian market activated carbon BAC demonstrate their competitive ability to remove TOC, including the "Moskok" TOC.

References

1. Pushpa Jha, Bhajan Dass (2020) Anlaysis of biomasses for their thermochemical transformation to biofuels. International Journal of Energy Production and Management 2(5):115–124
2. Ogbuefi PS, Nwaokafor P, Njoku IJ, Nwadinobi ED (2020) Proximate and ultimate composition of parboiled rice husk from local rice species and their suitability for use in bio-briquette production. Journal of Chemistry 1(7):8–14
3. Piriya RS, Jayabalakrishnan RM, Maheswari M, Boomiraj K, Oumabady S (2021) Coconut shell derived ZnCl2 activated carbon for malachite green dye removal. Journal of Water Science and Technology 5(83):1167–1182
4. Myanmar (2024) Britannica. https://www.britannica.com/place/Myanmar. Accessed 23 Jan 2024
5. Oil and gas in Myanmar Available via DIALOG. https://web.archive.org/web/20150415084151/; http://burma.total.com/myanmar-en/oil-and-gas-in-myanmar/oil-and-gas-in-myanmar-900130.html. Accessed 13 Jan 2023
6. Aung Min and Toshihero Kudo. New government initiative for industrial development in Myanmar. Available via DIALOG. https://www.burmalibrary.org/en/new-governments-initiatives-for-industrial-development-in-myanmar. Accessed 24 Sept 2015
7. Lae Win Swe. Overviews of water pollution in Myanmar. Available via DIALOG. https://pcasia.org/pic/wp-content/uploads/simple-file-list/20191014_Overview-of-Water-Pollution-in-Myanmar.pdf. Accessed 18 Dec 2023
8. Wine Phyu Phyu K. Industrial zone wastewater shows rising pollution. Myanmar Times. Available via DIALOG. https://www.mmtimes.com/business/18303-industrial-zone-waste-watershows-rising-pollution.html. Accessed 02 Feb 2023
9. Ren Z, Jia B, Zhang G (2020) Study on adsorption of ammonia nitrogen by iron-loaded activated carbon from low temperature wastewater. Journal chemosphere 2:262–275
10. Jafarisani M, Khavar AHC, Mahjoub AR et al (2020) Enhanced visible-light-driven photocatalytic degradation of emerging water contaminants by a modified zinc oxide-based photocatalyst; In-vivo and in-vitro toxicity evaluation of wastewater and PCO-treated water. Separation and purification Technology 243:1–14

11. Kishibayev KK, Serafin J, Tokpayev RR, Khavaza TN, Atchabarova AA, Abduakhytova DA, Ibraimov ZT, Srenscek-Nazzal J (2021) Physical and chemical properties of activated carbon synthesized from plant wastes and shungite for CO_2 capture. Journal of Environmental Chemical Engineering 9(6):106798. https://doi.org/10.1016/j.jece.2021.106798
12. Chunxia Ding, Can Ye, Wei Zhu, Guangdong Zeng, Xuemei Yao (2023) Engineered hydrochar from waste reed straw for peroxymonosulfate activation to degrade quinclorac and improve solanaceae plants growth. Journal of environmental management 347:119090. https://doi.org/10.1016/j.jenvman.2023.119090
13. Ramalingam G, Priya AK, Gnanasekaran L, Rajendran S, Hoang TKA (2024) Biomass and waste derived silica, activated carbon and ammonia-based materials for energy-related applications. Fuel 355:129490. https://doi.org/10.1016/j.fuel.2023.129490
14. Ramesh T, Rajalakshmi N, Dhathathreyan KS (2015) Activated carbons derived from tamarind seeds for hydrogen storage. Journal of Energy Storage 4:89–95
15. Guannan Qui, MingxinGuo (2010) Quality of poultry litter-derived granular activated carbon. Bioresource Technology 101(1):379–386
16. Hailay Girmay, Sarah Tewolde-Berhan, Hadgu Hishe, Zemede Asfaw, Morgan Ruelle, Alison Power (2020) Use and management of tamarind (Tamarindus indica L., Fabaceae) local morphotypes by communities in Tigray. Northern Ethiopia, Trees Livelihoods 29(2):81–98
17. Gayathri R, Thirumarimurugan M, Kannadasan T (2013) Removal of chromium (VI) ions from aqueous solution using tamarind seeds as an adsorbent. International Journal of Pharmaceutical and Chemical Sciences 2(2):984–991
18. Saw Win Myint, Nistratov AV, Klushin VN (2021) Nature of transformation of the elemental composition of target products for processing tamarind fruit peels into carbon adsorbents. III All-Russian Student Conference with International Participation "Geo-ecology: Theory and Practice", Moscow, November 19–20, 2021, pp 541–550
19. Sumrit Mopoung, Phansiri Moonsri, Wanwimon Palas, Sataporn Khumpai (2015) Characterization and properties of activated carbon prepared from tamarind seeds by KOH activation for Fe III adsorption from aqueous solution. The scientific world journal 2015(1):1–9. https://doi.org/10.1155/2015/415961
20. Nurul AJ, Hashim IN (2018) Preparation of activated carbon from tamarind seeds and methylene blue (MB) removal. Journal of fundamental and applied science 9:102–114
21. Thurumal V, Dhamodharan K, Ravi G Cleaner production of tamarind fruit shell into biomass derived porous 3D-activated carbon nanosheets by CVD technique for supercapacitor applications. Journal of chemosphere 282:131033. https://doi.org/10.1016/j.chemosphere.2021.131033
22. Saw Win Myint, Zhukov AV, Klushin VN (2020) Thermographic study of tamarind fruit components. Collection of articles of the scientific and practical conference with international participation "Ekologicheskaya, promyshlennaya i energeticheskaya bezopasnosti—2020", Sevastopol, SSU, pp 527–531
23. Saw Win Myint, Nistratov AV, Klushin VN (2020) Optimization of Conditions for Pyrolysis of Tamarind Fruit Shells. Proceedings of the 8th All-Russian Conference "Actual Issues of Chemical Technology and Environmental Protection", Cheboksary, 16–17, pp 70–71
24. Saw Win Myint, Nistratov AV, Klushin VN (2022) Active coal from tamarind fruit peel—waste food production in Myanmar. Federal State Unitary Enterprise "Federal Environmental Operator" Kirov Atomic Energy Information Center Institute of Biology of the Komi Scientific Center of the Ural Branch of the Russian Academy of Sciences Institute of Chemistry of the Komi Research Center of the Ural Branch of the Russian Academy of Sciences, pp 195–163
25. Saw Win Myint, Zaw Ye Naing, Min Thu et al (2020) Inexpensive resources of Myanmar as a source of carbon adsorbents. International Journal of Modern Agriculture 9(3):342–350
26. Saw Win Myint, Nistratov AV, Klushin VN (2021) Nature of transformation of the elemental composition of target products of tamarind fruit peel processing to carbon adsorbents. Student Conference and "Geoecology: Theory and Practice", Moscow, November 19–20, pp 124–125
27. Saw Win Myint, Nistratov AV, Klushin VN et al (2021) Kinetics and equilibrium of adsorption of volatile organic solvent vapors from their vapor-air mixtures by activated coal obtained

from the peel of tamarind fruits. Collection of articles of the scientific and practical conference "Ecological, Industrial and Energy Safety—2021", Sevastopol, SSU, pp 583–587
28. Zubakhin NP, Klushin VN, Starostin KG, Nistratov AV (2015) Purification of Coke-Plant Waste by Carbon Adsorbents. Coke and Chemistry 58(2):75–78
29. Mukhin VM, Nomenclature catalog "Active carbons. Elastic sorbents, catalysts, desiccants and chemical absorbers based on them." Under the general editorship. Publishing House "Ore and Metals", Moscow, 2003, p 278
30. Why fish die in our reservoirs, Federal Center for Animal Health Protection (FSBI "ARRIAH"). Available via DIALOG. https://tmvl.ru/poleznovsem/vazhno-znat/iz-za-chego-gibnet-ryba-v-nashikh-vodoemakh/

Specification of Nitrogen and Phosphorus Income into the Neva Bay and the Northeastern Part of the Finnish Gulf Ecosystems

V. Tretyakov, S. Klubov, A. Nikulina, V. Dmitriev, and A. Vichare

Abstract There is considered precise specification of total nitrogen and phosphorus income in the Neva Bay and the northeastern part of the Finnish Gulf ecosystems. The bay is the estuary located within the administrative border of St. Petersburg, the biggest conurbation on the Baltic Sea shore. The Neva Bay and the Finnish Gulf have significant fishery and recreation value. Anthropogenic eutrophication is the main danger to the Baltic Sea ecosystem. However, the danger is widespread, extending beyond the Finnish Gulf. Therefore, the presented methodology can be widely used for the evaluation of the biogenic elements incomes from the catchment areas of watercourses without monitoring points. The paper presents the evaluation methodology. It includes research of watersheds with and without data of the nitrogen and phosphorus outflow from them. The research consists of ascertainment of the watershed boundaries, determination of the watershed spatial structures, calculation of the biogenic elements unit discharges, cluster analysis and calculation of the nitrogen and phosphorus outflow from the catchment areas without the nitrogen and phosphorus outflow monitoring. The cluster analysis is used for the assessment of similarities and differences between the watersheds of watercourses with and without monitoring points. The results show a significant role of the nitrogen and phosphorus income from the watersheds without monitoring points.

Keywords Anthropogenic eutrophication · Catchment areas · Unit discharges

V. Tretyakov (✉) · S. Klubov · A. Nikulina · V. Dmitriev · A. Vichare
Saint Petersburg State University, VO 10-Line 33-35, St. Petersburg 199178, Russia
e-mail: v_yu_tretyakov@mail.ru

V. Tretyakov · S. Klubov
Russian State Hydrometeorological University, 11, Rizhsky Ave, St. Petersburg 190103, Russia

S. Klubov
State Budgetary Institution of Additional Education, Palace of Child Youth Art "At the Voznesensky Bridge" of the Admiralteyskiy District, 26, Grazhdanskaya Str, St. Petersburg 190031, Russia

L. V. Radionova and D. V. Ulrikh (eds.), *Advances in Ecology and Environmental Engineering*, Springer Proceedings in Earth and Environmental Sciences,
https://doi.org/10.1007/978-3-031-64423-8_20

1 Introduction

Modernity is named as "the global ecological crisis". Its cause consists of the biosphere poisoning by toxic chemical substances produced by humankind. Aquatic ecosystem deterioration has one more reason, that is anthropogenic eutrophication. It includes income into the aquatic ecosystems and additional amounts of the nitrogen and phosphorus of anthropogenic origin. Of course, chemical compounds of the elements are not toxic substances; otherwise, the elements are needed for all organisms. However, the overabundant income of the elements, especially phosphorus, can lead to an ecosystem crash, disappearance of valuable fish fauna, the bloom of blue-green algae, and the degeneration of the aquatic resources quality. The Baltic Sea catchment area is characterized by a high degree of urbanization, a good deal of agricultural and rural areas, and agricultural enterprises. These objects are intensive sources of nitrogen and phosphorus. Therefore, according to the HELCOM design [1], eutrophication is considered as the main danger for the Baltic Sea ecosystems. However, the problem is actual for aquatic objects in various climatic zones. Thus, there is a noted contamination of the Danube River urbanized watershed rivers in Romania [2]. There was an increase in the Cin'huaj River contamination after the river flowing through Nanjing City (China) [3]. Mexican scientists deduced that the rivers within the urban zones are open-air drainage ditches [4]. There has been noted a high degree of Western African river contamination. It is connected with the diffused outflow from the agricultural and urbanized areas [5]. It should be emphasized that many anthropogenic and natural factors that work on unit discharges of the total nitrogen and phosphorus outflow. The factors include the landscape structure of the catchment area, the land-use pattern, the watershed geological structure, the relief roughness, the surface slope, the density of the watershed river network, the watershed soil factors (soil types, oxidation–reduction potential, and acid–base conditions), and the watershed climate [6, 7]. High values of the phosphorus and nitrogen outflow unit discharges are promoted by the significant intensity of soil erosion and the light texture of the soil. Otherwise, the nitrogen and phosphorus are retained within clay soils. The soils prevent the nitrogen and phosphorus outflow from the watersheds into the aquatic ecosystems. The biogenic elements are trapped by vegetation, which works as a natural biogeochemical barrier. Macrophytes and coastal wetlands have a significant capability for nitrogen and phosphorus absorption. Plants contain far more of the biogenic elements than upper biogenous soil horizons. Nitrogen and phosphorus are involved in the substance circuit after the plants decompose. Compounds of the elements are washed out from leaves, needles, litter and live matter [8–10]. Accumulation of nitrogen and phosphorus takes place due to infiltration of atmospheric precipitation, surface runoff, the leaching of dead organic substances, and fixation of atmospheric nitrogen by microbiota. In addition, there is the anthropogenic outflow of nitrogen and phosphorus. Thus, the elements are contained in waste, fertilizer components, and contaminants from detergents [9]. The biogenic load on the aquatic ecosystems can be strengthened in the case of big amounts of organic substance within the soil [7, 8]. Vegetation can influence nitrogen and phosphorus outflow in many ways. Thus, the unit discharge in forests usually has minimal values. At the same time, nitrogen and phosphorus concentrations in the snow at young flood are often higher than the concentrations in the exposed district [9]. The various compounds of nitrogen and phosphorus have different mobility within the soil. Nitrates within the soil move more freely in comparison with phosphates, as far as the soil particles absorb the phosphates. Therefore, the phosphates have tended to move within the particles. Outflow of the phosphorus is characterized by maximum intensity during hurricanes. The bulk of the nitrogen migration is groundwater outflow, whereas the phosphorus outflow above all outflows within the surface runoff and meltwater from the watershed [11, 12]. In addition, there is the atmospheric income of the biogenic elements in soil with precipitation. The maximal precipitation leads to maximal surface and delayed runoff, and, in turn, to maximal outflow of the nitrogen and phosphorus. Therefore, the maximal unit discharges are observed during maximal precipitation periods [7]. Within-year variability of the biogenic elements has a maximum during the spring high water period and a slightly sloping decrease in the summer. The minimal concentrations of the total phosphorus take place in winter. The maximal concentrations are usually observed in spring when there is an increase in the soil temperature. The increase leads to an increase in the organic phosphorus mineralization rate [8]. Some increase in the phosphorus concentrations can be observed in the summer second half. As the possible cause of it, we can consider the income of the phosphorus from bottom sediments. The process is named water body secondary pollution. Extreme precipitation in wetter years increases outflow of the nutrients from the catchment area. The nitrogen in this case is mainly moved in dissolved inorganic compounds, and the phosphorus is transported in suspended substances. The last particularity indicates the origin of the transferred phosphorus. It means that the substance is mainly the result of soil erosion. The watershed morphometric features play an important role. Thus, phosphorus can be retained within bowl-shaped depressions. Nitrogen accumulation usually takes place within wide plains characterized by numerous shallow aquatic objects [11]. Connected with river drainages, wetlands promote retaining of the nitrogen and phosphorus within the watersheds [6, 13]. Climatic factors essentially influence nutrient outflow. Thus, the temperature increases from 1 °C up to 35 °C is responsible for the mineralization and the nitrate concentration in the groundwater increases up to two times [9]. Urbanized areas are characterized by high rates of nutrient unit discharges. The phenomenon can be explained by the outflow of storm water from impenetrable surfaces and sewage effluent. In addition, there is the impact of consumer waste [6]. Besides that, there are the following causes of the nutrient's income into surface water and groundwater, such as the usage of detergents and fertilizers, feeding of food-producing animals, and the nitrogen and phosphorus extraction of minerals [8, 14].

2 Materials and Methods

Ascertainment of the watershed boundaries is processed by some GIS technologies hydrological functions, which are applied to the watershed surfaces digital terrain models. The GIS technologies provide determination of such catchment area parameters as types of the surface, exposures, slopes, and multifactorial analysis of the digital data. The usage of GIS allows us to produce spatial-time models and the obtained data visualization [15, 16]. Manual drawings of the catchment area polygons in some GIS also can be used for the watershed localizations on an electronic chart. In the last case, a vector or raster map was used containing the altitude data. The watershed's subsequent refinement can be processed on the base of global terrain models such as the Shuttle Radar Topography Mission (SRTM). Maps of land use and soil cover are usually verified with the use of satellite images, for example, Landsat 8 Operational Land Imager (OLI) remote sensing data. We can produce by GIS technologies the watershed morphometric indicators, such as average length of overland flow, mean slope, ruggedness number, infiltration number, stream frequency, drainage texture, drainage density, dissection index, mean bifurcation ratio, circularity ratio, elongation ratio, and the watershed form factor. These parameters are extremely significant in erosion susceptibility evaluation. It is necessary because the high intensity of the erosion leads to a high rate of nitrogen and phosphorus outflow. The Fuzzy Analytical Hierarchy Process (AHP) is used as a feasible and proficient technique for prioritization of the watershed parameters [17]. The nutrient outflow rates depend on the susceptibility of the watersheds to surface runoff formation, transfer, and accumulation. There is presented a GIS-based method for susceptibility mapping [18, 19]. The method is based on the Interest route indicator parameter (IRIP). It is a pluvial runoff intensity indicator. The method allows the production and use of various charts for the identification of areas, that are susceptible to the formation, transfer, and accumulation of surface runoff. The charts must be able to reflect three stages of surface runoff that are its generation, transmission, and accumulation. Assessment of surface runoff occurrence is based on multiple landscape indicators, which can be obtained by using such data as land-use and soil maps, global terrain models, for example, Digital Elevation Model (DEM), and terrestrial linear networks. The assessment can use topographic, hydrological, meteorological, soil, and agronomic criteria [20]. The indicators can be maintained as raster matrix cell values or attribute table values. The range of the indicator values is from 0 up to 1. The values close to one are located within the areas, that are more susceptible to the occurrence of surface runoff. Multi-criteria assessment methodology of determining the possibility in the watershed area to prevent aquatic ecosystem contamination due to nutrient outflow is being swiftly advanced. The watersheds can be ranked according to the assessment results. Watersheds are researched as sources of contamination with the usage of multi-criteria analysis approaches, which use GIS methods and remote sensing data [21]. Thus, the Koyna River watershed in India was divided into 19 subwatersheds. They were prioritized by the DEM application. The drainage intensity, stream frequency, stream density, bifurcation ratio, texture ratio, and overland flow were used as geomorphological parameters. It is the intensity of drainage is considered as the most significant factor of the morphometric characteristics, which can be used for the sub-watershed prioritization. Regarding the sub-watersheds prioritization procedure, there is an opinion that both the morphometric parameters and the hydrological characteristics, such as stream power index (SPI), topographic wetness index (TWI), and sediment power index (STI) [22] should be taken into consideration. The uncertainties associated with hydrological and morphometric variables must be reduced.

Of course, the terrain models can be made on the ground of global models such as ASTER GDEM and SRTM. However, the models can have too many faults within timberlands and urbanized areas due to the reflection of the satellite emission from tree and shrub vegetation, anthropogenic industrial objects, roofs of multistory buildings, and so on. Therefore, for the digital terrain model creation, we carried out vectorization of topographic contours in a raster image of the topographic map of St. Petersburg and Leningrad Region issued in 2001 with a scale 1:200,000. The model was produced in ArcGIS by the method of the "Universal kriging". It is the raster matrix of the height values. The cell's dimension is equal to 100 per 100 m. Then the matrix was modified. A hundred meters wide buffer zones around the watercourses were prepared. The altitude values inside the zones were lessened by 10 m. It is necessary for easement of the river's watershed ascertainment in ArcGIS. The ascertainment was made by the tool "Watershed" of ArcGIS. The first stage of the ascertainment consists of the runoff directions creation by the "Flow Direction" tool. The nodes in points of the river outlets were manually located. In the case of rivers with monitoring points, the nodes coincide with the monitoring points. In the case of rivers without monitoring points, the nodes coincide with the river outfalls. All the catchment areas were produced by the "Watershed" tool upstream of the nodes. The tool creates new matrices with information about the cells belonging to partial watersheds. At the next stage of the technique, the cells belonging to the same partial catchment area must be converted into vector polygon by "Conversion tools". Then the polygons of all the partial watersheds are aggregated into the entire watersheds of the river-drainage systems. The Shingarka River and the Strelka River catchment areas are joined because the river systems are connected. In addition, there is dispersed discharge from the coastal area adjoined the Neva Bay. This area is included in the Shingarka and the Strelka Rivers joined watershed. The indispensable step of the research consists of the determination of surface types inside the watersheds. We used electronic maps of GIS-Lab—an informal community of Russian-speaking GIS/RS specialists (gis-lab.info) as the information source about surface types. The watershed layers were intersected with the GIS-Lab shapefiles. Then the watershed areas and ones of the watershed's parts occupied by various surface types were calculated. After that, there were calculated shares of the surface types inside the catchment areas.

Thus, we have the researched watersheds of watercourses without any monitoring points, and we want to evaluate the total nitrogen and phosphorus outflow from the watersheds into Neva Bay and the northeastern part of the Finnish Gulf. We attempt to use it for the evaluation of the nutrient outflow unit discharges of the

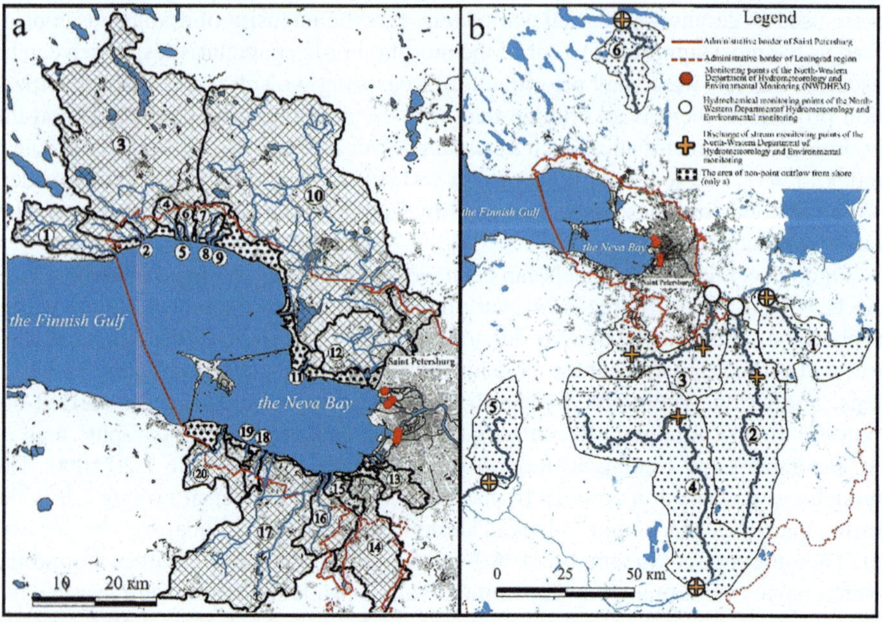

Fig. 1 Location of the river-drainage systems watersheds and the monitoring points: **a** the researched watersheds; **b** the analogous watersheds

closely located watersheds of the rivers with monitoring points. For this purpose, we compare the parameters of the above-mentioned groups of the watersheds by cluster analysis. Then we must determine analogue catchment areas of the rivers with the monitoring points for the watersheds without the points. Of course, the watersheds must have similar characteristics. Figure 1 demonstrates the location of the river-drainage systems entering Neva Bay and the northeastern part of the Finnish Gulf, and the river systems with monitoring points, the watersheds of the systems.

The researched watersheds in the Fig. 1 are designated by the numbers: (1) the Privetnaya River, (2) the Smolyachkov Stream, (3) the Chyornaya River, (4) the Ushkovskiy Stream, (5) the Vosmoy Stream, (6) the Bystraya River, (7) the Zelenogorskiy Stream, (8) the Shestoy Stream, (9) the Tretiy Stream, (10) the Sestroretskiy Razliv, (11) the Chyornaya Lakhta River, (12) the Lahtinskij Razliv, (13) the Ligovskij Channel and the Krasnenkaya River, (14) the Dudergofka River and the Dudergofskiy Channel, (15) the Strelka Stream, (16) the Kikenka River, (17) the Shingarka and Strelka Rivers, (18) the Troitskiy Stream, (19) the Kristatelka River, (20) the Karasta River watersheds. The analogous watersheds are designated by the numbers: (1) the Mga River, (2) the Tosna River, (3) the Izhora River, (4) the Oredezh River, (5) the Vruda River, (6) the Volchya River.

At the next stage of the exploration, we evaluate the unit discharges from the analogue watersheds. The final stage of the research consists of the evaluation of the total nitrogen and phosphorus outflow from the catchment areas without monitoring.

The watershed parameters, which are taken into consideration in the cluster analysis, are shares of the following types of surfaces: forests, wetlands, farmland areas, residential constructions, industrial zones, and lakes and reservoirs. The drainage network density is used as one of the parameters. Moreover, we want to include in the list of the cluster analysis parameters the average slopes of the watersheds. Ascertainment of the parameter can be processed in two ways. The first uses digital terrain models and GIS tools. The second proposes topographic map usage for the creation of strict lines between adjoining topographic contours. We can calculate the lengths of the lines. Altitude differences between the endpoints of the lines are known. Therefore, the slope of each line can be calculated by arithmetically dividing the altitude difference on the line length. Of course, for precise ascertainment of the parameter, we have to produce many of the lines. All the above-mentioned characteristics are used in the evaluation of distances between the catchment areas in the many-dimensional space of the features at the cluster analysis.

The Neva Bay residence time is very short. It is equal to 7 days. Therefore, it is necessary to ascertain the inter-year variability of the total nitrogen and phosphorus daily unit discharges. To do so, we used data from the Northern-Western Department of Hydrometeorology and Environmental Monitoring. There are data values about the analogue river runoff for each day throughout the year. Unfortunately, there are only a few values of the nutrient concentrations in the rivers per year. Nonetheless, we must have the values for each day of the year. Therefore, we have to interpolate the available values. The interpolation was made by a specially worked out program of Mathcad. There was used the linear interpolation. In that case, the results cannot exceed the range of the monitoring data values. The average daily unit discharges were calculated in Microsoft Excel tables. The river's runoff initial values for each day have dimensions in cubic meters per second. The values were recalculated in cubic meters per day (24 h). Then the results were multiplied by the total phosphorus and nitrogen daily concentrations, which were the interpolation results. Thus, we obtained the substance flowing through the points of the chemical monitoring per day (24 h). Then the results were divided by the watershed areas in square kilometers. In the result of the processing, we have within-year variability of the nutrient unit discharge from the analogous watersheds for each year. By processing of analogous watershed data of year series, we evaluated the generalized annual dynamics of the nutrient unit discharges with the confidence intervals. The modified Nash–Sutcliffe criteria are used for revealing similarities and dissimilarities amongst the generalized dynamics. The daily and total annual unit discharge is applied for evaluation of ones in catchment areas without the monitoring data of the nutrient outflow.

3 Results

Figure 2 and Table 1 present the spatial structures of the watersheds. The same numbers as in Fig. 1 designate the watersheds. Figure 3 demonstrates the cluster analysis results. The cluster analysis organized the watersheds into two groups.

The first includes the catchment areas of the Bystraya, Chyornaya, Karasta, Mga, Oredezh, Privetnaya, Tosna, Volchya, and Vruda Rivers; Shestoy, Smolyachkov, Tretiy, Ushkovskiy, and Vosmoy Streams; and the Sestroretskiy Razliv reservoir. The Mga, Oredezh, Tosna, Volchya, and Vruda Rivers in the group are the rivers with monitoring points that are the rivers of the analogous watersheds. The second group includes the watersheds of the Chyornaya Lakhta River, the Dudergofka River and the Dudergofskiy Channel, the Izhora River, the Kikenka River, the Kristatelka River, the Ligovskij Channel and Krasnenkaya River, and the Shingarka and the Strelka Rivers; the Strelka, Troitskiy, and Zelenogorskiy Streams; the Lahtinskij Razliv reservoir; the Northern shore of the Finnish Gulf, the Northern and Southern shores of the Neva Bay. The Izhora River in the group is only one river of the analogous watershed. The catchment areas of the analogous rivers have the following generalized annual values of the nutrient's unit discharges: of the Izhora River—2346.0 kgN/km^2 and 133.8 kgP/km^2, of the Mga River—578.8 kgN/km^2 and 132.0 kgP/km^2, of the Oredezh River—812.8 kgN/km^2 and 11.5 kgP/km^2, of the Tosna River—186.3 kgN/km^2 and 21.7 kgP/km^2, of the Volchya River—130.6 kgN/km^2 and 11.7 kgP/km^2, and the Vruda River—1376.1 kgN/km^2 and 8.8 kgP/km^2. The total area of the researched watersheds in the first group is equal to 1517.2 km^2, and in the second group—884.5 km^2. For evaluation of the nutrient outflow from the researched watersheds of the groups, we used the unit discharges in the analogous catchment areas of the same groups.

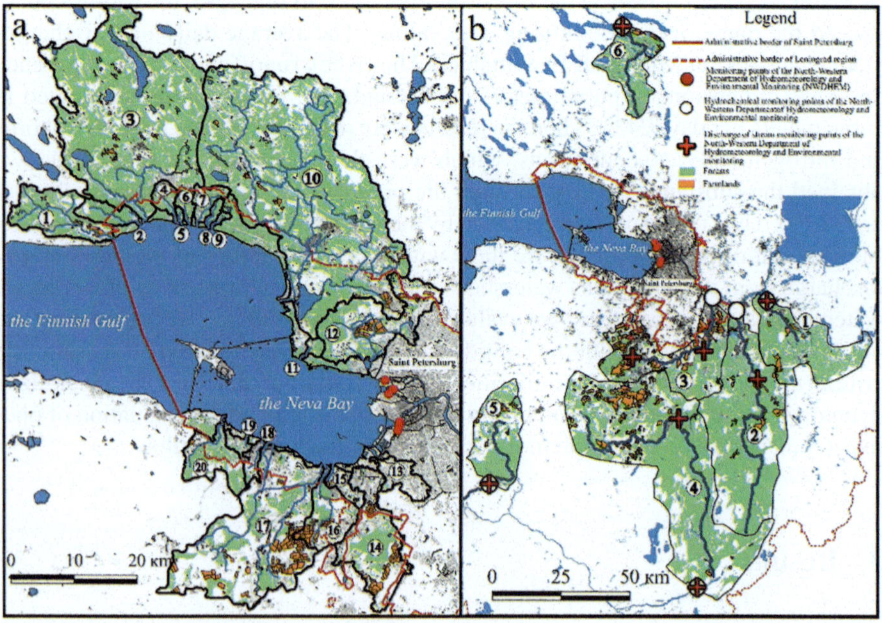

Fig. 2 Spatial structure of the watersheds: **a** the researched ones; **b** the analogous ones

Table 1 Spatial structure of the watersheds (1—forests, 2—wetlands, 3—farmland areas, 4—residential constructions, 5—industrial zones, 6—lakes and reservoirs, 7—drainage network density, 1/km)

Watersheds	Area, km^2	Share of the total watershed area, %						7
		1	2	3	4	5	6	
The researched watersheds								
Bystraya River	9.8	75.6	0.3	0.5	3.5	0.2	3.3	1.50
Chyornaya River	544.1	68.2	5.5	0.7	2.5	0.2	11.0	0.56
Chyornaya Lakhta River	7.6	61.7	0.0	0.0	17.6	0.7	0.1	0.37
Dudergofka River and Dudergofskiy Channel	126.7	27.7	0.1	10.0	10.7	4.7	1.0	1.12
Karasta River	59.6	60.6	3.2	0.8	3.4	0.7	4.0	1.00
Kikenka River	65.4	22.5	0.5	14.8	11.3	3.1	1.6	1.15
Kristatelka River	10.1	38.7	0.1	0.0	16.4	6.0	0.8	0.31
Lahtinskij Razliv	135.7	37.5	5.5	7.6	10.3	4.2	7.5	0.67
Ligovskij Channel and Krasnenkaya River	31.6	2.6	<0.1	0.0	27.9	14.6	0.9	0.62
Privetnaya River	83.6	68.6	5.8	3.4	4.1	<0.1	6.7	0.37
Sestroretskiy Razliv	766.1	60.7	9.7	1.2	4.8	0.4	12.5	0.58
Shestoy Stream	6.2	79.7	0.5	0.0	0.9	0.4	0.6	0.85
Shingarka and Strelka Rivers	324.7	27.8	3.4	9.1	9.5	2.1	4.7	0.75
Smolyachkov Stream	26.5	84.6	11.2	<0.1	0.2	0.1	11.0	0.69
Strelka Stream	19.0	11.6	0.9	6.1	16.0	5.8	4.3	0.83
Tretiy Stream	3.1	90.6	3.1	0.0	0.0	0.0	3.1	1.00
Troitskiy Stream	12.7	20.3	0.9	0.5	17.3	1.3	1.6	1.20
Ushkovskiy Stream	15.6	68.3	1.5	0.0	8.6	0.1	11.0	0.56
Vosmoy Stream	2.6	72.9	0.3	0.0	2.7	0.0	0.3	3.25
Zelenogorskiy Stream	19.9	62.3	0.7	0.0	21.1	1.8	1.0	0.56
Northern shore of the Finnish Gulf	68.7	63.3	1.6	0.0	14.6	2.0	3.0	0.31
Northern shore of the Neva Bay	22.3	33.2	5.8	0.1	21.5	0.7	13.0	0.29
Southern shore of the Neva Bay	40.1	25.3	9.2	0.0	7.9	17.2	14.2	1.32
The analogous watercourses watersheds								
Izhora River	864.7	36.9	3.1	13.9	7.0	2.8	3.9	0.90
Mga River	709.4	67.6	15.5	2.8	1.9	0.2	16.1	0.98
Oredezh River	2485.8	58.4	9.9	5.2	3.1	0.3	11.0	0.54
Tosna River	1637.7	69.5	17.5	2.8	1.9	0.4	18.3	0.75
Volchya River	417.3	71.1	3.5	3.0	4.0	0.1	7.6	0.59
Vruda River	537.3	62.2	7.9	3.3	1.1	<0.1	8.2	0.37

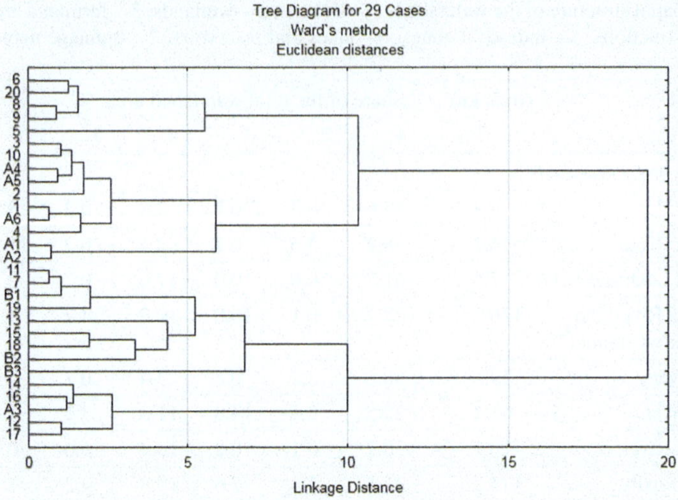

Fig. 3 The cluster analysis results. The numbers 1—20 as in Figs. 1 and 2 designate the watersheds. The symbols A1—A6 designate the analogous watersheds of the rivers: (1) Mga, (2) Tosna, (3) Izhora, (4) Oredezh, (5) Vruda, and (6) Volchya. The symbols B1—B3 designate the researched watersheds of (1) the northern shore of the Finnish Gulf, (2) the northern and (3) the southern shore of the Neva Bay

For the calculation of the nutrient outflow from the researched watersheds of the first group, we used average annual unit discharges in the group analogous watersheds. The average annual unit discharge of the total nitrogen is equal to 616.9 kgN/km^2, one of the total phosphorus—37.1 kgP/km^2. Therefore, the total nitrogen annual outflow from the researched watersheds of the first groups is equal to 936.0 metric tons, and the total annual phosphorus outflow—13.4 metric tons. The total nitrogen annual outflow from the researched catchment areas of the second group is equal to 2075.0 metric tons, and the total annual phosphorus outflow—118.3 metric tons. Thus, the annual average outflow of the total nitrogen from the researched catchment area into Neva Bay and the northeastern part of the Finnish Gulf within the administrative boundary of St. Petersburg is equal to 3011.0 metric tons, and the outflow of the total phosphorus—131.7 metric tons.

Atmospheric precipitation is an additional source of nutrients. Atmospheric unit discharge of the total nitrogen within the Leningrad Region is equal to 0.66 metric tons/km^2 per year, and of the total phosphorus—0.014 metric tons/km^2 per year [23]. The area of the Neva Bay and the northeastern part of the Finnish Gulf within St. Petersburg is equal to 843.3 km^2. Therefore, in the area, the annual atmospheric income of the total nitrogen on the water surface is equal to 556.6 metric tons, and the annual income of the phosphorus—11.8 metric tons. There are evaluations of the ammonium nitrogen and nitrate nitrogen annual income on the Finnish Gulf with atmospheric precipitation. The ammonium nitrogen income is equal to 0.265 metric tons/km^2 per year, and the nitrate nitrogen income—0.828 metric tons/km^2 per year

[24]. According to the values, the total annual income of the ammonium nitrogen from the atmosphere per year is equal to 223.5 metric tons, and the nitrate nitrogen income—698.3 metric tons.

4 Discussion

The results can add new information to knowledge about the outflow simulation of nutrients from the Finnish Gulf watershed according to the ILHM and ILLM models of the Institute of Limnology of the Russian Academy of Sciences [25]. Our research deals with smaller watersheds within St. Petersburg and its outskirts.

5 Conclusion

The total nitrogen and phosphorus average annual incomes into Neva Bay with the runoff of the Neva River and its spill streams for 1979–2019 were equal to 56,860 metric tons of nitrogen and 1716 tons of phosphorus [26]. Thus, there is non-registered nutrient income in Neva Bay and the northeastern part of the Finnish Gulf within the administrative boundary of St. Petersburg from the researched watersheds. The income is equal to 5.3% of the total nitrogen and 7.7% of the total phosphorus income from the Neva River and its spill stream runoff. The amounts do not include the nutrient income from any areas of St. Petersburg that adjoin the Neva Bay, for example, the Vasilyevsky Island. The authors intend to evaluate the income in future research. The income of the total nitrogen from atmospheric precipitation accounts for approximately 1.0% of the income from Neva River and its spill stream runoff, and the analogous income of the total phosphorus is 0.7%. It is evident, that the urbanization of the researched watersheds and the increase in residential construction areas within the watersheds can lead to an increase in nitrogen and phosphorus income from the St. Petersburg administrative area.

Acknowledgements The research was funded by the Russian Science Fund, grant 23-27-10011, and by the Saint Petersburg Science Fund.

References

1. Baltic Sea Action Plan 2021 update. Baltic Marine Environment Protection Commission. October 2021. https://helcom.fi/wp-content/uploads/2021/10/Baltic-Sea-Action-Plan-2021-update.pdf
2. Zaharia L et al (2016) Urbanization effects on the river systems in the Bucharest City region. Ecosystem Health and Sustainability 2(11):1–19. https://doi.org/10.1002/ehs2.1247

3. Zhiyi L (2015) Analysis on pollution factors of urban river. Journal of Geoscience and Environment Protection 3:9–16. https://doi.org/10.4236/gep.2015.310002
4. Caro-Borrero A, Carmona JJ, Mazari HM (2016) Evaluation of ecological quality in peri-urban rivers in Mexico: A proposal for identifying and validating reference sites using benthic macroinvertebrates as indicators. Journal of Limnology 75(s1):1–16. https://doi.org/10.4081/jlimnol.2015.1304
5. Kabore I et al (2018) Developing reference criteria for the ecological status West-African rivers. Environment Monitoring and Assessment 190(2):1–17. https://doi.org/10.1007/s10661-017-6360-1
6. Boardman E et al (2019) Fertilizer, landscape features and climate regulate phosphorus retention and river export in diverse Midwestern watersheds. Biogeochemistry 146:293–309. https://doi.org/10.1007/s10533-019-00623-z
7. Jun D et al (2023) Spatiotemporal variations and determinants of stream nitrogen and phosphorus concentrations from a watershed in the Three Gorges Reservoir Area, China. International Soil and Water Conservation Research 11(3):507–517. https://doi.org/10.1016/j.iswcr.2022.09.004
8. Verheyen D et al (2015) Dissolved phosphorus transport from soil to surface water in catchments with different land use. AMBIO A Journal of the Human Environment 44(2):228–240. https://doi.org/10.1007/s13280-014-0617-5
9. Dolgov SV, Koronkevich NI (2021) Current Features and Dynamics of Nutrient Balance in the Kud'ma River Basin. 2. Seasonal Removal of Nitrogen and Phosphorus. Water Resources 48(5):794–803. https://doi.org/10.1134/S0097807821050092
10. Ulen B et al (2019) Seasonal variation in nutrient retention in a free water surface constructed wetland monitored with flow-proportional sampling and optical sensors. Ecological Engineering 139. https://doi.org/10.1016/j.ecoleng.2019.105588
11. Goyette JO et al (2019) Differential influence of landscape features and climate on nitrogen and phosphorus transport throughout the watershed. Biogeochemistry 142:155–174. https://doi.org/10.1007/s10533-018-0526-y
12. Law JY et al (2020) Impact of stacked conservation practices on phosphorus and sediment export at the catchment scale. Journal of Environmental Quality 49(6):1552–1563. https://doi.org/10.1002/jeq2.20140
13. Hansen AT et al (2018) Contribution of wetlands to nitrate removal at the watershed scale. Nature Geoscience 11(2):127–132. https://doi.org/10.1038/s41561-017-0056-6
14. Yang J et al (2023) Dynamic characteristics of net anthropogenic phosphorus input to the upper Yangtze River Basin from 1989 to 2019: Focus on the phosphate ore rich area in China, Journal of Environmental Management 347. https://doi.org/10.1016/j.jenvman.2023.119140
15. Buryak Z (2022) Basin-Scale Approach to Integration of Agro- and Hydroecological Monitoring for Sustainable Environmental Management: A Case Study of Belgorod Oblast, European Russia. Sustainability 14(2):927. https://doi.org/10.3390/su14020927
16. Abdelkebir B et al (2021) Evaluating Low-Impact Development practice performance to reduce runoff volume in an urban watershed in Algeria. Arabian Journal of Geosciences 14(9). https://doi.org/10.1007/s12517-021-07178-0
17. Hembram TK, Saha S (2020) Prioritization of sub-watersheds for soil erosion based on morphometric attributes using fuzzy AHP and compound factor in Jainti River basin, Jharkhand, Eastern India. Environment, Development and Sustainability. 22(1):1241–1268. https://doi.org/10.1007/s10668-018-0247-3
18. Braud I et al (2020) A method to use proxy data of runoff-related impacts for the evaluation of a model mapping intense storm runoff hazard: application to the railway context. Natural Hazards and Earth System Sciences 20(4):947–966. https://doi.org/10.5194/nhess-20-947-2020
19. Rodrigues ALM et al (2021) Methodology for Allocation of Best Management Practices Integrated with the Urban Landscape. Water Resources Management 35(3):1353–1371. https://doi.org/10.1007/s11269-021-02791-w
20. Krois J, Schulte A (2014) GIS-based multi-criteria evaluation to identify potential sites for soil and water conservation techniques in the Ronquillo watershed, northern Peru. Applied Geography 51:131–142. https://doi.org/10.1016/j.apgeog.2014.04.006

21. Shelar RS et al (2022) Sub-watershed prioritization of Koyna river basin in India using multi criteria analytical hierarchical process, remote sensing and GIS techniques. Physics and Chemistry of the Earth Parts A/B/C 128. https://doi.org/10.1016/j.pce.2022.103219
22. Rahmati O et al (2019) SWPT: An automated GIS-based tool for prioritization of sub-watersheds based on morphometric and topo-hydrological factors. Geoscience Frontiers 10(6):2167–2175. https://doi.org/10.1016/j.gsf.2019.03.009
23. Minakova EA, Shlychkov AP (2019) Assessment of humid atmospheric throughfalls of biogenic elements inside of catchment area of Middle and Lower Volga. Ecological proceedings 7. Publications of young researchers. All-Russian post-teen scientific conference with international participation 1:308–310. https://doi.org/10.24411/9999-010A-2019-10077
24. Svistov PF et al (2023) Atlas diagramm himicheskogo sostava atmosfernyh osadkov (primenenie) (Atlas of atmospheric precipitation chemical composition chemical composition (application)). Pero, Moscow
25. Kondratyev SA et al (2020) Model-based assessment of nutrient load into water bodies from different landscape types. In: Wenkel K-O et al (ed) Landscape Modelling and Decision Support, Springer Nature Switzerland AG, pp 299–310. https://doi.org/10.1007/978-3-030-37421-1_15
26. Serebritskiy IA, Grigoryev IA (2019) Environment protection, natural management, and ecological safety supporting in St. Petersburg. Sezam-Print, St. Petersburg

Practical Application of Lead Dioxide Anode for Removal of Heavy Metals from Wastewater

Ch. M. Nyein, V. A. Brodsky, S. W. Jima, and C. Myo

Abstract The lack of cost-effective anodes with good catalytic activity and stability still remains one of the most important problems in electrochemical oxidation for organic wastewater treatment. Electrochemical oxidation based on hydroxyl radicals formed on lead dioxide has become the main process of advanced oxidation. Titanium-based lead dioxide electrodes play an increasingly important role in electrochemical oxidation process. Wastewater treatment is a complex task, requiring the use of different physical and chemical approaches depending on the type and number of pollutants. Currently, electrochemical methods of wastewater treatment are considered the most advanced and effective. Electrochemical methods widely applied to effectively remove impurities of various compositions and dispersions, and make adjustments to the physical and chemical characteristics of water. Removing heavy metal ions from wastewater is of paramount importance for environmental safety and human health. These methods aim to remove heavy metal ions from various wastewater sources. This work presents a study of electroflotation purification of model solutions on a lead dioxide anode for the removal of heavy metal ions. The efficiency and stability of lead dioxide anode was investigated in different types of salt solution in various concentrations. Scientific and practical recommendations on the use of lead dioxide anode for the electroflotation process of wastewater treatment are presented. It has been established that lead dioxide anode materials have sufficient efficiency and stability for electroflotation process. The degree of electroflotation extraction of heavy metal ions reached more than 90%.

Keywords Electrochemical oxidation · Lead dioxide electrode · Wastewater · Electro-flotation · Heavy metals

---

Ch. M. Nyein (✉) · V. A. Brodsky · S. W. Jima · C. Myo
Mendeleev University of Chemical Technology of Russia, Miusskaya Pl. 9, Moscow 125047, Russia
e-mail: mrx12693@gmail.com

L. V. Radionova and D. V. Ulrikh (eds.), *Advances in Ecology and Environmental Engineering*, Springer Proceedings in Earth and Environmental Sciences,
https://doi.org/10.1007/978-3-031-64423-8_21

1 Introduction

One of the insoluble electrodes, called lead dioxide electrode, which has the unique properties of high electrical conductivity, high chemical resistance in aggressive solutions, high overvoltage when releasing oxygen, is of interest to all researchers as a cheap anode material in electrochemical technology. Due to these properties, lead dioxide electrode has been used for various electrochemical processes such as removing of petroleum hydrocarbons [1], ozone production [2], water clarification by electrolytic oxidation [3], anodic oxidation for decomposition of organic pollutants [4], etc. According to the structure of the phase composition, lead dioxide has two different types of crystal structure: α-PbO_2 (orthorhombic structure of columbite) and β-PbO_2 (tetragonal structure of rutile. It is known that the crystal structure of PbO_2 deposits depends on the pH of the precipitation solution.α-PbO_2 obtained from bases, β-PbO_2 from acids. In this work α-PbO_2 electrode obtained by anodizing in an alkaline plumbite solution on a titanium substrate was used to remove heavy metal ions from wastewater by electroflotation. α-PbO_2 has a more compact structure compared to β-PbO_2. Consequently, in the α-PbO_2 modification crystal, the contact between sediment grains is better. Taking these advantages into account, α-PbO_2 electrode deposited from a plumbite solution on a titanium substrate was chosen to remove heavy metal ions from wastewater [5]. In line with the growth of various industrial processes such as coating and electroplating, batteries, pesticides, mining, viscose, metal washing processes, leather, fluidized bed bioreactors, textiles, metal smelting, petrochemicals, paper making, electrolysis, as well as human activities, the presence of heavy metals in water has become greater [6–8]. Due to the increase in heavy metals in water, this can pose a risk to human health and the ecosystem. The most popular heavy metals are lead (Pb), zinc (Zn), mercury (Hg), nickel (Ni), cadmium (Cd), copper (Cu), chromium (Cr) and arsenic (As). Heavy metals are not biodegradable and may be carcinogenic; thus, the presence of these metals in water in an inappropriate quantity can lead to serious health problems in living organisms [9, 10]. Table 1 shows the maximum contaminant level (MCL) standards for heavy metal ions established by the US Environmental Protection Agency [11].

There are different alternative methods for removing heavy metals from wastewater, such as physicochemical methods, chemical precipitation, coagulation and flocculation, electrochemical treatment, ion exchange, membrane filtration, electrodialysis, biological methods, etc. [12]. Among these methods, electroflotation process was used in this work due to the following advantages:

- The electric field gradient between the electrodes promotes flocculation and flotation even without the presence of any surfactant.
- Gas production (release), flotation time and other operating conditions can be quickly checked and controlled relatively easily.
- The equipment is reliable and safe to use.
- The ionic strength of the purified solution (suspension) is not a decisive parameter, as with other flotation methods.

Table 1 US maximum levels of heavy metal contamination in wastewater

Heavy metal	Toxicity	MCL (mg/l)
Arsenic	Skin manifestations associated with arsenic, cancer of internal organs, vascular diseases	0.05
Cadmium	Kidney damage, renal dysfunction, human carcinogen	0.01
Chromium	Headaches, diarrhea, nausea, vomiting, carcinogen	0.05
Copper	Liver damage, insomnia	0.25
Nickel	Dermatitis, nausea, chronic asthma, cough, human carcinogen	0.20
Zinc	Depression, lethargy, neurological symptoms and increased thirst	0.80
Lead	Damage to the fetal brain, kidney disease, circulatory system and nervous system	0.006
Mercury	Rheumatoid arthritis, kidney diseases, circulatory system, nervous system	0.00003

- By-product recovery is a potentially beneficial aspect of electroflotation processing, so proteins, fats, oils, organics, and toxic or valuable metals can be recovered from waste streams, especially when no other chemicals (such as surfactants) are added to facilitate the separation of pollutants [13].

Study of the potential recovery of heavy metals, especially copper, nickel, zinc, which are formed during metal coating, mining, smelting, battery production, tanneries, oil refining, paint production, pesticides, pigments, printing and photography industries, etc. was performed [14, 15]. It is known that to remove ions of these three heavy metals, we must first establish the formation of the most sparingly soluble precipitates ($Cu(OH)_2$, $Zn(OH)_2$, $Ni(OH)_2$) at pH 10 [16]. In this work, a sparingly soluble precipitate was obtained by adding sodium hydroxide.

The purpose of this work was to study the process of electroflotation extraction of a mixture of copper, nickel and zinc ions in the form of hydroxide precipitation under complexation conditions usingα-PbO_2 electrodes. The following tasks were set:

- Investigate the effect of various background salts in different concentrations on the efficiency of extraction of the dispersed phase of precipitation of Cu, Zn and Ni hydroxides.
- Create optimal conditions for the extraction of heavy metals from wastewater at various concentrations of background salts.
- Evaluate the performance of α-PbO_2 electrodes in the electroflotation recovery of heavy metal ions.

2 Experimental Method

The studies were carried out on model solutions containing Cu^{2+}, Ni^{2+} and Zn^{2+} at a concentration of 100 mg/l in the form of mixtures. Sodium chloride and sodium sulfate in different concentrations (10, 50, 200 g/l) were used as an electrolyte to create a constant ionic strength.

The purification of model wastewater was carried out by electroflotation using a laboratory installation consisting of a source of direct electric current HY 1803D, a non-flowing electroflotator with a volume of 500 ml with a cross-sectional area of 10cm^2 and a height of 80 cm. An insoluble α-PbO_2/Ti anode and a cathode from stainless steel 12X18H10T (NPF "RutTech"). The volumetric current density was 0.4 A/l and pH was 10 [17–20].

The efficiency of the electroflotation process was assessed by the degree of heavy metals ion recovery ∞ (%):

$$\infty = \frac{c_0 - c_f}{c_f}.100\% \quad (1)$$

where c_0, c_f are the initial and final concentrations of copper, nickel and zinc in the volume of solution, respectively (mg/1). The extraction of Zn^{2+}, Cu^{2+}, Ni^{2+} ions from an aqueous solution with and without salt was carried out in a non-flow electroflotator with insoluble anodes (Fig. 1).

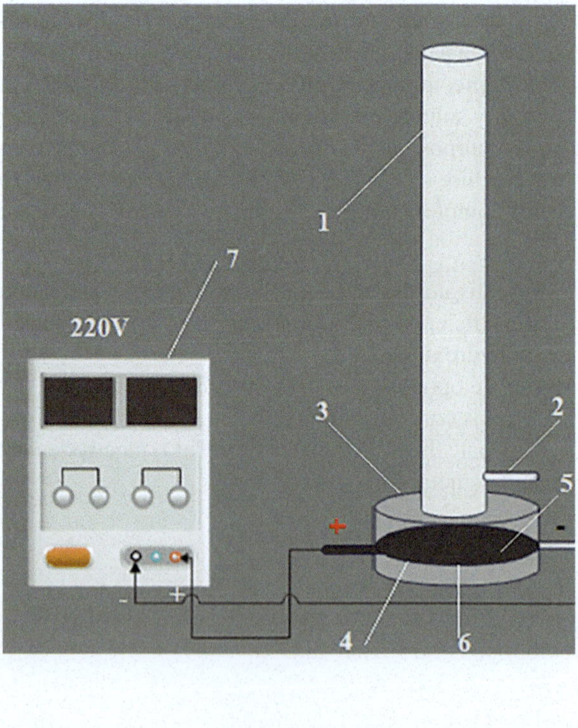

Fig. 1 Diagram of a laboratory batch electroflotation unit, 1—electroflotator column; 2—valve; 3—electrode block; 4—anode (PbO_2/Ti); 5—cathode; 6—rubber gasket; 7—DC source

The concentration of heavy metal ions were determined by the atomic absorption method using a Kvant-AFA spectrometer (KORTEK LLC). The objects of study were aqueous solutions of a mixture of copper, zinc and nickel ions.

The objective of this work is the disposal of liquid industrial waste, especially containing toxic ions of heavy and non-ferrous metals. The main source of such waste, which is discharged into surface water bodies together with waste technological solutions and washing waters, is galvanic production.

The purpose of the work is to determine the influence of electroflotation treatment on the efficiency of the process of cleaning solutions from metal ions by electroflotation and electroflotation methods with subsequent filtration. In this work, the resulting lead dioxide anode from plumbate electrolyte by electrodeposition was used to remove heavy metal ions from wastewater.

3 Results and Discussion

In this section, the finding of the study with respect to the effect of salt backgrounds at a constant current of 0.4A/l on the degree of extraction of poorly soluble metal compounds from aqueous solutions elaborated.

3.1 *Study of the Kinetics of Electroflotation Extraction of Poorly Soluble Compounds of Copper (II), Zinc (II), Nickel (II) as Part of a Multicomponent System with the Presence of a Background Electrolyte Na_2SO_4 ($c(Na_2SO_4) = 10$ and 200 g/l)*

The obtained data presented in Fig. 2 reflect how salt backgrounds affect the residual concentration of components and the degree of extraction α (%) of metal ions. The choice of pH = 10 as optimal was based on a study of the solubility of metal compounds in the system.

Under these conditions, the degree of extraction reaches 98–99% at 15 min of the process, while the lowest residual concentration of Zn^{2+} Ni^{2+} and Cu^{2+} ions is at $\tau = 15$ min, and is 0.13; 0.13; 0.0068 mg/l, respectively. With additional filtration, the degree of extraction of heavy metal ions reaches up to 99%.

The dependence of the degree of extraction of metal ions on time was also studied. Na_2SO_4 salt was used as an electrically conductive additive.

As depicted in Fig. 2, the data obtained show that these metals float equally well; the maximum degree of extraction for the three metals is achieved at 15 min after the start of the process.

The process of electroflotation extraction followed by filtration of poorly soluble compounds Cu^{2+}, Zn^{2+} and Ni^{2+} was studied. The studies were carried out in aqueous

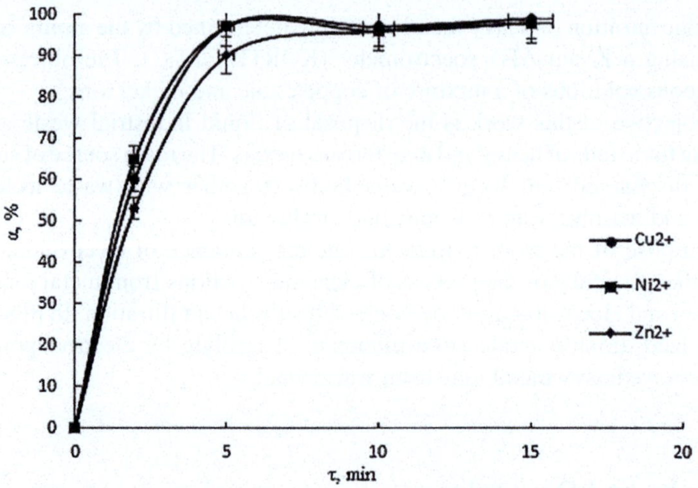

Fig. 2 Dependence of the degree of extraction of metal ions on the process time in the Cu(II)–Zn(II)–Ni(II) system. $C\sum Me = 100$ mg/l; C (Na_2SO_4) = 10 g/l; pH = 10; $i_v = 0.4$ A/l

solutions containing copper, zinc and nickel ions in a ratio of 1:1:1 at a total metal concentration of 100 mg/l in the presence of Na_2SO_4 with a concentration of 200 g/l, $i_v = 0.4$ A/l. The electroflotation process takes 15 min, followed by filtration.

The obtained data, presented in Fig. 3, demonstrate the relationship between the residual concentration and the degree of extraction of metal ions at a volumetric current density of 0.4 A/l.

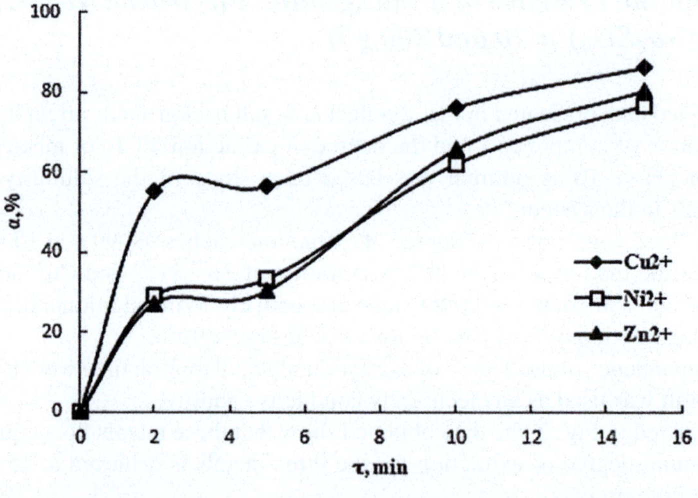

Fig. 3 Dependence of the degree of extraction of metal ions on the process time in the Cu(II)–Zn(II)–Ni(II) system. $C\sum Me = 100$ mg/l; $C(Na_2SO_4) = 200$ g/l; pH = 10; $i_v = 0.4$ A/l

Based on the data obtained, the efficiency of the electroflotation process is quite ineffective. Thus, the degree of extraction of Cu (II), Ni (II), Zn (II) at 15 min of the process is 86%, 77% and 80%, respectively.

The maximum degree of extraction for three metals was up to 96–99% with additional filtration. Figure 3 shows the metal ion cure efficiency over time at i_v = 0.4 A/l.

3.2 *Study of the Kinetics of Electroflotation Extraction of Poorly Soluble Compounds of Copper (II), Zinc (II), Nickel (II) as Part of a Multicomponent System in the Presence of a Background Electrolyte NaCl (C(NaCl) = 10 and 200 g/l)*

As shown in Fig. 4, the introduction of chloride ions into the solution has a beneficial effect on the electroflotation process. The presence of chlorine ions increases electrical conductivity, which increases the efficiency of electroflotation. This is because high electrical conductivity allows electrical current to pass through the solution more easily, resulting in better removal of suspended particles.

The kinetics of the electroflotation process was studied using solutions containing sodium chloride with a concentration of 10 g/l. To successfully remove copper, nickel and zinc in the form of sparingly soluble compounds, the sufficient time for the solution to remain in the electroflotation apparatus is at least 10 min. The foam layer has a bluish tint.

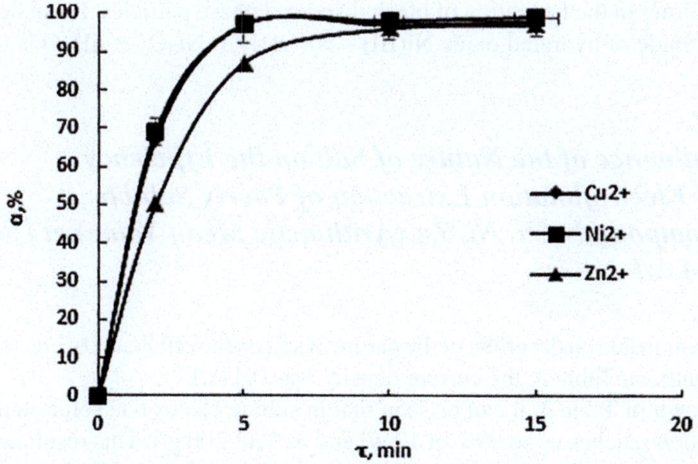

Fig.4 Dependence of the degree of extraction of metal ions on the process time in the Cu(II)-Zn(II)-Ni(II) system. C$\sum$Me = 100 mg/l; C(NaCl) = 10 g/l; pH = 10; i_v = 0.4 A/l

Table 2 Effect of $i_v = 0.4$ A/l on residual concentrations and the degree of extraction of poorly soluble metal compounds in the Ni(II)–Zn(II)–Cu(II) system in the presence of a background electrolyte $C_{NaCl} = 200$ g/l

τ (min)	Extraction degree "α"%(%)/Residual concentration (mg/l)		
	Cu	Ni	Zn
2	24/0.4155	28/0.3291	19/0.7569
5	91/0.0495	92/0.0352	90/0.0898
10	97/0.0114	98/0.0060	97/0.0207
15	99/0.0069	99/0.0044	98/0.0171
Filtration	99/0.0051	99/0.0005	98/0.0158

$C_{NaCl} = 200$ g/l; $C\sum Me = 100$ mg/l; $[C_{M1}:C_{M2}:C_{M3}] = 1:1:1$; pH = 10.0

As illustrated in Fig. 4 the maximum degree of extraction for three metals was 96–99%.

The next experiment was carried out at a volumetric current density of 0.4 A/l. The maximum process time is 15 min, followed by filtration. The total concentration of metal ions in the Cu(II)–Ni(II)–Zn(II) system was $C\sum Me = 100$ mg/l, background content C(NaCl) = 200 g/l.

Under these conditions, the highest degree of extraction of heavy metals reaches 98–99% at 15 min of the process. Additional filtrations did not bring significant changes as it has been revealed in Table 2.

The finding of the study depicted that the maximum degree of extraction of metal compounds occurs at 15 min of the process and is 98–99%. The further process of electroflotation does not have any significance, since the resulting dark brown foam layer is stable and no re-contamination has occurred. The change in color of the foam layer is associated with the presence of chloride ions, which oxidize nickel ions, resulting in the formation of black dispersed phase particles. It can be assumed that hydroxide or hydrated oxide Ni(III) – $Ni(OH)_3$ or $Ni_2O_3 \times nH_2O$ is formed.

3.3 *Influence of the Nature of Salt on the Efficiency of Electroflotation Extraction of Poorly Soluble Compounds Cu, Ni, Zn (Arithmetic Mean Value) at i_v = 0.4 A/l*

Experimental data on the effect of the nature of salt on the efficiency of electroflotation are presented in Table 3; the current density was 0.4 A/l.

As shown in Table 3, it can be seen that in sulfate electrolyte solution the degree of extraction reaches up to 99% at 10 g/l and 81% at 200 g/l. This result agrees with the study conducted with the titanium plate with oxide ruthenic titanium coating (ORTA) electrode in the study of [21]. However, in a direct chloride environment, the recovery rate was up to 98% at 10 g/l and 99% at 200 g/l. Thus, we can conclude

Table 3 Effect of $i_v = 0.4\ A/l$ on residual concentrations and the degree of extraction of poorly soluble metal compounds in the Ni(II)-Zn(II)-Cu(II) system with the presence of a background electrolytes C_{NaCl} $10\ g/l$– $200\ g/l$ and C_{Na2SO4} $10\ g/l$– $200\ g/l$

τ (min)	Extraction degree "α", (%)/Residual concentration (mg/l)			
	$i_v = 0.4$ A/l			
	$C_{(Na2SO4)} = 10$ g/l	$C_{(Na2SO4)} = 200$ g/l	$C_{(NaCl)} = 10$ g/l	$C_{(NaCl)} = 200$ g/l
	Ni(II)-Zn(II)-Cu(II)	Ni(II)-Zn(II)-Cu(II)	Ni(II)-Zn(II)-Cu(II)	Ni(II)-Zn(II)-Cu(II)
2	59/0.20	37/0.48	62/0.04	23/0.50
5	95/0.03	40/0.45	94/0.04	91/0.06
10	96/0.03	68/0.24	97/0.015	97/0.01
15	98/0.02	81/0.14	98/0.01	99/0.01
Filtration	99/0.01	98/0.02	99/0.01	99/0.01

$C\sum Me = 100$ mg/l; $[C_{M1}:C_{M2}:C_{M3}] = 1:1:1$; pH = 10.0

that in a chloride environment the efficiency of removing heavy metal ions is higher than in a sulfate environment.

It was revealed that the nature of the salt affects the particle size of insoluble compounds. Moreover, it was noted that according to the experimental results, if it is compared with the results by using oxide-ruthenium-titanium anode(ORTA) by using lead dioxide anode (PbO_2/Ti) the removal rate of heavy metal ions (Cu^{2+}, Ni^{2+},Zn^{2+}) from wastewater is also up to 98–99% [22].

Passing of an electric current through the NaCl electrolyte results in the release of chlorine gas at the anode, which can then dissolve and form hypochlorous acid, which in turn reacts with the OH^- ions present in the solution to form the hypochlorite ion ClO^-, being a strong oxidizing agent. Therefore, adding NaCl to a solution is beneficial for the formation of additional oxidizing agent in the bulk of the solution, which thereby increases the rate of oxidation of metal (II) ions.

4 Conclusions

The basic principles of electroflotation extraction of sparingly soluble compounds of zinc, copper and nickel from concentrated solutions of electrolytes in an electroflotator with insoluble electrodes have been studied. It was revealed that the process of joint electroflotation extraction of metal ions in the Cu(II)–Ni(II)–Zn(II) system is influenced by the pH value of the solution. The choice of pH = 10 as optimal was based on the minimum solubility of the resulting particles of the dispersed phase, which contributes to the maximum effect of extracting metal ions from model solutions. It has been established that the presence of background salts affects the speed and efficiency of the electroflotation process. An increase in the degree of extraction of metal ions is observed in the presence of Cl^-, which negatively affects the efficiency of electroflotation- SO_4^{2-} (at concentrations of 200 g/l). The degree of electroflotation extraction of sparingly soluble compounds of the studied metals (hydroxide, phosphate) increases at a low concentration of sodium sulfate (Na_2SO_4-10 g/l). However, in the presence of chloride ions it showed the best efficiency in removing suspended solids. Subsequent filtration makes it possible to achieve high degrees of electroflotation recovery (90–99%) for all studied metals and it was revealed that lead dioxide anode electrodes are efficient and stable at the same time.

Acknowledgements I would like to express my very great appreciation to Brodsky Vladimir Aleksandrovich—a candidate of chemical sciences, associate professor of the department of technology of inorganic substances and electrochemical processes of the D. I. Mendeleev Russian Chemical Technical University as my research supervisors, for his valuable and constructive suggestions during the planning and development of this research work. I would also like to thank Mr. Jima Solomon Workenekh—graduate student of the Department of Technology of Inorganic Substances and Electrochemical Processes of the D. I. Mendeleev Russian Chemical Technical University and Mr. Myo Swe—Ph.D., doctoral student for their advice and assistance in keeping my progress. My special thanks are extended to the administrator and organization of the Conference RusEcoCon 2024 for invitation. The work was carried out with financial support DI. Mendeleev (project No. Z-2020-004).

References

1. Boutheina Gargouri, Olfa Dridi Gargouri , Bochra Gargouri, Souhel Kallel Trabelsi , Ridha Abdelhedi , Mohamed Bouaziz (2014) Application of electrochemical technology for removing petroleum hydrocarbons from produced water using lead dioxide and boron-doped diamond electrodes. Chemosphere 117:309–315. https://doi.org/10.1016/j.chemosphere.2014.07.067
2. Xi Wang, Dandan Wu, Hang Ge, Luyang Wang, Xu Wu (2023) Shape-controlled synthesis of lead dioxide nanoparticles for enhanced electrocatalysis of electrochemical ozone production. J of Environmental Chemical Engineering 11:1–5. https://doi.org/10.1016/j.jece.2023.110248
3. Shmychkova O, Luk'yanenkoT, Dmirtikova L, Velichenko A (2019) Modified lead dioxide for organic wastewater treatment: physico-chemical properties and electro catalytic activity. J of the Serbian Chemical Society 84:1–13. https://doi.org/10.2298/JSC180712091S
4. Yingying Jiang, Haitao Zhao, Jie Liang et al (2020) Anodic oxidation for the degradation of organic pollutants: Anode materials, operating conditions and mechanisms. A mini review. Electrochemistry Communications 7:4–15. https://doi.org/10.1016/j.elecom.2020.106912
5. Xiaohong Li, Derek Pletcherb, Frank C. Walsha (2011) Electrodeposited lead dioxide coatings. ChemSoc Rev 7:3879–3894. https://doi.org/10.1039/c0cs00213e
6. Pugacheva IN, Karmanov AV, Zueva SB, De Michelis I, Ferella F, Molokanova LV, Vegliò F (2020) Heavy metal removal by cellulose-based textile waste product. J. Chem. & Chem. Tech 63:105–110. https://doi.org/10.6060/ivkkt.20206302.6098
7. Kakhramanov NT, Gadjieva RSh, Gahramanl JN, Arzumanova NB (2019) Sorption of heavy metals by multicomponent foam polymer sorbents. Khim. Khim. Tekhnol 62:110–117. https://doi.org/10.6060/ivkkt.20196205.5769
8. Yakovleva EV, Brudnik SV, Yakovlev AV et al (2022) Sorption of copper (II) cations from aqueous solutions by thermally reduced graphene oxide. ChemChemTech 65:35–42. https://doi.org/10.6060/ivkkt.20226505.6543
9. Naef AA Qasem, Ramy H Mohammed, Dahiru U Lawal (2021) Removal of heavy metal ions from wastewater: a comprehensive and critical review. Clean water 4(36):1. https://doi.org/10.1038/s41545-021-00127-0

10. Renu, Madhu Agarwal, Kailash Singh (2017) Methodologies for removal of heavy metal ions from wastewater: an overview. Interdisciplinary Environmental Review 18:125. https://doi.org/10.1504/IER.2017.10008828
11. Thien-Khanh Tran, Hoang-JyhLeu, Kuo-Feng Chiub, Chiu-Yue Linc (2017) Electrochemical Treatment of Heavy Metal-containing Wastewater with the Removal of COD and Heavy Metal Ions. Chin Chem Soc. 64:493–502. https://doi.org/10.1002/jccs.201600266
12. Gunatilake SK (2015) Methods of Removing Heavy Metals from Industrial Wastewater. Journal of Multidisciplinary Engineering Science Studies (JMESS) 1(1):12–18
13. Matis KA, Peleka EN (2010) Alternative Flotation Techniques for Wastewater Treatment: Focus on Electroflotation. Separation Science and Technology 45:2473. https://doi.org/10.1080/01496395.2010.508065
14. Kalavathy HM, Miranda LR (2010) Moringa oleifera - A solid phase extractant for the removal of copper, nickel and zinc from aqueous solutions. Chemical Engineering Journal 158:188–199. https://doi.org/10.1016/j.cej.2009.12.039
15. Parmar M, Thakur LS (2013) Heavy metal Cu, Ni and Zn: toxicity, health hazards and their removal techniques by low-cost adsorbents: a short overview. International Journal of Plant, Animal and Environmental Sciences 4(3):143–157
16. Jinhui Huang, Fang Yuan, Guangming Zeng, Xue Li, YanlingGu, Lixiu Shi, Wenchu Liu, Yahui Shi (2017) Influence of pH on heavy metal speciation and removal from wastewater using micellar enhanced ultrafiltration. Chemosphere 173:199–206. https://doi.org/10.1016/j.chemosphere.2016.12.137
17. Xenophontov BS, Bondarenko AV, Kapitonova SN (2012) Electroflotation wastewater treatment of microbiological productions. Santekhnika (Plumbing) 1:52–56
18. Xenophontov BS, Kapitonova SN, Bondarenko AV (2013) Electroflotation wastewater treatment of biotechnology productions. Vodoochystka (Water Purification) 8:19–24
19. Ksentini I, Ben Mansour L (2014) Modeling the hydrodynamic of an electroflotation column for the treatment of industrial wastewaters. Desalination and Water Treatment 56:1722–1727. https://doi.org/10.1080/19443994.20 14.950989
20. Bouyakoub AZ, Kacha S, Ouhib R, Bellebia S, Lartiges B (2010) Combined treatment of a textile effluent containing reactive dyes by coagulation-flocculation and electroflotation. Journal of Water Science 23:89–103. https://doi.org/10.7202/038927
21. Brodskiy VA, Malkova YuO, Soboleva VA (2021) The Impact of Surface Characteristics of Slightly Soluble Cu (II) and Ni (II) Compounds on their Electroflotation Extraction from Concentrated Solutions of Electrolytes. Earth and Environmental Science 814:1–5. https://doi.org/10.1088/1755-1315/814/1/012010
22. Zaw Htay Than, Pyae Aung, Thu Aung Hein, Kolesnikov AV (2020) Investigation of the process of electroflotation extraction of insoluble compounds aluminum (iii) in the presence of magnesium, calcium, barium and surfactants. Metal 2020:1–5

Thermal Treatment of Sewage Sludge

M. Dyagelev

Abstract This article provides the results of several series of experiments on the combustion of mechanically dehydrated sewage sludge as one of the ways to solve the problems of sludge storage, treatment and disposal. Solving this problem requires the development and implementation of new technologies and methods for the treatment, processing and disposal of sewage sludge, which would be more efficient and economically sound. Non-treated sewage sludge contains toxic substances such as heavy metals, organic pollutants and pathogenic microorganisms that can be dangerous to living organisms and ecosystems. In addition, sewage sludge accumulated at landfills occupies huge territories, which causes additional problems in terms of land use, since about 100 million tons of liquid sludge (2 million tons by dry weight) are generated annually at Russian wastewater treatment plants. The research on the thermal disposal of sludge were conducted at the sewage treatment plants of one of the cities in the region at a small experimental unit with vertical sludge supply and with a side point flame source. A number of experiments revealed that mechanically dehydrated sludge is able to burn without an external flame source at a humidity of 69.9%; with a decrease in the sludge humidity its calorific value increases. In the process of sludge burning ash is obtained with a volume at least 5 times smaller than the initial volume of sludge which makes it possible to increase manifold the utilization rate of sludge piles.

Keywords Sewage sludge · Thermal treatment · Combustion · Energy-saving and resource-saving systems · Sludge ash content

M. Dyagelev (✉)
Kalashnikov Izhevsk State Technical University, 7, Studencheskaja, Izhevsk 426069, Russia
e-mail: m.yu.dyagelev@istu.ru

L. V. Radionova and D. V. Ulrikh (eds.), *Advances in Ecology and Environmental Engineering*, Springer Proceedings in Earth and Environmental Sciences,
https://doi.org/10.1007/978-3-031-64423-8_22

1 Introduction

Sewage sludge (SS) is an uninfected and unstabilized mass containing a large amount of organic substances (up to 70% per dry mass), which, under favorable conditions, can quickly rot and produce unpleasant odors. SS contains toxic organic substances, pathogenic microflora and ions, i.e., they pose a great chemical and biological hazard to the environment, polluting the soil, water and atmosphere. The peculiarity of sewage sludge formed during biochemical wastewater treatment is a high content of heavy metals, sulfur-containing and chlorine-containing compounds in them, which complicates the processes of their disposal. 100 m^3 of wastewater contains from 15 to 20 kg of sludge dry matter while 12–15 tons of sludge per day are registered per 100 thousand inhabitants [1].

In the Russian Federation, up to 100 million tons of soil are currently formed with a humidity of 98% or about 2 million tons/year in dry matter. In Moscow, according to Mosvodokanal JSC, more than 9 million m^3 of sludge were formed in 2014 (200 thousand tons by dry mass). In St. Petersburg, about 160 thousand tons of SS are formed at the Northern, Central and Southwestern stations by dry mass [2]. In 2009, sludge at wastewater treatment plants in Europe amounted to about 12 million tons by dry mass, including 2.26 million tons in Germany and 1.3 million tons in France [3–9].

In the Russian Federation, quite frequently such types of waste are stored outside and are not subject to further treatment. According to official data alone, for example, in the Moscow region, more than 120 million tons of non-disposed sludge with a humidity of 98% have been accumulated, the total area of sludge pits exceeded 700 hectares. The bulk of sewage sludge is stored on sludge pits and dumps causing technological problems in the process of wastewater treatment. In 99% of cases, sludge piles are located next to wastewater treatment plants discharging into water bodies. In most wastewater treatment plants, sludge piles have not been cleaned in the last 10–15 years. Many piles are currently overflowing, which causes a huge problem of placing newly formed SS and overflow of the existing sites. The SS storage of on sludge pits is associated with environmental risks of contamination of surface and groundwater, soils, and vegetation, i.e., the existing conventional technologies do not meet modern environmental and technical requirements nor do they allow using the energy and resource potential of waste. Thus, SS treatment is one of the urgent, priority tasks aimed at reducing and preventing negative effects on environmental objects [10].

Numerous environmental problems in the Russian Federation include soil pollution as one of the most important. The difficulty consists not only in the technological part of the issue, but also in the legislative framework. Against increasingly stringent requirements for the quality of wastewater treatment, legislation does not regulate the issues of wastewater treatment and disposal. The result is secondary environmental pollution, in particular, due to indefinitely long storage [11–17].

The issue of dehydrated SS disposal was raised by Russian scientists as early as 1914, they were recommended to be used as fertilizers or fuel. Currently, considerable experience has been accumulated in the SS use [2]. Thus, SS treatment is one of the urgent, priority tasks aimed at reducing and preventing negative effects on environmental objects.

To solve this problem, various methods of SS treatment and further use are being developed and implemented in Russia and abroad. The main promising areas of sludge disposal are combustion, anaerobic digestion to produce biogas, pyrolysis and use in agriculture and soil reclamation [16, 18]. In this paper, the authors consider such a method of sludge disposal as combustion.

Combustion has always been a very expensive solution available mainly for the largest plants in the developed countries. These units are mainly used abroad in the Nordic countries, France, Japan, the USA in such megacities as Berlin, Munich, Paris, Vienna, London, Manchester, etc. [3, 19, 20]. A good illustration of the increasing use of thermal sludge neutralization is the example of treatment plants in Germany. About two thirds of SS is currently treated thermally in the form of co- or mono combustion, while only 24% of the total SS volume is used directly in agriculture. It should also be noted that after treatment sludge is also used for landscaping or reclamation which account jointly for about 12% of the total amount of SS removal.

According to Eurostat [21], the majority of EU countries have switched to thermal SS disposal: the Netherlands, Austria, Belgium, Poland, Czech Republic, etc. However, statistics on the volume of burnt sludge is ambiguous. For example, Fig. 1 shows that the amount of recycled sludge in the Netherlands from 2009 to 2020 decreased from 336 thousand tons per year to 294 thousand tons, and in Poland it increased almost 11 times over the same period.

Today several technologies are available for the SS thermal treatment. These technologies can be classified in different ways. This article overviews and compares

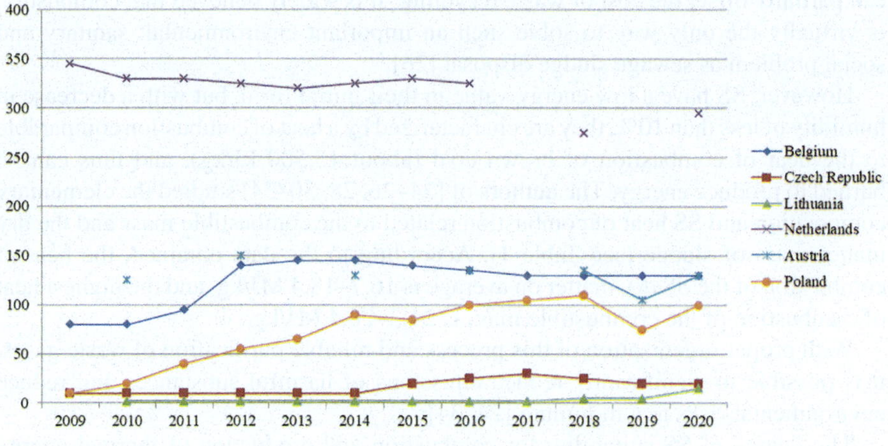

Fig. 1 The amount of burnt SS in terms of dry matter, thousand tons

thermal methods for the treatment of wastewater and provides the results of experiments to determine the possibility of SS combustion obtained from a mixture of household and industrial effluents from one of the cities of the Udmurt Republic (Russia).

2 Theoretical Basis

The main directions of SS thermal treatment are SS pyrolysis or gasification and oxidative SS neutralization (combustion) with waste gas heat recovery [22, 23]. Pyrolysis and gasification are promising from the point of view of energy efficiency, energy saving and the possibility of obtaining valuable secondary liquid organic products and synthesis gas, however, due to the low cost of electricity, pyrolysis and gasification are currently considered non-competitive in economic terms. Thus, the most common thermal method of SS eliminating is the combustion of pre-dehydrated or dried sludge sewage [3, 9, 24–29].

It is advisable to use waste combustion under the following conditions:

- shortage of waste disposal areas
- the necessary environmental measures are being observed
- the possibility of improving the existing SS treatment system
- SS have a high combustion heat.

The main advantage of SS burning is a significant reduction in their mass (by about 75%) and volume (up to 90%), which is especially important against a shortage of free space for the organization of new sludge beds. Many hazardous organic compounds decompose during combustion while the use of the resulting heat for electricity generation and ash and slag residues for the production of some materials can partially offset the cost of waste recycling. It is widely believed that combustion is virtually the only way to solve such an important environmental, sanitary and social problem as sewage sludge disposal [26].

However, SS have a low energy value in their initial form, but with a decrease in humidity of less than 10%, they are characterized by a heat of combustion comparable to the heat of combustion of brown coal (about 15.500 kJ/kg), and thus can be burned to produce energy. The authors of [24–26, 28, 30–34] studied the elementary composition and SS heat of combustion related to the combustible mass and the dry matter mass of sludge (see Table 1). According to the data obtained, the heat of combustion of the SS dry matter on average is 16.7–18.4 MJ/kg, and the highest heat of combustion of the combustible mass is 24.3–26.4 MJ/kg.

With proper organization of this process and reliable purification of waste gases, it is possible to significantly reduce emissions of harmful substances and reduce environmental risks to a minimum [24, 34].

The degree of SS suitability for combustion and production of thermal energy depends mainly on the content of organic substances and humidity in them. When analyzing the cost-effectiveness of the SS combustion method, it is commonly

Table 1 Analysis of the sludge composition of urban aeration stations

Sludge type	Ash content (%)	Heat of combustion (MJ/kg)		Volatile Vc yield (%)
		Combustible mass	Dry mass	
Wet mass from primary sedimentation tanks	21.9–51.0	20.9–27.9	16.5–20.6	84.7–95.0
Active sludge	24.9–36.8	23.1–25.5	15.2–18.5	81.5–89.4
Fermented mixture	24.6–42.4	24.3–27.9	14.0–22.0	82.3–86.2

assumed that this result can be achieved in industrial conditions at the concentrations of solids in sewage sludge not lower than 40%, i.e., with a decrease in its humidity to 60% [2, 3, 30]. In addition, the combustion of dehydrated SS is always preceded by an endothermic process of their thermal preparation, including heating of the material, evaporation of moisture and emission of volatile components. The heat spend for this process is quite high and, in some cases, may exceed the amount of heat released during the sludge combustion, i.e., additional fuel is required for combustion [26, 32, 33].

Before the combustion sewage sludge is subject to preparation, which includes sand removal, thickening, establishment of the required composition and water removal (dehydration) [31, 35–37]. Sand shall be removed to protect the process equipment from wear. In addition, due to this procedure, the heat of SS combustion increases due to an increase in the content of organic compounds. The SS thickening which is commonly associated with the processes such as gravitational settling, surfacing or centrifugation, is primarily intended to reduce the SS volume to be processed. The water from sewage sludge is removed by centrifugation, vacuum filtration, pressure filtration, extraction, sometimes using cross-flow microfiltration and electric dewatering.

The co-combustion of SS with other fuel types has a number of advantages that make it a promising direction [36]. Firstly, it is cost efficient as it allows effective use of the energy contained in such sludge. Secondly, co-combustion reduces the emission of carbon dioxide and other harmful substances into the atmosphere, which is important environmentally. Thirdly, it makes it possible to use the energy potential of biomass, which contributes to the development of renewable energy sources. However, before the co-combustion of SS compounds with other fuels, it is necessary to conduct special studies. The reason for this is that non-conventional fuels can contain harmful impurities, both from a technological and environmental point of view. Thus, before implementing such methods, it is necessary to guarantee the safety of this process and its compliance with environmental safety requirements. One of the successful examples of the SS co-combustion with other fuels is the case of the SS co-combustion (humidity not exceeding 10%) with coal in the Netherlands. As a result of the research, it was found that such a combination allows achieving the greatest efficiency in all indicators, such as environmental safety, cost-effectiveness and technical feasibility. Similar methods are being tested in other countries, such as

Germany, where the SS co-combustion with coal at thermal power plants is considered feasible, as well as in cement kilns [3], where high temperatures facilitate a complete destruction of organic SS compounds and their inclusion in the furnace charge. This allows for the removal of solid combustion products, which significantly reduces environmental risks and reduces the cost of the combustion process. Thus, the SS co-combustion with other fuel types is a promising direction, since it is cheap, environmentally acceptable, reduces the emission of harmful substances and allows efficient use of the biomass energy potential. However, before its implementation, special studies must be conducted to provide safety and compliance with environmental requirements.

The SS combustion can be done in various aggregate states: liquid, paste-like and dry. The SS combustion in the initial liquid state is associated with significant fuel consumption (natural gas, fuel oil) up to 300 kg of conventional fuel per 1 ton of liquid solid waste [1]. It is advisable to burn liquid wastewater at treatment plants of industrial companies in cases where their quantity does not exceed 5 m^3/h (water treatment plants with a capacity of up to 15,000 m^3/ day), when they contain toxic impurities, heavy metal salts, pesticides, etc. At wastewater treatment plants with a capacity of over 15,000 m^3/day, the organization of the SS combustion process in its liquid initial state is associated, alongside with a high demand for fuel, with an increase in the number of furnaces (whose aggregate capacity is limited) and, consequently, technological systems as a whole, which will entail significant operational and capital costs and the complication of operating conditions.

The solid waste combustion in a paste-like state is possible after the application of methods of mechanical dehydration of liquid SS using centrifuges, filter presses and vacuum filters. As a result of mechanical dewatering, the volume of soil is sharply reduced due to the removal of external moisture, a paste-like cake with a humidity of 60–80% is formed [23].

However, the most efficient method of SS combustion having almost an unlimited scope of application for modern wastewater treatment plants with a capacity of up to several million cubic meters per day, is the method of their combustion in a dry pulverized state. The implementation of this method is possible with the sequential implementation of mechanical dewatering and SS thermal drying (requires significant heat expenditure).

On the territory of Russia, most of the treatment facilities are equipped with mechanical dewatering workshops, thus, the experiments on the heat treatment of paste-like cake are of significant interest.

3 Materials and Methods

The research was conducted at one of the wastewater treatment plants of the Udmurt Republic. During the experiments, we used SS after mechanical dehydration (filter press); the generalized indicators of the composition of the studied mixture consisting

Table 2 Sludge indicators of urban wastewater treatment plants

Sludge type	Sludge amount per person/day		Humidity (%)	Ash content of the dry matter (%)
	By dry matter (g)	By volume of raw sludge (l)		
Raw sludge of primary sedimentation tanks	25–40	0.5–0.8	93–95	15–30
Raw activated sludge (thickened)	20–32	0.7–1.1	97	25–30
A mixture of sludge from primary sedimentation tanks and activated sludge (thickened)	45–70	0.6–2.3	93–97	20–30
Aerobically stabilized activated sludge (thickened)	15–25	0.3–0.6	97	30–35

of sludge from primary and secondary settling tanks are shown in Table 2, the density of the mechanically dehydrated sludge was 0.63 t/m^3.

A ready-made isobutane-butane-propane gas mixture (gas proportions in the mixture of 70, 25 and 5%) in a cylinder with a volume of 520 cm^3 and a mass of a liquefied mixture of 220 g, with a continuous combustion duration of 150 min and a calorific value of about 46.8 MJ/kg was used as a combustible material.

In the first series of experiments, SS samples were taken, with a humidity of 80% (humidity was measured with a moisture meter MA-35). A sample weighing 1 kg was put into an experimental unit (see Fig. 2) and was burned under the flame of a gas burner for 1 h. The temperature in the experimental unit was measured using a pyrometer DT-8896, in zone 4 (see Fig. 2) the average combustion temperature was 500 °C, in zone 5 it decreased to 100 °C. This series consisted of three experiments.

In the second series of experiments, the sewage sludge was dried in natural conditions (average daily temperature of 18 °C and air humidity of 34%) for 24 h, and a sample weighing 1 kg was subject to direct combustion inside the experimental unit for 1 h. This series also consisted of three experiments.

In the third series of experiments, the SS was dried naturally (average daily temperature 18 °C and air humidity 34%) for 120 h, and a sample weighing 1 kg was subject to direct combustion inside an experimental unit for 1 h. This series, similar to the previous ones, consisted of three experiments.

4 Results

The ash obtained by combustion was analyzed in terms of its chemical composition, the results are provided in Table 3. Due to the fact that the sewage sludge was obtained from mixed effluents (industrial effluents and household effluents), the sludge have a high content of heavy metals Pb, Cu, Ni, Cr, Zn, and Mn. However, the ash content

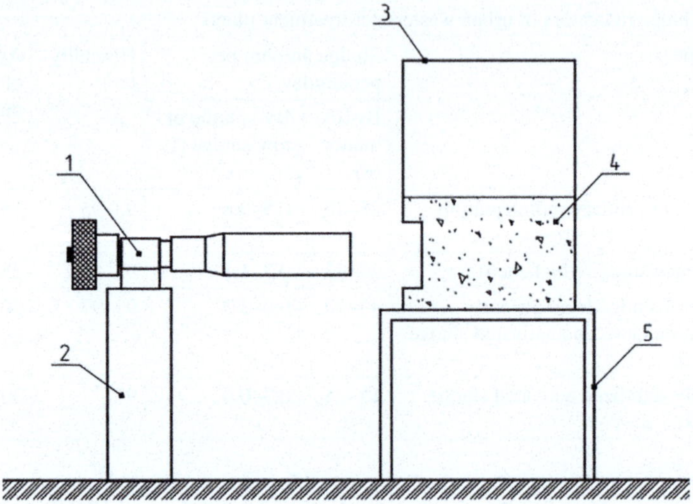

Fig. 2 The configuration of the experimental unit: 1—a burner, 2—a container with a combustible composition, 3—an incinerator, 4—sewage sludge, 5—a pedestal

of the sludge is not lower than the average ash content in the household wastes, thus, the average biomass content was 73.1%. During the combustion process, the mechanically dehydrated sludge began to smolder after 50–55 min, however, at the same time it lost significant mass. Thus, as a result of the first series of experiments, the sludge mass decreased by 5 times, and the ash weight was 200 g.

In the second series of experiments, after natural drying the sludge humidity decreased to 69.9%, and during combustion, the sludge began to smolder completely,

Table 3 The chemical composition of the sludge ash

Indicators	Units of measurement	The average value
Lead	mg/kg	20.32
Copper	mg/kg	148.91
Nickel	mg/kg	29.71
Chrome	mg/kg	677.46
Cadmium	mg/kg	0.42
Zinc	mg/kg	1068.28
Manganese	mg/kg	854.18
Ash content	% on dry matter	32.9
Mass fraction of total nitrogen	% on dry matter	4.39
Mass fraction of total phosphorus	% on dry matter	3.29
Hygroscopic humidity	%	3.52

and at 40–45 min the process of autothermal combustion started at a burning temperature of 300 °C. After drying and combustion, the sludge mass decreased by 8 times.

In the third series of experiments, the sludge humidity amounted to only 8%, thus, the process of autothermal combustion began 30 min following the exposure to direct fire.

5 Conclusions

When cleaning 100 m^3 of wastewater, from 15 to 20 kg of sludge dry matter is formed while from 12 to 15 tons of sludge per day are registered per 100 thousand inhabitants. At the constant storage of sludge on sludge pits (piles), it causes their overflow, which in turn entails an additional increase in the area of sludge piles or the withdrawal and disposal of previously transported sludge to free up existing piles.

Combustion is frequently used as one of the methods of sludge disposal, since from 50 to 70% of waste consists of activated sludge or biomass. Thus, a number of experiments have shown that mechanically dehydrated sludge, when dried for 24 h in natural conditions, is able to burn by itself if exposed to a direct fire source with a temperature of no more than 500 °C. The experiments also showed that 73.1% of the organic matter in the SS composition is sufficient for the reaction of exothermic combustion even at a sludge humidity of 69.9%. At the same time, combustion temperature can reach 300 °C.

During the combustion of mechanically dehydrated sludge in the first series of experiments, ash was obtained in the volume 5 times smaller than the initial volume of sludge while the combustion of the pre-dried SS in the second and third series of experiments allowed reducing the sludge volume by 8 times. Thus, the resulting ash increases the utilization rate of sludge piles several times, and the method of SS combustion can be recommended as one of the most promising for cities and municipalities in Russia.

References

1. Khoreva PV, Bernadiner IM (2017) Experimental determination of thermal characteristics of wastewater sludge for design of incineration plants. Ecology and Industry of Russia 21(3):32–34. https://doi.org/10.18412/1816-0395-2017-3-32-35
2. Bernadiner IM, Khoreva PV, Ekimova VA (2020) Determination of Thermal Technical Characteristics of Sewage Sludge. Journal of Physics: Conference Series 1683(5):052010. https://doi.org/10.1088/1742-6596/1683/5/052010
3. Schnell M, Horst T, Quicker P (2020) Thermal treatment of sewage sludge in Germany: A review. Journal of Environmental Management 263:110367. https://doi.org/10.1016/j.jenvman.2020.110367

4. Fytili D, Zabaniotou A (2008) Utilization of sewage sludge in EU application of old and new methods – A review. Renewable and Sustainable Energy Reviews 12(1):116–140. https://doi.org/10.1016/j.rser.2006.05.014
5. Mountouris A, Voutsas E, Tassios D (2008) Plasma gasification of sewage sludge: Process development and energy optimization. Energy Conversion and Management 49(8):2264–2271. https://doi.org/10.1016/j.enconman.2008.01.025
6. Ghanem A, Bados P, Estaun AR et al (2007) Concentrations and specific loads of glyphosate, diuron, atrazine, nonylphenol and metabolites thereof in French urban sewage sludge. Chemosphere 69(9):1368–1373. https://doi.org/10.1016/j.chemosphere.2007.05.022
7. Luostarinen S, Luste S, Sillanpää M (2009) Increased biogas production at wastewater treatment plants through co-digestion of sewage sludge with grease trap sludge from a meat processing plant. Bioresource Technology 100(1):79–85. https://doi.org/10.1016/j.biortech.2008.06.029
8. Romero-Güiza MS, Flotats X, Asiain-Mira R et al (2022) Enhancement of sewage sludge thickening and energy self-sufficiency with advanced process control tools in a full-scale wastewater treatment plant. Water Research 222:118924. https://doi.org/10.1016/j.watres.2022.118924
9. Hall JE (1995) Sewage sludge production, treatment and disposal in the European union. Journal of the Chartered Institution of Water and Environment Management 9(4):335−343.https://doi.org/10.1111/j.1747-6593.1995.tb00950.x
10. Gaponenkov IA, Fedorova OA, Krashevskaya AA (2015) Heat Treatment of Waste Water Sludges. Ecology and Industry of Russia 19(10):4–9. https://doi.org/10.18412/1816-0395-2015-10-4-9
11. Dyagelev M (2022) An Example of Practical Training on the Discipline "Wastewater Treatment" for Students in the Field of Construction Engineering. AIP Conference Proceedings 2647:030030. https://doi.org/10.1063/5.0104524
12. Augenstein D, Wise D, Dat N et al (1996) Composting of municipal sewage waste and sludge: Potential for fuel gas production in a developing country. Recourses, Conservation and Recycling 16:265−279. https://doi.org/10.1016/0921-3449(95)00060-7
13. Dyagelev MYu (2023) Improving the efficiency of biological treatment of industrial wastewater as part of urban wastewater. Theoretical and Applied Ecology 2:96–103. https://doi.org/10.25750/1995-4301-2023-2-096-103
14. Kharkina OV (2015) Effektivnaia ekspluatatsiia i raschet sooruzhenii biologicheskoi ochistki stochnykh vod (Efficient operation and calculation of biological wastewater treatment facilities). Panorama. Volgograd
15. Kirillov NA, Fadeeva NA (2015) Prospects for the use of sewage sludge to increase the productivity of low-humus soils. Ecological Bulletin of the North Caucasus 11(1):79–83
16. Jamalov RG, Karlsson A, Yushmanova SI et al (2021) Household waste management in the countries of Northern Europe and Russia. Municipal Academy 1:208–217. https://doi.org/10.52176/2304831X_2021_01_208
17. Longobardi A, Elizaryev AN, Nasyrova ES et al (2020) Spread of landfill leachate into groundwater. Theoretical and Applied Ecology 2:36–43. https://doi.org/10.25750/1995-4301-2020-2-036-043
18. Ruchkinova OI, Zvereva AN (2020) Methods of utilization of sludge from municipal treatment facilities. Modern technologies in construction. Theory and practice 1:192–196
19. Donatello S, Cheeseman CR (2013) Recycling and recovery routes for incinerated sewage sludge ash (ISSA): A review. Waste Management 33(11):2328-2340. https://doi.org/10.1016/j.wasman.2013.05.024
20. Chen Z, Hou Y, Liu M et al (2022) Thermodynamic and economic analyses of sewage sludge resource utilization systems integrating Drying, Incineration, and power generation processes. Applied Energy 327:120093. https://doi.org/10.1016/j.apenergy.2022.120093
21. Sewage sludge production and disposal from urban wastewater (in dry sub-stance (d.s)) (2022) https://ec.europa.eu/eurostat/databrowser/view/TEN00030__custom_3980828/default/table?lang=en. Accessed 21 Dec 2023
22. Xu D, Han X, Chen H et al (2020) New insights into impact of thermal hydrolysis pretreatment temperature and time on sewage sludge: Structure and composition of sewage sludge from

sewage treatment plant. Environmental Research 191:110122. https://doi.org/10.1016/j.envres.2020.110122
23. Ngo PL, Udugama IA, Gernaey KV (2021) Mechanisms, status, and challenges of thermal hydrolysis and advanced thermal hydrolysis processes in sewage sludge treatment. Chemosphere 281:130890. https://doi.org/10.1016/j.chemosphere.2021.130890
24. Hu Q, Dai Y, Wang C-H (2020) Steam co-gasification of horticultural waste and sewage sludge: Product distribution, synergistic analysis and optimization. Bioresource Technology 301:122780. https://doi.org/10.1016/j.biortech.2020.122780
25. Peltola P, Ruottu L, Larkimo M et al (2023) A novel dual circulating fluidized bed technology for thermal treatment of municipal sewage sludge with recovery of nutrients and energy. Waste Management 155:329–337. https://doi.org/10.1016/j.wasman.2022.11.017
26. Bonu R, Anand N, Palani SG (2023) Impact of thermal pre-treatment on anaerobic co-digestion of sewage sludge and landfill leachate. Materials Today Proceedings 72(1):99–103. https://doi.org/10.1016/j.matpr.2022.06.130
27. Zhu Y, Zhai Y, Li S et al (2022) Thermal treatment of sewage sludge: A comparative review of the conversion principle, recovery methods and bioavailability-predicting of phosphorus. Chemosphere 291(3):133053. https://doi.org/10.1016/j.chemosphere.2021.133053
28. Chang H, Zhao Y, Zhao S et al (2022) Review of inventory data for the thermal treatment of sewage sludge. Waste Management 146:106–118. https://doi.org/10.1016/j.wasman.2022.05.002
29. Castro-Fernandez A, Taboada-Santos A, Balboa S et al (2023) Thermal hydrolysis pre-treatment has no positive influence on volatile fatty acids production from sewage sludge. Bioresource Technology 376:128839. https://doi.org/10.1016/j.biortech.2023.128839
30. Bąk K, Ziółkowski P, Frost J et al (2023) Comparative study of a combined heat and power plant retrofitted by CO_2 capture during the combustion of syngas from sewage sludge gasification versus zero-emission combustion of hydrogen produced using renewables. International Journal of Hydrogen Energy 48(99):39625–39640. https://doi.org/10.1016/j.ijhydene.2023.07.322
31. Ortner B, Schmidberger C, Gerhardter H et al (2023) Application of computationally inexpensive CFD model in steady-state and transient simulations of pulverized sewage sludge combustion. Advanced Powder Technology 34(12):104260. https://doi.org/10.1016/j.apt.2023.104260
32. Ogugua PC, Su H, Wang E (2024) Synergistic blending of biomass, sewage sludge, and coal for enhanced bioenergy production: Exploring residue combinations and optimizing thermal conversion parameters. Journal of Environmental Management 352:120035. https://doi.org/10.1016/j.jenvman.2024.120035
33. Xia Y, Liu M, Zhao Y et al (2023) Hydration mechanism and environmental impacts of blended cements containing co-combustion ash of sewage sludge and rice husk: Compared with blended cements containing sewage sludge ash. Science of The Total Environment 864:161116. https://doi.org/10.1016/j.scitotenv.2022.161116
34. Quan LM, Kamyab H, Yuzir A et al (2022) Review of the application of gasification and combustion technology and waste-to-energy technologies in sewage sludge treatment. Fuel 316:123199. https://doi.org/10.1016/j.fuel.2022.123199
35. Jadlovec M, Výtisk J, Honus S et al (2023) Pollutants production, energy recovery and environmental impact of sewage sludge co-incineration with biomass pellets. Environmental Technology & Innovation 32:103400. https://doi.org/10.1016/j.eti.2023.103400
36. Xing J, Tang Q, Gan M et al (2023) Co-treatment of municipal solid waste incineration fly ash (MSWI FA) and municipal sludge: A innovative method to improve sludge dewatering with fly ash dechlorination. Journal of Environmental Management 332:117403. https://doi.org/10.1016/j.jenvman.2023.117403
37. Zhang X, Ye P, Wu Y (2022) Enhanced technology for sewage sludge advanced dewatering from an engineering practice perspective: A review. Journal of Environmental Management 321:115938. https://doi.org/10.1016/j.jenvman.2022.115938

System Analysis and Processes Modeling of Waste Water Treatment and Automatic Control of the Indicators Taking into Account an Emergency Probability

O. Shikulskaya, G. Abuova, M. Shikulskiy, O. Nikulin, and A. Burenin

Abstract The article revealed the problem of the impossibility of ensuring the required water quality in the event of an emergency, or other unforeseen situations. The choice of tools for solving the problem is justified: System analysis and functional modeling in IDEF0 notation. A functional model was developed and a systematic analysis of the wastewater treatment technology of one of the enterprises of the gas processing industry was carried out. The choice of the facility was due to the high requirements for the treatment of wastewater of the enterprise due to the strong influence of processed products on the environment of the surrounding area. Analysis of the model showed the inability of the used wastewater treatment system to respond in a timely manner to the sudden release of contaminated water in an extremely situation. Based on the identified shortcomings of the applied technological scheme, the TO-BE model of the modernization of the used technological scheme of purification was developed, which will allow the life of the population in case of an emergency.

Keywords Wastewater treatment · Emergency · Systems analysis · Automation · Model · Process · Diagram

O. Shikulskaya (✉) · G. Abuova · M. Shikulskiy · O. Nikulin · A. Burenin
Astrakhan State University of Architecture and Civil Engineering, 18, Tatishcheva Str., Astrakhan 414056, Russia
e-mail: shikul@mail.ru

L. V. Radionova and D. V. Ulrikh (eds.), *Advances in Ecology and Environmental Engineering*, Springer Proceedings in Earth and Environmental Sciences,
https://doi.org/10.1007/978-3-031-64423-8_23

1 Introduction

The quality of water consumed affects the life and health of the population. Currently, the problem of water quantity and quality is acute in the world [1–5]. Due to the technical progress and rapid growth of industry, there is a sharp deterioration in the state of water resources. In these conditions, the problem of treatment of industrial wastewater at urban-type treatment plants is acute, since the quantity and quality of industrial wastewater can affect the efficiency of wastewater treatment and the possibility of reuse.

Currently, a huge number of methods and technologies for water treatment are known [6–18]. The effectiveness of their use depends on the amount and composition of pollutants in each specific case. The choice of the process treatment scheme in the design of water treatment systems should be due to many factors, both environmental and economic. Water treatment is very expensive, and municipal authorities do not always have enough financial and material resources to ensure the required water quality and sewage treatment. The probability of emergencies and man-made disasters [19] significantly aggravates the problem not only by a sharp increase in pollutants, but also by introducing an uncertainty factor in their composition, which is difficult to foresee when designing treatment facilities. Therefore, in order to select optimal solutions for wastewater treatment technology in the context of environmental and financial problems, taking into account the uncertainty factor, it is necessary to use a system analysis at the pre-design study stage. Currently, intelligent design support methods are used in this area [16–23]. But the considered works do not take into account the likelihood of emergencies and man-made disasters. Therefore, it is necessary to expand and deepen research in this direction, taking into account the specifics of specific conditions. This is due to the relevance and scientific significance of the presented study.

The scientific problem, the solution of which is described in the article, is the development and study of a model of mixed wastewater treatment processes in modern conditions.

The purpose of the study is to model the wastewater treatment technology taking into account the whole variety of factors of the modern stage, taking into account the specific conditions for the pre-design study stage.

Setting of the task: to analyze, on the basis of a systematic approach, the technology for treating wastewater of an enterprise with hazardous production waste and develop a model for improving the waste water treatment technology used on it.

The scientific novelty of the study consists in taking into account in the design of water treatment systems the uncertainty factors of the qualitative and quantitative composition of water pollutants arising from emergencies and man-made disasters, based on a systematic approach and modeling.

The practical value of work is to ensure vital activities in an emergency, improve the health and quality of life of the population by improving the quality of water.

2 Materials and Methods

System analysis and technology of functional modeling in IDEF0 notation are used as tools for solving the identified problem.

Systems analysis is a way to solve complex problems. It is a set of studies aimed at identifying general trends and factors in the development of the organization and developing measures to improve the management system and all production and economic activities of the organization. System analysis can also be applied not to the entire organization, but to any processes, in particular, in this case, it is used to study technological processes.

The IDEF0 methodology is used to create a functional model that displays the structure and functions of the system, as well as the flows of information and material objects transformed by these functions.

For an existing system, the methodology can be used to analyze executable system functions as well as to document the mechanisms by which they are executed.

IDEF0 notation allows you to model system functions of work, actions, operations, processes, functional connections and data information and objects that ensure the integration of system complexes. The developed models are a complete and interrelated description of the system functioning.

The influence of the external environment of the system can also be the object of modeling and research.

The use of a single language to represent the functioning of the system and the external environment allows you to obtain process models that reflect the consumer's point of view.

Existing procedures for discussing IDEF0 models allow the analyst and customer of the design work, the industrial consumer, to reach consensus and mutual understanding.

3 Results and Discussion

The authors developed two models: AS-IS—for the system analysis of the applied technology and TO-BE—a model of the modernized cleaning technology based on the analysis of the first model.

3.1 System Analysis of Mixed Wastewater Treatment Technology at Gas Processing Plant

As a source, the authors developed a functional model of mixed wastewater treatment processes for one of the branches of Gazprom energo LLC located in Russia (AS-IS model). The choice of the facility is due to the high requirements for the treatment of wastewater of the enterprise due to the strong influence of processed products on the environment of the surrounding area. The node tree diagram of this model is shown in Fig. 1.

In normal everyday production conditions, the wastewater treatment system operates efficiently, ensuring compliance with regulatory requirements. It implements the trend of secondary use of treated water, which will save water [23]. During its design, the modern principles of improving the quality of treatment for the purpose of recycling treated water for irrigation were laid, namely, the construction of additional treatment facilities for the purpose of treating mixed wastewater (industrial and biological and rainwater) and dividing the technological process into stages and stages.

At the 1st stage of industrial wastewater treatment, a 3-stage treatment scheme is used, the basis of which is mechanical treatment with mandatory control of pollution indicators at the inlet. In this case, depending on the specific conditions, gravity devices of various structures are used, and in order to increase the treatment effect, preliminary or subsequent treatment of wastewater (oil traps, sumps) can be carried out. At the 3rd stage, in addition to settling, the purification technology includes flotation.

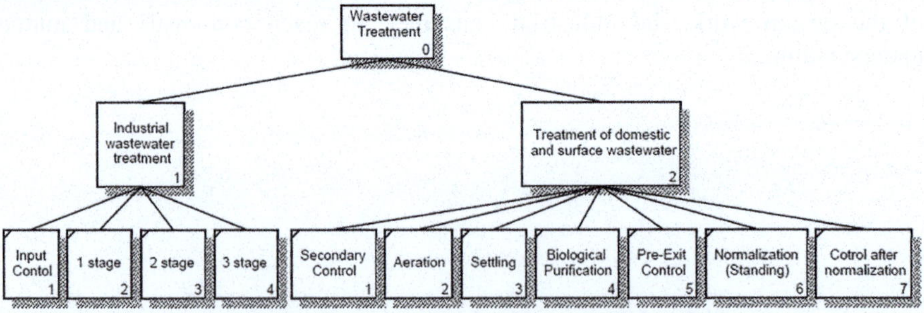

Fig. 1 Node tree diagram of the mixed wastewater treatment model (AS-IS)

The 2nd treatment stage of biological and rainwater effluents includes 7 processes of treatment with input parameters control to determine the efficiency of industrial waste water treatment at the 1st stage and output parameters control. Stages are used aerotanks, sedimentation tanks, oxythenks, control capacities. After analyzing the output indicators, the runoff will go to watering the planted vegetation. If the controlled parameters are exceeded, the effluents are sent for re-purification.

However, the implemented model is not focused on the likelihood of an emergency or a technological disaster. A system analysis of the AS-IS model revealed the disadvantages of the technological process: the impossibility to divert water with a high concentration of pollutants not known in advance in terms of composition in order to separate them from the total wastewater flows, and the probability of untimely detection of significant deterioration due to manual water control technology at regular intervals, which will result in severe contamination of the entire stream and in the shutdown of the water treatment process to identify the composition of the contaminants and the choice of water treatment methods. A TO-BE model was developed to address the identified deficiencies.

3.2 *TO-BE Waste Water Treatment Model at Gas Processing Plant*

The upgrade process is described by the TO-BE model in a number of diagrams.

The node tree diagram of the model TO-BE as a whole and its fragments is shown in Fig. 2.

The context diagram of waste water treatment model with automated control is shown in Fig. 3

The Context diagram decomposition is shown in Fig. 4.

The industrial wastewater treatment process decomposition diagram is shown in Fig. 5.

The decomposition diagram of the additional pretreatment process is shown in Fig. 6.

The decomposition diagram of the treatment of domestic and surface wastewater process is shown in Fig. 7.

Process modernization is reflected in the TO-BE model in the node tree diagram (Fig. 2) and the decomposition diagram of the industrial wastewater treatment (Fig. 5) by highlighting the supplemented process with background and color, the decomposition diagram of the process "Additional pre-treatment" (Fig. 6), which was not in the AS-IS model (Fig. 1).

In addition, in diagrams (Figs. 5, 7) red arrows of the mechanism indicating the automated monitoring system are displayed.

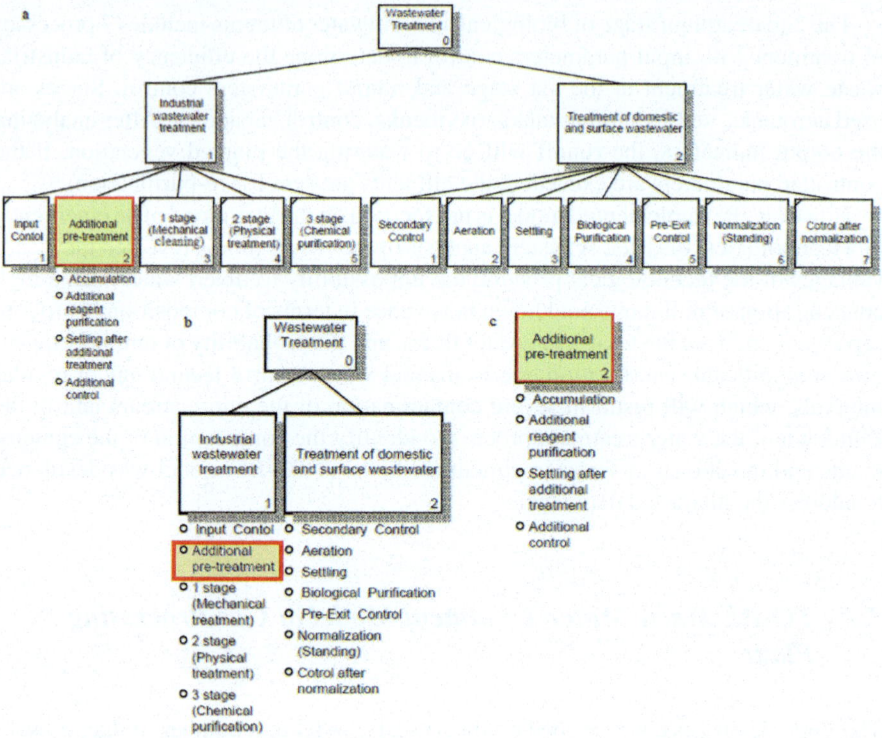

Fig. 2 Mixed wastewater treatment model node tree diagram with automated control (TO-BE) **a** 1–4 levels (all levels); **b** 1–3 levels; **c** 4–5 levels

Fig. 3 Context diagram of waste water treatment model with automated control

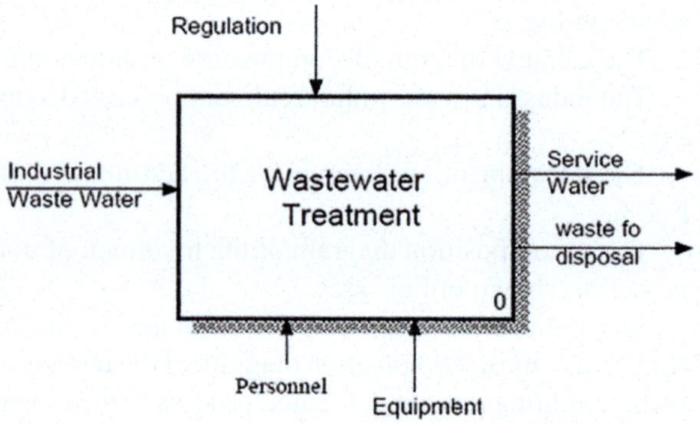

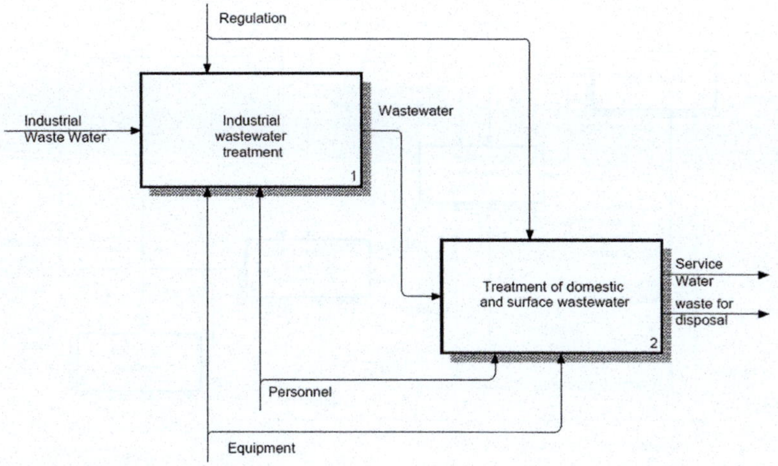

Fig. 4 Context diagram decomposition

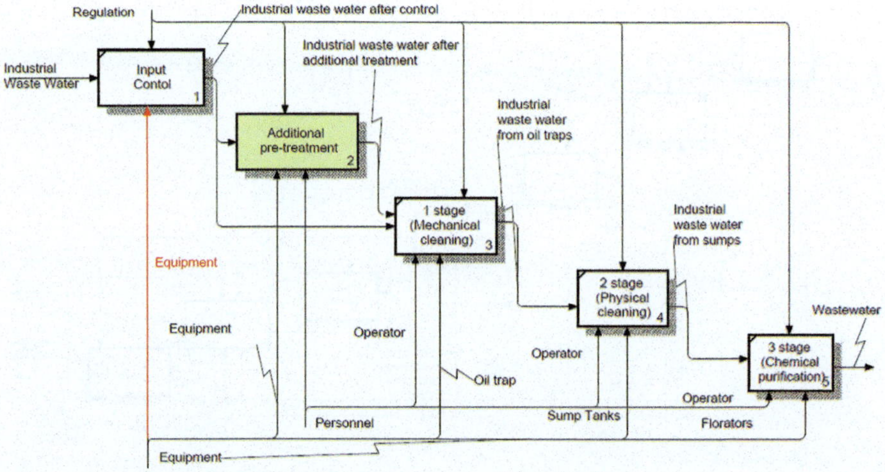

Fig. 5 Industrial wastewater treatment process decomposition diagram

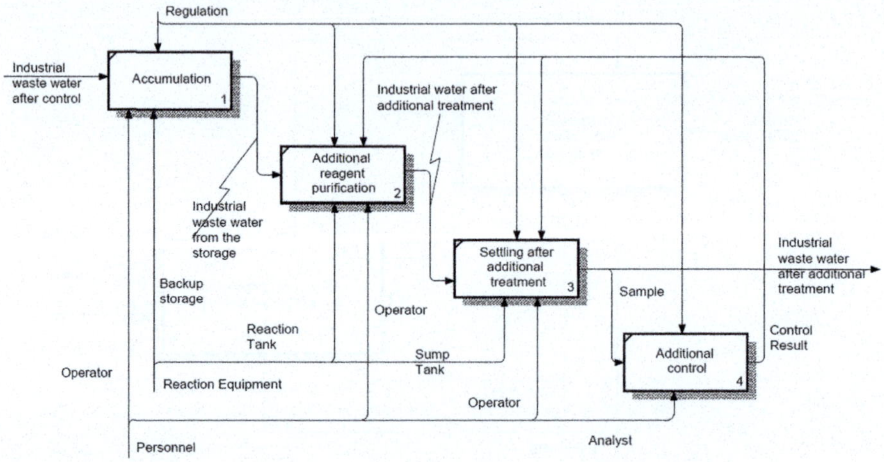

Fig. 6 Decomposition diagram of the additional pretreatment process

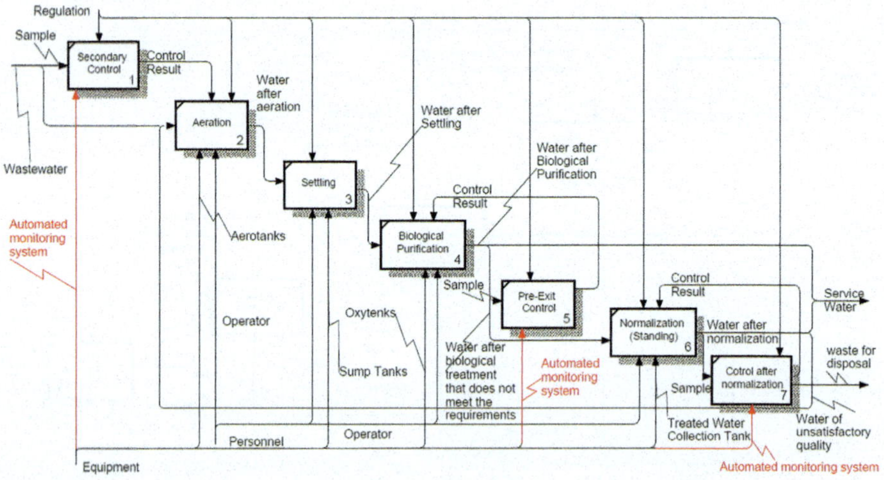

Fig. 7 Decomposition diagram of the treatment of domestic and surface wastewater process

4 Conclusion

To reduce the risk of a sudden sharp change in the environmental situation due to unforeseen problems of an emergency nature (emergency, technosphere disaster, etc.), the authors conducted a systematic analysis of the wastewater treatment technology used at one of the enterprises in the gas processing industry.

The facility choice was due to the high requirements for the wastewater treatment of the enterprise due to the strong influence of processed products on the environment of the surrounding area.

Analysis of the developed AS-IS model showed that despite the effective treatment of wastewater in everyday conditions, the enterprise is not ready to quickly respond to a sudden release of contaminated water in an different extremely situations.

This is due to the following circumstances. As a result of melting snow or rain, chemicals, the remains of spilled various fuels, etc. can get into the sewer. In case of complete or partial destruction of any dangerous enterprise and the formation of a volley discharge of chemical toxic substances, the excess of the indicators can be monitored at the input control, but additional analysis is required to specify the reason for this and determine the method of neutralization. For this purpose, backup storage must be provided at the inlet.

In addition, analysis of the developed AS-IS model showed that the least effective processes of the applied technology are control processes at different wastwater treatment stages due to the use of manual technology. Control is performed by personnel in accordance with the specified time interval. The timely dosing of the reagent in the desired concentration during the purification process depends on the time interval. However, if wastewater with a high concentration of contaminants enters the receiving chamber in the period of time between sampling, the operator does not detect this excess, which affects the quality of treatment.

A TO-BE model was developed to address these deficiencies. The design and modernization of treatment facilities at especially dangerous enterprises will ensure the vital activity in an emergency, as well as affect the duration and quality of the population life.

References

1. Water resources across Europe – confronting water scarcity and drought (2009) EEA Report 2:11
2. Kumar Reddy DH, Lee SM (2012) Water Pollution and Treatment Technologies. J Environ Anal Toxicol 2(e103):2–4

3. Schwarzenbach RP, Escher BI, Fenner K, Hofstetter TB, Johnson CA, Von Gunten U, Wehrli B (2006) The challenge of micropollutants in aquatic systems. Science 313:1072–1077
4. Grace KC Ding (2023) Wastewater Treatment, Reused and Recycling – A Potential Source of Water Supply, Reference Module in Earth Systems and Environmental Sciences, Elsevier. https://doi.org/10.1016/B978-0-323-90386-8.00062-0
5. Ikanina E (2011) Industrial pollution of the hydrosphere and problems of water treatment in Russia. International Conference on Sustainable Cities. IOP Conf. Series: Earth and Environmental Science 177(2018):012011. IOP Publishing. https://doi.org/10.1088/1755-1315/177/1/012011
6. Viessman W and Mark J (2005) Water Supply and Pollution Control. Upper Saddle River. Pearson Prentice Hall, NJ
7. Peter AC, Catalina AP, Matthew F, Stephen C (2012) Slow-sand water filter: Design, implementation, accessibility and sustainability in developing countries. Med Sci Monit 18(7):RA105–17
8. Chaurasia S, Tiwari A (2016) A review on traditional water purification methods used in Rural Area Indian. Journal of Environmental Protection 36:43–48
9. Murat E, Serkan A, Ercan G, Ebubekir Y (2017) Textile Materials in Liquid Filtration Practices: Current Status and Perspectives in Water and Wastewater Treatment. Textiles for Advanced Applications. https://doi.org/10.5772/intechopen.69462
10. Lee A, Elam J and Darling SB (2016) Membrane materials for water purification: Design, development, and application. Environmental Science: Water Research & Technology 2:17–42
11. Mukhtar Nuhu Yahyaa, Hüseyin Gökçekuş, Dilber Uzun Ozsahin (2020) Performance Analysis of the Water Treatment Reverse Osmosis Plant. Jurnal Kejuruteraan 32(2):221–230. https://doi.org/10.17576/jkukm-2020-32(2)-06
12. Dhote Jayashree, Ingole Sangita, Chavhan Dr Arvind (2012) Review on Waste Water Treatment Technologies. International Journal of Engineering Research and Technology 1(05):01–10
13. Epelle EI, Okoye PU, Roddy S, Gunes B, Okolie JA (2024) Wastewater Treatment Technologies. Encyclopedia. https://encyclopedia.pub/entry/37133. Accessed 22 January 2024
14. Letterman, Raymond D (1999) Water Quality and Treatment: A Handbook of Community Water Supplies. American Water Works Association. McGraw-Hill, New York
15. Viessman W, Mark J (2005). Water Supply and Pollution Control. Upper Saddle River. Pearson Prentice Hall, NJ
16. Jin Quan, Lingzhong Kong, Xiaohui Lei, Mingna Wang (2018) Emergency Operations of Sudden Water Pollution Accidents in the Middle. https://doi.org/10.5772/intechopen.82021
17. An Integrated Framework for Treatment and Management of Produced Water (2009) Technical assessment of produced water treatment technologies. 1st EDITION. RPSEA Project 07122–12
18. Luis Arismendy, Carlos Cárdenas, Diego Gómez, Aymer Maturana, Ricardo Mejía, Christian G Quintero M (2020) Intelligent System for the Predictive Analysis of an Industrial Wastewater Treatment Process. Sustainability 12(16):6348 https://doi.org/10.3390/su12166348
19. Shikulskaya OM, Urechko MA, Usynina AE (2018) Information and analytical providing for research of reservoirs' state in conditions of anthropogenic pollution. IOP Conf. Series: Materials Science and Engineering 451:012195
20. Shikulskaya O, Boronina L, Yurechko M, Petrova I, Shikulskiy M (2015) Information intelligent model of the aquatic ecosystem state Identification under the heavy metals influence. The International Conference on Information, Intelligence, Systems and Applications IISA2015, 7388098
21. Shikulskaya OM, Boronina LV, Abuova GB, Usynina AE (2020) Electronic classifier of natural water treatment technologies. IOP Conference Series: Materials Science and Engineeringm 775(1):012098

22. Zhurba MG, Nechayev AP, Ivleva GA, Vostova WM, Rodina IC, Vanin B, Kozina AK (2000) Classifiers of technologies of natural water purification Zhurbs M.G. (Moscow), p 118
23. Englande AJ, Krenkel P, Shamas J (2015) Wastewater Treatment &Water Reclamation. Reference Module in Earth Systems and Environmental Sciences, Elsevier. https://doi.org/10.1016/B978-0-12-409548-9.09508-7

Effectiveness of Ultrafiltration (UF) and Reverse Osmosis (RO) in the Purification of Groundwater with a High Content of Iron and Manganese

Lin Maung Maung, G. G. Kagramanov, and Htet Aung

Abstract The article discusses the removal of iron and manganese ions from groundwater using the ultrafiltration (UF) and reverse osmosis (RO) process. The influence of technological parameters, such as operating pressure and temperature, on the characteristics of the UF and RO membranes has been studied. It is shown that high values of operating pressure and temperature lead not only to an increase in flux, but also to a decrease in the rejection of the UF membrane. This study provides useful data on the use of ultrafiltration as a pre-treatment for reverse osmosis process in the purification of groundwater for removing iron and manganese.

Keywords Groundwater · Oxidation · Membrane separation · Reverse osmosis · Ultrafiltration · Iron · Manganese

1 Introduction

The scarcity of freshwater resources and the need for additional water supply are critically important in many arid regions of the world and will become increasingly important in the future. The main water resources for domestic, communal, irrigation and economic needs are natural water resources—surface and underground. Fresh water from surface water and groundwater accounts for only 0.01% of the total volume of water resources on earth [1]. Most of the world's countries depend on surface waters for daily needs, including for drinking water sources. This is due to the fact that surface water can relatively be found on the surface of the earth in rivers, ponds, lakes or other fresh water (unsalted) sources. However, surface waters contain a wide range of pollutants: suspended particles, organic compounds, inorganic salts

L. Maung Maung (✉) · G. G. Kagramanov · H. Aung
Mendeleev University of Chemical Technology of Russia, 9, Miusskaya Sq., Moscow 125047, Russia
e-mail: linmg51@gmail.com

L. V. Radionova and D. V. Ulrikh (eds.), *Advances in Ecology and Environmental Engineering*, Springer Proceedings in Earth and Environmental Sciences,
https://doi.org/10.1007/978-3-031-64423-8_24

(dissolved ions), hardness salts, bacteria, etc. Therefore, groundwater plays a vital role in providing clean and adequate drinking water worldwide [1].

The high content of iron and manganese in groundwater is the most urgent task in the purification of groundwater to obtain drinking-quality water. Ultrafiltration (UF) effectively eliminates the microbiological risks associated with groundwater and is widely used to meet the ever-increasing needs of the world suffering from a shortage of clean water. The UF process is the best and most reliable pre-treatment method before the reverse osmosis desalination stage [2]. Reverse osmosis (RO) system plays a major role in both water treatment and the chemical industry, especially in the production of purified drinking water.

Ultrafiltration allows the purification of all suspended solids and biological impurities from water for reverse osmosis desalination plants. However, the traditional UF process based on size exclusion provides a low purification efficiency of Fe^{2+} and Mn^{2+}, since their molecular sizes were much smaller than the pore size of the membrane. Iron (Fe^{2+}) and manganese (Mn^{2+}) ions penetrate into the pores of the membrane to form iron and manganese oxides and would be deposited inside the pores of the membrane.

As a result, irreversible hydraulic contamination of the membrane occurs and a significant decrease in water productivity. Most methods of removing iron (Fe^{2+}) and manganese (Mn^{2+}) from water are based on pretreatment—aeration or reagent oxidation of their divalent forms to insoluble forms with further filtration of precipitation. Oxygen from air, sodium hypochlorite, ozone, chlorine dioxide, etc. are used as oxidizing agents (oxidizers). The rate of oxidation of these metals (iron, manganese) depends on the pH value of solution and the type of reagent [3–5]. After UF pre-treatment, the hydraulic characteristics of the RO plant increase when high-quality water is supplied to it. Another advantage of UF pre-treatment compared to conventional pre-treatment for RO is low footprint and low energy consumption.

2 Experimental Data

The main impurities present in groundwater are dissolved manganese and iron compounds. The concentrations of iron and manganese ions in groundwater significantly exceed the WHO limit values [6]. With an iron content of more than 0.3 mg/L and manganese of more than 0.1 mg/L, such water is unacceptable for drinking, domestic and communal purposes. Groundwater purification experiments were carried out on model solutions, the composition of which is presented in following Table 1.

TDS—total dissolved solids; TH—total hardness; EC—electroconductivity; WHO—World Health Organization

The TDS contents of groundwater—7.9 g/L and 10 g/L respectively. Therefore, groundwater from Myanmar can be called brackish water [3].

In this research work, all experiments were carried out using a laboratory UF and RO unit, the diagram of which is shown in Fig. 1. Experiments were carried

Table 1 Physico-chemical parameters of groundwater from Myanmar

Parameters	Groundwater compositions		WHO limits
	1st component	2nd component	
Fe^{2+}, mg/L	14	10	0,3
Mn^{2+}, mg/L	11,5	8,5	0,1
Ca^{2+}, mg/L	226,4	731,4	180
Mg^{2+}, mg/L	434,5	407,1	40
Na^{+}, mg/L	2075	2117	120
K^{+}, mg/L	10,95	10,85	50
$NO_3{}^{-}$, mg/L	354	305,1	50
Cl^{-}, mg/L	2738	3855	250,0
$SO_4{}^{2-}$, mg/L	2065	2642	250,0
*TDS, mg/L	7901	10,066	600
*TH, mg/L	2353	3500	100
*EC, μS/cm	10,204	12,710	2500
pH	7,35	7,48	6,5–8,5

out on the following membrane modules—ultrafiltration: hollow fiber membrane modules based on poly-sulfone membranes, characterized by a pore size of 20 nm (0.02 μm) and molecular weight cutoffs ≈100 kDa; reverse osmosis: spiral-wound module, model—Desal TFM-75 NT. All modules were placed in standard housings. All experiments were carried out at a temperature of 20 ± 2 °C and an operating pressure of 4.0 ± 1 bar in a laboratory unit. Measurements of the concentration of iron and manganese ions in purified water were carried out using the method of atomic absorption spectroscopy (AAS) on a spectrophotometer. For measuring the temperature and pH of the solution, a thermometer (±0.2 °C) and a pH meter (±0.01 pH) were used.

3 Results and Discussions

Most methods of water purification from iron and manganese are based on the process of oxidation by aeration or the use of chemical reagents with further mechanical filtration [7]. Iron and manganese are present in groundwater mainly in the form of dissolved compounds of divalent iron (Fe^{2+}) and manganese (Mn^{2+}) [8].

To convert them into insoluble compounds, various oxidizing agents are used: sodium hypochlorite (NaClO) oxidizes iron (Fe^{2+}) to insoluble iron hydroxide (Fe^{3+}), and potassium permanganate ($KMnO_4$) oxidizes manganese (Mn^{2+}) to manganese oxide (Mn^{4+}).

The processes of converting the forms of these compounds occurring in water treatment plants are described by the following chemical reactions:

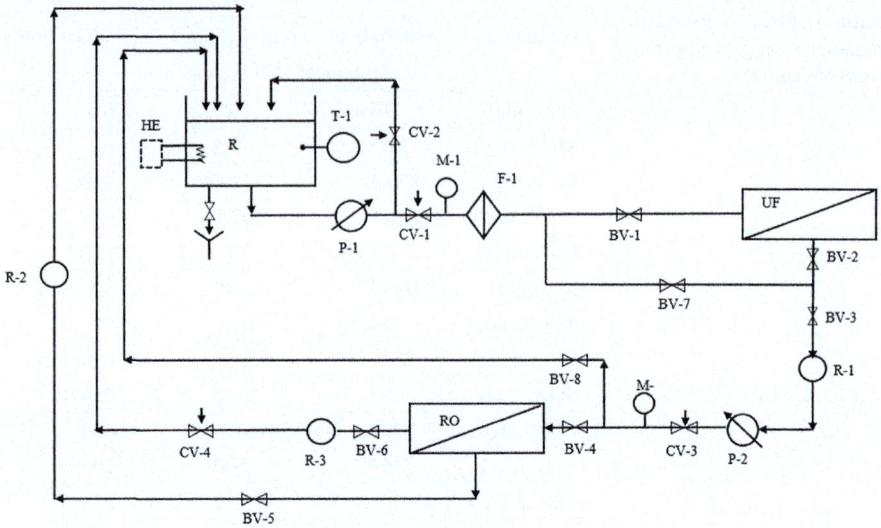

Fig. 1 Flow diagram of UF and RO laboratory unit: R—raw water tank; HE—heat exchanger; T-1—thermometer; P-1, P-2—pumps; F-1—coarse filter; UF—ultrafiltration membrane apparatus; CV—control valves, BV—ball valves; R-1—rotameter; M-1—manometer; RO—reverse osmosis membrane apparatus

$$Fe^{2+} \rightarrow Fe^{3+} + e \tag{1}$$

$$4Fe(HCO_3)_2 + O_2 + 2H_2O \rightarrow 4Fe(OH)_{3\downarrow} + 8CO_2 \tag{2}$$

$$Mn^{2+} \rightarrow Mn^{4+} + 2e \tag{3}$$

$$2Mn(HCO_3)_2 + O_2 \rightarrow 2MnO_{2\downarrow} + 2H_2O + 4CO_2 \tag{4}$$

After oxidation process, water is supplied to the ultrafiltration stage for pre-treatment of reverse osmosis desalination with removing iron and manganese colloids. UF membranes have a pore size from 10 to 50 nm and effectively trap not only colloidal particles and fine impurities, but also high-molecular substances, organic pollutants and bacteria, while the salt composition of water does not change [9–11].

The focus of this study was to evaluate the effectiveness of ultrafiltration in removing iron and manganese from groundwater sources for process optimization.

A study of the influence of operating characteristics (temperature and operating pressure) on the efficiency of the UF purification (flux and rejection) was carried out on hollow fiber UF membrane materials.

It was noted that an increase in operating pressure leads to an increase in flux due to an increase in the driving force–the pressure difference on both sides of the membrane. In Fig. 2a, b, the rejection of the UF membrane is estimated at different operating pressures. The rejection value was determined by the specific electroconductivity TDS value of the solution. It was observed that increasing the operating pressure leads to higher rejection towards the iron and manganese contaminants.

Temperature also significantly affects the purification efficiency of water containing iron and manganese by ultrafiltration process [9]. In Fig. 3a, b, increasing the temperature has a positive effect on the characteristics of the UF membrane by

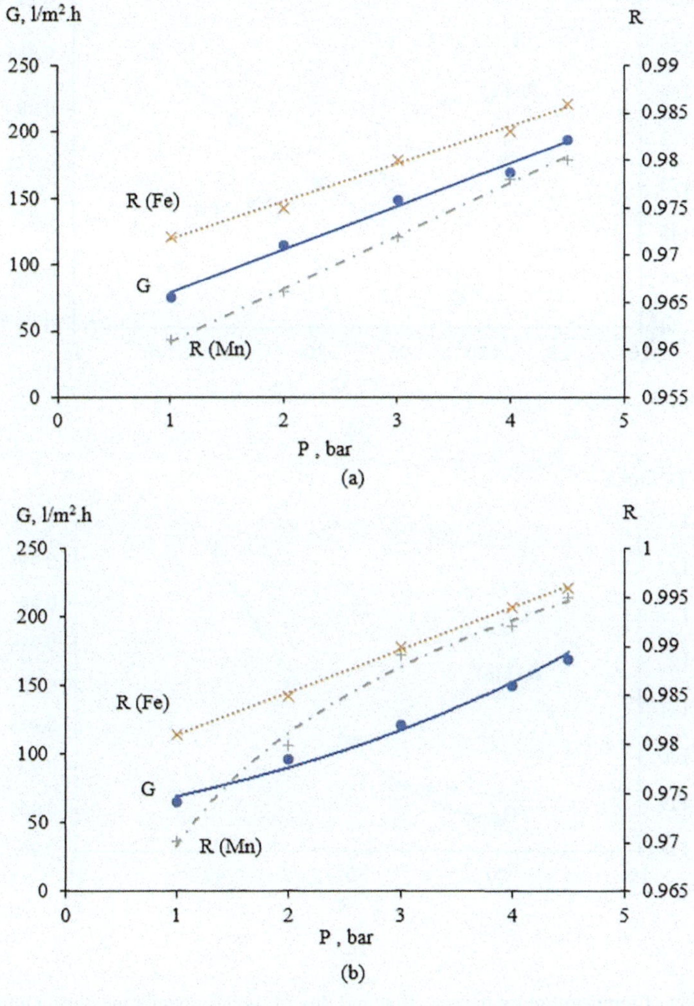

Fig. 2 Effect of operating pressure on the rejection and flux of the UF membrane during purification of groundwater sample **a** 1st component and **b** 2nd component T = 20°C ± 1; pH = 6,8 ± 0,5; TDS (1st component)—7, 9 g/L; TDS (2nd component)–10 g/L

increasing the flux and rejection of the UF membrane. However, very high temperatures (>40 °C) can potentially lead to degradation of the membrane polymer, which in turn leads to a decrease in rejection. It is important to determine the optimal temperature range, which provides a balance between rejection of membrane and energy consumption.

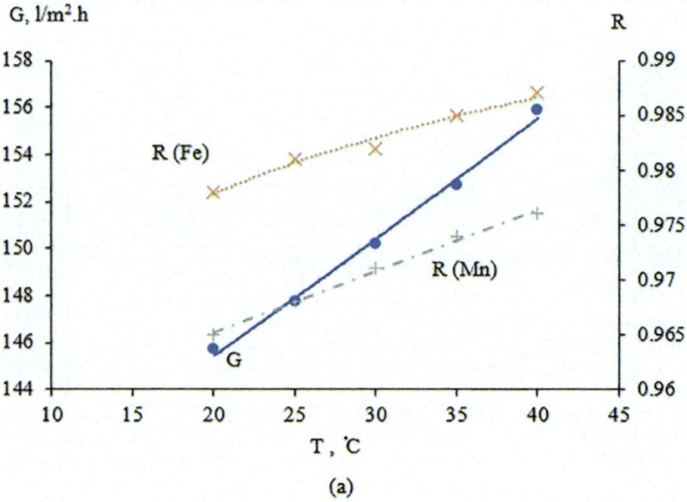

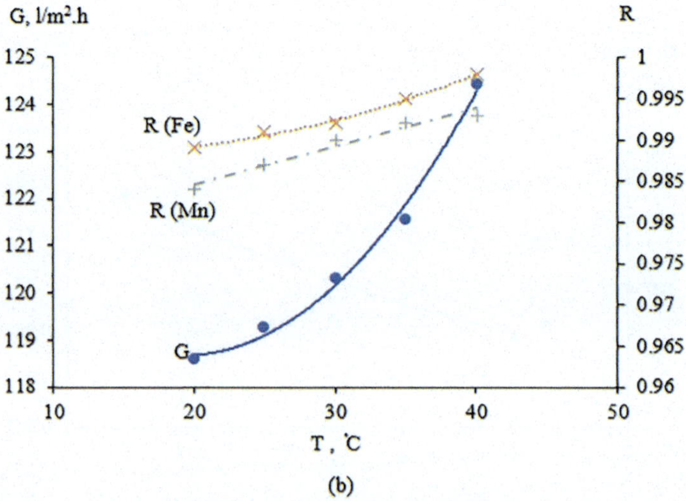

Fig. 3 Effect of temperature on the rejection and flux of the UF membrane during purification of groundwater sample **a** 1st component and **b** 2nd component P = 3 bar; pH = 6, 8 ± 0, 5; TDS (1st component)—7, 9 g/L; TDS (2nd component)–s10 g/L

After UF purification, the water is supplied to the reverse osmosis stage in order to reduce the total salinity (TDS) content and, in particular, the content of hardness salts (calcium and magnesium ions).

Reverse osmosis is the most energy efficient process compared to other desalination processes, which is mainly due to the absence of a phase transition in the purification process [12]. The main advantages of RO are a high degree of purification (up to WHO standards), low costs of chemicals and electricity, especially compared to distillation, compact equipment, ease of management and control of the purification process and the possibility of full automation of processes.

In the process of RO purification, water is divided into two streams: permeate, which is desalinated (purified) water, and concentrate, a concentrated stream in which the concentration of salts is four times higher than in the feed water [13]. In Tables 2 and 3 show that RO purification provides remarkable rejection value (≥97–98%) for ions of salts, but requires high efficiency pre-treatment system of water, including the UF stage. The obtaining purified water meets the drinking water quality requirements established by the World Health Organization (WHO).

In Fig. 4 shows the main technological process used in the purification of natural groundwater with a high content of iron and manganese compounds.

Sodium hypochlorite and potassium permanganate are added into the feed water from tanks 1 and 2, then the water is supplied to the ultrafiltration stage 3 for pre-treatment before reverse osmotic desalination. Before being fed to the reverse osmosis stage, sodium pyrosulfite is dosed into the water to protect the membranes from

Table 2 Effect of operating pressure on the characteristics of RO membrane

P, bar	G, l/m^2.h	Rejection, R (%)				
		Fe^{3+}	Mn^{4+}	Mg^{2+}	Ca^{2+}	Na^+
1	75,27	98,39	97,33	97,36	98,24	94,87
2	115,19	98,49	97,51	97,56	98,38	94,99
3	148,49	98,56	97,68	97,79	98,45	95,23
4	170,06	98,75	97,83	97,95	98,49	95,49
4,5	194,35	98,94	97,94	98,18	98,77	96,37

Table 3 Effect of temperature on the characteristics of RO membrane

T, C	G, l/m^2.h	Rejection, R (%)				
		Fe^{3+}	Mn^{4+}	Mg^{2+}	Ca^{2+}	Na^+
20	145,72	98,54	97,63	97,10	98,50	95,02
25	147,75	98,55	97,66	97,21	98,58	95,12
30	150,17	98,58	97,71	97,32	98,66	95,24
35	152,69	98,61	97,74	97,43	98,78	95,44
40	155,91	98,63	97,99	97,56	98,92	95,51

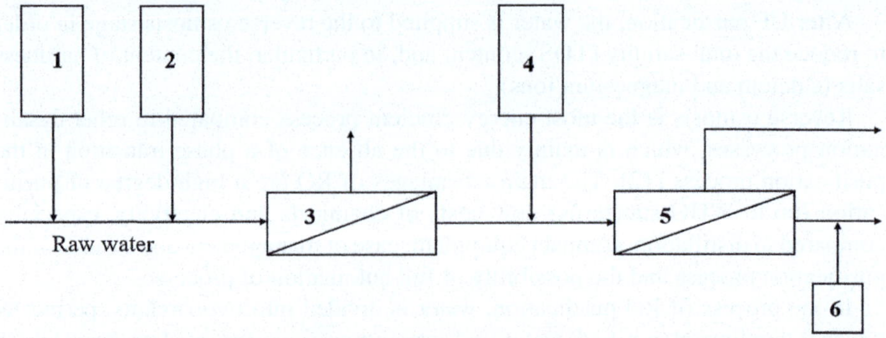

Fig. 4 Schematic flow diagram for natural groundwater purification, 1—sodium hypochlorite tank; 2—potassium permanganate tank; 3—UF purification stage; 4—sodium pyrosulfite tank; 5—RO purification stage; 6—reagent for chlorination

oxidative degradation. After that, the water enters the reverse osmosis stage. OO permeate is chlorinated from container 6 before being supplied to the network.

The capacity of purified water in this system is 11–18 m^3/h at an operating pressure of 10–16 bar. The degree of desalination of the membrane element is at least 99%.

4 Conclusion

The influence of the main technological parameters on the separation efficiency of UF and RO membranes during the purification of brackish groundwater with a high content of iron and manganese has been studied. The obtained results emphasize the importance of optimizing the operating parameters of the membrane separation process—pressure and temperature, in order to achieve higher values of flux and rejection.

The ultrafiltration process provides fine purification of water from suspended solids and colloidal particles. The use of UF process for pre-treatment of RO unit is a promising and reliable way to adequately produce drinking water. Reverse osmosis purification system provides good high quality post UF treatment after obtaining purified drinking water that meets WHO standards.

In further studies, it is planned to conduct the specific experiments of UF membranes and study the stability of the UF-RO purification system under various operating parameters and changing water quality characteristics. The obtained results can be used by engineers in the design of groundwater treatment plants to obtain drinking-quality water.

References

1. The Human Right to Water and Sanitation (2010) https://www.un.org/es/comun/docs/?symbol=A/RES/64/292. Accessed 8 Jan 2024
2. Sobsey MD, Stauber CE, Casanova LM, Brown JM, Elliott MA, Farrah SR (2008) Point of use household drinking water filtration: a practical, effective solution for providing sustained access to safe drinking water in the developing world. Environmental Science & Technology 42(12):4261–4267. https://doi.org/10.1021/es702746n
3. Kavcar P, Sofuoglu A, Sofuoglu SC (2009) A Health risk assessment for exposure to trace metals via drinking water ingestion pathway. International Journal of Hygiene and Environmental Health 212(2):216–227. https://doi.org/10.1016/j.ijheh.2008.05.002
4. Dytnersky YI (1995) Basic processes and devices of chemical technology. Part 2. Mass transfer processes and devices. Chemistry, Moscow
5. Izabela Krupińska (2020) Impact of the Oxidant Type on the Efficiency of the Oxidation and Removal of Iron Compounds from Groundwater Containing Humic Substances. Molecules 25(15):3380. https://doi.org/10.3390/molecules25153380
6. Guidelines for drinking-water quality. 4-th Ed. site World Health Organization (2011) https://www.who.int/publications/i/item/9789241548151. Accessed 12 Jan 2024
7. Lin Maung Maung, Htet Aung, Saw Thurein, Parusov DV, Kagramanov GG, Farnosova EN (2021) Methodological foundation and assessment of artesian water treatment technology in the Republic of the Union of Myanmar. Ecology and Industry of Russia 25(8):34–39. https://doi.org/10.18412/1816-0395-2021-8-34-39
8. Rusydi AF, Onodera SL, Saito M et al (2021) Vulnerability of groundwater to iron and manganese contamination in the coastal alluvial plain of a developing Indonesian city. SN. Applied Science (3):399. https://doi.org/10.1007/s42452-021-04385-y
9. Dytnersky YI (1986) Baromembrane processes. Theory and calculation. Chemistry, Moscow
10. Rajindar Singh, Nicholas P. Hankins (2016) Introduction to Membrane Processes for Water Treatment. In: Emerging Membrane Technology for Sustainable Water Treatment, Elsevier Science, p 15–52. https://doi.org/10.1016/B978-0-444-63312-5.00002-4
11. Perfilev AV, Chirikov AYu, Badulin YuM, Yudakov AA, Slesarenko VV (2014) Research of Demanganation Process of Artesian Well Water. Russian Journal of General Chemistry 84(13):2588–2591. https://doi.org/10.1134/S1070363214130155
12. Khaled Walha, Raja Ben Amar, Loubna Firdaous, Francis Quemeneur, Pascal Jaouen (2006) Brackish groundwater treatment by nanofiltration, reverse osmosis and electrodialysis in Tunisha: performance and cost comparison. Desalination 207:95–106. https://doi.org/10.1016/j.desal.2006.03.583
13. Xianhui Li, David Hasson, Raphael Semiat, Hilla Shemer (2019) Intermediate concentrate demineralization techniques for enhanced brackish water reverse osmosis water recovery—A review. Desalination 466:24–35. https://doi.org/10.1016/j.desal.2019.05.004

Using Industrial Waste for Treatment of Urban Surface Runoff

I. Arkanova, L. Fedorova, E. Terekhova, and A. Podporin

Abstract A radical solution to modern environmental problems caused by the negative impact from human technological activity is possible through the use of zero-waste and low-waste technologies. In this case, industrial waste is used as valuable types of secondary material and energy resources that are generated in large industrial centers, where there are conceptual opportunities for their reuse. Agricultural waste is used as sorbents: corn husk, modified leaves near the cob after cleaning; foam blocks (building material made from a type of cellular concrete). The sorbate is surface wastewater from urban territories of the city of Chelyabinsk. The main pollutants of stormwater and snowmelt runoff in the city of Chelyabinsk, especially near surface sources, such as Smolino Lake and the Miass River (which are reservoirs for fishing purposes), have been identified. Under dynamic conditions, the use of crushed leaf corn husk is more effective for removing iron ions (76.5%); the removal of suspended solids by crushed foam concrete is 83.5%; and the combination of these sorption materials makes it possible to reduce the iron content by 70.6% and suspended solids by 95.7%. It makes it possible to propose a combined filter medium for treatment of urban surface runoff. The studied waste from the agro-industrial complex and construction production are environmentally friendly and effective sorption materials that allow purification from iron ions and suspended solids, which is necessary to protect surface sources from pollution by stormwater and snowmelt runoff.

Keywords Surface runoff · Stormwater and snowmelt runoff · Corn husk · Foam concrete block · Heavy metals · Suspended solids · Water bodies · Sorption

---

I. Arkanova (✉) · L. Fedorova
South Ural State University, 76, Lenin Prospekt, Chelyabinsk 454080, Russia
e-mail: waterbas@mail.ru

E. Terekhova · A. Podporin
Saint Petersburg State University of Architecture and Civil Engineering, 4, 2nd Krasnoarmeiskaya Str., St Petersburg 190005, Russia

L. V. Radionova and D. V. Ulrikh (eds.), *Advances in Ecology and Environmental Engineering*, Springer Proceedings in Earth and Environmental Sciences,
https://doi.org/10.1007/978-3-031-64423-8_25

1 Introduction

Everyday and technological activities of human beings inevitably lead to generation of various types of waste that have a negative impact on the environment.

The formation of surface runoff depends on the terrain conditions, and the flow rate depends on the size of drainage basin and the nature of the use of its territory. During precipitation, a significant portion of it either evaporates, is absorbed, or remains on the surface on which it fell. Average annual evaporation varies from 200 mm in cold and humid climates to 2000 mm in hot, dry areas. Its maximum intensity can reach 0.3 mm/h. Although the intensity of evaporation is small compared to the intensity of precipitation (for example, a light rainstorm can have an intensity of more than 10 mm/h), evaporation continues even after the rain stops. Therefore, total losses from large watersheds and watersheds with long time of runoff concentration can be significant.

Some of the rainfall is retained by vegetation. The possible total interception of precipitation by trees is in the range of 2–10 mm. Precipitation is retained and accumulated in surface depressions. Observations show that the amount of retained water layer, as if lost to runoff, is 5 mm on sand, 4 mm on lawns and 3 mm on clay surfaces. In general, losses vary from 1 mm on paved surfaces to 10 mm in gardens.

Water falling on porous surfaces seeps into them at Uf speed, which depends on the water permeability of the soil surface and the initial moisture content. During the period of rainfall, the intensity of infiltration decreases as the soil pores are filled with water and the water surface rises. The intensity of infiltration over 1 h can be with sufficient accuracy taken as: 0.2–2 mm/h for clay, 2–10 mm/h for loam, and 12–25 mm/h for sandy soils. In the presence of vegetation cover, these values can increase to 200 mm/h in a seeded field with sandy soil.

For the conditions of the Ural region, where the generation of a significant amount of surface (stormwater and snowmelt) runoff from urban areas and the territories of industrial enterprises and a high degree of its pollution is typical, a need for protection of water bodies from pollution by these waters arises.

Pollution of the Miass River, which is one of the largest waterways in the Chelyabinsk Region, occurs due to the disposal of storm water from urban areas and industrial storm water from the territories of industrial enterprises. The river is mainly fed by melting snow and rainwater.

Lake Smolino, located within the city of Chelyabinsk, also experiences significant anthropogenic pressure. The formation of the modern water regime is associated not only with the discharge of technical waste water from surrounding enterprises, but also with the discharge of surface runoff since a significant part of the catchment area has hard surfaces. The lake has significantly changed its natural characteristics, so there is a need to improve the operation of stormwater drainage.

The generation of waste is a consequence of the process of using natural resources by humans to meet their needs.

A special area of environmental protection activities has emerged: "Management of Production and Consumption Waste" [1].

Waste management policies should primarily focus on reducing of the generated waste and developing methods for maximizing their use (recycling). Most wastes are sources of secondary material resources.

According to the source of generation, waste is classified into: production waste from various sectors of the national economy (metallurgical, mechanical engineering, chemical, agricultural, construction, food industries, etc.); transport waste, and municipal waste.

The Russian Federation is implementing a comprehensive waste management strategy, which provides for the prevention of hazardous effects on human health and the environment from waste, as well as the involvement of components contained in waste into the national economy as secondary natural resources [2].

Waste processing consists of carrying out technological operations that lead to changes in its composition, physical, chemical, and biological properties. When recycling waste, various methods of its treatment are combined: physical, chemical, biological, or hybrid ones.

In physical processes, only the shape, size, and state of aggregation change while maintaining the qualitative chemical composition.

Mechanical processes dominate, for example, during crushing, grinding, particle size separation, mixing of waste and forming products from it.

For waste of organic and inorganic origin that cannot be used directly, the following types of mechanical processing are used: grinding, sorting by size or fractionation, coarsening, and mixing.

After mechanical processing, waste is transformed into products ready for further use.

Grinding can range from coarse to ultra-fine according to the degree of grinding, that is, from 2 to 1000 microns. In this case, various types of crushers and mills can be used.

Vibration sifters are widely used to disperse various granular materials.

Construction waste includes residues that are generated after the construction of real estate objects. Construction waste is classified by type of material: concrete products, reinforced concrete blocks, wood, plastic, fragments of ceramic products, etc. Increasing construction volumes correspondingly increases the amount of waste. The feasibility of recycling lies in saving money on the purchase of new materials, reducing the amount of waste and reducing land plot areas required for its placement, and reducing the burden on nature.

Waste from the production and operation of concrete and reinforced concrete products and structures is dust, sludge from various types of raw materials used in production, inevitable breakage, defects and products themselves that have expired service life.

For the disposal of construction waste, stationary, collapsible and mobile installations and equipment have been created.

Large volumes of agricultural production produce a very significant amount of waste.

Of interest is agricultural waste (organic waste from crop production, livestock farming and processing industries). The total amount of agricultural waste reaches

630–650 million tons. 35.6% is crop waste, 4.7% is waste from the processing industry.

Waste from corn production is classified as solid waste in terms of its state of aggregation; as large-tonnage waste in terms of material consumption (over 100 thousand tons per year); and safe waste in terms of its impact on the environment. It can be used in feed production, as bedding for animals, fertilizers and soil protective agents, as well as in bioenergy and decorative and applied crafts [3].

The share of using crop waste in bioenergy sector is increasing. To produce solid, liquid or gaseous fuel, straw, husks, stems and leaves of agricultural plants, corn cobs and shells of corn grains, as well as other plant materials are used.

A promising field is the use of crop waste in the construction industry and is called eco-technology [4].

In the context of the aggravation of the environmental situation, the use of polymer packaging, biodegradable packaging, containers and disposable tableware, and bags made of natural polymers are of interest; the raw materials for production are corn, rice, and sugar cane waste [5].

Currently, plant waste has found application in agriculture (feed additives) [3, 6], to improve soil quality [7, 8], in the production of fertilizers, granular carbon-containing products and liquid biofuels by compacting pre-crushed raw materials, accelerated hydrolysis, and pyrolysis with further activation [9]. Buckwheat husks, sunflower husks and sawdust are promising raw materials for obtaining various sorption materials for wastewater purification from oil and petroleum products [10–14].

Various types of agricultural products are grown in the Ural region: buckwheat, oats, wheat, corn, rapeseed, etc. Therefore, it is relevant for the region to use waste of mineral and organic origin from construction production and the agro-industrial complex as sorbents for the purification of urban surface runoff and for the purification of melt runoff during snow melting from accumulation sites before discharge into water bodies. The Chelyabinsk Region has significantly smaller volumes of water resources than other regions in the country (per capita water is 16 times less than the Russian average). The region ranks the 8th in our country in terms of the volumes of polluted wastewater discharged into water bodies (3% of the national level).

Unlike organic waste, heavy metals are not biodegradable and must be removed before being released into the ecosystem.

Possibilities of replacing complex and expensive purification methods such as chemical sedimentation, ion exchange, membrane separation, etc. with cheaper and more accessible sorption technologies are considered [12, 15, 16]. Adsorption is an effective physicochemical process of removing heavy metal ions from aqueous solutions [17–20].

A developing field is the elaboration of technologies for the creation and use of new sorption materials made from various wastes, which allows minimizing the burden on the environment and at the same time saving natural resources since waste becomes a secondary raw material. The original waste is modified or combined for higher sorption properties and possibility of using it for purification of multicomponent wastewater.

Many studies on the sorption of heavy metal ions have been carried out on various unprocessed plant materials, such as corn leaves, teak leaf powder, rice husk ash and neem tree bark, corn cobs, sunflower husks [16, 21, 22], straw, coconut and pine nut shells [4, 23].

The goal of this study is to evaluate the effectiveness of using two types of waste to purify urban surface runoff from suspended solids and iron ions: corn husks from the agricultural sector, and defective foam concrete blocks from the construction industry.

2 Materials and Methods

Waste from the agricultural sector (corn husk) and construction industry (foam concrete blocks) were used as sorbents.

Corn husk is modified (dried and crushed) corn leaves collected in the fields of the Chelyabinsk Region.

Corn is a species of annual herbaceous plant of the grass family. In terms of its feeding and productive qualities, it is one of the most valuable agricultural crops in the world.

Foam blocks are a building material made from a type of cellular concrete called foam concrete, which is made from ordinary cement mortar, sand and water with the addition of a foaming agent.

Bentonite is a clay with a seventy percent content of montmorillonite, that is, a highly dispersed layered aluminosilicate. Its main feature is its chemical-crystalline structure, which is formed by the presence of ion-exchange cations on the surface of bentonite, which determines its physical and chemical properties as a mineral. When bentonite is lowered into an aqueous solution, water penetrates through the interlayer space of montmorillonite, which causes hydration of the latter, resulting in swelling. With further dilution, bentonite is able to form a viscous suspension, in which thixotropic properties are clearly manifested. In addition, montmorillonite has a fairly high adsorption property.

The appearance of samples is shown in Fig. 1.

The structure and composition of the surface of sorbents were studied on a JEOL JSM—6460LV electron microscope with an attachment for micro-X-ray spectral analysis. The preparation of sorbent samples was carried out using a JEOL JFC–1600 installation for sputtering conductive platinum layers. Microphotographs of corn husk samples and foam concrete particles at different resolutions are presented in Fig. 2a–d.

Analysis of the samples showed that the main chemical elements in their composition are environmentally friendly carbon, oxygen, and calcium.

The results of X-ray microspectral analysis of the surface of the samples are given in Table 1.

a b

Fig. 1 Appearance of samples: **a** corn husk; **b** defective foam concrete

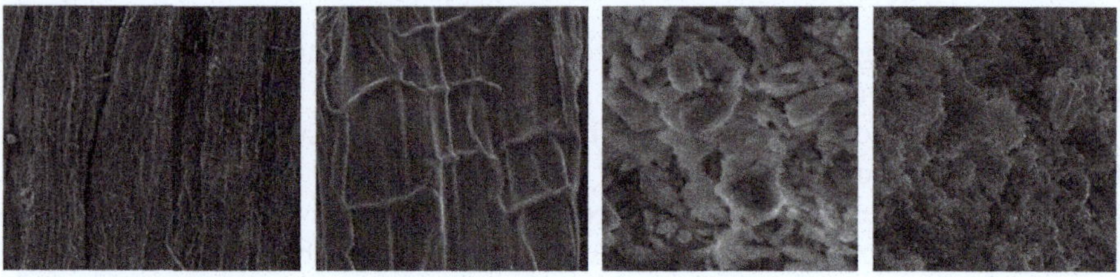

a b c d

Fig. 2 Microphotographs of samples **a, b** corn husk; **c, d** defective foam concrete

Table 1 Results of micro-X-ray spectral analysis of sample surfaces, %

Sorbent	C	O	K	Mg	Ca	Si
Defective foam concrete	8.15	52.10	0.15	0.53	21.54	1.06
Corn husk	48.12	50.52	1.00	–	–	–

To determine the effectiveness of wastewater treatment, surface water (stormwater and snowmelt runoff) from urban territories of the city of Chelyabinsk was used as a sorbate. Sampling and drainage points were chosen taking into account the differences in the territory: industrial, residential, and transport territories.

Chemical analysis of iron content was carried out using atomic emission spectrometry with an OPTIMA 2100 DV device. The pH value was set on a microprocessor pH meter to 150 MI. The results are presented in Table 2.

Table 2 shows a high degree of pollution of surface runoff from the transport areas of the city of Chelyabinsk.

Surface runoff (stormwater and snowmelt runoff) is characterized by a fairly wide range of concentrations [24–28]. The selected sorbate samples agree well with the data [29].

Table 2 Chemical composition of sorbate

Types of territory	pH	Content, mg/l	
		Fe	Suspended solids
Industrial territories	7.2	1.7	230
Residential constructions	7.7	0.9	420
Transport areas	8.0	0.2	2430

Studies of the interaction in the "sorbent-sorbate" system were carried out in dynamic mode on a special installation, where surface runoff was filtered through a layer of sorbent (Fig. 3).

The change in sorbate during interaction with sorbents was determined by atomic emission spectrometry method with the help of an OPTIMA 2100 DV device (Perkin Elmer).

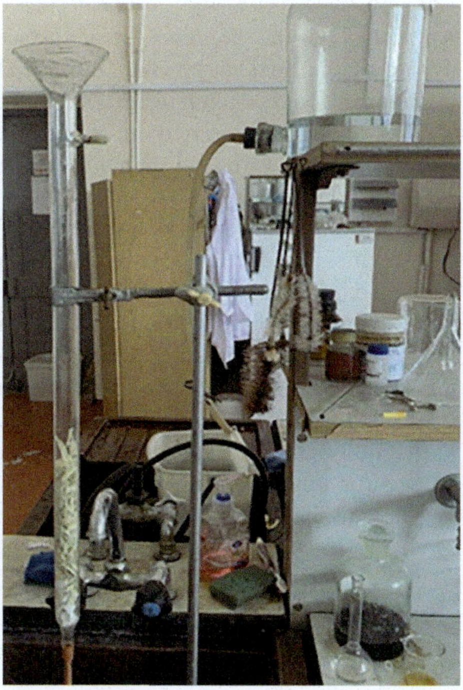

Fig. 3 Installation for filtering sorbate in dynamic mode

Table 3 Content of iron ions and suspended solids

Filter medium	Initial concentration, mg/l		2 h, mg/l		4 h, mg/l		6 h, mg/l	
	Fe	Suspended solids	Fe	Suspended solids	Fe2	Suspended solids	Fe	Suspended solids
Corn husk	1.7	230	0.3	–	0.3	–	0.4	–
Foam concrete	1.7	230	1.55	66.9	1.53	48.4	1.5	37.8
Corn husk (layer 1), foam concrete, (layer 2)	1.7	230	0.3	12.2	0.4	11.4	0.5	9.9

3 Research Results

An analysis of the effectiveness of surface runoff purification with sorbents under dynamic conditions is shown in Table 3. Dried and crushed corn husk and 1–3 mm fractions of foam concrete were used in the experiments.

The studies were carried out in laboratory conditions, the initial concentrations of contaminants were analyzed in each experiment for corn husk, for foam concrete and for a combined sorption medium in a ratio of 30:70 over the time period of 6 h.

As follows from Table 3, the crushed foam concrete solely does not contribute to a significant reduction in iron ions, but quite effectively retains suspended solids. The effect of cleaning suspended solids with foam concrete is associated with the compaction of the filter medium under the influence of water flow. This was confirmed when compiling a combined medium: a corn husk and crushed foam concrete. At the same time, adding of shredded corn husk showed a reduction in iron ions and suspended solids.

As follows from the data in Table 3, the use of a combined sorption medium helps reduce iron ions by an average of 70.6%, and suspended solids by 95.7%

The crushed corn husk has a loose structure. During the experiment it did not swell, and the sorbing surface area remained unchanged.

4 Conclusion and Discussion

The structure and composition of the surface of waste from agro-industrial complex (crushed corn husks) and construction industry (crushed and sifted defective foam concrete) have been studied.

The data obtained confirmed the environmental safety of the materials used. The main elements in their composition are carbon, oxygen, potassium, and calcium.

The effectiveness of sorption of iron ions and suspended solids from surface runoff in urban territories of Chelyabinsk has been determined.

The effectiveness of the proposed sorbents on real urban wastewater was studied for the first time.

Some literature sources contain data on the use of plant raw materials as sorbents [15, 21, 23]. V.V. Ulyanova has proven in her work [4] that purification of wastewater from heavy metals is possible on the basis of sludge from a ceramic workshop and waste from agricultural processing enterprises. It has been shown that high effectiveness of 97–99% is achieved by combining sorption materials.

Under dynamic conditions, the maximum effect of purification from suspended solids by 95.7% and from iron ions by 70.6% was achieved on a combined sorption medium consisting of crushed corn husk and crushed foam concrete.

The conducted studies have demonstrated that when choosing a sorbent, one should take into account the chemical composition of the wastewater and select an appropriate scheme based on the use of environmentally friendly sorbents–waste from the agro-industrial complex and the construction industry.

The use of combined sorbents, which make it possible to use the advantages of the studied materials with maximum effectiveness, is promising.

When using industrial waste as sorbents for wastewater treatment, economic costs are reduced to the cost of crushing, grinding, and sieving these materials [30].

The results obtained show that the studied waste serves as a promising economically safe sorption material, allowing to extract iron ions and reduce the content of suspended solids in the surface runoff of urban areas before being discharged into water bodies.

The use of sorption technology is also possible in the case of temporary storage of removed snow in open areas. Open areas are arranged with a waterproof coating for the temporary storage of snow removed from urban and enterprises' territories. To purify snowmelt runoff, formed during spring snowmelt, from storage and accumulation sites, it is also possible to organize sorption technology using industrial waste to prevent the volley discharge of such water into water bodies.

It is necessary to eliminate the exceeding of the standards for permissible impact on the environment in terms of pollution of surface and ground waters and soils of adjacent territories. To do this, it is necessary to intercept the snowmelt runoff by drainage and then clean it before discharging it into a water body.

References

1. Russian Federation. Laws. On Production and Consumption Waste: Federal Law No. 89-ФЗ: [adopted by the State Duma on May 22, 1998: approved by the Federation Council on June 10, 1998]—Consultant Plus: official website. https://www.consultant.ru/. Accessed 23 Jan 2024
2. Lipaev AA, Lipaev SA (2021) Management of Production and Consumption Waste. Infra-Engineering, Moscow, Vologda, p 408
3. Lesnov AP, Leontiev AP, Nikitenkov AI (2009) Production of Feed from Low-value Plant Raw Materials for Cattle. Effective Animal Husbandry 9:44–45

4. Ulianova VV (2015) Purification of Wastewater from Heavy Metal Ions with Sorbents Based on Modified Waste from Ceramic Production and Agricultural Processing.: author's abstract. Saratov, p 16
5. Zavorokhina NV, Semukhin AS (2022) Prospects for Using Agricultural Waste for the Production of Biodegradable Packaging. Industry and Agriculture 8(49):5–9
6. Lesnov AP, Leontyev AP, Kovalev AN (2011) Low-value Plant Raw Materials in Biotechnologies of Feed Production. APK-YuG 5:40–43
7. Buzyreva NV, Tashlykova AN, Vasina MV (2018) Natural Sorbents for Cleaning Water Bodies and Soil from Oil Products. Omsk State Technical University 1:348–351
8. Ryazantseva NV (2016) The Use of Straw for Mulching the Soil Surface in Corn Crops. Agrotechnological Processes within the Framework of Import Substitution: International science-to-practice conference dedicated to 85th anniversary of the birth of the honored worker of higher education of the Russian Federation, pp 63–64
9. Spitsyn AA, Belousov II, Tursunov TB, Hen VA (2018) Thermochemical Conversion of Crushed Compacted Plant Biomass. News of the St. Petersburg Forestry Academy 224:256–272. https://doi.org/10.21266/2079-4304.2018.224.256-272
10. Sirotkina LV, Novoselova NIu (2005) Materials for Adsorption Purification of Water from Oil and Petroleum Products. Chemistry for Sustainable Development 13(3):359–377
11. Gromyko NV (2015) New Plant Sorbents for Cleaning Reservoirs from Accidental Oil Spills and Oil-based Products: immediate research. Young Scientist 10(90):192–195
12. Safarov AKh, Proskuriakova EA (2020) Selection of Organic Sorbents for Wastewater Treatment from Oil and Petroleum Products. Enigma 22-2:157–160
13. Eremin IS, Zaitseva EA, Rossolova AS, Voronina KE (2020) Sorption Properties of Carbon Adsorbent Based on Corn Stalks. Problems and Prospects for Sustainable Development of the Agro-industrial Complex: collection of articles based on the proceedings of the 2nd All-Russian Science-to-Practice Conference with international participation, pp 254–261
14. Brishov SB, Kozybaev AK (2023) Effectiveness of Wastewater Treatment from Petroleum Products Using Sorbents Based on Natural Origin. Innovation in Science and Technology. Collection of articles based on the proceedings of the 12th International Science-to-Practice Conference 1:46–52
15. Somin VA, Komarova LF, Kutalova AV (2020) Research on the Use of Buckwheat Husks for Water Softening. News of Higher Educational Institutions. Applied Chemistry and Biotechnology 10(2):213–222. https://doi.org/10.21285/2227-2925-2020-10-2-213-222
16. Chirova VS, Sobgaida NL, Rzazade FA (2015) Sorbents Based on Agricultural Waste for Wastewater Treatment. Bulletin of Kazan Technological University 18(20):263–266
17. Bagrov VV, Kamrukov AS, Kostritsa VN, Zubkov AA (2017) Natural Sorbents and Their Use for Wastewater Treatment. Water Magazine 6:38–43
18. Khlytina AA, Matiushin AA (2017) Search for Effective Sorbents of Plant Origin by Determining Their Specific Absorption. Modern Aspects of the Use of Plant Raw Materials and Raw Materials of Natural Origin in Medicine, pp 236–238
19. Shaikhiev IG, Almazova GA, Kalinin LV (2015) The Use of Natural Mineral Sorbents for the Purification of Storm Water Generated During the Production of Heat-insulating Building Materials. Bulletin of Technological University 18.3(13):195–197
20. Zaitseva EA, Eremin IS (2021) Sorption Properties of Materials Made from Agricultural Waste. Ecological, economic and technological aspects of sustainable development of the Republic of Belarus and the Russian Federation. Collection of articles of the 3rd International Science and Technology Conference 2:62–65
21. Gromyko NV, Usmanova LM (2015) Prospects for Using Waste from Processing Agricultural Crops to Purify Natural Waters from Oil Pollution. Magistracy Bulletin. Chemical Sciences, pp 4–6
22. Safarov AKh, Nurislamova DA, Iagafarova GG, Akchurina LR, Mikulik DI (2022) Promising Sorbents Based on Agricultural Waste. Ecology (in chemistry and petrochemistry). The problem of collecting, preparing and transporting oil and petroleum products 2(136):129–138. https://doi.org/10.17122/ntj-oil-2022-2-129-138

23. Sverguzova SV, Bely VA (2019) Plant Materials as Raw Materials for the Production of Sorbents. Safety, and Protection of the Natural Environment: Fundamental and Applied Research: All-Russian Scientific Conference: Collection of Reports. Part 2 (Belgorod, October 14–18, 2019). Publishing House of Belgorod State Technological University named after V.G. Shukhov, Belgorod, pp 119–125
24. Na Nagara V, Sarkar D, Datta R (2022) Phosphorus and Heavy Metals Removal from Stormwater Runoff Using Granulated Industrial Waste for Retrofitting Catch Basins. Molecules 27:7169
25. Davis AP, Shokouhian M, Sharma H, Minami C, Winogradoff D (2003) Water Quality Improvement through Bioretention: Lead, Copper, and Zinc Removal. Water Environ. Res. 75:73–82
26. USEPA. Stormwater O&M Fact Sheet: Catch Basin Cleaning, EPA 832-F-99–011; Washington, DC, USA, USEPA, 1999
27. Okaikue-Woodi FEK, Cherukumilli K, Ray JR (2020) A Critical Review of Contaminant Removal by Conventional and Emerging Media for Urban Stormwater Treatment in the United States. Water Res. 187:116434
28. Stagge JH, Davis AP, Jamil E, Kim H (2012) Performance of Grass Swales for Improving Water Quality from Highway Runoff. Water Res. 46:6731–6742
29. Alekseev MI, Kurganov AM (2000) Organization of the Drainage of Surface Runoff (Snowmelt and Rainfall) from Urban Areas: Textbook. ACB Publishing House, Saint Petersburg State University of Architecture and Civil Engineering, Moscow, SPb, p 352
30. Bulaev VG, Gushchina NV, Kuvshinov DIu (2020) Disposal of Construction Waste. Management of Technosphere 2:259–270

Research on Photocatalysts Based on Fly Ash Cenospheres from CHP Plants Used for Purifying Water from Organic Pollutants

T. Lonzinger, M. Bryukhov, S. Makarova, and D. Ulrikh

Abstract This article examines the results of a research on photocatalysts based on fly ash cenospheres from CHP plants used when purifying water from organic pollutants. The goal of the work is to obtain photocatalysts based on fly ash cenospheres from CHP plants, for water purification from organic compounds. Structural and chemical parameters of cenospheres have been studied. It has been determined that cenospheres are highly porous materials with a low density of 0.7 kg/m^3. Under the influence of mechanical stress, destruction of cenospheres occurs along the least durable elements—internal pores. The thickness of the outer shell of the spheres ranges from 5.6 to 14.4 microns. The destruction of the walls of the spheres is complicated by defects of closed pores in their structure, which increase the path of the main crack. With a mass of cenospheres in an aqueous solution 4 times less than generally accepted concentrations, high effectiveness of the photocatalytic coating was achieved in the decomposition of the methylene blue redox indicator. The obtained experimental data show that the photocatalytic coating on cenospheres reduces the concentration of methylene blue by 25% within two hours of irradiation. The photocatalytic activity of cenospheres towards cyclic hydrocarbons has been assessed using the example of the destruction of phenol. It has been determined that the presence of cenospheres on the water surface significantly increases the rate of phenol destruction during photocatalytic oxidation.

Keywords Fly ash · Cenospheres · Photocatalytic decomposition · Organic pollutants

T. Lonzinger (✉) · M. Bryukhov · D. Ulrikh
South Ural State University, 76, Lenin Prospekt, Chelyabinsk 454080, Russia
e-mail: lonzingertm@susu.ru

M. Bryukhov · S. Makarova · D. Ulrikh
Saint Petersburg State University of Architecture and Civil Engineering, 4, 2nd Krasnoarmeiskaya Str., St Petersburg 190005, Russia

L. V. Radionova and D. V. Ulrikh (eds.), *Advances in Ecology and Environmental Engineering*, Springer Proceedings in Earth and Environmental Sciences,
https://doi.org/10.1007/978-3-031-64423-8_26

1 Introduction

One of the global governmental tasks, that contribute to the implementation of the goal of the energy sector development in the Russian Federation and are stipulated in the Energy Strategy of the Russian Federation for the period until 2035, is the reduction of the negative impact from the fuel and energy complex on the environment. Therefore, special attention is paid to the problem of managing waste from coal generation. It is intensified by the tightening of environmental legislation with the transition to the principles of the best available technologies, the greening of coal energy, the depletion of the capacity of ash and slag dumps of coal CHP plants, etc. Disposing of ash, slag and ash and slag mixtures generated during the combustion of coal, peat and shale, is one of the main problems of modern solid fuel CHP plants in Russia [1].

During the combustion of coal fuel at CHP plants, mineral residue, consisting of slag and fly ash, or, as it is also called, flue ash, is generated in large quantities. Fly ash mainly consists of solid spheres (density of ~2300–2600 kg/m^3) and cenospheres (density of < 1400 kg/m^3), and also contains irregularly shaped mineral particles and unburned carbon. Depending on combustion method, coal source, and more, the particle size of fly ash varies in the range from 1 to more than 200 microns, yet hollow cenospheres and particles of unburned carbon are of large size [2]. The main components of ash are oxides of silicon, aluminum and iron. In addition to the main oxides, fly ash also includes: CaO, MgO, SO_3, SO_2, Na_2O, K_2O, TiO_2, MnO, P_2O_5, and other oxides. In Russia, in 2019 alone, CHP plants operating on coal fuel generated about 25 million tons of ash and slag, and only 10% of the annual output was involved in circulation. The mass of accumulated waste at ash dumps equals 1.5 billion tons, and this mass will exceed 2 billion tons with further accumulation by 2030 [3, 4].

In Western countries, the level of ash and slag waste recycling has been increased up to 50–100%. In Germany, ash dumps are prohibited by law and almost 100% of the annual output of ash and slag waste is used in the building materials industry. In the USA, Great Britain and Poland, the level of use of ash and slag waste reaches 70% [5, 6]. The adoption of a number of laws at the state level in China has made it possible to increase the use of ash and slag waste up to 70–80%.

The fields of use of fly ash are constantly expanding. Fly ash is applied in construction industry for the production of cement, concrete, gypsum, ceramic products, and thermal insulation materials. [4, 7, 8, 9, 10]. In road construction, ash and slag dumps for the production of cement-soil and filling mixtures are also known to be used [11–13]. Data have been published on the use of fly ash in the refractory industry without pre-treatment. In the work [14], the authors distinguish fly ash as an alternative material for the synthesis of pure cordierite with a low synthesis temperature and improved mechanical properties. Technologies for producing refractory and insulating materials using fly ash are described in the works [14–18].

At present after special treatment, fly ash is used as catalyst, catalyst carriers or adsorbents for purifying water and gases from heavy metals. Fly ash particles have a developed surface, and such characteristics as morphology and chemical composition make it possible to use it in the manufacture of filters for water purification [19–22].

The goal of the work is to obtain photocatalysts based on fly ash cenospheres from CHP plants for water purification from organic compounds.

Currently it is impossible to solve the problem of water purification and disinfection without using new technologies. The applied methods (such as sorption, filtration, osmosis, chlorination, flotation, etc.) mostly just move pollutants from one object to another, rather than removing them completely. Photocatalytic oxidation is one of the newest methods used to remove organic pollutants from water and disinfect it. The photocatalytic method belongs to the destructive type of purification methods, but it does not require specially prepared oxidizing agents. Oxidation occurs due to air oxygen dissolved in water. Unlike the currently used destructive methods of water purification using strong oxidizing agents such as chlorine and ozone that are undesirable for release into the environment, the photocatalytic approach appears to be much safer. This method is environmentally friendly since its use does not leave any waste requiring burial, neutralization or disposal and does not emit hazardous gases into the atmosphere [23, 24]. From an economic point of view, photocatalysis can compete with the cheapest sorption purification methods almost equally [25, 26].

It has been proven that almost any organic compound can be oxidized (mineralized) to CO_2 and H_2O on the surface of titanium oxides. [27]. The use of titanium oxides in the form of suspensions for water purification is not effective enough since photocatalytic reactions are initiated by ultraviolet radiation. The use of "floating" cenospheres coated with a photocatalyst solves the problem of activating the photocatalyst directly by sunlight. "Floating" photocatalysts are especially effective for purifying the surface of water from organic compounds.

The generally accepted method for producing photocatalyst films is sol-gel technology. The works [27–29] examine chemical and technological methods for producing TiO_2 films on the surface of hollow cenospheres, which are fly ash that is formed during fuel combustion. Sol-gel technology is considered to be labor-intensive and involves the use of toxic organic reagents. Other methods for producing a photocatalyst using cenospheres as a carrier are not used in world practice.

2 Research Results

When performing this work, the author's method has been used to obtain a titanium-oxides-based photocatalytic coating firmly fixed on the surface of cenospheres (method of activated pyrohydrolysis of titanium chloride on the finished surface of cenospheres) [30].

The research on the structure and composition of the photocatalytic coating on cenospheres has been conducted by electron microscopic and micro-X-ray spectral analysis using a JEOL JSM 6460 LV microscope.

The photocatalytic effect has been assessed on a laboratory installation with a DRB 841 lamp (TU 16,535,659), allowing to create ultraviolet radiation with a wavelength of 203–315 nm.

Determining physical and chemical characteristics of cenospheres. Physical characteristics of a material depend equally on its chemical composition and structure. Microphotograph 1 shows the cenosphere and its cross-section, the structure of the outer shell and the cenosphere with a photocatalytic coating. Table 1 shows the composition of the cenosphere surface before and after application of the photocatalytic coating. In order to study the size and nature of the distribution of pores in the spheres, cross sections of granules have been examined using an electron microscope after mechanical action.

The data given in the Table show that the cenosphere consists of aluminosilicates with minor admixtures of titanium and iron oxides. The specifics of structure of cenospheres lies in hard glassy fused surface and a high, predominantly closed porosity of the shell, therefore the bulk weight of the spheres does not exceed 0.6–0.7 kg/m^3. Under the influence of mechanical stress, the destruction of cenospheres occurs along the least durable elements: internal pores. The thickness of the outer shell of the spheres ranges from 5.6 to 14.4 microns. The destruction of the walls of the spheres is complicated by defects in their structure: closed pores, which increase the path of the main crack. The microphotograph clearly shows the strong outer shell of the sphere with destroyed internal pores. Closed internal pores in the walls of the cenospheres have a regular spherical shape with a diameter of 2–6 microns. Physical characteristics of the cenospheres and, above all, mechanical strength are determined by the structure of the outer shell, which includes evenly distributed spherical closed pores.

When conducting the study, cenospheres with a photocatalytic coating have been used. The microphotograph of a coated sphere and the surface composition can be seen in the Fig. 1 and Table 1. On the surface of the cenosphere there is a film coating consisting of titanium oxides, the content of which in terms of TiO_2 is on average 7.49–10.58%. Data from previous studies show that to achieve a photocatalytic effect, the surface of the material must contain at least 1.5–2.0% titanium oxide in the form of anatase, therefore experimental samples must have high photochemical activity.

The most hazardous pollutants in potable and waste water are organic compounds of artificial and natural origins. The effectiveness of decomposition of organic compounds in water by photocatalysts can most clearly be assessed using the example of the decomposition of the methylene blue redox indicator. In order to examine the

Table 1 Chemical composition of the cenosphere surface

Element	O	Al	Si	Ti	Fe
Content of elements on the uncoated surface, wt.%	47.0	23.5	27.9	0.6	0.5
Content of elements on the surface with photocatalytic coating, wt.%	56.3	13.7	15.5	6.8	0.4

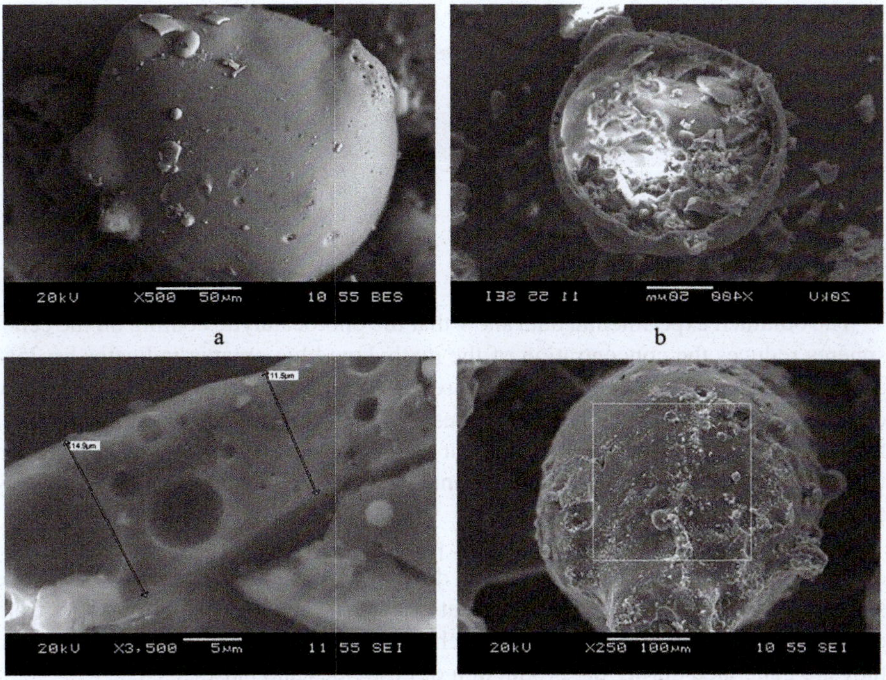

a b c d

Fig. 1 Microphotographs of the cenosphere: **a** cenosphere, **b** cross-section of the cenosphere, **c** structure of the cenosphere shell, **d** cenosphere with photocatalytic coating

structure of cells of living organisms (such as bacteria, algae), solutions of methylene blue with a concentration of 0.15% were used. The effectiveness of decomposition of methylene blue by cenospheres with a photocatalytic coating was assessed using a laboratory installation that provided exposure to ultraviolet irradiation of the objects under study with a wavelength of 203–315 nm. The method of conducting the experiment included preparing methylene blue as a working solution, which was put in Petri dishes in the amount of 50 ml. Then 1.5 wt.% photocatalytic coated cenospheres were added. The number of cenospheres was reduced by 4 times compared to similar methods, which usually use a 6% catalyst concentration, that is, the tests were carried out under harsh conditions. In order to study the kinetics of the methylene blue decomposition, solution samples were taken at certain time intervals and the color intensity was examined by photocolorimetric analysis on an FEK-M device.

Figure 2 shows the results of the study of the methylene blue decomposition by cenospheres with a photocatalytic coating. The reference samples used in each experiment were the methylene blue stock solution. The decrease in the concentration of the methylene blue indicator upon exposure to ultraviolet radiation is a combined result of the ultraviolet rays effect and the photocatalytic effect of cenospheres. The use of reference samples made it possible to isolate the effect of ultraviolet radiation on the decomposition of methylene blue.

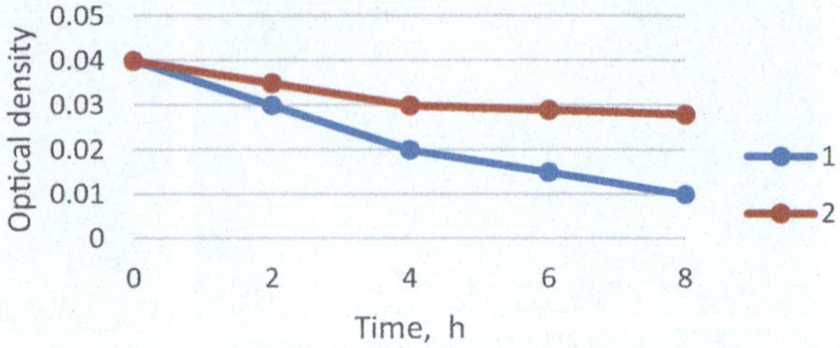

Fig. 2 Kinetics of changes in the optical density of methylene blue: 1–solution with cenospheres, 2–solution without cenospheres

The obtained experimental data show that the photocatalytic coating on the cenospheres reduces the concentration of the methylene blue organic indicator by 25% after 2 h of irradiation; after 8 h of exposure of the cenospheres to the test solution, the concentration of methylene blue decreases by 75%, which is 2 times higher than for the ultraviolet version irradiation without photocatalyst (30% decrease). Moreover, in an aqueous solution, the number of cenospheres is 4 times less than the generally accepted concentrations.

In relation to cyclic hydrocarbons, the photocatalytic activity of cenospheres was evaluated using the example of phenol destruction. Phenols are considered to be among of the most common pollutants that enter surface waters with wastewater from oil refining, oil shale processing, timber chemical, coke, and aniline paint industries; they are generated during the production of pesticides and plastics, etc.

In the wastewater from these enterprises, the content of phenols in very diverse combinations can exceed 10–20 g/dm^3. In addition to phenols of technogenic origin, low-molecular-weight phenols that are formed during the biochemical decomposition and transformation of organic substances (microbiological or hydrolytic destruction of high-molecular humic and fulvic acids) can also be found in natural water. Phenols in water lead to the formation of chlorine and nitro-substituted phenols, which are even more hazardous for human health.

The conducted analysis of the literature data on the influence of phenol and its compounds on the state of water bodies, and of the regulatory documents on maximum permissible concentrations of phenol made it possible to select the concentration of aqueous solutions of phenol for photocatalytic destruction. In the examined solutions, the concentration of phenol equaled 0.5 mg/l. The change in the phenol concentration was monitored using the method of sorption-photometric determination. The interaction of phenol with 3-methylbenzothiazoline hydrazone occurred with a formation of a colored product. The measurement was carried out on a FEK-3 photocolorimeter with a wavelength of 490–520 nm. Subsequent sorption was conducted on a membrane filter.

When conducting experiments to determine the photocatalytic activity of cenospheres, a solution of phenol in water with cenospheres located on the surface was exposed to ultraviolet light with a wavelength of 300 nm in a closed chamber for 6 h. The ratio of the number of cenospheres in the studied phenol solution was varied. For comparison, experiments were carried out without using cenospheres. The obtained

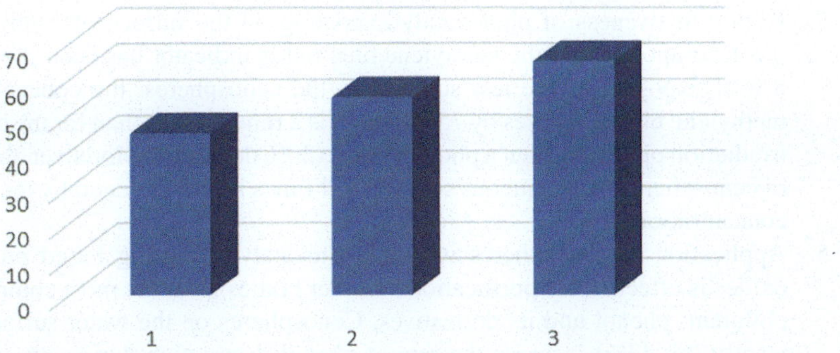

Fig. 3 Change in the degree of phenol destruction depending on the content of cenospheres on the surface of the aqueous solution: 1–phenol solution without cenospheres, 2–phenol solution with 1 wt. % cenospheres, 3–phenol solution with 2 wt. % cenospheres

results are presented in Fig. 3. The chart shows that cenospheres with a photocatalytic coating increase the degree of phenol destruction by 30–43%. The higher the concentration of cenospheres, the more effective the destruction process is.

An increase in the concentration of cenospheres on the surface of a phenol solution up to 2% in water leads to an increase in the number of phenol molecules that decompose under the influence of photocatalytic oxidation. After a 24 h irradiation, the degree of phenol destruction equaled 81.2%, which was 43% higher than for a solution without cenospheres.

The conducted experiments showed that the use of cenospheres with a photocatalytic coating based on titanium oxides was effective in cleaning water bodies from the most common toxic pollutant: phenol and its derivatives.

3 Conclusion

1. Research on the composition and structure of cenospheres, a component of fly ash from CHP plants, has been conducted.
2. It has been established that cenospheres are highly porous materials with a low density of 0.7 g/cm^3. The destruction of cenospheres under the influence of mechanical stress occurs along the least durable elements–internal pores. The thickness of the outer shell of the spheres ranges from 5.6 to 14.4 microns. The destruction of the walls of the spheres is complicated by defects of closed pores in their structure, which increase the path of the main crack.
3. When conducting research on purification of water from organic compounds, cenospheres have been used, coated (using a previously developed technology) with a photocatalytic film, consisting of titanium oxides, the content of which in terms of TiO_2 averages in 7.49–10.58%.

4. High effectiveness of photocatalytic coating on the surface of cenospheres in the decomposition of the methylene blue redox indicator has been shown. After 8 h of exposure of the test solution to the cenospheres, the concentration of methylene blue decreases by 75%, which is 2 times higher than for the ultraviolet irradiation option without a photocatalyst (30% decrease). Moreover, the number of cenospheres in an aqueous solution is 4 times less than generally the accepted concentrations.
5. Application of cenospheres with a photocatalytic coating based on titanium oxides is effective for purification of water bodies from the most common toxic pollutant: phenol and its derivatives. Cenospheres on the water surface significantly (by 43%) increase the rate of phenol destruction during photocatalytic oxidation.
6. Research data make it possible to expand the scope of application of CHP plant waste, cenospheres, through the use of a more effective technology of applying photocatalytic coating, which is environmentally friendly and less labor-intensive compared to the well-known sol-gel technology that uses toxic organic reagents.

References

1. Khudyakova LI, Zalutskiy AV, Paleev PL (2019) Use of Ash and Slag Waste of Thermal Power Plants. XXI Century. Technosphere Safety 4(3):290–306
2. Xu G, Shi X (2018) Characteristics and Applications of Fly Ash as a Sustainable Construction Material: A State-of-the-art Review. Resources Conservation and Recycling 136:95–101
3. Denisov GA (2020) Ash and Slag in the Construction Materials Industry. New Chemical Technologies. Analytical Portal of the Chemical Industry
4. Ash and Slag: An Unsolved Problem. Energy and Industry of Russia: newspaper, March 2019, No. 05 (361). https://www.eprussia.ru/epr/361/1492205.html
5. Delitsin ML, Ryabov YuV, Vlasov FS (2014) Possible Technologies for Ash Utilization. Energy Saving 2:60–66
6. Hemalatha T, Ramaswamy A (2017) A Review on Fly Ash Characteristics - Towards Promoting High Volume Utilization in Developing Sustainable Concrete. Journal of Cleaner Production 147:546–559
7. Feuerborn H–J, Müller B, Walter E (2012) Use of Calcareous Fly Ash in Germany. Proceedings of the EUROCOALASH 2012 Conference, September 25–27, Thessaloniki
8. Gerasimova NP (2016) Fly Ash as a Raw Material for Precast Concrete Block Production When Solving the Environmental Problem of CHP Ash and Slag Waste Disposal. Proceedings of Irkutsk State Technical University 6(113):122–127
9. Sobolev K, Vivian IF, Saha R (2014) The Effect of Fly Ash on the Rheological Properties of Bituminous Materials. Fuel 116:471–477
10. Sett R (2017) Flyash: Characteristics, Problems and Possible Utilization. Advances in Applied Science Research 8(3):32–50
11. Weilian G, Werner L (2010) Geopolymer Composite Binders with Given Properties for Cement and Concrete. Patent of the Russian Federation 2011134840/03. publ. 27.02.13
12. Han L, Wang J, Liu Z (2020) Synthesis of Fly Ash–based Self–supported Zeolites Foam Geopolymer Via Saturated Steam Treatment. Journal of Hazardous Materials 393:12468
13. Pichugin EA (2019) Analytical Review of the Experience of Involving Ash Slag Waste of Thermal Power Plants in Economic Circulation in the Russian Federation. Problems of Regional Ecology 4:77–87

14. Putilin EI, Tsvetkov VS (2003) Application of Fly Ash and Ash and Slag Mixtures in the Construction of Highways. Review of Russian and Foreign Experience in the Use of Waste from Burning Solid Fuels at Thermal Power Plants. SOUZDORNII, Moscow
15. Senthil Kumar M, Vanmathi M, Senguttuvan G (2019) Fly Ash Constituent–silica and Alumina Role in the Synthesis and Characterization of Cordieritebased Ceramics. Silicon 11:2599–2611
16. Nguyen M, Sokolář R (2020) Impact of Fly Ash as a Raw Material on the Properties of Refractory Forsterite-spinel Ceramics. Minerals 10(9):835–846
17. Luo Y, Ma S, Liu C (2017) Effect of Particle Size and Alkali Activation on Coal Fly Ash and Their Role in Sintered Ceramic Tiles. Journal of the European Ceramic Society 37(4):1847–1856
18. Braganza SR, Zimmer A, Bergmann SP (2008) The Use of Coal Ash in the Production of Insulating Refractories. New Refractories 6:60–63
19. Kim M, Ko H, Kwon T (2020) Development of Novel Refractory Ceramic Continuous Fibers of Fly Ash and Comparison of Mechanical Properties with Those of E-glass Fibers Using the Weibull Distribution. Ceramics International 46(9):13255–13262
20. Jing Z, Li YY, Cao S (2012) Performance of Double–layer Biofilter Packed with Coal Fly Ash Ceramic Granules in Treating Highly Polluted River Water. Bioresource Technology 120:212–217
21. Mushtaq F, Zahid M, Ahmad Bhatti I (2019) Possible Applications of Coal Fly Ash in Wastewater Treatment. Journal of Environmental Management 240:27–40
22. Nguyen TC, Loganathan P, Nguyen TV (2017) Adsorptive Removal of Five Heavy Metals from Water Using Blast Furnace Slag and Fly Ash. Environmental Science and Pollution Research 25(21):20430–20438
23. Atun G, Ayar N, Kurtoğlu AE (2019) A Comparison of Sorptive Removal of Anthraquinone and Azo Dyes Using Fly Ash from Single and Binary Solutions. Journal of Hazardous Materials 371:94–107
24. Savinov EN (2000) Photocatalytic Methods for Water and Air Purification. Soros Educational Journal 6(11):52–56
25. Praveen K Surolia, Rajesh J Tayade, Raksh V Jasra (2010) TiO_2-Coated Cenospheres as Catalysts for Photocatalytic Degradation of Methylene Blue, *p*-Nitroaniline, *n*-Decane, and *n*-Tridecane under Solar Irradiation. Ind. Eng. Chem. Res. 49(19):8908–8919
26. Chen Shufu, Cao Gengyu (2005) Photocatalytic Degradation of Organophosphorus Pesticides Using Floating Photocatalyst Tio_2 Sio_2/Beads by Sunlight. Solar Energy 79(1):1–9
27. Wei-hai Pang, Nai-yun Gao, Yang Deng, Yu-lin Tang (2009) Novel Photocatalytic Reactor for Degradation of DDT in Water and Its Optimization Model. Journal of Zhejiang University–Science A, volume 10 (5), Springer Journals 1
28. Addamo M, Augugliaro V, Di Paolo A, Garcia-Lopez E, Loddo V, Marci V, Palmisano L (2008) Photocatalytic Thin Films of Tio_2 Formed by a Sol-Gel Process Using Titanium Tetraisopropoxide as the Precursor. Thin Solid Films 516:3802–3807
29. Bing Wang, Qin Li, Wei Wang, Ying Li and Jianping Zhai (2011) Preparation and Characterization of Fe^{3+}–doped TiO_2 on Fly Ash Cenosphere for Photocatalytic Application. Applied Surface Science 257(8):3473–3479
30. Patent. Photocatalytic Element. Patent for the Application for a Utility Model of the Russian Federation 115686. 27.08.13

Use of Expanded Clay for the Treatment of Acidic Wastewater from Mining Enterprises

A. Samodolov and A. Kudryavtsev

Abstract The extraction of non-ferrous metals such as copper, zinc, and gold leads to the production of a huge amount of processing waste. The danger to the environment posed by mining and metallurgical enterprises processing copper pyrite ores lies in the pollution of surface water bodies with technogenic acidic wastewater. When purifying acidic wastewater from heavy metals, sorption technologies play a pivotal role and their development requires new types of sorbents. Expanded clay, a lightweight porous material that looks like granules with a burnt brown–red surface and dark pores inside, is chosen for the study. This article discusses the composition of expanded clay and the results of research on its sorption capacity at various temperatures under static conditions. In dynamic mode, data on the effect of filtration speed on cleaning effectiveness are obtained. It has been established that the effectiveness of wastewater treatment when using expanded clay varies from 70 to 90% under static conditions and from 3 to 85% under dynamic conditions, depending on the nature of metal cations in the effluent, sorbate temperature, and filtration rate. A recommendation for the use of expanded clay as part of a composite sorbent is given.

Keywords Technogenic water · Expanded clay · Mining and processing enterprise · Sorption · Heavy metals · Pollutants · Filter medium

A. Samodolov (✉)
South Ural State University, 76, Lenin Prospekt, Chelyabinsk 454080, Russia
e-mail: samodolov@mail.ru

A. Kudryavtsev
Saint Petersburg State University of Architecture and Civil Engineering, 4, 2nd Krasnoarmeiskaya Str., St Petersburg 190005, Russia

L. V. Radionova and D. V. Ulrikh (eds.), *Advances in Ecology and Environmental Engineering*, Springer Proceedings in Earth and Environmental Sciences,
https://doi.org/10.1007/978-3-031-64423-8_27

1 Introduction

The extraction of non-ferrous metals such as copper, zinc, and gold leads to generation of a huge amount of production waste. If not managed correctly, these types of waste can cause irreversible damage to the environment. The danger to the environment posed by mining and metallurgical enterprises processing copper pyrite ores lies in the pollution of surface water bodies with technogenic acidic wastewater, called *acid mine drainage* (AMD) in world practice [1–8]. Such runoff is a source of environmental pollution with heavy metals. The vital activity of the biota of aquatic ecosystems, which are a transit medium for unbalanced flows, is either suppressed by the action of heavy metals or does not have a mechanism for their assimilation [9, 10]. This problem is relevant for developed and developing countries around the world.

In the eastern United States, more than 10,000 km of streams and more than 72,000 ha of lakes and reservoirs were negatively impacted by AMD before 1990 [11].

In Madagascar, according to a 2019 study, water downstream from the Rio Tinto mining site contained high concentrations of uranium and lead that could harm local residents who rely on water from a nearby lake and river for drinking. The study, commissioned by the Andrew Lees Trust, confirmed that uranium concentrations downstream of the QIT-Madagascar Minerals mining site were 350 times higher than those upstream, and lead concentrations were 9.8 times higher. A 2019 report from Columbia University found elevated concentrations of heavy metals in rivers and streams near the Porgera mining site in Papua New Guinea. The mine site, operated as a joint venture of Barrick Gold Corp and Zijin, discharged tailings directly into rivers with government permission. A Columbia study confirmed that levels of toxins such as cadmium, lead, nickel, arsenic and zinc exceed those indicated in national and international potable water standards.

In the Russian Federation, the main contribution to hydrosphere pollution is made by enterprises of the coal mining industry, ferrous and non-ferrous metallurgy, which are concentrated in the European part, in Siberia and the Urals. Particular attention is paid to the control of pollution of water bodies in the regions of Siberia, where the main reserves of fresh water are concentrated (in Lake Baikal) on the border of the Irkutsk Region and the Republic of Buryatia. Unfortunately, as a result of the activities of mining enterprises, the state of water bodies in Buryatia is causing concern. For example, for the period from 1999 to 2015, OAO Buryatzoloto discharged 25% of wastewater without treatment, and 75% was insufficiently treated. Sources of pollution of surface and groundwater are: mine waters; recycled water from the tailings of enrichment shops; recycled water from the tailings pond of the hydrometallurgy workshop; and surface runoff from residential areas and industrial sites [12]. Sources of water pollution are not only active mines, but also abandoned mines. In June 2017, orange mine water emerging from the adit and flowing into the Modon-Kul River was observed from the unexploited Zapadnaya mine in the Zakamensky District of the Republic of Buryatia. The water samples were found to exceed standard values for cadmium content by 99 times, manganese by 91 times, iron by 78 times, and lead by 20 times [13].

In the Urals, the highest degree of pollution of water bodies was recorded in rivers flowing in the area of the Denezhkin Kamen nature reserve in the north of the Sverdlovsk Region. They were found to contain high levels of heavy metals. Tamsher turned out to be the dirtiest river: in its waters the concentration of copper exceeds the norm by 60 thousand times, zinc by 6 thousand times, and manganese by 3 thousand times. A similar situation is observed in other rivers. The streams from the dumps of the Shemursky quarry exceed standard values by a hundred to a thousand times. Here, the ecological disaster zone with destroyed forests has an area of more than 600 square kilometers [14].

When purifying acidic wastewater from heavy metals, sorption technologies play a pivotal role and their development requires new types of sorbents. Expanded clay, a lightweight porous material that looks like granules with a burnt brown–red surface and dark pores inside, was chosen for the study. [15].

2 Expanded Clay

Expanded clay is a pressure-resistant material, which does not undergo biological degradation, is a heat and sound insulating material, non-toxic one, and has a neutral pH and sorption properties. Research has shown that expanded clay pre-treated with steam activation at a temperature of 800–900 °C has higher effectiveness in removing any pollutants that can clog the pores and expand microporosity to meso- and macroporosity. Expanded clay is produced from clay by firing at high temperatures of 1100–1300 °C in rotary kilns, each of which consists of a long, large-diameter steel cylinder inclined to the horizon at an angle of about 5°. Inside, in the firing zone, the kiln is lined with refractory bricks, which, when the kiln rotates, heat up to the required temperature and "fry" the clay pellets until the required degree of expansion is achieved. The length and configuration of the kiln depends in part on the composition of the clay and the duration of time required to "condition" the clay granules in a preheater to a temperature of about 650 °C so that they do not crumble before being converted into pyroplastics [16].

2.1 Physical and Chemical Properties of Expanded Clay

To study the physical and chemical properties of the expanded clay surface, an OPTIMA 2100DV atomic emission spectrometer with inductively coupled plasma and a JEOL JSM-6460 LV electron microscope with an attachment for micro-X-ray spectral analysis were used. Electron microscopic analysis of expanded clay with varying degrees of magnification is presented in Fig. 1.

Fig. 1 Microphotograph of the expanded clay surface and the results of micro-X-ray spectral analysis (photo by the authors)

The photograph of expanded clay is in the upper left part of the Figure. The right side of the Figure shows a microphotograph of the expanded clay surface. A study of a microphotograph of the expanded clay surface indicates to a porous structure of the sample under study. The lower part of the Figure shows micro-X-ray spectral analysis of the sample under study.

The main elements that make up the expanded clay sample under study are summarized in Table 1. As can be seen from the Table, the main chemical elements in expanded clay are oxygen, magnesium, aluminum, silicon, calcium, and iron.

Table 1 Chemical composition of expanded clay, %

Element	O	Na	Mg	Al	Si	K	Ca	Ti	Fe
Element content, wt. %	46.07	0.59	1.52	10.02	28.96	1.90	3.09	0.52	7.32
	44.37	0.75	2.00	11.62	22.30	2.42	2.44	0.61	13.49

2.2 *World Experience in Using Expanded Clay*

An analysis of world literature indicates that the use of expanded clay is very diverse. For example, expanded clay is used to optimize the treatment of wastewater contaminated with dyes. To treat colored wastewater, a biological method with the application of enzymes and microorganisms is used. Bioremediation with the help of white rot fungi produces biodegradation of dyes, then these fungi are immobilized using expanded clay. [17].

Expanded clay is used as a sorbent to remove polycyclic aromatic hydrocarbons (PAH) from water. PAH are generally resistant to biodegradation and are not easily removed by traditional physicochemical methods such as flocculation, coagulation, sedimentation, filtration or ozonation. Methods such as chemical or biological oxidation, ion exchange, electrolysis, and solvent extraction can be difficult to use and are sometimes ineffective at removing trace amounts of contaminants. However, sorption has been proven to be effective in removing persistent organic pollutants. Expanded clay is used for these purposes instead of traditional activated carbon [18].

Expanded clay can be used to remove nitrogen from nitrifying and denitrifying bacteria in an aquaculture system, where the floating expanded clay provides a basic structure to immobilize nitrifiers and denitrifiers, additionally the expanded clay is coated with alginate to provide the necessary nutrients to maintain bacterial cohesion and growth [19].

Expanded clay can also be used in the free zone of road infrastructure (back filling along roads) to purify storm water [20].

Literature analysis shows that expanded clay is actively used in research to purify water from various pollutants; however, the bulk of articles on the use of expanded clay are devoted to building materials, where expanded clay is used as a filler for lightweight concrete [21–24], and as a basis for road construction in soils with limited bearing capacity [25].

3 Results of Experiment on the Effectiveness of Acidic Wastewater Treatment with Expanded Clay

The effectiveness of sorption of heavy metals by expanded clay was assessed using a mixture of dump waters from the Novo-Shemurskoye and Shemurskoye polymetallic ore deposits in the Northern Urals as a sorbate. The effluent is highly acidic. The pH value is 2.38, so simultaneously with the removal of pollutants, the problem of neutralizing the wastewater arises. The composition of the sorbate is presented in Table 2.

As can be seen from the Table, the main pollutants in the source water are aluminum, cobalt, copper, iron, lead, and zinc. The interaction of the sorbent (expanded clay) with the sorbate was studied using static and dynamic methods.

Table 2 Chemical composition of the sorbate

Element	Al	Co	Cu	Fe	Pb	Zn
Element content, mg/l	216.29	2.43	59.95	614.41	0.98	106.7

The effectiveness of purification of the source wastewater with expanded clay under static conditions depending on temperature and time is presented in Fig. 2.

As can be seen from the charts, after cleaning the source wastewater with expanded clay, the concentration of pollutants decreased by an average of 70% to 90%, and it is worth noting that the cleaning effectiveness increased with an increase in temperature from 10 to 20 °C.

The obtained dependences make it possible to determine the mechanism of sorption of heavy metal cations using expanded clay. The porous structure of expanded clay should ensure the occurrence of the sorption process according to the type of physical adsorption. But physical adsorption is an exothermic process that occurs spontaneously. Sorbed ions tend to occupy the entire surface of the adsorbent; this is prevented by the process opposite to adsorption–desorption (endothermic process). Therefore, the higher the temperature, the less physical adsorption should be. The improvement in the effectiveness of sorption of heavy metal cations with expanded clay with increasing temperature suggests that the process occurs not only through the mechanism of physical adsorption, but also chemical processes are present, so when using expanded clay, pollutants will be firmly retained on the surface. With this

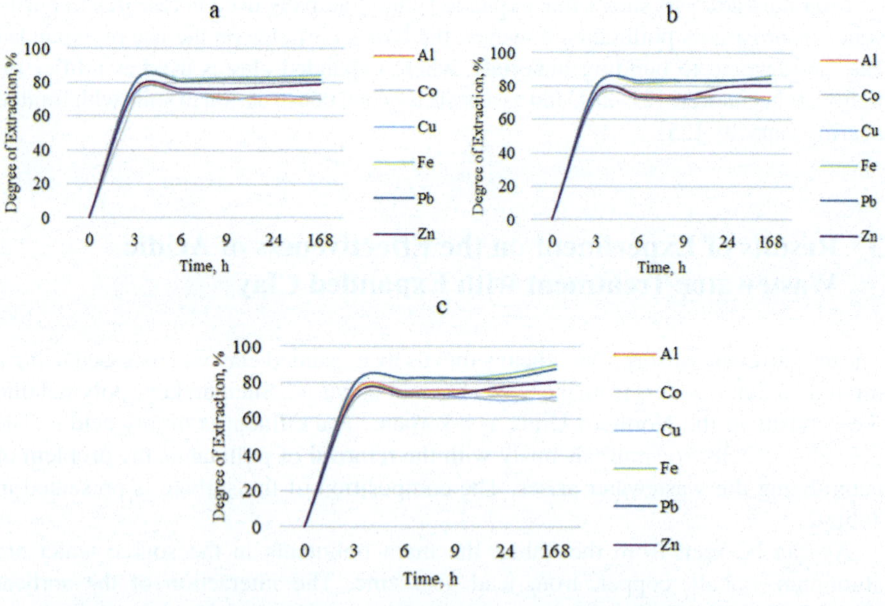

Fig. 2 Effectiveness of purification of pollutants with expanded clay under static conditions depending on temperature and time: **a** 0 °C; **b** 10 °C; c–20 °C (authors' data)

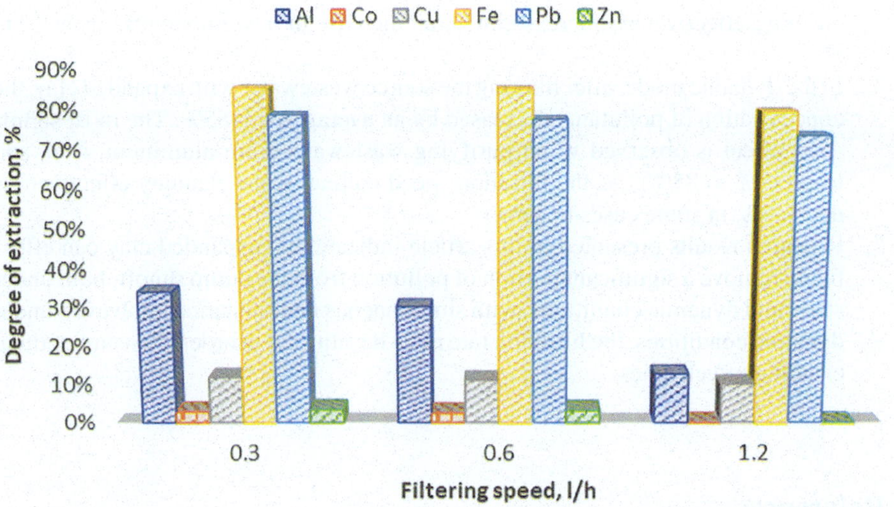

Fig. 3 Effectiveness of pollutant purification with expanded clay under dynamic conditions (authors' data)

mechanism, the danger of a burst release of pollutants into the sorbate after reaching the saturation limit of the expanded clay surface centers is eliminated.

The effectiveness of purification of the source wastewater with expanded clay under dynamic conditions depending on the filtration rate is presented in Fig. 3.

As can be seen from the presented charts, after filtering the source wastewater with expanded clay, the concentration of pollutants decreased by an average of 3–85%. The most significant effect is observed when treating wastewater from aluminum, iron, and lead (13–85%). As the filtration speed increases, the cleaning effectiveness decreases, in some cases to zero.

The research results presented in this article indicate that expanded clay can effectively remove a significant portion of pollution from the dump runoff, both under static and dynamic conditions, with simultaneous neutralization. When treating wastewater, the authors recommend using these types of waste mainly as part of composite sorbents.

4 Conclusions

1. Possibility of using expanded clay to remove heavy metals from acidic wastewater from the mines of the Northern Urals has been studied.
2. Composition and structure of expanded clay have been studied.
3. Sorption capacity of expanded clay under static conditions depending on the temperature and type of pollutant has been determined. The concentration of pollutants decreased by an average of 70–90%. And it is worth noting that the

cleaning effectiveness increased with an increase in the temperature from 10 to 20 °C.

4. In the dynamic mode, after filtering the source wastewater with expanded clay, the concentration of pollutants decreased by an average of 3–85%. The most significant effect is observed when purifying wastewater from aluminum, iron, and lead (13% to 85%). As the filtration speed increases, the cleaning effectiveness decreases, in some cases to zero.
5. Research results presented in this article indicate that expanded clay can effectively remove a significant portion of pollution from the dump runoff, both under static and dynamic conditions, with simultaneous neutralization. However, under dynamic conditions, the filtration rate must be adjusted in order to ensure optimal cleaning effectiveness.

References

1. Ugya AY, Ajibade FO, Ajibade TF (2018) Water Pollution Resulting From Mining Activity: An Overview. Proceeding of the 2018 Annual Conference of the School of Engineering and Engineering Technology (SEET): Need-Driven Engineering Research for Entreprenuerial Development in Developing Countries, pp 17–19
2. Akabzaa TM, Banoeng - Yakubo BK, Seyire JS (2007) Impact of Mining Activities on Water in the Vicinity of the Obuasi Mine. West African Journal of Applied Ecology 11:101–109
3. Bench DW (2003) PCBs, Mining, and Water Pollution. Mine Design, Operations and Closure Conference, Polson, Montana, April 27–May 1, 2003
4. Vorobyev AV, Karginov KG, Ananikyan SA, Odintsova ES (2002) Evaluation of the Environmental Impact of Mining Industry Enterprises. Ekologicheskaya ekspertiza (Environmental Expert Review) 3:96–104
5. Akcil A, Koldas S (2006) Acid Mine Drainage (AMD): Causes, Treatment and Case Studies. Journal of Cleaner Production 14:1139–1145
6. Skousen J, Ziemkiewicz P, McDonald L (2018) Acid Mine Drainage Formation, Control and Treatment: Approaches and Strategies. The Extractive Industries and Society. https://doi.org/10.1016/j.exis.2018.09.008
7. Akcil A, Koldas K (2006) Acid Mine Drainage (AMD): Causes, Treatment and Case Studies. J. Cleaner Prod. 14(1):139–1146. https://doi.org/10.1016/j.jclepro.2004.09.006
8. Skousen J (2014) Overview of Acid Mine Drainage Treatment with Chemicals. Acid Mine Drainage, Rock Drainage, and Acid Sulfate Soils: Causes, Assessment, Prediction, Prevention, and Remediation, no 29. https://doi.org/10.1002/9781118749197.ch29
9. Burtsev LI, Galchenko IuP, Sabyanin GV (2003) Methodological Approaches to Normalizing the Level of Technogenic Changes in the Hydrosphere During Underground Development of Subsoil. Environmental Safety in the Development of Deposits, IPKON, Moscow, 226–230
10. Elpatievskaia VP (1997) Interaction of Sub-dump Waters of Polymetallic Deposits with Waters of Local River Runoff: Using the Example of the Southern Far East. Geography and Natural Resources 2:57–62
11. Herlihy A, Kaufmann P, Mitch M, Brown D (1990) Regional Estimates of Acid Mine Drainage Impact on Streams in the Mid-Atlantic and Southeastern United States. Water Air Soil Poll. 50:91–107
12. Bazarova SB. Impact of Mining Enterprises on the Regional Ecosystem and Assessment of the Effectiveness of Their Environmental Activities. Regional Economics and Management: Electronic Scientific Journal. https://eee-region.ru/article/1008/. Accessed 14 Feb 2023

13. Astrakhantseva AIu, Timofeeva SS (2022) Mining Enterprises of the Republic of Buryatia: Impact on Atmospheric Air and Environmental Risks. Mining Information and Analytical Bulletin 5(2):198–208
14. Bodrov V. How Ural Mining and Metallurgical Company (UMMC) is Destroying the Ural Rivers. Water, Dumps, Pyrites. Federal Press: [website]. https://fedpress.ru/article/2614512. Accessed 14 Feb 2023
15. Expanded Clay Production—What Expanded Clay is Made From. https://gruntovozov.ru/chasto-zadavayemiye-voprosy/proizvodstvo-keramzita/. Accessed 14 Feb 2023
16. Owens PL, Newman JB (2003) Lightweight Aggregate Manufacture. Advanced Concrete Technology, Butterworth-Heinemann, pp 1–12. https://doi.org/10.1016/B978-075065686-3/50283-4
17. Dede Heri Yuli Yanto, Rayi Mishellia Chempaka, Oktan Dwi Nurhayat, Bambang Dwi Argo, Takashi Watanabe, Yusuf Wibisono, Yung-Tse Hung (2023) Optimization of Dye-contaminated Wastewater Treatment by Fungal Mycelial-light Expanded Clay Aggregate Composite. Environmental Research 231(2):116207. https://doi.org/10.1016/j.envres.2023.116207
18. Marian Asantewah Nkansah, Alfred A. Christy, Tanja Barth, George William Francis (2012) The Use of Lightweight Expanded Clay Aggregate (LECA) as Sorbent for PAHs Removal from Water. Journal of Hazardous Materials 217–218:360–365. https://doi.org/10.1016/j.jhazmat.2012.03.038
19. Thi Minh Nguyen, Phuong Thu Ha, Thi Thu Huong Le, Ke Son Phan, Thi Nhi Cong Le, Thi Thu Trang Mai, Phuong Ha Hoang (2022) Modification of Expanded Clay Carrier for Enhancing the Immobilization and Nitrogen Removal Capacity of Nitrifying and Denitrifying Bacteria in the Aquaculture System. Journal of Bioscience and Bioengineering 134(1):41–47. https://doi.org/10.1016/j.jbiosc.2022.04.006
20. Tonoy K. Das, Onja D. Raoelison, Hamid Rehman, Yuhui Zhang, Wendy Chau, Lisa Thamiz, Michael K. Stenstrom, Sanjay K. Mohanty (2023) Use of Expanded Shale, Clay, and Slate Aggregates and Biochar in the Clear Zone of Road Infrastructures for Sustainable Treatment of Stormwater. Journal of Cleaner Production 428:139443. https://doi.org/10.1016/j.jclepro.2023.139443
21. Bogas JA, Gomes A, Pereira MFC (2012) Self-compacting Lightweight Concrete Produced with Expanded Clay Aggregate. Construction and Building Materials 35:1013–1022. https://doi.org/10.1016/j.conbuildmat.2012.04.111
22. Rashad AM (2018) Lightweight Expanded Clay Aggregate as a Building Material—An Overview. Construction and Building Materials 170:757–775. https://doi.org/10.1016/j.conbuildmat.2018.03.009
23. Elías Roces, Mauro Muñiz-Menéndez, Jesús González-Galindo, José Estaire (2021) Lightweight Expanded Clay Aggregate Properties based on Laboratory Testing. Construction and Building Materials 313:125486. https://doi.org/10.1016/j.conbuildmat.2021.125486
24. BrabinWinsley JB, Muthukannan M (2022) A Study on Strength Parameters of Concrete with Expanded Fly Ash Clay Aggregate. Materials Today: Proceedings 56(6):3484–3489. https://doi.org/10.1016/j.matpr.2021.11.128
25. Salisa Chaiyaput, Jiratchaya Ayawanna, Pornkasem Jongpradist, Hatairat Poorahong, Raksiri Sukkarak, Pitthaya Jamsawang (2023) Application of a Cement–clay–air Foam Mixture as a Lightweight Embankment Material for Construction on Soft Clay. Case Studies in Construction Materials 18:e02188. https://doi.org/10.1016/j.cscm.2023.e02188

Chrysotile for Removing Heavy Metals from AMD

A. Samodolov, D. Ulrikh, and Yu. Romanova

Abstract The article discusses the possibility of using chrysotile to remove various pollutants, in particular heavy metals, from acid mine drainage (AMD). The purpose of this study is to develop and justify new technologies for purifying wastewater from heavy metals (using the example of Al, Co, Cu, Fe, Pb and Zn). The authors study the physicochemical properties and composition of chrysotile. The structure and surface composition of the material under study are examined using a JEOL JSM-6460 LV electron microscope with an attachment for micro-X-ray spectral analysis, and the chemical composition of the sorbate and its changes depending on the research conditions are studied using an OPTIMA 2100DV atomic emission spectrometer with inductively coupled plasma. The fibrous structure of the chrysotile sample under study consisting of fibers 2–3.5 mm in length and 0.06–0.08 μm in diameter is determined using electron microscopic analysis. Its sorption capacity is determined under static conditions at various temperatures. Data on the effect of filtration speed on the purification effectiveness are obtained in dynamic mode. It has been established that the effectiveness of wastewater treatment when using chrysotile varies from 71 to 100% under static conditions and from 35 to 100% under dynamic conditions, depending on the nature of metal cations in the effluent, sorbate temperature, and filtration rate. The authors note that the value of the hydrogen index during dynamic testing increases from a value of 2.38 to values of 4.11–4.20 at filtration rates of 0.3–0.6 l/h, and to a value of 5.84 at a filtration rate of 0.78 l/h.

Keywords Technogenic dump water · Chrysotile · Mining and processing enterprise · Sorption · Heavy metals · Pollutants · Filter medium

A. Samodolov (✉) · D. Ulrikh
South Ural State University, 76, Lenin Prospekt, Chelyabinsk 454080, Russia
e-mail: samodolov@mail.ru

D. Ulrikh · Yu. Romanova
Saint Petersburg State University of Architecture and Civil Engineering, 4, 2nd Krasnoarmeiskaya Str., St. Petersburg 190005, Russia

L. V. Radionova and D. V. Ulrikh (eds.), *Advances in Ecology and Environmental Engineering*, Springer Proceedings in Earth and Environmental Sciences,
https://doi.org/10.1007/978-3-031-64423-8_28

1 Introduction

Extraction of mineral resources plays a vital role in the development of the national economies of many countries around the world due to the development of a consumer society, and therefore the demand for raw materials is constantly increasing. Mining industry enterprises are no exception in this production chain and are also increasing production volumes, having an extremely undesirable impact on the environment. Environmental problems arising as a result of the work of mining enterprises include disturbance of the natural topography, degradation and chemical pollution of land, pollution of the atmosphere and water bodies with heavy metals, organic and inorganic waste, which in turn has a negative effect on terrestrial and aquatic ecosystems. The consequence of such mining is the formation of surface and underground acidic wastewater as a result of drainage of acid mines, lowering of groundwater levels, subsidence of soil, disruption of the hydrological cycle and sedimentation. The outflow of acidic wastewater from mines, commonly referred to as acid mine drainage (AMD), is considered one of the most serious threats to water resources in the mining industry. AMDs are capable of causing long-term destructive effects on rivers, streams, and aquatic life [1–7]. The negative environmental impacts from AMD have been described by researchers in Asia [8–10], New Zealand [11, 12], Europe [13–16], South America [17, 18], Canada [19, 20], the USA [21–23] Madagascar, and Papua New Guinea [24]. This problem is relevant for developed and developing countries around the world.

In the eastern United States, more than 10,000 km of streams and more than 72,000 ha of lakes and reservoirs were negatively impacted by AMD before 1990 [25].

In Madagascar, according to a 2019 study, water downstream from the Rio Tinto mining site contained high concentrations of uranium and lead that could harm local residents who rely on water from a nearby lake and river for drinking. The study, commissioned by the Andrew Lees Trust, confirmed that uranium concentrations downstream of the QIT-Madagascar Minerals mining site were 350 times higher than those upstream, and lead concentrations were 9.8 times higher. A 2019 report from Columbia University found elevated concentrations of heavy metals in rivers and streams near the Porgera mining site in Papua New Guinea. The mine site, operated as a joint venture of Barrick Gold Corp and Zijin, discharged tailings directly into rivers with government permission. A Columbia study confirmed that levels of toxins such as cadmium, lead, nickel, arsenic and zinc exceed those indicated in national and international drinking water standards.

In the Russian Federation, the main contribution to hydrosphere pollution is made by enterprises of the coal mining industry, ferrous and non-ferrous metallurgy, which are concentrated in the European part, in Siberia and the Urals. Particular attention is paid to the control of pollution of water bodies in the regions of Siberia, where the main reserves of fresh water are concentrated (in Lake Baikal) on the border of the Irkutsk Region and the Republic of Buryatia. Unfortunately, as a result of the activities of mining enterprises, the state of water bodies in Buryatia is causing concern. For example, for the period from 1999 to 2015, OAO Buryatzoloto discharged 25% of wastewater without treatment, and 75% was insufficiently treated. Sources of pollution of surface and groundwater are: mine waters; recycled water from the tailings of enrichment shops; recycled water from the tailings pond of the hydrometallurgy workshop; and surface runoff from residential areas and industrial sites [26]. Sources of water pollution are not only active mines, but also abandoned mines. In June 2017, orange mine water emerging from the adit and flowing into the Modon-Kul River was observed from the unexploited Zapadnaya mine in the Zakamensky District of the Republic of Buryatia. The water samples were found to exceed standard values for cadmium content by 99 times, manganese by 91 times, iron by 78 times, and lead by 20 times [27].

In the Urals, the highest degree of pollution of water bodies was recorded in rivers flowing in the area of the Denezhkin Kamen nature reserve in the north of the Sverdlovsk Region. They were found to contain high levels of heavy metals. Tamsher turned out to be the dirtiest river: in its waters the concentration of copper exceeds the norm by 60 thousand times, zinc by 6 thousand times, and manganese by 3 thousand times. A similar situation is observed in other rivers. The streams from the dumps of the Shemursky quarry exceed standard values by a hundred to a thousand times. Here, the ecological disaster zone with destroyed forests has an area of more than 600 km^2 [28].

When AMDs enter surface water bodies, they have a biological effect on river and lake organisms due to changes in the acidity and chemical composition of the habitat, the presence of heavy metals, deterioration of the organoleptic properties of water bodies (increased turbidity and color), orange or yellow coloration of river sediments, and disruption of nutrient cycling. In areas adjacent to mining enterprises, reservoirs become unsuitable for domestic, agricultural and industrial purposes.

2 Methods

Treatment of wastewater containing AMD can be carried out in various ways, differing in effectiveness and economic costs. Many of them are not widely used due to the high cost of reagents, low effectiveness of filter materials or the complexity of the equipment, so the development of new methods for treating such wastewater is still relevant today. To reduce costs and increase the effectiveness of acid wastewater treatment, in this work we propose to use a sorption treatment method based on a combined filter load, where the fibrous mineral chrysotile is used as one of the components.

Chrysotile was first described by von Kobell. The word *chrysotile* comes from a combination of Greek words *golden* and *fibrous*. Chrysotile is a fibrous mineral of the silicate class, serpentine group, with a chemical composition close to $Mg_3Si_2O_5(OH)_4$.

Chrysotile asbestos or *white asbestos* is a hydrous magnesium silicate, found in nature in the form of a crystal consisting of hollow fibril tubes with a diameter of 2.6×10^{-5} mm and a length of up to 2–3 cm. The mineral is capable of splitting into the finest chrysotile fibers with a thickness of up to 0.5 μm. Chrysotile fibers have a tensile density of more than 3000 MPa, alkali resistance from 9.1 to 10.3 pH, density from 2.4 to 2.6 g/cm^3, specific surface area of 20 m^2/g and capacity for adsorption.

Chrysotile asbestos has some exceptional properties that no other mineral or synthetic fibers have: it is stronger than steel, non-corrosive, fire-resistant (thermal insulating) and non-conductive; it is much cheaper than synthetic fibers such as polyvinyl acetate (PVA); and it requires less energy during production [29].

The composition and properties of the material depend on the deposit where it is mined. Asbestos, widely used in construction, has carcinogenic properties. The fibrogenicity and carcinogenicity of fibers of different types of asbestos are different. The types of asbestos traditionally mined and used in Europe are highly hazardous. However, long-fiber chrysotile asbestos produced in Russia has significantly lower toxicity levels (tens of times lower). It belongs to Class 4 in terms of the degree of negative environmental impact on the environment (low-hazard).

To determine a more detailed component composition in percentage terms, it is necessary to be guided by the results of a chemical analysis of the material used, in accordance with POT PM-010-2000 Interindustry Rules for Labor Protection in the Production of Asbestos and Asbestos-containing Materials and Products. A JEOL JSM-6460 LV electron microscope with an attachment for micro-X-ray spectral analysis was used to study the structure and composition of the surface of the material. The chemical composition of the sorbate and its changes depending on the test conditions were studied using an OPTIMA 2100DV atomic emission spectrometer with inductively coupled plasma.

Chrysotile from the Orsko-Khalilovskoye deposit was used in this work. The article is part of the dissertation research by Aleksandr P. Samodolov.

3 Results and Discussion

The results of electron microscopic analysis of chrysotile are presented in Fig. 1. The microphotograph clearly shows the fibrous structure of the chrysotile sample under study. The size of the fibers is shown: length is 2–3.5 mm, diameter is 0.06–0.08 μm.

The results of X-ray microspectral analysis of the chrysotile surface are presented in Table 1.

As can be seen from the Table, the main chemical elements in chrysotile are oxygen, magnesium, silicon, carbon, iron, and calcium.

The effectiveness of sorption of heavy metals by chrysotile was assessed using a mixture of dump waters from the Novo-Shemurskoye and Shemurskoye polymetallic ore deposits in the Northern Urals as a sorbate. The effluent is highly acidic. The pH value is 2.38, so simultaneously with the removal of pollutants, the problem of neutralizing the wastewater arises. The composition of the sorbate is presented in Table 2.

The effectiveness of purification of the source wastewater with chrysotile under static conditions depending on temperature and time is presented in Fig. 2. As can be

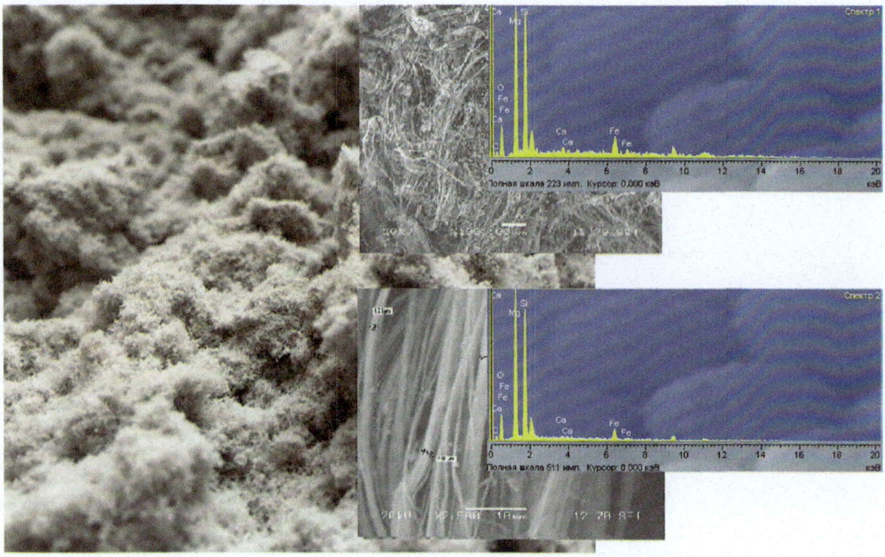

Fig. 1 Microphotograph of the chrysotile surface and the results of micro-X-ray spectral analysis (photo by the authors)

Table 1 Results of micro-X-ray spectral analysis of chrysotile

Element	C	O	Mg	Si	Ca	Fe
Element content (wt%)	11.9	43.9	21.8	17.5	0.5	4.4

Table 2 Chemical composition of the sorbate

Element	Al	Co	Cu	Fe	Pb	Zn
Element content (mg/l)	216.29	2.43	59.95	614.41	0.98	106.7

seen from the Table, the main pollutants in the source water are aluminum, cobalt, copper, iron, lead, and zinc. The interaction of the sorbent (chrysotile) with the sorbate was studied using static and dynamic methods.

The effectiveness of purification of the source wastewater with chrysotile under static conditions depending on temperature and time is presented in Fig. 2.

As can be seen from the charts, after purification of the source wastewater with chrysotile, the concentration of pollutants decreased by an average of 85–90%. It is worth noting that the purification effectiveness increased with an increase in temperature from 10 to 20 °C. At temperatures of 10–20 °C, it was possible to reduce the content of such metals as aluminum, cobalt, copper, iron, and zinc in the wastewater almost to zero, and the treatment effectiveness was almost 100% accordingly.

The obtained dependences make it possible to determine the mechanism of sorption of heavy metal cations by chrysotile. The high specific surface area of chrysotile

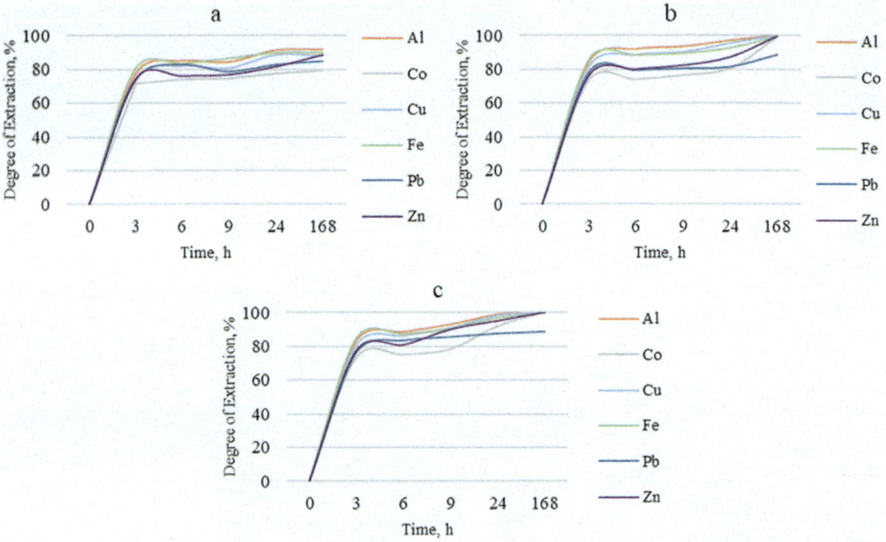

Fig. 2 Effectiveness of purification of pollutants with chrysotile under static conditions depending on temperature and time: **a**—0 °C; **b**—10 °C; c—20 °C (authors' data)

should ensure that the sorption process occurs according to the type of physical adsorption. But physical adsorption is an exothermic process that occurs spontaneously. Sorbed ions tend to occupy the entire surface of the adsorbent; this is prevented by the process opposite to adsorption—desorption (endothermic process). Therefore, the higher the temperature, the less physical adsorption should be. The increase in the effectiveness of sorption of heavy metal cations by chrysotile with increasing temperature suggests that the process occurs not only through the mechanism of physical adsorption, but also chemical processes are present, therefore, when using chrysotile, pollutants will be firmly retained on the surface. With this mechanism, the danger of a burst release of pollutants into the sorbate after reaching the saturation limit of the surface centres of chrysotile is eliminated.

The effectiveness of purification of the source wastewater with chrysotile under dynamic conditions depending on the filtration rate is presented in Fig. 3.

As can be seen from the presented charts, after filtering the source wastewater with chrysotile, the concentration of pollutants decreased by an average of 70–80%. The most significant effect is observed when removing aluminum, copper, iron, and lead from wastewater (96–100%), the concentrations of which differ by 3–650 times. Therefore, chrysotile can be used for the sorption of heavy metals from wastewater of complex chemical composition, where pollutants are present in very high and low concentrations. It has been observed that the purification effectiveness under dynamic conditions increases as the filtration speed increases. At the maximum achieved filtration rate of 0.78 l/h, it was possible to reduce the content of such metals as aluminum, copper, iron, and lead in water almost to zero, and the purification effectiveness was almost 100% correspondingly. This distinguishes chrysotile from

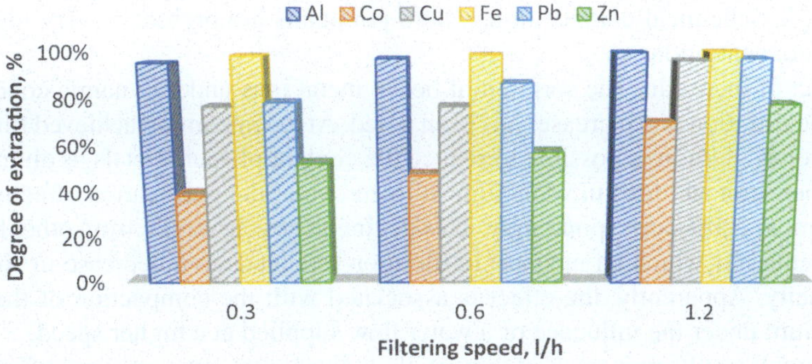

Fig. 3 Effectiveness of purification of pollutants with chrysotile under dynamic conditions (authors' data)

other known sorbents, for which an increase in filtration rate leads to a decrease in sorption capacity. Apparently, the effect is associated with the compaction of the filter medium under the influence of a water flow supplied at a higher speed. In a denser load, the number of sorbent centers interacting with pollutant ions increases, and the purification effectiveness increases accordingly. In the process of sorption of heavy metals by chrysotile, the runoff is neutralized. This is another positive effect from using this material. The value of the hydrogen index during the dynamic test mode increases from 2.38 to values of 4.11 to 4.20 at filtration rates of 0.3 to 0.6 l/h, and to a value of 5.84 at a filtration rate of 0.78 l/h.

The research results presented in this article indicate that chrysotile can effectively remove a significant portion of pollution from the dump runoff, both under static and dynamic conditions, with simultaneous neutralization.

4 Conclusions

1. Possibility of using chrysotile to remove heavy metals in acidic wastewater from the mines of the Northern Urals has been studied.
2. Composition and structure of chrysotile have been studied.
3. Sorption capacity of chrysotile under static conditions depending on the temperature and type of pollutant has been determined. The concentration of pollutants (aluminum, cobalt, copper, iron, and zinc) during the sorption process decreased by an average of 85–90%. With an increase in temperature from 10 to 20°C, the cleaning effectiveness increased to 100%.
4. In the dynamic mode, after filtering the source effluent with chrysotile, the concentration of pollutants decreased by an average of 70–80%. The most significant effect is observed when purifying wastewater from aluminum, copper, iron, and lead (96–100%), the concentrations of which differ by 3–650 times. Therefore, chrysotile can be used for the sorption of heavy metals from wastewater of

complex chemical composition, where pollutants are present in very high and low concentrations.

5. Effect of increasing the sorption of heavy metal ions under dynamic conditions as the filtration rate increases has been noted. At the maximum achieved filtration rate of 0.78 l/h, it is possible to reduce the content of such metals as aluminum, copper, iron and lead in water almost to zero, and the purification effectiveness is almost 100% correspondingly. This distinguishes chrysotile from other known sorbents, for which an increase in filtration rate leads to a decrease in sorption capacity. Apparently, the effect is associated with the compaction of the filter medium under the influence of a water flow supplied at a higher speed.
6. Research results presented in this article indicate that chrysotile can effectively remove a significant portion of contaminants from the dump runoff, both under static and dynamic conditions, with simultaneous neutralization.

References

1. Ugya AY, Ajibade FO, Ajibade TF (2018) Water Pollution Resulting From Mining Activity: An Overview. Proceeding of the 2018 Annual Conference of the School of Engineering and Engineering Technology (SEET): Need-Driven Engineering Research for Entreprenuerial Development in Developing Countries, pp 17−19
2. Akabzaa TM, Banoeng-Yakubo BK, Seyire JS (2007) Impact of Mining Activities on Water in the Vicinity of the Obuasi Mine. West African Journal of Applied Ecology 11:101–109
3. Bench DW (2003) PCBs, Mining, and Water Pollution. Mine Design, Operations and Closure Conference, Polson, Montana, April 27–May 1, 2003
4. Vorobyev AV, Karginov KG, Ananikyan SA, Odintsova ES (2002) Evaluation of the Environmental Impact of Mining Industry Enterprises. Ekologicheskaya ekspertiza (Environmental Expert Review) 3:96–104
5. Akcil A, Koldas S (2006) Acid Mine Drainage (AMD): Causes, Treatment and Case Studies. Journal of Cleaner Production 14:1139–1145
6. Skousen J, Ziemkiewicz P, McDonald L (2018) Acid Mine Drainage Formation, Control and Treatment: Approaches and Strategies. The Extractive Industries and Society. https://doi.org/10.1016/j.exis.2018.09.008
7. Skousen J (2014) Overview of Acid Mine Drainage Treatment with Chemicals. Acid Mine Drainage, Rock Drainage, and Acid Sulfate Soils: Causes, Assessment, Prediction, Prevention, and Remediation, no 29. https://doi.org/10.1002/9781118749197.ch29
8. David CP (2003) Establishing the Impact of Acid Mine Drainage Through Metal Bioaccumulation and Taxa Richness of Benthic Insects in a Tropical Asian Stream (The Philippines). Environmental Toxicology and Chemistry. 22(12):2952−2959
9. Wei TT, Yu Y, Hu ZQ, Cao YB, Gao Y, Yang YQ, Wang XJ, Wang PJ (2013) Research Progress of Acid Mine Drainage Treatment Technology in China. Applied Mechanics and Materials. 409–410:214–220
10. Yang JE, Kim HJ, Ok YS et al (2007) Treatment of Abandoned Coal Mine Discharged Waters using Lime Wastes. Geosciences Journal. 11(2):111−114
11. Uster B, Trumm D, Pope J, Weber P, O'Sullivan A, Weisener C, Diloreto Z (2014) Waste Mussel Shells to Treat Acid Mine Drainage: A New Zealand Initiative. American society of mining and reclamation, pp 23−27

12. Winterbourn M, McDiffett W, Eppley S (2000) Aluminium and Iron Burdens of Aquatic Biota in New Zealand Streams Contaminated by Acid Mine Drainage: Effects of Trophic Level. Science of The Total Environment 254(1):45–54
13. Antivachis D, Chatzitheodoridis E, Skarpelis N, Komnitsas K (2016) Secondary Sulphate Minerals in a Cyprus-type Ore Deposit, Apliki, Cyprus: Mineralogy and Its Implications Regarding the Chemistry of Pit Lake Water. Mine Water Environ 36:226–238
14. Balci N, Demirel C (2018) Prediction of Acid Mine Drainage (AMD) and Metal Release Sources at the Kure Copper Mine Site, NW Turkey. Mine Water Env. 37:56–73
15. Casiot C, Egal M, Elbaz-Poulichet F, Bruneel O, Bancon-Montigny C, Cordier M, Gomez E, Aliaume C (2009) Hydrological and Geochemical Control of Metals and Arsenic in a Mediterranean River Contaminated by Acid Mine Drainage: Preliminary Assessment of Impact on Fish (Leuciscus Cephalus). Appl. Geochem. 24:787–799
16. Gray N, Delaney E (2008) Comparison of Benthic Macroinvertebrate Indices for the Assessment of the Impact of Acid Mine Drainage on an Irish River below an Abandoned Cu-S Mine. Environ. Poll. 155:31–40
17. Strosnider W, Llanos Lopez F, Nairn R (2011) Acid Mine Drainage at Cerro Rico de Potosi II: Severe Degradation of the Upper Rio Pilcomayo Watershed. Environ. Earth Sci. 64:911–923
18. Strosnider W, Llanos Lopez F, Nairn R (2011) Acid Mine Drainage at Cerro Rico de Potosi I: Unabated High-strength Discharges Reflect a Five-century Legacy of Mining. Environ. Earth Sci. 64:899–910
19. Neculita C, Zagury GJ, Bussiere B (2007) Passive Treatment of Acid Mine Drainage in Bioreactors using Sulfate-reducing Bacteria. J. Env. Qual. 36:1–16
20. Sracek O, Choquette M, Gelinas P, Lefebvre R, Nicholson RV (2004) Geochemical Characterization of Acid Mine Drainage from a Waste Rock Pile, Mine Doyon, Quebec, Canada. Journal of Contaminant Hydrology 69(1–2):45–71
21. Cherry D, Currie R, Soucek D, Latimer D, Grent G (2001) An Integrative Assessment of a Watershed Impacted by Abandoned Mined Land Discharges. Environ. Pollut 111:377–388
22. Kleinmann R (1989) Acid Mine Drainage. Eng. Mining J. 190:161–168
23. Soucek D, Cherry D, Currie R, Latimer H, Trent G (2000) Laboratory to Field Validation in an Integrative Assessment of an Acid Mine Drainage—Impacted Watershed. Environ. Tox. Chem. 19:1036–1043
24. Harmful Impacts of Mining. Responsible Mining Foundation: [website]. https://www.responsibleminingfoundation.org/ru/harmful-impacts-of-mining/. Accessed 14 Feb 2023
25. Herlihy A, Kaufmann P, Mitch M, Brown D (1990) Regional Estimates of Acid Mine Drainage Impact on Streams in the Mid-Atlantic and Southeastern United States. Water Air Soil Poll. 50:91–107
26. Bazarova SB. Impact of Mining Enterprises on the Regional Ecosystem and Assessment of the Effectiveness of Their Environmental Activities. Regional Economics and Management: Electronic Scientific Journal: [website]. https://eee-region.ru/article/1008/. Accessed 14 Feb 2023
27. Astrakhantseva AIu, Timofeeva SS (2022) Mining Enterprises of the Republic of Buryatia: Impact on Atmospheric Air and Environmental Risks. Mining Information and Analytical Bulletin. 5–2:198–208
28. Bodrov V. How Ural Mining and Metallurgical Company (UMMC) Is Destroying the Ural Rivers. Water, Dumps, Pyrites. Federal Press: [website]. https://fedpress.ru/article/2614512. Accessed 14 Feb 2023
29. What Is Chrysotile? Chrysotile Association: [website]. https://chrysotile.ru/page/what-is-chrysotile/. Accessed 14 Feb 2023

Using of Phytogenic Waste Sorbents for Purification of Urban Surface Wastewater from Aluminum

O. Samodolova, D. Ulrikh, and M. Lazurina

Abstract It is no secret that the condition of various ecological systems in the present-day world leaves much room for improvement. In addition to food issues, the growing population on our planet causes a number ecological problems, including contamination of the environment with different anthropogenic pollutants generated as a result of the human activity. There is a major problem with obtaining of clean potable water due to the pollution of water bodies, and the pollution processes have not ceased to this day. Among such factors is the polluted surface wastewater from urban territories that enters water reservoirs. Running over the impervious urban surfaces this wastewater collects all the pollutants along its way and transports these to a water body. Aluminum is one of such pollutants. After entering the human body, this microelement causes negative effects and leads to various diseases. In order to prevent this substance's further entering water bodies, we propose to use sorption, and phytogenic production waste—as sorbent. This is the easiest, most effective and relatively cheap approach to treatment of water mediums. This article presents a comprehensive study of the possibility of using four types of phytogenic waste: hemp hulls; carbonized straw; carbonized walnut shells; and granulated buckwheat husks—as sorbents for purification of urban surface wastewater from aluminum. The composition of the materials has been analyzed, and microanalysis using a JEOL JSM-6460 LV scanning electron microscope has been performed. The sorption capacity of the potential sorbents in various conditions has been studied.

Keywords Waste water · Sorbent · Vegetable waste · Aluminum

O. Samodolova (✉) · D. Ulrikh
South Ural State University, 76, Lenin Prospekt, Chelyabinsk 454080, Russia
e-mail: Samodolova@mail.ru

D. Ulrikh · M. Lazurina
Saint Petersburg State University of Architecture and Civil Engineering, 4, 2nd Krasnoarmeiskaya Str, St. Petersburg 190005, Russia

L. V. Radionova and D. V. Ulrikh (eds.), *Advances in Ecology and Environmental Engineering*, Springer Proceedings in Earth and Environmental Sciences,
https://doi.org/10.1007/978-3-031-64423-8_29

1 Introduction

It is no secret that the state of environment in various ecological systems leaves much room for improvement. In addition to food problems, the growing population on our planet causes a number ecological ones, for example: contamination of the environment, including both the air and water bodies. Degradation of water bodies compromises the population's standards of living and causes problems, both obvious ones, like potable water contamination, and hidden ones, like food contamination (fish). A significant factor contributing to the deterioration of the condition of these mediums is the urban surface wastewater, which on its way to a water body runs over impervious urban surfaces and collects all the pollutants, including aluminum.

Human body is a flawless self-regulating system. To maintain its normal functioning, it needs food, water and a complex of various vitamins and minerals. Unfortunately, besides the "vital" elements, there are substances, the concentration of which in our body must be limited to an acceptable level, or they must be prevented from entering it altogether as they can cause irreparable damage to human health. Aluminum is one of such substances. Aluminum microelement does not play any important role or function in our body, therefore its presence in any part of our body is not normal. Entering our body with food and water through the digestive system, it causes different irreversible changes inflicting damage to the central nervous system and leads to cognitive deterioration (Alzheimer's disease) and other neurodegenerative diseases. The accumulation of aluminum in the brain over a long period of time results in defects of short-term memory, working memory and data processing speed due to the chronic exposure. Also, substantial pathological changes are observed in the mitochondria ultrastructure and functions [1, 2]. When it enters our body together with another neurotropic chemical agent, arsenic, it can cause disturbance in the bioenergetics of nerve cells [3]. In the bone tissue, aluminum disrupts its mineralization and increases the general load on the skeletal system. It is also hazardous for people with kidney failure as the renal dysfunction leads to the increase in the accumulation of aluminum [4, 5].

In 2022, following the calculations, the UN press service reported that the world population had reached 8 billion people. According to the organization's further assessment, the world population would reach 9 billion people by 2037, and 10 billion people by 2058 [6]. In a situation like this, the global community has to boost the food production, including the share of crop products. Increasing the manufacture of the end-products (food) from this raw material would lead to the generation of gigantic amounts of waste and would bring on a number of problems, including the ecological ones, as well as difficulties related to waste disposal since for the most part such waste is buried in the ground (best case scenario), or left in the fields. An attendant problem is the lack of sufficient amounts of potable water of the required quality. Over the recent four decades, the use of water in the world has been increasing by approximately one percent every year. Despite the measures being taken with regards to the rational use of resources, this trend will persist up until

2050. This is caused by such factors as the increasing population and the social-and-economic development [7]. According to the UN's assessment, more than a quarter of the world population (over 2 billion people) have no adequate access to potable water. India has the largest potable water shortage in the world. And although the government of this country is trying to improve the situation, 163 million people still need help. Next come Ethiopia (where the population spends huge amounts of money on water) and Nigeria (the most populous state on the African continent). Approximately 60 million people experience problems with potable water in each of these countries. In China, residents of remote regions suffer from a shortage of drinking water. In Indonesia, the problem is caused by pollution of inland waters; and in Pakistan, by competition for limited water resources with the neighboring India.

The leaders among countries with sufficient supplies of fresh, clean water are Brazil, Russia and Canada. Nevertheless, even in Russia, about 40% of residents use water of low quality [8].

At present, it is possible to solve part of these problems by implementing a comprehensive approach and the principles of a closed-loop economy: using the waste generated from food production to make sorbents for purification of surface wastewater from pollutants, before these enter water bodies. Today there are a number or research studies that prove the effective use of crop production waste as sorbents to purify water from aluminum. The following can be used as sorbents for wastewater treatment. Guava leaves allow removing up to 90% of aluminum upon contact of the sorbent with the sorbate for 120 min, in which case the optimal pH value for the adsorption of metal ions is pH = 6.5, and the best temperature is 40 °C [9]. Samples of barley grains pericarp modified with a 1% solution of sulfuric acid allow reducing the content of aluminum ions by 66.26%, and are also recommended for use as a filtration and adsorption medium for purification of water from a petrochemical enterprise at the stage of mechanical purification [10]. Activated carbon from pine nut shells (treated with a solution of orthophosphoric acid at the ratio of 8% of H_3PO_4 per weight of raw materials), carbonized and activated in a muffle furnace at the temperature of 300 °C, can be effectively used as a sorption material, and the degree of purification from aluminum will depend on the initial concentration of the metal in the wastewater (0.05 mg/dm^3—90%, 1.00 mg/dm^3—70%) [11]. Phosphorus-containing product isolated from rice flour of rice varieties of the Far Eastern selection, makes it possible to extract aluminum ions from an aqueous solution by 97% already after 10 min; and after 30 min, equilibrium is established in the distribution of metal ions between the solution and the phosphorus-containing product (the degree of extraction for An^{3+}ions reaches 99%) [12]. Buckwheat husks treated with a solution of orthophosphoric acid at the ratio of 8% of H_3PO_4 per weight of raw material, carbonized and activated in a muffle furnace at a temperature of 300 ± 50 °C, if used as a sorbent, allow extracting from 13 to 73% of aluminum from water, and the extraction effectiveness increases with the increasing metal concentration in the source raw materials (0.1 mg/l—13%, 0.5 mg/l—63%, 1.0 mg/l—73%) [13].

Various reagents for the purification of wastewater from aluminum can be obtained from agricultural waste. A reagent produced from barley grains pericarp allows obtaining results for the purification of wastewater from aluminum comparable to traditional reagents [14, 15]. A reagent obtained as a result of alkaline pulping from wheat grains pericarp using the soda cooking method makes it possible to achieve maximum permissible concentration (MPC) values for aluminum ions, and at low initial concentrations to achieve discharge standards for MPC as well (the achieved effectiveness results are comparable to those of NaOH and at a lower cost, too) [16–18]. Reagents obtained from wheat grains make it possible to also purify wastewater from nickel ions (98.84%) and zinc (83.36%). When using a reagent obtained from barley grains pericarp, it is possible to extract copper by 91.95%, and when using a reagent obtained from oat grains pericarp, it is possible to purify wastewater from iron ions by 87.17% [19, 20].

The goal of this research is to determine the most effective sorbent from among the agroindustrial complex waste (hemp hulls; straw; walnut shells; buckwheat husks) for the purposes of purifying surface wastewater from aluminum.

2 Materials and Methods

The objects of the research were a system of four sorbents (hemp hulls; straw heat-treated at 80 °C during 10 min; walnut shells carbonized at 300 °C during 15 min; granulated buckwheat husks) and the aluminum-polluted surface wastewater from urban territories of the city of Chelyabinsk (sorbate).

Hemp hulls is a residual product (waste) obtained after cleaning hemp seeds. They look like hemispherical and slightly elongated hard shells of brown color.

Straw are dry stems obtained after threshing grain crops, in this case, wheat. For the study, the straw was pre-chopped and heat-treated at a temperature of 80 °C for 10 min.

The walnut shells are the hard outer shells of the nuts that protect the kernels. Before the study, walnut shells were crushed and exposed to heat treatment (carbonization) for 15 min at the temperature of 300 °C to destroy pathogenic microorganisms and improve the sorption characteristics of the sorption material.

Buckwheat husks are the dry outer shells of buckwheat grains, consisting of elastic convex petals. In the study we used husks that had undergone granulation, that is, the process of grinding, moistening, and pressing at the temperature of 120 °C. They look like cylindrical granules with a smooth surface.

All materials under study represent large-tonnage waste from the agro-industrial complex, which is annually renewable in huge quantities.

Figure 1 shows what the sorbents look like.

The surface wastewater from urban territories of the city of Chelyabinsk was in-taken according to the conventional techniques. The concentration of aluminum in the urban surface wastewater was 2.707 mg/l, with the maximum permissible concentration applied for discharge into a fishery water body being 0.04 mg/ dm^3.

Fig. 1 Sorbents appearance, from left to right: hemp hulls; straw; walnut shells; buckwheat husks

The effectiveness of the sorption process was evaluated in static and dynamic conditions. In static conditions, we used the limited volume method with the 1:20 solid phase—liquid ratio. The system temperature varied from 0 to 20 °C. The exposure time equaled 3–168 h. The study under static conditions is presented in Fig. 2.

The dynamic conditions were created using a specialized unit to filter the sorbate through the sorbent at a set speed. The initially set speeds equaled 0.3, 0.6, 1.2, and 15 l/h.

To perform the microanalysis in the course of the research, we used a JEOL JSM-6460 LV scanning electron microscope with an X-ray microspectroscopy attachment, and OPTIMA 2100DV Perkin Elmer emission spectrometer with inductively coupled plasma (USA). Extra purified water produced in a Simplicity UV purification system (France) was used as the background solution, and we used a pH-150MI microprocessor pH meter.

Microrelief study of the samples is given in Fig. 3.

Fig. 2 Research in static conditions: **a** hemp hulls; **b** straw; **c** walnut shells; **d** buckwheat husks

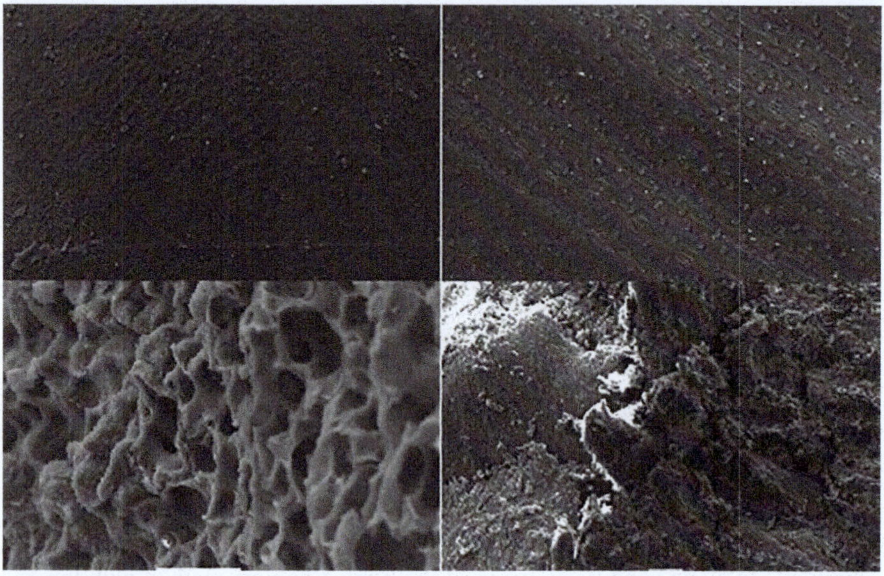

Fig. 3 X-ray microanalysis data from left to right: hemp hulls; straw; walnut shells; buckwheat husks

Table 1 Results of X-ray microanalysis of the samples (%)

Sorbent	C	O	K	Mg	Ca	Si
Hemp hulls*	64.38	34.99	0.35	0.13	–	0.15
Straw*	40.96	40.22	1.98	–	0.57	16.27
Walnut shells*	55.83	43.97	1.92	–	–	–
Buckwheat husks	54.26	44.25	0.77	0.42	0.30	–

* *Note* Average of 2 spectra

The percentage correlation of the averaged elemental composition of the samples is presented in Table 1. It shows that the main elements constituting the samples are carbon, oxygen and potassium.

3 Research Results

Table 2 presents the results of research on the effectiveness of sorption process in the sorbent-wastewater (stormwater) system in static conditions at various temperature modes.

Analysis of the effectiveness of aluminum removal from the urban surface wastewater has shown that the maximum removal of the pollutant equals 94%, and this result

Table 2 Results of research on the effectiveness of sorption process in static conditions, (%)

No.	Sorbent	Degree of pollutants removal from the sample at various t °C, (%)					
		t = 0 °C		t = 10 °C		t = 20 °C	
		Min	Max	Min	Max	Min	Max
1	Hemp hulls	41	88	35	94	53	94
2	Straw	18	76	6	94	82	88
3	Buckwheat husks	41	82	59	94	82	94
4	Walnut shells	76	88	71	82	41	82

can be achieved using three of the sorbents under study: hemp hulls, straw and buckwheat husks. The sorption process for all three materials is most effective at the temperature of 10 °C, over the period of 7 days. But hemp hulls and buckwheat husks make it possible to achieve the same result at the temperature of 20 °C, with the same exposure time of 7 days.

A less effective sorption process in static conditions was registered when using walnut shells, with the maximum effect (88%) achieved over the exposure time of 7 days, but at the temperature of 0 °C.

The results are given in Figs. 4, 5 and 6.

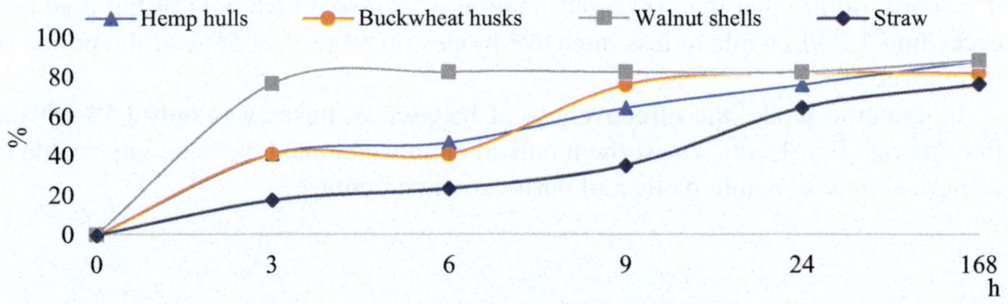

Fig. 4 Chart of changes in the indicators at the temperature of 0 °C

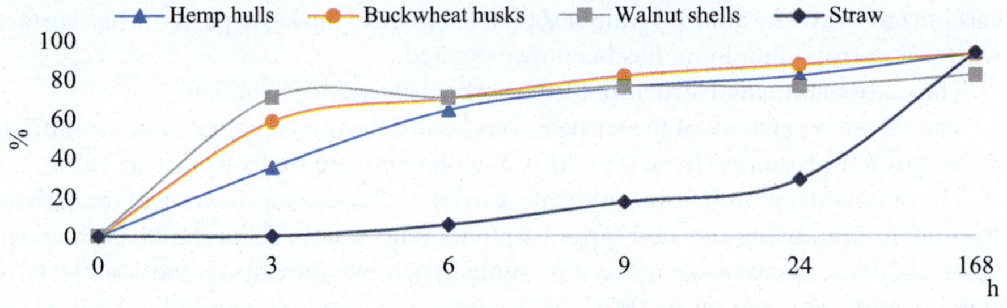

Fig. 5 Chart of changes in the indicators at the temperature of 10 °C

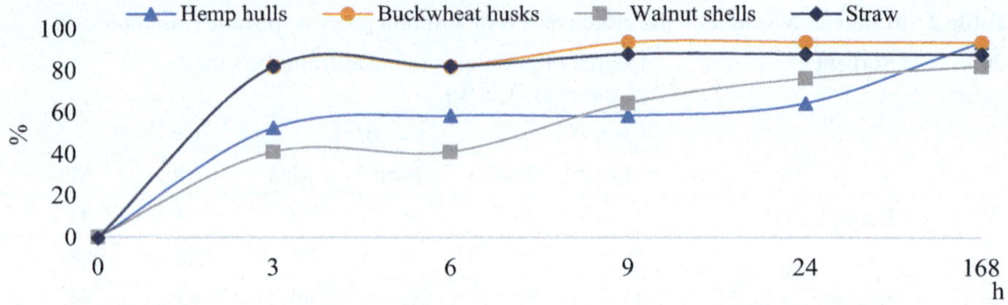

Fig. 6 Chart of changes in the indicators at the temperature of 20 °C

The research in dynamic conditions has shown that it is more preferable to use hemp hulls, straw and walnut shells as sorbents for purification of wastewater from aluminum. Using these materials, the degree of wastewater purification from this pollutant reaches 100%. But each sorbent should be used taking into consideration its specific features. Thus, carbonized straw is the most undemanding material. Namely this sorbent has allowed to maintain all the set conditions of the experiment, and its effectiveness equals 100% at any flow rate of the filtrate. Using hemp hulls as sorbent to remove aluminum from the sorbate, allows to achieve the effectiveness of 100%, but the maximum flow rate of the filtrate in this case equals 83 ml/min (4.98 l/h). The maximum flow rate of the filtrate when using walnut shells equaled 13.8 l/h. And it is worth noting that the 100% purification is achieved here only at the flow rate exceeding 1.2 l/h, while at less intensive modes no more than 56% of the pollutant is removed.

In dynamic mode, the effectiveness of buckwheat husks was only 65%, at the filtering rate of 0.15 l/h. The sorbent bulked up in water and it became impossible to achieve a more dynamic mode and purification percentage.

4 Conclusions and Discussion

The possibility of using waste from the agro-industrial complex (hemp hulls; carbonized straw; carbonized walnut shells; buckwheat husks) to purify urban surface wastewater from aluminum has been investigated.

The composition and structure of the sorbents have been studied.

The sorption capacity of the sorbents has been determined under static conditions at various temperatures (from 0 to 20 °C), with exposure times from 1 to 168 h.

The obtained research results indicate at a quite effective purification of the sorbate from aluminum using the said types of phytogenic waste. Hemp hulls, carbonized straw and buckwheat husks make it possible to achieve the effectiveness of 94% (in static conditions), and up to 100% in dynamic conditions (hemp hulls, carbonized straw, walnut shells). The less effective results in static conditions were registered for walnut shells (maximum effect of 88%), and in dynamic conditions—for granulated buckwheat husks (65%).

The sorbents have a number of advantages: low cost, annual replenishment, and environmental safety.

Therefore, hemp hulls, straw, buckwheat husks and walnut shells can be classified as quite promising and highly-effective sorbents. Using these sorbents for purification of polluted wastewater (including from aluminum) will help solve the problem of generation of this type of waste in case of further increase in the food production at the global scale.

References

1. Skupnevskiy SV, Ivanov DV (2023) The Effect of Aluminum and Its Compounds on the Functions of Human Organs and Tissues (review article) Journal of New Medical Technologies, Edition 1
2. Ju QN (2023) Overview of the Relationship Between Aluminum Exposure and Human Health. Neurotoxicity of Aluminum 7:1–32. https://doi.org/10.1007/978-981-99-1592-7_1
3. Flurik SV, Dremza IK (2022) Mechanisms of Mitochondrial Dysfunction of Neurons under Exposure to Arsenic and Aluminum (review). Journal "Vestnik Vitebskogo Gosudarstvennogo Meditsinskogo Universiteta" ("Vitebsk Medical Journal") 2
4. Eliseeva T (2022) Aluminum (Al): Its Significance for Body and Health, and What Its Sources. Journal of Healthy Nutrition and Dietology 20
5. Trace Aluminum, Potentially Toxic (2023) Available via DIALOG. https://news.un.org/ru/story/2023/03/1438972. Accessed 14 Dec 2023
6. World Population Has Exceeded 8 Billion People (2023) Available via DIALOG. https://www.kommersant.ru/doc/5667917. Accessed 14 Dec 2023
7. UN Report: We Can Overcome Water Crises Only by Joint Effort (2023) Available via DIALOG. https://news.un.org/ru/story/2023/03/1438972. Accessed 14 Dec 2023
8. World Thirst: What Caused Water Shortage in Different Countries and What Can We Do about It? (2023) Available via DIALOG. https://tass.ru/opinions/11935383. Accessed 14 Dec 2023
9. Mustafa A, Kolkaila SA, May AAE (2023) Removal of Aluminum (III) from Water by Adsorption on the Surface of Natural Compound. Journal of Environmental Treatment Techniques 11 (2):100–105. https://doi.org/10.47277/JETT/11(2)105
10. Nikolaeva ES (2020) Purification of Wastewater from Petrochemical Productions Using Modified Grain Crop Waste Environment. Safety and Protection: Fundamental and Applied Research: Collected Papers of the All-Russian Science Conference, Belgorod, 19–23 October 2020, pp 192–194
11. Labuzova OM, Noskova TV, Ilyina EG (2015) Obtaining Activated Carbon from Pine Nut Shells to Remove Aluminum Ions from Water. Lomonosov Readings in Altai: Fundamental Problems of Science and Education: Collected Papers of the International Conference, Barnaul, 20–24 October 2015, pp 1515–1518
12. Kovekhova AV, Arefyeva OD, Makarenko NV (2018) EXTRACTION OF Al3+ IONS. Chemistry of vegetable raw materials 4:281–288. doi:https://doi.org/10.14258/jcprm.2018043761
13. Merinova OM (2015) Obtaining Activated Carbon from Buckwheat Husks to Remove Aluminum from Water. International Student Science Bulletin 3-4:530–531
14. Miftakhova AR, Magizova EF, Stepanova SV, Shaykhiev IG (2018) Purification of Water from Aluminum Ions Using a Reagent Obtained from Barley Grains Pericarp. Relevant Issues of Environment Protection: Collected Papers of the All-Russian Science and Technology Conference, Belgorod, 17–19 September 2018, pp 140–144

15. Magizova EF, Shaykhiev IG, Stepanova SV (2019) Purification of Water from Aluminum (III) Ions Using Wastewater after Soda Pulping of Grains Pericarp. Environment Safety and Protection: Fundamental and Applied Research: Collected Papers of the All-Russian Science Conference, Belgorod, 14–18 October 2019, 182–186
16. Miftakhova AR, Magizova EF, Stepanova SV, Shaykhiev IG (2019) Purification of Water from Aluminum Ions Using an Innovative Reagent Obtained from Wheat Grains Pericarp. Innovative Approaches to Solving Modern Problems of the Rational Use of Natural Resources and Environment Protection: Collected Papers of the International Science and Technology Conference, Alushta, 03–07 June 2019, pp 231–235
17. Magizova EF, Stepanova SV (2018) Purification of Water from Aluminum Ions Using an Alternative Reagent Obtained from Wheat Grains Pericarp. Chemistry and Engineering Ecology - XVIII: Collected Papers of the International Science Conference, Kazan, 27–29 September 2018, pp 70–72
18. Miftakhova AR (2018) Research on Purification of Water from Aluminum Ions Using an Alternative Reagent. Nanomaterials and Nanotechnologies: Problems and Prospects: Proceedings of the 8th International Youth Science Conference, Saratov, 03–05 November 2018, pp 191–195
19. Magizova EF, Stepanova SV, Shaykhiev IG (2018) Searching for the Ways to Purify Wastewater from Heavy Metal Ions Using Reagents Obtained from Grain Crop Waste. Chemistry and Engineering Ecology - XVIII: Collected Papers of the International Science Conference, Kazan, 2729 September 2018, pp 189–191
20. Stepanova SV, Alekseeva AA (2022) Purification of Model Waters from Aluminum Ions by Wastewater from the Soda Pulping of Cellulose from the Shells of Cereals. Theoretical and Applied Ecology 1:124–128. https://doi.org/10.25750/1995-4301-2022-1-124-128

Using Medicinal Plants for Purification of Urban Surface Wastewater

O. Samodolova and A. Avsyukevich

Abstract At present, there is quite a significant contamination of water bodies with various pollutants, and this process has not ceased to this day. Among the most serious ones are heavy metals, which enter water bodies with urban surface wastewater, including from impervious surfaces. It is possible to prevent this from happening by using modern bioengineering facilities. One of the components in these facilities are bioaccumulator plants. Today, the search for such plants is still relevant, but they must meet a number of requirements: they must be pollution-tolerant, adapted to a region's climate, aesthetically attractive (what is important for urban territories), and they must be cheap. Namely that is why wild mint (Mentha arvensis L.) medicinal plant drew our attention, as it meets all the above-mentioned requirements. In this article, we study the plant's structure and composition. The interaction within the plant-sorbate system has been examined in laboratory conditions using the limited volume method. During the study, the time of contact of the plant with the sorbate has been varied from 1 to 72 h, while the ambient temperature has been 20 °C. The goal of this research is to study the bioaccumulating capacity of wild mint when used for purification of urban surface wastewater from heavy metals (on the example of Al, Cr, Cu, Fe, Mn and Zn). The obtained research results show that the concentration of pollutants has been reduced by 40–100%. The authors recommend to use wild mint in combination with other bioaccumulator plants to achieve synergistic effect.

Keywords Urban surface wastewater · Bioaccumulator plant · Heavy metals · Pollutants · Mentha Arvensis (wild mint)

O. Samodolova (✉)
South Ural State University, 76, Lenin Prospekt, Chelyabinsk 454080, Russia
e-mail: Samodolova@mail.ru

A. Avsyukevich
Saint Petersburg State University of Architecture and Civil Engineering, 4, 2nd Krasnoarmeiskaya Str., St. Petersburg 190005, Russia

L. V. Radionova and D. V. Ulrikh (eds.), *Advances in Ecology and Environmental Engineering*, Springer Proceedings in Earth and Environmental Sciences,
https://doi.org/10.1007/978-3-031-64423-8_30

1 Introduction

When heavy metals enter various water bodies, they pose a serious threat and contribute to further degradation of water reservoirs. One of the modern methods of reducing the amount of pollutants entering water bodies is the use of bioengineering facilities installed on the way of urban surface wastewater polluted with heavy metals. Bioengineering facilities effectively remove heavy metals. One of the components in these facilities are bioaccumulator plants. Such plants are widely used for purification from these pollutants in industrial and contaminated areas in various countries. Both one specimen, or whole fields of bioaccumulator plants can be used. Development of bio-facilities is currently in progress, and the search for plants meeting these requirements is still underway. Namely that is why the *Labiatae* (*Lamiaceae*) family of plants drew our attention.

The *Labiatae* family is massive and abundant. It includes almost 8 thousand species. Their specific features are the active phytogenic compounds and essential oils. At various stages of the human civilization, the *Lamiaceae* have been playing an important role: suffice it to recall lavender, rosemary, sage and marjoram. Wild mint also belongs to this family. About 40 species of mint have been described. They are mostly found in humid areas in the temperate zones. In addition to these species, there are plenty of hybrids and varieties. In the temperate zones in Russia, wild mint (*Mentha arvensis* L.), horse mint (*Mentha longifolia*) and water mint (*Mentha aquatica*) mostly grow.

Wild mint (*Mentha arvensis* L.) is an aromatic wilding perennial plant belonging to the *Lamiaceae* or *Labiatae* family. Popular names are: horse mint, black mint, deaf mint, ground ivy, Polish mint, Chinese mint, and deadnettle. Wild mint does not have the strong and characteristic aroma and fresh flavor of peppermint. The first data on its use date very far back: it was mentioned in cuneiform tablets of the Assyrians, and was found in tombs of Egyptian pharaohs. The ancient Greeks named this plant in honor of Mentha nymph (the love of Pluto, ruler of the underworld, whose jealous wife turned this poor creature into a plant with a subtle and cool aroma) [1].

It was first cultivated in the ancient times in Europe, was spread to Japan by the nineteenth century, and later—to China and other Asian countries (according to ethnomedical records) [2].

The Eurasian species grows in Europe, Western and Central Asia, in the Caucasus, and as far as India and Nepal. In Russia it is found in the North Caucasus (Pre-Caucasian area, (Krasnodar Territory (south), Republic of Adygeya, Stavropol Territory, Karachay-Cherkess Republic, Kabardino-Balkarian Republic, Republic of North Ossetia, Chechen Republic, Republic of Ingushetia), in the European part, in the Western and Eastern Siberia. It is also widely spread in Latvia, Lithuania, Estonia, Byelorussia, Ukraine, Kazakhstan, Uzbekistan, Kirghizia, Tajikistan, Turkmenistan, Georgia (Racha-Lechkhumi and Kvemo Svaneti, Imereti, Samtskhe–Javakheti, Shida Kartli, Mtskheta-Mtianeti, Kvemo Kartli and Tbilisi, Kakheti), Azerbaijan (except for Talish and the Nakhchivan Autonomous Republic) [3]. The sites where this plant grows vary: watersides of natural reservoirs, fields, marshes, loose woodland, and building-surrounding grounds. In the Russian Federation, wild mint is included in the register of perennial weeds, but it actually found wide application in folk medicine and as a flavoring agent.

Wild mint has a very high concentration of essential oils, their main component being menthol (50–80%) [4]. It also contains vitamin C, carotene, flavonoids, terpenes, organic acids, bitter principles, tanning principles, microelements, and estrogen.

Flower heads are most rich in essential oils (4–6%). Leaves may contain up to 2.5% of oils. Stems have almost no oils. The main oil components are oxygen derivatives of monocyclic terpenes; menthol, pulegone, menthofuran, as well as ethers of menthol with acetic and isovalerianic acids. In addition, mint oil contains terpenes: monocyclic ones (1-limonene, alpha-phellandrene) and bicyclic ones (alpha- and beta-pinenes). Mint leaves also contain ursolic and oleanic acids (up to 0.5%), carotene (up to 40 mg), hesperidin, betaine, and other substances [5].

The plant is 70–100 cm high, and in certain cases may be short-growing—down to 15 cm. It has a straightly standing branchy stem. Common features: downiness, downwards-oriented cilia, green or reddish color. The straightly standing stems may become less straight in certain natural conditions of their growing, though the plant preserves its characteristic hollow structure. The size of leaves may vary and can be 2–7 cm long and 1–3 cm wide. They also have slight downiness and are covered in cilia, but can be of various shapes: ovoid, oblong, close to oval or elongated, or reminding of an ellipsoid. Common features: meandering jaggedness along the edges, cilia, and arrow-headed tip. Color: all shades of green, depending on the age and habitat. Small blossom clusters are located in ball-shaped verticillasters, and can also vary in color: from purple to pinkish, or even violet in the flower cup, the shape of which is similar to a small bellflower with a wide crown. Flowers are located on a bare or hairy floral stem. There are no specific differences in eremus types. Coenobium is monoseeded in each of the four constituting parts and looks a like small nut. All 4 parts develop harmoniously, are round-shaped, slightly blunt on top, and form the so-called schizocarp. This is a strong couch grass rooting quite deeply sometimes. Nature created this mechanism so that the plant could fix itself in wet soil, spread on the adjacent territories and withstand aggressive destruction [6, 7]. Figure 1 shows what the plant looks like [8].

Experts advise to set these plants near shrubs or give them more space so that they could reproduce by root propagules. A great solution would be to plant them under trees, where the *Labiatae* would have enough shade that they like. All that wild mint needs for comfortable growing are the diffused lighting or lacy shade, good humidity and free space.

This plant is widely used in pharmacology and traditional medicine, cooking and cosmetology. The mint essential oil is very often used in medicine as stomachic medication and pain killer. Menthol is used to cure running nose, migraines and freeze burns; it is used as an antisepsic and pain killer, and also used for the production of validol (menthol valerate). Mint leaves help to cure migraines and neuralgic disorders. The aerial parts of plants are used as antitussives, expectorants, as well as to cure elevated heart rate, nausea, vomiting, and allergies; they are also

Fig. 1 Appearance of wild mint [8]

used as appetite stimulators and medications for treating hyperpeptic gastritis, gastric ulcers [9], gastro-intestinal and hepatic colics, and as an astringent remedy. In the Tibetan medicine these plants are used to cure pulmonary tuberculosis and gastric diseases (internal application), and as pain killer for spasms, rheumatic and arthroses pains, and as anti-inflammatory agent for skin diseases (external application). Mint is included in the tea medications (appetite stimulating, gastric, antiflatulent, sudorific, bile-expelling, and sedative ones) and in compositions for taking baths [1, 10–12]. Leaf extract is used for treating rheumatoid arthritis, infections, as well as against mosquito bites [13].

Research has been performed at the University of Kashmir aimed at evaluating the antioxidant parameters of *Mentha arvensis* to study how it can actively improve memory and help fight against the Alzheimer's disease [14].

The research results show that *Mentha arvensis* extract has a great pharmacological potential, especially with consideration to its antibacterial, antidiabetic, and cytotoxical effects. In particular, luteolin and rosemarinic acid have been recognized as the main candidates for potential bioactivity with acceptable properties that are similar to medicines [15].

This plant is also used to salt away cabbage heads, and add flavor to sauces, drinks and kvas (bread drink). Leaves are added as dressing to salads, meat and fish dishes, baked foods (sponge cakes with adding of Dutch cocoa powder and different varieties of mint [16]), vegetable and fruit dishes, soups, borshch, dumplings, as well as to milk to prevent it from souring. Menthol from wild mint extract is a health nutritional supplement that fights against resistant pathogens (used for preservation and production of healthy foods [17]). Essential oil is used as flavoring agent in perfume manufacturing for toothpastes, powders, essences, elixirs, and eau de toilette. Nectar-bearing flowers [1] are recommended to be used as natural repellent when storing wheat grain (to repel grain storage pests in various conditions) [18].

Wild mint is also used in landscape design for decoration of watersides of big reservoirs, when growing with limitation—and for small compositions.

Scientists from India have published a research that confirms the possibility of using wild mint for remediation of contaminated soil (purify it from cadmium) to ensure sustainable agriculture and food products safety [19].

Research conducted to evaluate the concentration of zinc in the plants growing in the flood plain of the Irtysh River (contamination resulted from the activity of Kazzinc industrial company) has shown that plants of the *Labiatae* family, including wild mint, are capable of actively accumulating zinc. Thus, the concentration of zinc in the wild mint growing in steppificated chestnut meadow floodplain soil equaled 15.9 mg/kg [20].

The unquestionable benefits of this plant are a weighty argument in favor of using it: flower-bearing stem is hugely attractive for insects and facilitates the pollination of plants growing in a land plot or a near-house vegetable garden. This plant can also be used as a background decorative crop thanks to its impressive green matter that can emphasize other plants, which have long stems and big flowers.

The goal of this work is to study the capability of wild mint for bioaccumulation and evaluate its use as a bioaccumulator plant for removing heavy metals from urban surface wastewater.

2 Materials and Methods

The object of the research is a system consisting of wild mint (bioaccumulator plant) and surface wastewater from urban territories of the city of Chelyabinsk (sorbate).

The surface wastewater from urban territories of the city of Chelyabinsk was in-taken according to the conventional techniques. The chemical composition of a mixed sample of wastewater (stormwater) of the city of Chelyabinsk is presented in Table 1.

Effectiveness of the sorption process was assessed in static conditions. Wild mint (without a sub-sample) was placed in a glass and poured over with the sorbate. The samples for the chemical analysis were taken according to a schedule. The temperature during conducting of the research was 20 °C. The time of exposure was from 1 to 72 h.

Table 1 Chemical composition of the sorbate

Element	Al	Cr	Cu	Fe	Mn	Zn	pH
Content in urban surface wastewater (mg/l)	2.707	0.01	0.028	2.624	0.307	0.286	6.43

To perform the research, we used a JEOL JSM-6460 LV scanning electron microscope with an X-ray microspectroscopy attachment, and OPTIMA 2100DV Perkin Elmer emission spectrometer with inductively coupled plasma (USA). Extra purified water produced in a Simplicity UV purification system (France) was used as the background solution, and we used a pH-150MI microprocessor pH meter.

Figure 2 features microphotographs and results of the electron-microscopic analysis of the wild mint's vegetative organs (parts of the plant), at various magnification ranges.

Figure 2a shows microphotographs and data of the X-ray microanalysis of the plant's root with a 100-fold magnification. Figure 2b depicts the plant's stem (140-fold magnification), and Fig. 2c—the leaf with a 100-fold magnification. The bottom part of the figure presents the X-ray microanalysis of the sample under study.

The main elements constituting the source wild mint sample under study (as per the plant's parts) are consolidated into Table 2. As one many see in this Table, the main chemical elements in wild mint are carbon, oxygen, potassium, silica, aluminum, calcium, chlorine, magnesium, iron and sulphur.

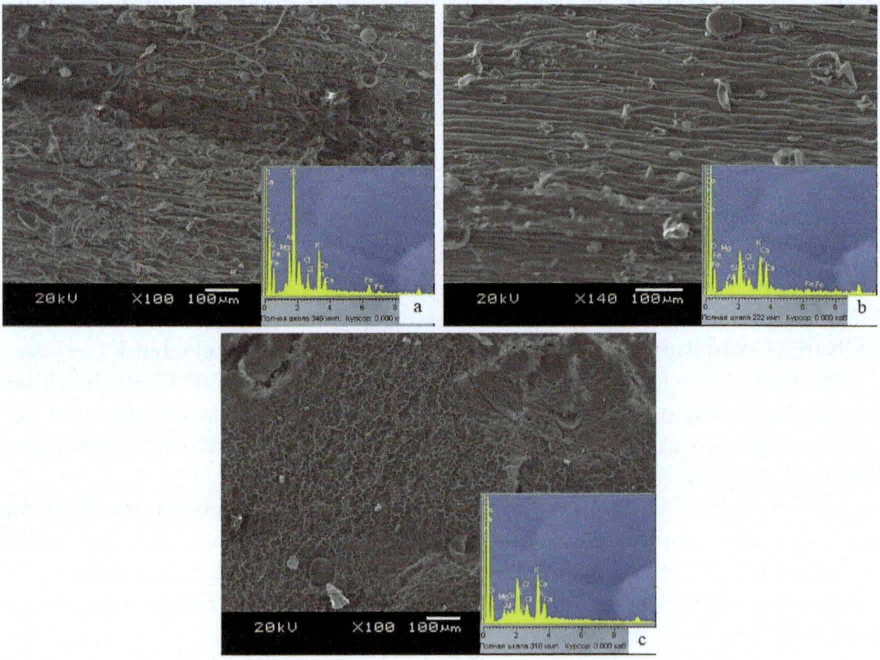

Fig. 2 Wild mint microphotographs and data of the X-ray microanalysis (photo by the authors): **a**—root, **b**—stem, **c**—leaf

Table 2 Results of X-ray microanalysis of the sample (%)

Element	C	O	Mg	Al	Si	S	Cl	K	Ca	Fe
Average element's concentration in the root (wt%)	47.6	42.8	0.5	1.35	3.8	0	0.6	2.0	0.8	0.6
Average element's concentration in the stem, (wt%)	55.7	37.9	0.8	0.30	0.7	0.4	0.5	2.1	1.3	0.3
Average element's concentration in the leaves, (wt%)	52.5	41.41	0.69	0.27	0.31	0.31	0.61	2.83	1.07	0

3 Research Results

The results obtained in the course of the research on removing heavy metals from the sorbate with the help of the plant are presented in Fig. 3.

As one may see in this Figure, after treatment of the source wastewater using the plants under study, the heavy metals concentration was reduced by 40–100%. The plants were less effective for sorbing aluminum, with the maximum effect equaling 59% over 1 h, and 41% till the completion of the study.

The next element with the achieved degree of purification from it is zinc, which is reliably removed by 50% regardless of the time of contact of wild mint with the sorbate. Copper removal also does not depend on the time of contact and equals 80% over the entire course of the experiment.

We managed to achieve the effectiveness exceeding 90% during purification from iron, manganese and chromium. The removal of iron is most effective during the first hours (95.7%), and next the degree of purification slightly decreases (91%). Very stable results were registered for manganese: regardless of the time of contact of the sorbate with the bioaccumulator plant, the effectiveness equals 99.9%. The

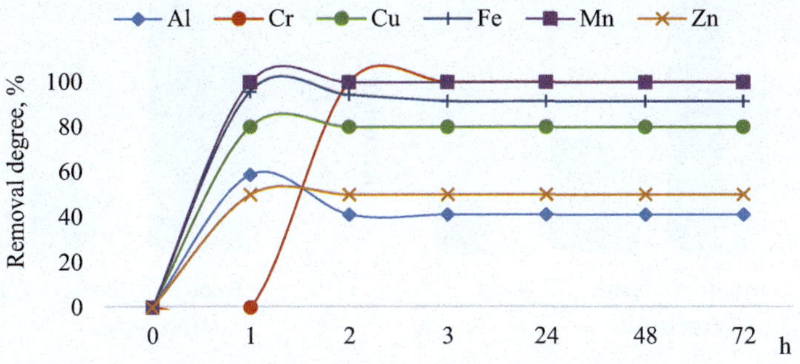

Fig. 3 Effectiveness of wild mint for pollutants removal in static conditions (authors' data)

maximum effect (100%) was achieved during purification from chromium, but the required time of contact is more than 2 h.

Figure 4 shows the dependence of the pH index on the time of contact in the plant-sorbate system.

As the chart shows, after contact of the source wastewater with the plant, the pH index slightly increases—3.2%.

The microscopic analysis of the wild mint reference sample proved the accumulation of part of the removed pollutants (and namely, iron and aluminum) in various organs of the plant. Comparison of the data on the source and reference samples is given in Fig. 5.

According to Fig. 5a, all vegetative organs of wild mint accumulate iron with more than 40% in the stem and 55% in the root. And we can see the accumulation of iron in the leaf in the reference sample despite the absence of iron in the leaf in the source sample.

Figure 5b is an illustration of a similar situation with aluminum, but this pollutant is accumulated not as effectively: root—6%, stem—23%, and (interesting point) no changes in the metal content in the leaf.

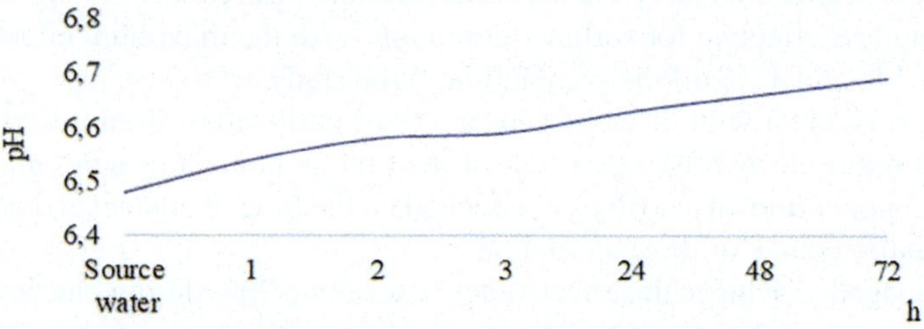

Fig. 4 Changes in the pH with the change in the time of contact of the plant (wild mint) with the sorbate (wastewater) (authors' data)

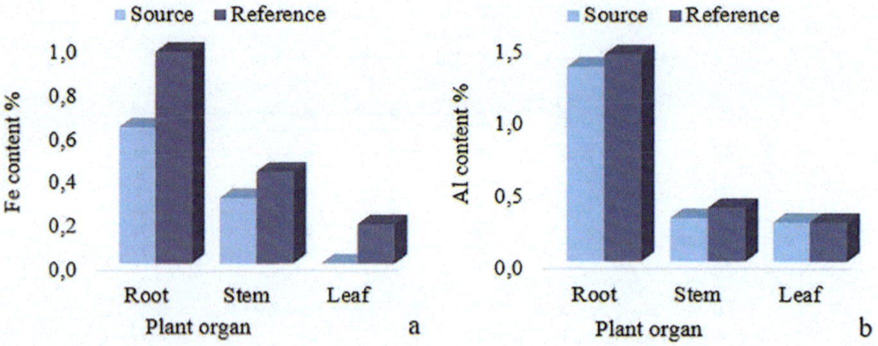

Fig. 5 Results of the X-ray microanalysis of the reference sample (%): **a**—iron content, **b**—aluminum content

The results obtained in this research indicate that wild mint quite effectively removes heavy metal ions from urban surface wastewater. The authors recommend to use this plant in combination with other plants during the wastewater treatment. This will help create a modern and effective bioengineering facility that has a number of competitive advantages: the plant is tolerant to the metals under study, is capable of simultaneously sorbing and accumulating several different heavy metals, grows in the majority of European territories, and is aesthetically attractive and cheap.

4 Conclusion

1. Possibility of using wild mint (*Mentha arvensis* L.) medicinal plant for purification of urban surface wastewater from heavy metals (aluminum, chromium, copper, iron, manganese, and zinc) has been studied.
2. Said plant's composition and structure have been examined.
3. Wild mint' sorption capacity in static conditions has been determined at the temperature of 20 °C and exposure time from 1 to 72 h. The pollutants concentration has been reduced by the average of 40–100%.
4. Research results presented in this article indicate that wild mint allows to quite effectively remove a significant part of pollutants from urban surface wastewater.
5. Said plant has a number of competitive advantages: it is tolerant to the metals under study, is capable of simultaneously sorbing and accumulating several different heavy metals, grows in the majority of European territories, and is aesthetically attractive and cheap.
6. Wild mint is recommended to be used in combination with other bioaccumulator plants to achieve synergistic effect.

References

1. Wild Mint (Mentha arvensis L.). https://karpolya.ru/page/mjata-polevaja-lugovaja-mentha-arvensis-l. Accessed 20 Jan 2024
2. Dar M, Mir P, Raja W, Rasool S (2021) Mentha Arvensis (L.): an Insight on the Pharmacological and Phytochemical Profile. Science of Spices & Culinary Herbs, pp 1–33. https://doi.org/10.2174/9789814998154121050005
3. Wild Mint (Mentha arvensis). https://universityagro.ru. mentha-arvensis/. Accessed 20 Jan 2024
4. Iosebidze TI, Ubiria MN, Kuridze MG (2019) Genetic Features of Mentha Arvenis L and Technology of Its Cultivation in Georgia. 2nd International Science Conference on the Role of Metabolomics in Improving the Biotechnological Means of Production, in the field of Metabolomics and Quality of Life, Moscow: All-Russian Scientific Research Institute of Medicinal and Aromatic Plants, pp 416–419, Moscow, June 06–07, 2019
5. National Projects in the Russian Federation. Our Russian Mint – Wild Mint or Meadow Mint (MENTHA ARVENSIS). https://xn----7sbmeb4apchekgg5a1ki.xn--p1ai/%D0%BC%D1%8F%D1%82%D0%B0-%D0%BB%D1%83%D0%B3%D0%BE%D0%B2%D0%B0%D1%8F/. Accessed 20 Jan 2024

6. Wild Mint. https://stroy-podskazka.ru/myata/o-polevoj/. Accessed 20 Jan 2024
7. Wild Mint. How to Gather and Use? https://botanicoved.com/poleznoe/travy/myata/vidy-myata/polevaya.html Accessed 20 Jan 2024
8. Wild Mint, pictures. https://triptonkosti.ru/2-kartinki/myata-polevaya-kartinki.html. Accessed 20 Jan 2024
9. Saleem M, Yadav S, Kumar R (2021) Evaluation of antiulcer activity of Mentha arvensis. IJCRT. 8(8):464–491
10. Wei H, Kong S, Jayaraman V, Selvaraj D (2023) Mentha arvensis and Mentha × piperita-Vital Herbs with Myriads of Pharmaceutical Benefits. Horticulturae 9(2):224. https://doi.org/10.3390/horticulturae9020224
11. Jan HA, Sher H, Hussain W, Rahman I (2021) Mentha arvensis L. Mentha longifolia (L.) L. Mentha royleana Benth. Mentha viridis (L.) L. Lamiaceae. Ethnobotany of the Himalayas, pp 1265–1277. https://doi.org/10.1007/978-3-030-45597-2_151-1
12. Thakur S, Walia B, Chaudhary G (2021) Mentha arvensis (Pudina): A Review based upon its Medicinal Properties. Research Journal of Pharmacognosy and Phytochemistry, pp 143–148. https://doi.org/10.52711/0975-4385.2021.00024
13. Hamsa CB, Aishwarya B, Roopa D, Manirupesh K (2022) A contemporary review on pharmacognosy and pharmacological activities of mentha arvensis. International Journal of Pharmacognosy and Chemistry, pp 107–112. https://doi.org/10.46796/ijpc.v3i3.344
14. Ameen N, Shafi S (2017) Evaluation of In Vitro Anti-Oxidant Activity of Mentha arvensis Linn in Memory Enhancing Study. International Journal of Pharmacognosy and Phytochemical Research, 9. https://doi.org/10.25258/phyto.v9i08.9627
15. Faisal S, Hamza Tariq M, Ullah R, Zafar S (2023) Exploring the antibacterial, antidiabetic, and anticancer potential of Mentha arvensis extract through in-silico and in-vitro analysis. BMC Complementary Medicine and Therapies, 23(1). https://doi.org/10.1186/s12906-023-04072-y
16. Sik B, Kovács K, Hanczné Lakatos E, Kapcsándi V (2023) Increasing the functionality of sponge cakes by mint, and cocoa powder addition. Heliyon 9:20029. https://doi.org/10.1016/j.heliyon.2023.e20029
17. Asghar M, Khattak M, Arshad B, Zaman W, Ayaz A, Rasheed S, Shah A, Ullah F, Saqib S (2022) Bioactive potential of cultivated Mentha arvensis L. for preservation and production of health-oriented food. Journal of Animal and Plant Sciences 32:835–844. https://doi.org/10.36899/JAPS.2022.3.0484
18. Fedotova ZA, Zhuravlyov AP, Beregnaya GA (2012) Using Plant Repellents at Storage Grain Wheats. Vestnik of Ulyanovsk State Academy of Agriculture 2(18):34–39
19. Mahanto S, Kumar I, Sharma RK (2024) Cadmium Accumulation, Growth Performance and Antioxidant Properties of Bacterial Consortium Inoculated Mentha arvensis L. Grown in Polluted Soil. Natl. Acad. Sci. Lett. https://doi.org/10.1007/s40009-023-01375-z
20. Klimova EV (2003) Zinc in the Plants Growing in the Flood Plain of the Irtysh River [Contamination as a Result of the Activity of Kazzinc Industrial Company]. Environmental Safety in Agro-industrial Complex. Abstract Journal 4:849

Comparative Assessment of Implementing Modern Ash Extraction Technologies at Russian Coal-Fired Thermal Power Plants

P. V. Roslyakov and O. E. Kondratieva

Abstract Fly ash emissions from flue gases for coal-fired thermal power plants are of great importance from the point of view of atmospheric air pollution. Various ash removal technologies are used to reduce them. The purpose of this work was to assess the possibility of using various types of ash collectors to achieve the established technological indicators of ash emissions from solid fuels. For this the technical and environmental characteristics of the most effective ash removal technologies—electrostatic precipitators and bag filters—were compared. A comparative technical and economic assessment of their use for boilers of cogeneration and condensing thermal power plants burning the main Russian coals has been carried out. The costs of their implementation have been determined, as well as the cost of removing one ton of ash from various coals. Recommendations on the use of electrostatic precipitators and bag filters at thermal power plants for various types of coals have been formed.

Keywords Environmental protection · Marker pollutants · Coal ash · Ash collection · Emission indicators · Electrostatic precipitator · Bag filter · Power plant

1 Introduction

Heat and power engineering is traditionally one of the industrial sectors with the greatest negative impact on the environment. The main marker pollutants released into the atmospheric air during the combustion of solid fuels at coal-fired thermal power plants are solid fuel ash, sulfur dioxide SO_2, nitrogen oxides NO_X and carbon monoxide CO [1, 2]. Emissions of these marker pollutants into the atmosphere are subject to continuous monitoring and accounting and should not exceed the technological indicators established for them [3]. To effectively solve of these problems, a

P. V. Roslyakov (✉) · O. E. Kondratieva
National Research University, "MEI", 14, Krasnokazarmennaya Str., Moscow 111250, Russia
e-mail: roslyakovpbv@mpei.ru

L. V. Radionova and D. V. Ulrikh (eds.), *Advances in Ecology and Environmental Engineering*, Springer Proceedings in Earth and Environmental Sciences,
https://doi.org/10.1007/978-3-031-64423-8_31

new Russian environmental legislation has been developed [2–6] aimed at reducing the negative impact on the environment.

For coal-fired thermal power plants emissions of fly ash with flue gases are of great importance from the point of view of atmospheric air pollution. In order to reduce them, various ash removal technologies are used. However, some of them no longer allow to provide the required technological indicators of ash emissions [7, 8]. This requires the modernization or replacement of outdated and inefficient equipment at thermal power plants and the transition to the best available technologies (BAT) [9].

2 Purposes and Objectives of the Study

Currently, 79 coal-fired thermal power plants with an installed electric capacity of 39.95 GW are in operation in Russia, which is 16.2% of the total installed capacity of power plants in Russia [10]. Currently, various types of dry and wet ash collectors, electrostatic precipitators and bag filters are used to clean flue gases from solid fuel ash at coal-fired thermal power plants [11, 12]

The purpose of this work was to assess the possibility of using different types of ash collectors to meet the established technological indicators of solid fuel ash emissions [1, 7] during the combustion of main Russian energy coals. Special attention was paid to the issues of meeting the technological indicators (TI) of solid fuel ash emissions during the combustion of the main Russian brown, hard coal and anthracite in boilers with a steam capacity of 420 t/h at cogeneration heat plant (CHP) and power units with an electric capacity of 300 MW at condensing power plant (CPP), as well as computational studies on the feasibility assessment of the introduction of bag filters and electrostatic precipitators.

3 Research Methodology

The main coals burned at Russian thermal power plants are Ekibastuzsky, Kuznetsky, Kansk-Achinskie and Donetsky. They form about 60% of the total annual supply of coal fuel [13]. When choosing the technology of ash extraction at coal-fired thermal power plants, it is necessary to take into account the composition of the burned fossil fuels, the properties and mass concentration of fly ash in the flue gases at the entrance to the ash collector. According to the ash content in the fuel per working weight (A^r, %) and the calculated content of fly ash in flue gases (C_{ash}, mg/m^3), all coals can be conditionally divided into low-ash ($A^r < 10\%$, $C_{ash} < 20{,}000$ mg/m^3), medium-ash ($10 \leq A^r \leq 20\%$, $20{,}000 \leq C_{ash} \leq 30{,}000$ mg/m^3) and high-ash ($A^r > 20\%$, $C_{ash} >$ 30,000 mg/m^3) [14].

It should be noted that the initial ash content in flue gases depends only on the ash content of the fuel itself and does not depend on its combustion modes. However, the higher the ash content of solid fuels, the greater the required ash recovery efficiency to perform technological ash emission indicators.

In connection with the approval of new technological indicators for solid fuel ash emissions and updating the list of BAT ash collection, coal energy companies in the coming years will have to modernize or replace installed ash collectors to meet new environmental requirements for energy equipment [1]. This will undoubtedly contribute to the development of domestic environmental engineering and the fulfillment of the tasks of the national project "Ecology" [15] and the federal project "Clean Air" [16] in the energy sector.

An analysis of the efficiency of flue gas ash purification showed that bag filters provide the best results of cleaning gases from solid particles (emissions of no more than 20–40 mg/m^3) for all coals. Electrostatic precipitators make it possible to achieve ash emissions of less than 150 mg/m^3 for most coals, with the exception of anthracites and the most high-ash brown and hard coals [17].

The multicyclones currently in operation are no longer able to provide the established technological indicators, and therefore must be replaced with more efficient ash extraction devices. Limited use of wet ash collection devices with Venturi coagulant and emulsifiers is possible only for boilers with a steam capacity of no more than 640 t/h when burning low- and medium-ash coals, provided that their efficiency is at least 98% [18]. The positive side of their application is the partial desulfurization of flue gases, the efficiency of which increases when capturing ash with a high content of CaO (more than 30–35%) [19].

For the technical and economic assessment of the use of EP and BF at thermal power plants in order to meet the technological indicators of coal ash emissions, computational studies were conducted for the main Russian hard and brown coals: Ekibastuzsky, Kuznetsky, Irsha-Borodinsky, as well as Donetsk anthracite. The studies were carried out for the most common boilers at cogeneration heat plant (CHP) with a steam capacity of 420 t/h and boilers of power units with an electric capacity of 300 MW of condensation power plant (CPP).

4 Estimated Cost Studies for the Implementation of Electrostatic Precipitators and Bag Filters

The comparison of the technical and environmental characteristics of the different ash collection technologies currently used at Russian thermal power plants is shown in Table 1. The most effective modern ash collection technologies at thermal power plants are electrostatic precipitators (EP), bag filters (BF) and combined storage devices. The possible degree of reduction in emissions of particles smaller than 10 microns reaches 99–99.9 when using EP and 98–99.99 + when using BF.

The calculated initial values of the concentration of fly ash (up to ash collector), as well as the required degree of ash removal to achieve technological indicators of ash emission in accordance with [1], are presented in Table. 2 The initial ash content

Table 1 Comparison of technical characteristics of different types of ash collectors

Type of ash collectors	Multicyclones	Venturi wet scrubbers	Emulsifiers	Electrostatic precipitators (EP)	Bag filters (BF)
Flue gas cleaning efficiency	до 85–92%	до 95–97%	до 96–98%	до 99–99,9%	до 99–99,99%
Cleaning of large volumes of flue gases	Limitations on gas consumption	Limitations on gas consumption	Limitations on gas consumption	Yes	Yes
Dependence of the cleanning degree on the ash electrical resistance	No	No	No	Yes	No
Overall dimensions	Moderate	Moderate	Moderate	Large	Large
Effective capture of particles up to 0.1 microns in size	No	Yes	Yes	No	Yes
The ability to ensure an ash content of no more than 250 mg/m^3	It cannot be recommended as a BAT	Very limited, only for the low-ash coals	Very limited, only for the low-ash coals	For coals with an ash content of less than 30%	For all coals without limitations
The ability to ensure an ash content of no more than 150 mg/m^3	It cannot be recommended as BAT	It cannot be recommended as BAT	It cannot be recommended as BAT	For coals with an ash content of less than 20%	For all coals without limitations

in flue gases depends only on the ash content of the fuel itself and does not depend on its combustion modes. However, the higher the ash content of solid fuels, the greater the required ash recovery efficiency to meet the technological indicators.

It is obvious that it is possible to ensure the necessary degrees of reduction of coal gas emissions only when using electrostatic precipitator or bag filters. Therefore, when analyzing the transition of Russian thermal power plants to BAT, modern serial EGAV type electrostatic precipitators (EP) were selected [12, 20]. As the ash content in flue gases after bag filters (BF), based on the operational characteristics of bag filters already operating in Russia, a concentration of 20 mg/m^3 was adopted, regardless of the type of coal [21, 22].

With an increase in the steam capacity of boilers, the volume consumption of flue gases increases, therefore, in order to ensure high efficiency of ash collection, two electrostatic precipitators are usually installed in parallel behind high-power

Table 2 Calculated mass concentrations of fly ash, established technological indicators of ash emissions and the necessary degrees of ash emission reduction for different types of coals

Coals	Ash content of coal,%/ Initial ash concentration before ash collector, mg/m^3	CHP, Boiler unit, steam capacity 420 t/h		CPP, Power unit, electric capacity 300 MW		EP type/EP efficiency, %	Ash concentration in exhausted gases after EP/BF, mg/m^3
		TI of fly ash emissions, mg/m^3	Required degree of ash removal, %	TI of fly ash emissions, mg/m^3	Required degree of ash removal, %		
Irsha-Borodinsky	7,4/8433	250	99,67	130	99,80	EGAV2-66–9-5–5/ 99,83	130/20
Kuznetsky	18,2/20903	160	99,23	70	99,71	EGB1M2-56–12-6–4/ 99,67	70/20
Ekibastuzsky	45,6/76108	150	98,22	38	99,41	EGBM2-48–12-6–5/ 99,55%	38/20
Donezsky anthracite	36,0/36976	250	99,32	63	99,59	EGAV2-56–12-6–4/ 99,83%	63/20

boilers at condensation power plants, regardless of the characteristics of the coal being burned. As can be seen, the installation of serial electrostatic precipitators selected for the studied coals allows reducing ash emissions to values even below established technological indicators (TI) [1] (Table 2).

The calculation of capital and operating costs for the introduction of EF and BF at thermal power plants as BAT was carried out in accordance with GOST R 113.38.02–2019 [23] with an inflation forecast of 5% (Table 3). Analysis of the calculation results allows us to draw the following conclusions.

It is obvious that the total costs of implementing the EF and BF increase proportionally with increasing boiler capacity and, accordingly, flue gas rate. Moreover, for both types of ash collectors, the share of operating costs (calculated for 15 years) significantly exceeds capital costs.

The significant predominance of operating costs over capital costs is due to the fact that when calculating operating costs, not only the boiler capacity is taken into account, but also the inflation coefficient, which can increase the cost of equipment several times. At the same time, the increase in inflation has a stronger effect on the cost of BF introducing, than on the cost of EP introducing.

The operating costs of using bag filters are higher than similar costs when using EP due to the need for periodic (approximately once every three years) complete replacement of expensive bags that are not yet manufactured in Russia. In the case of localization of the production of filter materials and bags in Russia, operating and total costs should significantly decrease.

An increase in inflation above the calculated value will have a very negative impact on the pace of BAT implementation at Russian thermal power plants, which should be taken into account when planning modernization and replacement of equipment. For example, an increase in inflation from 5 to 13.5% will lead to an increase in the total cost of introducing electrostatic precipitators and bag filters by 67.8–76.6%, primarily due to a significant increase in operating costs. At the same time, the total cost of implementing bag filters is 22.8–28.4% higher than the cost of implementing electrostatic precipitators with an inflation forecast of 5% and by 33.4–37.7% with

Table 3 Costs for the implementation of EP and BF for boilers of CHP and CPP

Ash collectors type	Capital expenditures, million rubles (min–max)	Operating costs, million rubles (min–max)	Total costs, million rubles (min–max)
Cogeneration heat power (CHP), Boiler unit, steam capacity 420 t/h			
Electrostatic precipitator (EP)	208–377	464,44–593,45	672,44–970,45
Bag filter (BF)	166,4–301,6	696,66–890,18	863,06–1191,78
Condensation power plant (CPP), Power unit, electric capacity 300 MW			
Electrostatic precipitator (EP)	480–870	1071,79–1369,51	1551,79–2239,51
Bag filter (BF)	384–696	1607,68–2054,26	1991,685–2750,26

an inflation forecast of 13.5%. Unfortunately, this can slow down the implementation of the most effective BAT of ash collection—bag filters.

For a comprehensive comparative assessment of the implementation of EP and BF for the ash removal under different coals combustion, an estimate was made of the unit cost of ash removal of various coals (RUB/t) in these ash collection units. The estimated cost of ash removal under combustion of various solid fuels (RUB/t) in EP and BF is given in Table 4.

For this purpose, the annual mass of the removed ash in ash collectors G_{ash}, t/year, was determined as:

$$G_{ash} = V_g \cdot (C_{ash} - C_{end}) \cdot 10^{-9} \cdot T \quad (1)$$

where V_g is the volume flow rate of flue gases during coal combustion, m^3/h; C_{ash} is the calculated initial ash concentration in flue gases before ash collector, mg/m^3; C_{end} is the calculated ash concentration in the exhaust gases behind ash collector, mg/m^3 (see Table 2); $T = 8760$ h/year—the annual campaign of the equipment. The results of calculations of the mass of the removed ash for the considered coals at the CHP and CPP are shown in Table. 4.

Due to the fact that modern EGAV type electrostatic precipitators, although inferior to bag filters, can achieve high ash collection efficiency (ash content behind EP 63–130 mg/m^3), the difference between the annual mass of ash removed from the same coal by bag filters and electrostatic precipitators is small and amounts to only a few hundred tons per year (Table 4). The smallest difference occurs for the ash of low-ash Irsha-Borodinsky brown coal, and the largest for high-ash Ekibastuzsky coal. Therefore, when burning high-ash coals at thermal power plants, such as Ekibastuzky and others, the efficiency of using bag filters increases.

Table 4 Estimated annual mass of coal ash removed at ash collector and the unit cost of their removal

Coals	CHP			CPP		
	The annual weight of the remoted ash, t/year		Unit cost of ash removal in the EF/BF, RUB/ton of ash (min/max)	The annual weight of the remoted ash, t/year		Unit cost of ash removal in the EF/BF, RUB/ton of ash (min/max)
	EP	BF		EP	BF	
Ekibastuzsky	359,782,38	360,303,27	125–180/160–221	833,388,07	834,594,64	125–180/159–220
Kuznetsky	91,150,72	91,369,49	492–710/630–870	212,033,68	212,542,57	488–705/625–863
Irsha-Borodinsky	40,656,85	40,744,02	1103–1592/1413–1950	95,117,70	95,321,65	1088–1570/1393–1924
Donezsky anthracite	262,315,67	262,621,24	171–247/219–303	599,613,19	600,311,68	173–249/222–306

It is recommended to use bag filters primarily when burning high-ash coals, especially, for example, such as Ekibastuzsky, the ash of which is characterized by high electrical resistivity, which is why it is poorly captured in the EF. In turn, electrostatic precipitators should be mainly used for burning low-ash and medium-ash coals, since the difference in the mass of captured ash in the EP and BF for the same coal is small (Table. 4), and the cost of implementing the EF is significantly less (Table 3).

Assessment of unit costs of coal ash removal in the EP and BF UC, RUB/ton, (see Table. 4) was performed according to the expression:

$$UC = \Sigma Exp/(G_{ash} \cdot t) \quad (2)$$

where ΣExp is total capital and operating costs calculated in accordance with GOST R 113.38.02–2019 [23] (see Table 3); G_{ash}—annual weight of the remoted ash, ton/year (Table 4); t—the service life of the equipment (15 years).

The calculated results of the specific cost indicators of coal ash removal in the EF and BF, taking into account the fifteen-year service life, showed that the cost of removing 1 ton of ash for the considered coals at the CHP ranged from 160 to 3565 rubles/t when using bag filters and from 125 to 2670 rubles/t when using EF. Similar costs for CPP amounted to 159–3517 rubles/ton for bag filters and 125–2634 rubles/ton for EP (Table 4). The costs of removing one ton of ash for the considered coals are in fairly wide ranges and depend primarily on the ash content of coal and the magnitude of projected inflation. At the same time, the type of ash collector does not play such a significant role.

A significant difference in the estimated unit costs for ash removal of different coals is due to very different values of the initial ash content of these coals (from 7.4 to 45.6%). The higher it is, the greater the mass of the ash to be removed, which means that the unit cost of its removal is lower under comparable conditions. For this reason, for low-ash coals such as Irsha-Borodinsky, the specific cost of ash removal in the EP and BF is the highest compared to high-ash Ekibastuzsky and Donetsk anthracite.

Therefore, for low-ash coals, the introduction of other cheaper types of ash collectors should be considered, for example, wet scrubbers with a Venturi coagulator or emulsifiers, provided, however, that they ensure compliance with the established technological indicators of ash emissions.

At the same time, the unit cost of the ash removed differs little for the considered boilers of CHP and CPP operating on the same coal (Table 4). Thus, the unit cost of removing 1 ton of ash is determined only by the total cost of implementing ash extraction technology and the ash content of the burned coals, but does not depend on the capacity of the boiler plant.

The estimated payback period for the capital costs of the introduction of EP and BF, calculated according to the current standards of payment for 1 ton of captured ash [24], for both types of ash collectors is more than 26 years. This is almost twice the service life of fixed assets. In the case of commercial disposal of captured ash, the payback period can be significantly reduced.

However, the estimated payback estimates in the case of equipment modernization are not so unambiguous. It should be borne in mind that the introduction of BAT undoubtedly improves the quality and safety of life of the population, which implies taking into account important social aspects (reducing morbidity, increasing the age of life).

5 Conclusion

The use of highly efficient BAT at coal thermal power plants–electrostatic precipitator s and bag filters–make it possible to reliably achieve technological indicators for cleaning exhausted gases from solid particles presented in ITS 38-2022.

The implementation of new, more stringent technological indicators of fly ash emissions will require Russian coal-fired thermal power plants to introduce modern efficient ash collectors—mainly electrostatic precipitators and bag filters, the degree of ash capture in which exceeds 99.5%. Less effective ash collectors, primarily multi-cyclones, and in some cases wet ash collectors can no longer be used to effectively clean flue gases from coal-fired thermal power plants to established regulatory levels.

It is recommended to use bag filters primarily when burning high-ash coals, especially, for example, such as Ekibastuzsky, the ash of which is characterized by high electrical resistivity, which is why it is poorly captured in the EP. In turn, electrostatic precipitators should be mainly used for burning low-ash and medium-ash coals, since the difference in the mass of captured ash in the EP and BF for the same coal is small and the cost of implementing the EF is significantly less.

It is obvious that the total costs of implementing the EP and BF increase proportionally with increasing boiler capacity and, accordingly, flue gas consumption. Moreover, for both types of ash collectors the share of operating costs (calculated for 15 years) significantly exceeds capital costs.

The main problems of switching to BAT of ash extraction, in addition to their high cost, are: for bag filters—the lack of their domestic mass production, and for electrostatic precipitators—dependence on the electrical resistivity of ash and the complexity of upgrading already operating devices.

The total cost of implementing bag filters is about 26–30% higher than the cost of implementing EF, which is explained by the high operating costs of BF. The unit cost of removing 1 ton of ash for the considered coals was approximately 30% more when using the BF (160–3565 rubles/ton) than in the EF (125–2670 rubles/ton) for both CHP and CPP. At the same time, the higher the ash content of the coals, the lower the unit cost of removing 1 ton of ash in ash collectors.

The unit cost of the ash removed differs little for the considered boilers of CHP and CPP operating on the same coal. However, it strongly depends on the level of inflation, the growth of which significantly increases operating costs.

Bag filters provide lower ash emissions into the atmospheric air compared to EF under comparable conditions. However, their production in Russia has not yet been fully localized. So their use of BF at Russian thermal power plants carries certain risks. If the production of bag filters is localized at enterprises of the Russian environmental engineering, their operating cost will significantly decrease. This will increase their competitiveness and accelerate the introduction at Russian thermal power plants.

The increase in inflation significantly affects the cost of introducing modern and efficient equipment at thermal power plants.

In order to stimulate the introduction of BAT at thermal power plants, it is necessary to review the rates of payment for emissions in the direction of their increase, taking into account real inflation and the financial possibilities of their repayment within the framework of tariffs for produced energy. The new rates of payment for ash emissions should ensure for thermal power plants the payback of the total costs of introducing BAT ash collection within 15 years.

References

1. Fuel Combustion at Large Plants with the Goal of Energy Production: Informational and Technical Handbook on Best Available Techniques, ITS 38–2022, Byuro NDT, Moscow, 2022
2. Meshalkin VP, Roslyakov PV, Guseva TV, Dovi VJ (2021) New technological indicators of solid fuel ash and sulfur dioxide emissions for thermal power plants and the best available gas purification technologies. Ecology and industry of Russia 8:40–46. https://doi.org/10.18412/1816-0395-2021-8-40-46
3. Decree of the Government of the Russian Federation No. 2909-R dated 20.10.2023 "On approval of the list of pollutants for which the following are measures of state regulation in the field of environmental protection"
4. On Protection of the Environment, Federal Law of the Russian Federation No. 7-FZ of January 10, 2002, With Changes from March 26, 2022
5. On the protection of atmospheric air. Federal Law of the Russian Federation. 96-FZ of May 04, 1999
6. Roslyakov PV, Kondrat'eva OE, Skobelev DO, Guseva TV, Mikaelsson A (2020) Normative and Legal Regulation for Implementing the Best Available Technologies in Russian Heat-and-Power Engineering. Thermal Engineering 67(10):733–740. https://doi.org/10.1134/S0040601520100092
7. Order of the Ministry of Natural Resources and Ecology of the Russian Federation No. 248 dated 25.04.2023 "On approval of a regulatory document in the field of environmental protection "Technological indicators of the best available fuel combustion technologies at large installations for the energy production"
8. Roslyakov PV, Cherkassky EV, Guseva TV, Tikhonova IO, Lundholm M (2021) Technological rationing of thermal power engineering facilities: the best available technologies and norms of general action. Thermal Engineering 68(10):735–742. https://doi.org/10.1134/S0040601521100050
9. Decree of the Government of the Russian Federation No. 398-r dated 19.03.2014 "On a set of measures aimed at abandoning the use of outdated and inefficient technologies, switching to the principles of the best available technologies and the introduction of modern technologies"
10. The Unified Energy System of Russia. [Electronic source]. https://www.so-ups.ru/functioning/ees/ups2022. Accessed 15 Dec 2023
11. Roslyakov P.V (2018) Ash Collectors of Thermal Power Plants. Mosk. Energ. Inst., Moscow
12. Electrofilters. http://www.kondor-eco.ru/download/egb1.pdf. Accessed 12 Nov 2023
13. Alekhnovich AN (2016) Energy coals. Cicero, Chelyabinsk

14. Roslyakov PV, Guseva TV, Mikaelsson A (2021) Substantiation of the choice of new technological indicators of ash and sulfur oxides emission for Russian thermal power plants and best available techniques that provide them. Elektr. Stn. 8:2–13. https://doi.org/10.34831/EP.2021.1081.8.001
15. Passport of the National Project "Ecology". http://www.mnr.gov.ru/activity/directions/natsionalnyy_proekt_ekologiya/. Accessed 10 Oct 2022
16. Passport of the Federal Project "Clean Air" (Annex to the Record of the Meeting of the Project Committee on the National Project "Ecology" No. 3 dated December 21, 2018). https://www.mnr.gov.ru/activity/clean-air/. Accessed 10 Oct 2022
17. Sanaev YuI, Chekalov LV (2016) The specific electrical resistance of dust (ash) in electric filters. CJSC "Condor-Eco", Semibratovo
18. Alekhnovich AN (2018) Ash capture at domestic thermal power plants. Cicero, Chelyabinsk
19. Roslyakov PV (2007) Methods of environmental protection. Publishing House of Moscow Power Engineering Institute, Moscow
20. Rogalev ND, Prokhorov VB et al (2021) Environmental protection technologies at thermal power plants. Publishing House MEI, Moscow
21. Roslyakov PV, Kondratieva OE, Baranov VV (2020) Introduction of the best available technologies for desulfurization and ash collection at Russian thermal power plants. Elektricheskie. Stantsii 6:13–20. https://doi.org/10.34831/EP.2020.1067.6.003
22. Roslyakov PV, Kondratieva OE, Almgren AR, Sivtseva SA, Burchenko VD (2021) Technical and economic problems and risks of introducing BAT at Russian thermal power plants. Energetik 1:15–20
23. GOST R 113.38.02-2019 "Methodological recommendations for estimating the costs of electric power industry enterprises to reduce emissions of pollutants in order for them to achieve technological indicators of the best available technologies". https://docs.cntd.ru/document/1200167475. Accessed 17 Nov 2020
24. Decree of the Government of the Russian Federation dated 1.03/2020 No. 274 "On the application of payment rates for negative environmental impact in 2022". http://www.consultant.ru/document/cons_doc_LAW_410708/. Accessed 12 Nov 2023

Environmental Assessment of the Effect of Fuel Composition on Operating Conditions and Air Emissions from Motor Vehicles

Sh. V. Buzikov, O. M. Buzikova, and M. V. Motovilova

Abstract Due to the growth of vehicles, the number of emissions of exhaust gases and various harmful substances into the air is increasing. This leads to a deterioration in the life and health of the population. The authors are also interested not only in reducing emissions of pollutants into the air, which can be calculated by various complex indicators, but also in considering the efficiency of the engine and improving performance under various conditions. The work carried out experiments using different fuel mixtures with a diesel engine at different loads. The purpose of the study is to determine the effect of combined fuel on the environmental efficiency of the diesel engine. As a result of the conducted studies, the values of the optimal injection advance angles of the combined fuel with the content of rapeseed oil from 0 to 60% and the combined fuel with the content of ethanol from 0 to 40% were equal to 26° PCV to WDC and 22° PCV to WDC at which the values of the specific effective consumption of combined fuel ranged from 259 to 321 g/kwh and from 283 to 323 g/kwh, respectively.

Keywords Automobile fuel · Diesel · Diesel engine testing · Engine characteristics · Environmental assessment · Atmospheric pollution · Fuel and hardware supply

1 Introduction

Due to the growth of vehicles, the number of emissions of exhaust gases and various harmful substances into the air is increasing. This leads to a deterioration in the life and health of the population. Which in turn stimulates the development of alternative fuels and the development of the environmental agenda of the world community [1–5].

Sh. V. Buzikov (✉) · O. M. Buzikova · M. V. Motovilova
Vyatka State University, 36, Moskovskaya Street, Kirov 610000, Russia
e-mail: usr10012@vyatsu.ru

L. V. Radionova and D. V. Ulrikh (eds.), *Advances in Ecology and Environmental Engineering*, Springer Proceedings in Earth and Environmental Sciences,
https://doi.org/10.1007/978-3-031-64423-8_32

The transportation industry plays a critical role in promoting integration and construction between cities and villages. However, motor vehicle emissions also pose urban air pollution problems, and heavy diesel vehicle emissions have a major impact on the urban environment.

Currently, diesel cars still play an indispensable role in the traditional energy sector of the world and Russia. Exhaust emissions from diesel vehicles contain hydrocarbons, carbon monoxide, nitrogen oxides and particulate matter. The negative impact of these elements on the biosphere has been proven, pollutants can lead to foggy weather, photochemical smog and greenhouse effect; endangering human health; damage the environmental environment. Diesel vehicles, particularly diesel trucks, have become a top priority in tackling pollution. Accurate assessment of pollutant emission characteristics is essential for reducing emissions and their associated effects on the environment and public health.

Many authors propose various combinations of fuel for vehicles in order to reduce emissions into the air and reduce the environmental load, both on humans and on the biosphere as a whole [6–10].

In addition, the researchers are interested not only in reducing pollutant emissions into the air, which can be calculated by various complex indicators, but also in considering the efficiency of the engine and improving performance under various conditions [11–15].

Exhaust emissions from diesel cars tend to deteriorate with increased mileage due to engine wear and exhaust cleaning systems.

Emissions from heavy diesel trucks (HDDTs) have a significant impact on air quality, human health and climate change, and seriously threaten the achievement of the UN Sustainable Development Goals. There have been a number of emissions control measures around the world aimed at reducing emissions of pollutants from HDDT.

Some researchers have proven that real heavy-duty diesel (HDT) trucks emit far more excess nitrogen oxides (NO_x) and black carbon (BC) pollutants than regulatory limits. It is important to systematically assess road NO_x and BC emission levels to reduce HDT emissions.

2 Materials and Methods

The work carried out experiments using different fuel mixtures with a diesel engine at different loads.

The purpose of the study is to determine the effect of combined fuel on the environmental efficiency of the diesel engine.

The objectives of the study are:

- determination of effective performance indicators of the diesel engine;
- assessment of the level of environmental impact of combined fuel on atmospheric emissions;

- determination of load characteristics of diesel engine operation at combined fuel.

The impact of the composition of the combined fuel containing rapeseed oil and ethanol on the effective and environmental performance of the diesel engine was assessed by the load characteristics taken at the nominal speed mode at 1800 min^{-1} and the maximum torque mode 1400 min^{-1} [16].

3 Results

The environmental indicators of the diesel engine in connection with the modern environmental situation are of great importance for the world community.

As a result of the obtained data, Fig. 1a shows the environmental indicators of the diesel engine operation with the dimension 4ChN 11.0/12.5 when operating on combined fuel with the content of rapeseed oil 1800 min^{-1} and the optimal value 26° PCV to WDC [17, 18].

From the dependencies shown in Fig. 1a, it can be seen that the nature of curves of carbon emission levels (C) both during diesel operation on pure diesel fuel and during operation on combined fuel with rapeseed oil content is the same. With an increase from 0.2 to 1.0 MPa, carbon (C) emission with diesel exhaust gases during

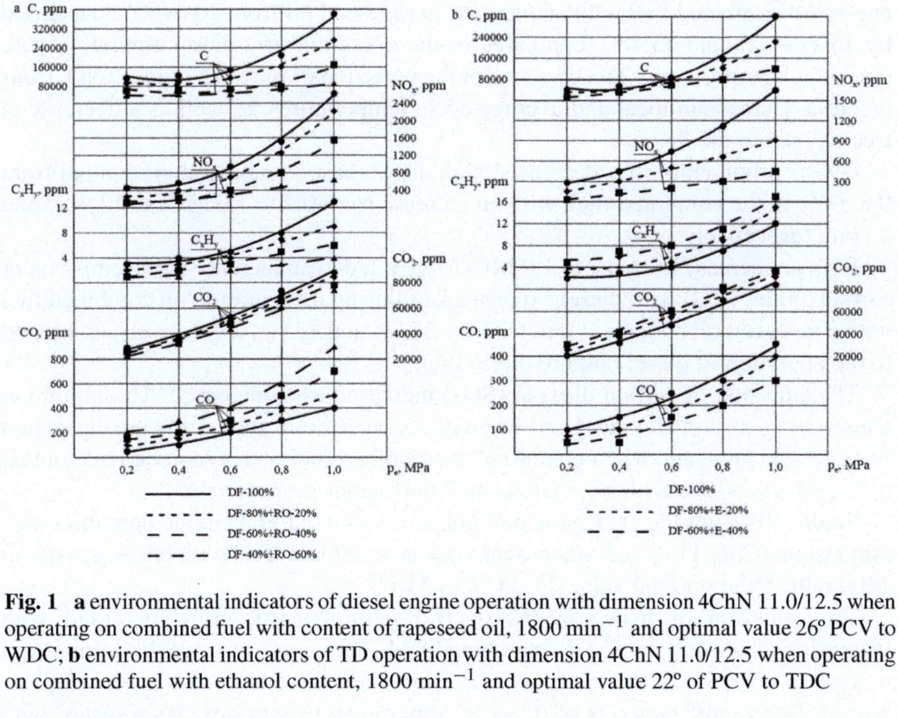

Fig. 1 **a** environmental indicators of diesel engine operation with dimension 4ChN 11.0/12.5 when operating on combined fuel with content of rapeseed oil, 1800 min^{-1} and optimal value 26° PCV to WDC; **b** environmental indicators of TD operation with dimension 4ChN 11.0/12.5 when operating on combined fuel with ethanol content, 1800 min^{-1} and optimal value 22° of PCV to TDC

operation on diesel fuel and combined fuel with rapeseed oil content increased. The emission of carbon (C) with the exhaust gases of the diesel engine was also reduced when the diesel engine was running on combined fuel with a rapeseed oil content of 20, 40 and 60%, respectively.

With an increase in the content of rapeseed oil in the combined fuel, carbon emission (C) with an increase of 0.2–1.0 MPa decreased more intensively, due to a more complete oxidation of atomic carbon. Thus, the increase to 0.6 MPa had an impact on carbon emissions (C) with exhaust gases and amounted to 147,000 ppm when the diesel engine was operating on pure diesel fuel, and when the diesel engine was operating on combined fuel with a rapeseed oil content of 20, 40 and 60%, respectively, 113,000, 67,000 and 47,000 ppm.

With an increase to 1.0 MPa, the emission of carbon (C) with exhaust gases was 380,300 ppm when the diesel engine was operating on pure diesel fuel, and when the diesel engine was operating on combined fuel with a rapeseed oil content of 20, 40 and 60% 270,000, 168,000 and 90,000 ppm, respectively.

The emission of nitrogen oxides (NO_x) during the operation of a diesel engine on pure diesel fuel and combined fuel with a content of rapeseed oil increased with an increase.

So, at 1.0 MPa, the emission values of nitrogen oxides (NO_x) during diesel operation on pure diesel fuel were 2630 ppm, and during diesel operation on combined fuel with a rapeseed oil content of 20, 40 and 60%, respectively, 2225, 1545 and 645 ppm. The emission of nitrogen oxides (NO_x) during the operation of a diesel engine on combined fuel with an increase in rapeseed oil from 20 to 60% decreased by 15.4, 41.23 and 75.48% compared to the operation of a diesel engine on pure diesel fuel, respectively. The decrease in the emission of nitrogen oxides (NO_x) was due to a decrease in local and average cycle temperatures, as well as a decrease in free oxygen in the flame.

Hydrocarbon emissions decreased with an increase in rapeseed oil content from 0 to 60% in the combined fuel, with an increase from 0.2 to 1.0 MPa to 13, 9, 6 and 4 ppm, respectively.

With an increase from 0.2 to 1.0 MPa MPa, it led to an increase in the emission of carbon oxides (CO) with diesel exhaust gases during its operation on combined fuel with a rapeseed oil content of 0 to 60% by 25, 75 and 125%, respectively, compared to the operation of diesel on pure diesel fuel.

The emission of carbon dioxide (SO_2) increased as it increased. The minimum emissions of carbon dioxide (SO_2) when operating a diesel engine on pure diesel fuel were 87,700 ppm, and when operating on combined fuel with a rapeseed oil content of 20, 40 and 60%—84,000, 78,000 and 74,000 ppm, respectively.

Figure 1b shows the environmental indicators of the diesel engine operation with dimension 4ChN 11.0/12.5 when operating on combined fuel with ethanol content, 1800 min^{-1} and optimal value 22° PCV to TDC.

The emission of hydrocarbons (C_xH_y) with exhaust gases increased with increasing ethanol content in the combined fuel. The acceleration of the combustion process, a decrease in the average cycle temperature and, as a result, the temperature of the exhaust gases, is of decisive importance in this case. As a result, with an increase from 0.2 to 1.0 MPa and during diesel operation on pure diesel fuel, hydrocarbon emissions (C_xH_y) increased from 2 to 13 ppm, and when operating on combined fuel with an ethanol content of 20 and 40% from 3 to 15 ppm and from 5 to 17 ppm, respectively.

As can be seen (Fig. 1b), the emission of nitrogen oxides (NO_x) with exhaust gases at 0.6 MPa decreased by 41.14–73.11% when the diesel engine was running on combined fuel with an ethanol content of 20–40%, respectively. At 1.0 MPa, the tendency to decrease the emission of nitrogen oxides (NO_x) remained at 45.03–42.79%, respectively. According to the thermal nature of the formation of nitrogen oxides (NO_x), their emission is directly dependent on the content of free oxygen in the flame, provided that its temperature is sufficiently high. The decrease in local and average cycle temperatures in the case of combined fuel with ethanol content directly causes a decrease in the emission of nitrogen oxides (x). At the same time, the emission of nitrogen oxides (NO_x) is in direct dependence on the characteristics, since the temperature of the diesel exhaust gases rises in parallel with the characteristics.

The nature of the carbon monoxide (CO) and carbon dioxide (SO_2) emission curves is the same, both when the diesel engine is operating on pure diesel fuel and on combined fuel with an ethanol content of 20 and 40%. As emissions of carbon monoxide (CO) and carbon dioxide (SO_2) with diesel exhaust gases increased, they changed. Thus, the operation of a diesel engine on combined fuel with an ethanol content of 20–40%, at 0.2–1.0 MPa, led to an increase in SO_2 emissions from 8.41 to 15.93% and from 3.71 to 16.48%, respectively. The decrease in CO emission during diesel engine operation at 0.2–1.0 MPa and diesel engine operation on combined fuel with ethanol content of 20–40% was 11.11–33.33% and 20–50%, respectively.

According to the classical scheme of CO and SO_2 formation, all carbon formed in the flame is oxidized to CO, and only then, in the presence of oxygen, goes into the SO_2 state. Reducing emissions at high temperatures causes under oxidation of the SO_2. With an increase in the ethanol content of the combined fuel, the temperature of the exhaust gases decreased throughout the range.

The carbon (C) content in the exhaust gases was also reduced when the diesel engine was running on combat-fired fuel with ethanol content. Thus, an increase from 0.2 to 1.0 MPa entailed an increase in carbon emissions (C) from 40,000 to 320,000 when the diesel engine was operating on pure diesel fuel, while the operation of the diesel engine on combined fuel was accompanied by lower carbon emissions (C) in the entire range. The decrease was caused by a lower tendency to smoke ethanol compared to diesel fuel.

Thus, the decrease in carbon emissions (C) with exhaust gases during the operation of a diesel engine on combined fuel with an ethanol content of 20% was 25, 27.27, 10.34, 22.93 and 30.53% in the entire range from 0.2 to 1.0 MPa compared to the operation of a diesel engine on pure diesel fuel. Similarly, the decrease in carbon emissions (C) with exhaust gases during the operation of a diesel engine on combined fuel with an ethanol content of 40% was 72.5, 45.45, 34.48, 50.32 and 62.5% in the entire range from 0.2 to 1.0 MPa compared to the operation of a diesel engine on pure diesel fuel.

The increase in C emission with exhaust gases with an increase from 0.2 to 1.0 MPa both during diesel operation on pure diesel fuel and on combined fuel with ethanol content was caused by a lack of oxygen in the air required for oxidation of the combined fuel injected into the cylinder. Complete combustion of the combined fuel was not provided, with a sharp increase in the temperature of the gases in the cylinder. Under conditions of high temperature and lack of air oxygen, unburned combined fuel was pyrolyzed to form C.

Figure 2a, b shows the environmental parameters of the diesel engine with a dimension of 4ChN 11.0/12.5 when operating on combined fuel with a content of rapeseed oil, 1400 min^{-1} and the optimal value 26° of PCV to TDC.

The nature of the change in environmental indicators, in general, did not change. Based on the dependencies, carbon emissions (C) (Fig. 2a) both when the diesel engine is running on pure diesel fuel and combined fuel with rapeseed oil content had the same tendency to increase throughout the operating interval. When a diesel engine runs on combined fuel, carbon (C) emissions with exhaust gases decreased more intensively with an increase in rapeseed oil content from 20 to 60%, due to a more complete oxidation of atomic carbon.

Thus, the increase in the indicator to 0.6 MPa had an impact on carbon emissions (C) with exhaust gases and amounted to: for diesel fuel—107,000 ppm, and for combined fuel with a rapeseed oil content of 20, 40 and 60%—92,000, 76,000 and 56,000 ppm, respectively. With an increase to 1.2 MPa, the value of carbon emissions (C) with exhaust gases amounted to: for diesel fuel—380,300 ppm, and for

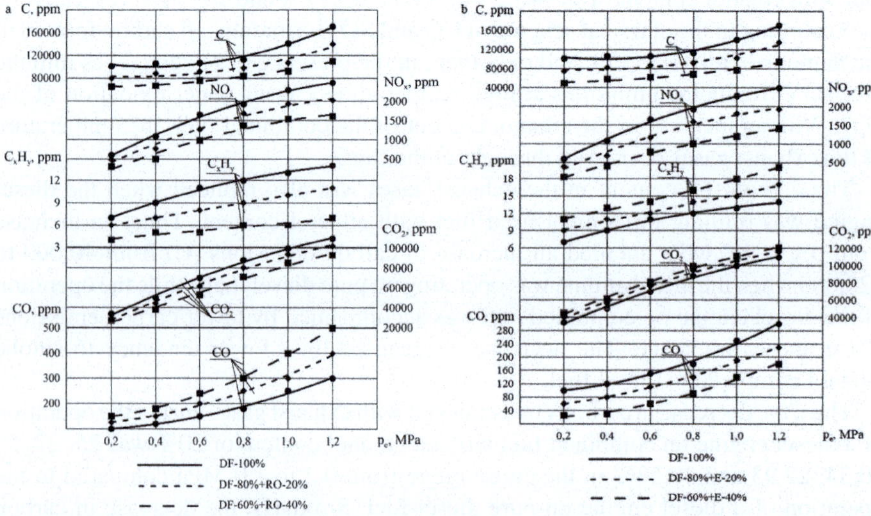

Fig. 2 **a** environmental indicators of diesel engine operation with dimension 4ChN 11.0/12.5 when operating on combined fuel with content of rapeseed oil, 1400 min^{-1} and optimal value 26° PCV to WDC; **b** environmental indicators of diesel engine operation with dimension 4ChN 11.0/12.5 when operating on combined fuel with ethanol content, 1400 min^{-1} and optimal value 22° of PCV to TDC

combined fuel with a content of rapeseed oil of 20, 40 and 60%—270,000, 168,000 and 90,000 ppm, respectively.

The change in carbon emission (C) in the range from 0.6 to 1.2 MPa was up to 171,000 ml^{-1} for diesel fuel, 140,000, 110,000, 77,000 ml^{-1} for rapeseed oil, depending on its content in the combined fuel. However, when analyzing the obtained data, it can be seen that the carbon emission (C) when operating on a combined fuel with a rapeseed oil content of 20% is less by 18.13%, and on a combined fuel with a rapeseed oil content of 40 and 60%—less by 35.67 and 54.97%, respectively.

The emission of nitrogen oxides (NO_x) during the operation of a diesel engine on combined fuel with a content of rapeseed oil 20. 40 and 60% amounted to—1926. 1531 and 1231 ppm, respectively. The emission of nitrogen oxides (NO_x) during the operation of a diesel engine on combined fuel with a content of rapeseed oil of 20, 40 and 60% decreased by 15.93, 33.17 and 46.26% compared to the operation of a diesel engine on pure diesel fuel. The reduction in nitrogen oxide (NO_x) emissions is most likely due to a decrease in local and average cycle temperatures, as well as a decrease in free oxygen in the flame.

The emission of hydrocarbons (C_xH_y) with a decrease in the content of rapeseed oil in combined fuel in the entire range from 0.2 to 1.2 MPa increased: during the operation of the diesel engine on pure diesel fuel to 13 ppm, during the operation of the diesel engine on combined fuel with a content of rapeseed oil of 20, 40 and 60%—to 10, 6 and 3 ppm, respectively. However, the emission of hydrocarbons (C_xH_y) during the operation of a diesel engine on combined fuel with a content of rapeseed oil of 20, 40 and 60% decreased by 23.08, 53.85 and 76.92% compared to the operation of a diesel engine on pure diesel fuel.

The operation of a diesel engine on combined fuel with a content of rapeseed oil was accompanied by an increase in the emission of carbon monoxide (CO) with exhaust gases. With an increase from 0.2 to 1.0 MPa, it led to an increase in the emission of carbon monoxide (CO) with diesel exhaust gases on combined fuel with a content of rapeseed oil of 20. 40 and 60% and amounted to 250. 400 and 500 ppm, respectively.

The emission of carbon dioxide (SO_2) increased with an increase and had the following values: when operating a diesel engine on pure diesel fuel—99,600 ppm, when operating a diesel engine on combined fuel with a content of rapeseed oil of 20, 40 and 60%—93,700, 85,200 and 75,800 ppm, respectively.

The dependencies (Fig. 2b) show the environmental indicators of the diesel engine with a dimension of 4ChN 11.0/12.5 when operating on combined fuel with ethanol content, 1400 min^{-1} and the optimal value 22° of PCV to TDC.

Analysis of experimental data showed that the operation of a diesel engine on combined fuel with ethanol content was accompanied by a decrease in the emission of nitrogen oxides (NO_x), carbon (C) and carbon monoxide (CO) over the entire range from 0.2 to 1.2 MPa, compared to the operation of a diesel engine on pure diesel fuel.

The emission of nitrogen oxides (NO_x) with exhaust gases, in accordance with the nature of the relationship arrangement (Fig. 2b), increased throughout the interval.

Emission during the operation of a diesel engine on combined fuel with an ethanol content of 20 and 40% at 0.6 MPa decreased by 29.09 and 52.73%, respectively.

At 1.2 MPa, NO_x emission increased to 2490 ppm when the diesel engine was operating on pure diesel fuel, to 1820 and 1200 ppm, respectively, when the diesel engine was operating on combined fuel with an ethanol content of 20 and 40%, respectively. The decrease in local and average cycle temperatures in the case of combined fuel with ethanol content directly causes a decrease in the emission of nitrogen oxides (NO_x). At the same time, the emission of nitrogen oxides (NO_x) is directly dependent on the temperature of the exhaust gases.

Thus, carbon emissions (C) in the entire range from 0.2 to 1.2 MPa decreased when the diesel engine was running on combined fuel with an ethanol content of 20% by 25.74, 17.02, 18.69, 17.7, 27.66 and 21.05%, in combined fuel diesel operation with ethanol content of 40% by 50.46, 42, 42.06, 39.82, 45.39 and 44.44%, respectively, compared to pure diesel operation.

The increase in carbon emissions (C) with the exhaust gases with an increase in temperature, both when operating on diesel fuel and on combined fuel, was primarily due to the lack of oxygen in the air necessary for oxidizing the combined fuel injected into the cylinder. Complete combustion of the combined fuel is not ensured, with a sharp increase in the temperature of the gases in the cylinder. Under conditions of high temperature and lack of air oxygen, unburned combined fuel will undergo pyrolysis to produce carbon (C).

Hydrocarbon emissions increased over the entire range. The increase in the emission of hydrocarbons (C_xH_y) during the operation of a diesel engine on combined fuel with an ethanol content of 20–40% was 15.38–38.46% compared to the operation of a diesel engine on pure diesel fuel. In this case, the acceleration of the combustion process with the addition of ethanol, a decrease in the averaged cycle temperature and, as a result, the temperature of the exhaust gases, is of decisive importance.

When the diesel engine operates on pure diesel fuel and 0.2 MPa, the emission of carbon monoxide (CO) and carbon dioxide (CO_2), respectively, was 100–33,900 ppm, and at 1.2 MPa, respectively, 300–171,000 ppm. Thus, when the diesel engine operates in the range from 0.2 to 1.2 MPa, the increase in carbon dioxide (SO_2) emissions was from 5.5 to 12% when the diesel engine operates on combined fuel with an ethanol content of 20% and from 4.32 to 12.53% when the diesel engine operates on combined fuel with an ethanol content of 40% compared to the diesel engine operates on pure diesel fuel.

The decrease in values at high levels causes under oxidation of carbon dioxide (CO_2). The decrease in carbon monoxide (CO) emissions during diesel fuel operation in the range from 0.2 to 1.2 MPa was from 22.22 to 40.0% for diesel fuel operation on combined fuel with ethanol content of 20% and from 21.74 to 62.5% for diesel fuel operation on combined fuel with ethanol content of 40% compared to diesel fuel operation on pure diesel fuel.

4 Discussion

Vehicle exhaust gases are an important anthropogenic source of atmospheric carbon aerosols; of which emission coefficients and stable isotope composition of carbon are important initial data. Different combinations of fuels have been selected to perform dynamometric tests using different test cycles and under cold/hot start conditions.

The transport sector's contribution to global emissions is a known fact, however, the mitigation path to achieving nationally defined carbon reduction targets is often not specified.

Air pollution is recognized as a new environmental risk factor for neurological diseases. Large-scale epidemiological studies have linked traffic-related particulate matter to cognitive impairment and increased incidence of neurodegenerative diseases.

Recent WHO recommendations include systematic measurements of environmental particle concentration and soot (BC) concentrations.

In future emission control and source distribution, attention should be paid to changes in emission factors and isotope signatures caused by the use of different types and combinations of fuel and the aging of vehicles.

As a result of the conducted studies, the values of the optimal injection advance angles of the combined fuel with the content of rapeseed oil from 0 to 60% and the combined fuel with the content of ethanol from 0 to 40% were equal to 26° PCV to WDC and 22° PCV to WDC at which the values of the specific effective consumption of combined fuel ranged from 259 to 321 g/kwh and from 283 to 323 g/kwh, respectively.

In this work, the possibility of using combined fuel (rapeseed oil and ethanol) as a renewable and alternative fuel source was investigated. Various fuel mixtures containing 20, 40 and 60% rapeseed oil containing 20, 40 and 60% ethanol with diesel fuel were prepared and the effect of the combined fuel on exhaust emissions and some operating parameters under different load conditions using a diesel engine were tested. The measured parameters of the engine operation, as a rule, showed a slight increase in the temperature of exhaust gases and specific fuel consumption [19–23].

The experiments also show that with an increase in the temperature of the exhaust gases in the cylinder, the substances do not pass the full combustion cycle and the emission from the diesel engine increases during operation on combined fuel. Thus, the optimal performance of the diesel engine is crucial in obtaining a combination and actual concentration of exhaust gases that can reduce or increase the negative impact on the atmosphere and the biosphere as a whole.

References

1. Jacob A, Ashok B, Lawrence J, Soosairaj AS, Anandan J, Elango M (2023) Exploring the potential of third-generation microalgae bio-alcohol and biodiesel in arresting particulate smoke emissions and greenhouse gases using CART. Environ Sci Pollut Res Int. 30(10):27650–27669. https://doi.org/10.1007/s11356-022-24110-5
2. Apicella B, Mancaruso E, Russo C, Tregrossi A, Oliano MM, Ciajolo A, Vaglieco BM (2020) Effect of after-treatment systems on particulate matter emissions in diesel engine exhaust. Exp Thermal Fluid Sci 116:110107. https://doi.org/10.1016/j.expthermflusci.2020.110107
3. Atmanli A, Yilmaz N (2018) A comparative analysis of n-butanol/diesel and 1-pentanol/diesel blends in a compression ignition engine. Fuel 234:161–169. https://doi.org/10.1016/j.fuel.2018.07.015
4. Ayodhya AS, Narayanappa KG (2018) An overview of after-treatment systems for diesel engines. Environ Sci Pollut Res 25(35):35034–35047. https://doi.org/10.1007/s11356-018-3487-8
5. Chen C, Yao A, Yao C, Qu G (2020) Experimental study of the active and passive regeneration procedures of a diesel particulate filter in a diesel methanol dual fuel engine. Fuel 264:116801. https://doi.org/10.1016/j.fuel.2019.116801
6. Sharma AK, Sharma PK, Chintala V, Khatri N, Patel A (2020) Environment-Friendly Biodiesel/Diesel Blends for Improving the Exhaust Emission and Engine Performance to Reduce the Pollutants Emitted from Transportation Fleets. Int J Environ Res Public Health 17(11):3896. https://doi.org/10.3390/ijerph17113896
7. Patel A, Arora N, Sartaj K, Pruthi V, Pruthi PA (2016) Sustainable biodiesel production from oleaginous yeasts utilizing hydrolysates of various non-edible lignocellulosic biomasses. Renew. Sustain. Energy Rev. J Mol Med 62:836–855. https://doi.org/10.1016/j.rser.2016.05.014
8. Scarlat N, Dallemand JF, Monforti-Ferrario F, Banja M, Motola V (2015) Renewable energy policy framework and bioenergy contribution in the European Union—An overview from National Renewable Energy Action Plans and Progress Reports. Renew. Sustain. Energy Rev. 51:969–985. https://doi.org/10.1016/j.rser.2015.06.062
9. Scarlat N, Dallemand JF, Monforti-Ferrario F, Nita V (2015) The role of biomass and bioenergy in a future bioeconomy: Policies and facts. Environ. Dev. 15:3–34. https://doi.org/10.1016/j.envdev.2015.03.006
10. Purohit P, Dhar S (2015) Biofuel Roadmap for India, Promoting Low Carbon Transport in India. 2015 UNEP DTU Partnership (United Nations Environment Programme) Available via DIALOG: https://backend.orbit.dtu.dk/ws/portalfiles/portal/120569025/Biofuel_Roa. Accessed 1 October 2015
11. Abu-Hamdeh NH, Alnefaie KA (2015) A comparative study of almond biodiesel-diesel blends for diesel engine in terms of performance and emissions. Biomed Res Int. 2015:529808. https://doi.org/10.1155/2015/529808
12. Zhu R, Wang X, Miao H, Yang X, Huang Z (2011) Effect of dimethoxy-methane and exhaust gas recirculation on combustion and emission characteristics of a direct injection diesel engine. 90(5):1731–1737. https://doi.org/10.1016/j.fuel.2011.01.035
13. Srivastava A, Prasad R (2000) Triglycerides-based diesel fuels. Renewable & Sustainable Energy Reviews 4(2):111–133. https://doi.org/10.1016/S1364-0321(99)00013-1
14. Ali Y, Hanna MA (1994) Alternative diesel fuels from vegetable oils. Bioresource Technology 50(2):153–163. https://doi.org/10.1016/0960-8524(94)90068-X
15. Altin R, Çetinkaya S, Yücesu HS (2001) Potential of using vegetable oil fuels as fuel for diesel engines. Energy Conversion and Management 42(5):529–538. https://doi.org/10.1016/S0196-8904(00)00080-7
16. State standard (1988) Tractor and combine diesel engines. Bench test methods (with Rev.1)
17. Lapidus AL, Bavykin VM, Smolensky EA, Chuvaeva IV (2008) Dependence of cetane numbers of hydrocarbons on the structure of molecules. Reports of the Academy of Sciences

18. Lapidus AL, Smolensky EA, Bavykin VM et al (2008) Models for calculating and predicting octane and cetane numbers of individual hydrocarbons. Petrochemistry
19. Vellaiyan S, Subbiah A, Chockalingam P (2018) Combustion, performance, and emission analysis of diesel engine fueled with water-biodiesel emulsion fuel and nanoadditive. Environ Sci Pollut Res Int. (33):33478–33489. https://doi.org/10.1007/s11356-018-3216-3
20. Bünger J, Krahl J, Schröder O, Schmidt L, Westphal GA (2012) Potential hazards associated with combustion of bio-derived versus petroleum-derived diesel fuel. Crit Rev Toxicol. 42(9):732–50. https://doi.org/10.3109/10408444.2012.710194
21. Aatola H, Larmi M, Sarjovaara T, Mikkonen S (2008) NExBTL - Biodiesel fuel of the second generation. SAE Technical Paper
22. Ackland ML, Zou L, Freestone D, van de Waasenburg S, Michalczyk AA (2007) Diesel exhaust particulate matter induces multinucleate cells and zinc transporter-dependent apoptosis in human airway cells. Immunol Cell Biol. 85:617–622
23. Alam M, Song J, Acharya R, Boehman A, Miller K (2004) Combustion and emissions performance of low sulfur, ultra low sulfur and biodiesel blends in a DI diesel engine. SAE Technical Paper. 2004:01–3024

The Concept for Recycling Waste from Multi-Apartment Residential Buildings

T. Zhilina, K. Afonin, and A. Zagorskaya

Abstract The article analyzes the problem of waste disposal from multi-apartment residential buildings using the example of global and domestic Russian practice. Graphs of waste generation and management in various countries of the world are presented. The article examines current data on the volume of waste disposal at landfills in the Russian Federation and analyzes the dynamics of accumulation of production and consumption waste over the past decade. The authors of the article are developing a concept for recycling organic waste from apartment buildings at the place of their generation. A scheme for fermentation of waste to produce biomethane and a humus-like product is considered. The economic efficiency of introducing fermentation plants for a mixture of waste from apartment buildings has been assessed. A scheme for joint waste processing has been developed. The composition of the substrate from food waste and domestic wastewater sludge is given. It is concluded that there is a synergistic effect from the introduction of a waste recycling scheme.

Keywords Waste recycling · Biomethane · Fermentation · Separate accumulation · Waste collection · Bioreactor

1 Introduction

Despite the introduction of modern technologies and updated approaches to organizing waste management, the recycling of all types of production and consumption waste still remains a difficult problem not only in Russia, but also abroad. In most countries of the world, the main method of waste management is incineration or disposal in landfills (Fig. 1). Among the leading countries in progressive waste management are Sweden, South Korea and Germany [1]. However, data from the Global Waste Index [1] show that not everything is actually recycled and the bulk of waste in these countries continues to be incinerated (Fig. 2).

T. Zhilina (✉) · K. Afonin · A. Zagorskaya
Industrial University of Tyumen, 2, Lunacharsky St., Tyumen 625002, Russia
e-mail: ts-z@yandex.ru

L. V. Radionova and D. V. Ulrikh (eds.), *Advances in Ecology and Environmental Engineering*, Springer Proceedings in Earth and Environmental Sciences,
https://doi.org/10.1007/978-3-031-64423-8_33

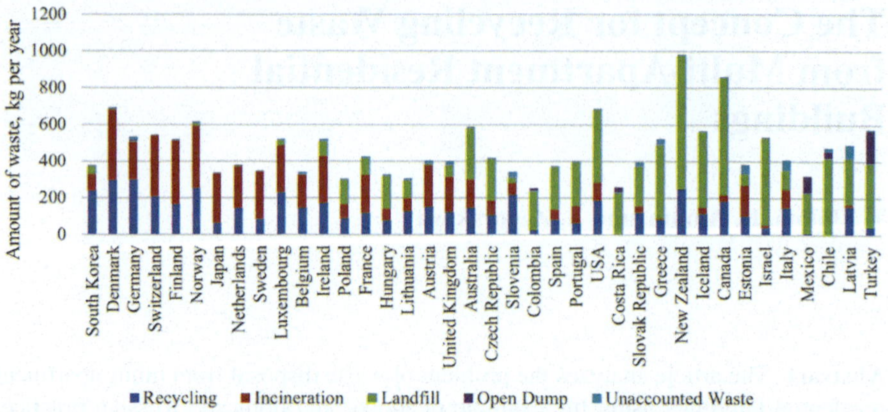

Fig. 1 Waste generation and management in different countries of the world in 2022, kg of waste per person per year (according to Sensoneo, Brastislava, Slovakia, 2023)

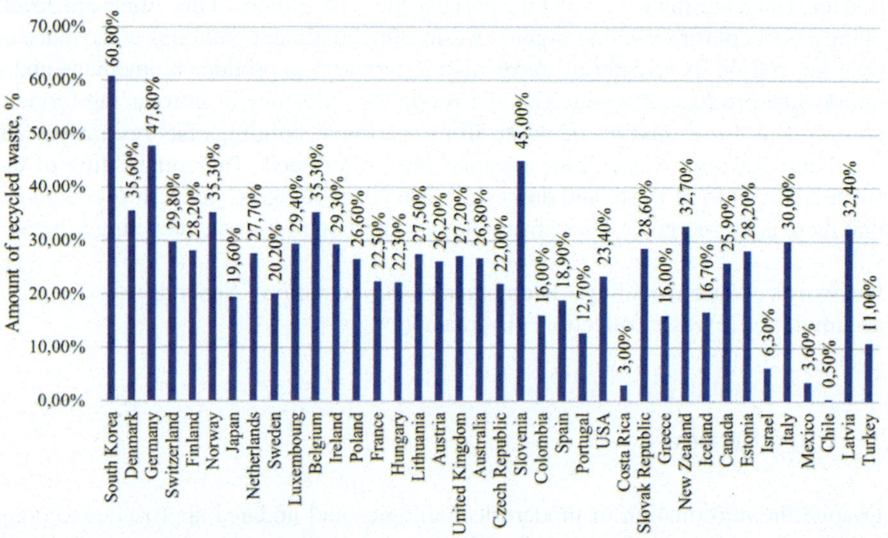

Fig. 2 Analysis of waste management in various countries of the world in 2022, (according to Sensoneo, Brastislava, Slovakia, 2023)

Waste management as an industry in the Russian Federation has existed since 2018, with the start of the so-called "garbage reform". Before its start, industrial waste was widely used as secondary resources in the production of building materials and structures. However, the return of production and consumption waste to the technological process did not exceed 10–15% of the total volume. The main facility for waste disposal in Russia, both industrial and municipal, was and still is landfills. Their dominance is associated with the presence of free, unused territories remote from populated areas. The waste disposal facilities made it possible to accept industrial waste of the third, fourth and fifth hazard classes, provided that they constitute no more than 30% of their payload. Under ideal conditions, waste at solid waste landfills fermented naturally, with small capital and operating investments for 4 years, and after this period of time it was possible to obtain technosoil suitable for the reclamation of quarries and other workings formed during open-pit mining. Unfortunately, in a significant part of the territory of the Russian Federation, the duration of seasons with temperatures below zero is 6 months or more, which does not allow colonies of decomposers, formed mainly by thermophilic and mesophilic bacteria, to fully ferment waste. In addition, insufficient isolation from the environment, poor-quality or completely absent pre-sorting of waste, and a low level of fire safety have led to the fact that landfills have turned from environmental and bioprotective structures into a serious source of air pollution, soil, surface and groundwater. Existing as an alternative waste disposal measure, incinerators are located in areas where climatic, environmental or economic constraints prevent waste from being disposed of in landfills. Existing as an alternative waste disposal measure, incinerators in Russia are located in areas where climatic, environmental or economic restrictions prevent waste from being disposed of in landfills. At the moment, only 7 waste incineration plants operate in the Russian Federation, so the dominant method of waste disposal is still centralized sorting with subsequent placement at various facilities (landfills, sludge pits, drill pits, tailings dumps, etc.). [2–5]. The total volume of Russian waste disposal facilities owned by municipalities in 2022 amounted to 90.2 thousand hectares, and large mining and processing companies reached 193.2 thousand hectares [2, 6].

The dynamics of waste generation over the past 20 years (Fig. 3.) shows a significant increase in the volume of waste generated in industry and in residential areas, which may indicate industrial growth and an improvement in the well-being of the population. However, this fact does not reduce the need for rational waste management, which makes it possible to reduce the area of waste disposal and disposal facilities, which in turn will lead to a reduction in anthropogenic load in adjacent areas and an undoubted improvement in environmental quality.

The ratio of waste disposed and reused in technological cycles in 2022 was approximately 55–45%, which confirms the prevalence of landfills over other disposal technologies [7]. And, as a result, it shows the need for inexpensive technologies for reusing waste as secondary material or energy resources. The low involvement of waste in recycling processes is also due to the fact that at this stage the system of separate accumulation and collection in Russia has been partially implemented, and in some regions, it is completely absent [8–10]

2 Materials and Methods

This article discusses the concept of separate accumulation of waste from apartment buildings with subsequent methanogenesis and recycle of the resulting products at the site of generation.

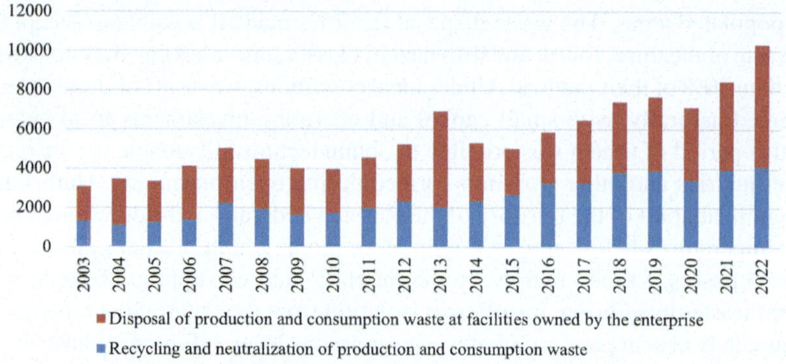

Fig. 3 Dynamics of waste generation in the Russian Federation, thousand tons of waste per year (according to Rosstat, 2023)

Modeling of the conditions for the formation of residential waste and its composition was carried out using the example of the city of Tyumen. All calculations were made based on the amount of waste typical for the south of the Tyumen region. The composition of the fermentable part, based on the type of municipal waste, is shown in the Fig. 4.

This waste management scheme assumes that part of the waste from apartment buildings, about 70% of the total, is accumulated separately on adjacent sites and transported to a waste sorting plant for deeper separation according to its specifics (Fig. 5). Food waste is crushed using a disposer and, together with household wastewater, is sent for treatment, which includes dewatering and drying. Thus, a substrate of food waste and sewage sludge will contain 4% fats, 82% carbohydrates and 14% proteins, with a total moisture content of about 60%, which are acceptable conditions for the onset of methanogenesis. [11–13]. The volume of the reactor for waste fermentation is calculated based on the total amount of fermented waste (Fig. 6) and the unevenness of their formation (Fig. 7).

Since high humidity significantly inhibits the activity of anaerobic bacteria, the reactor includes a drying system and removal of moisture formed during the early

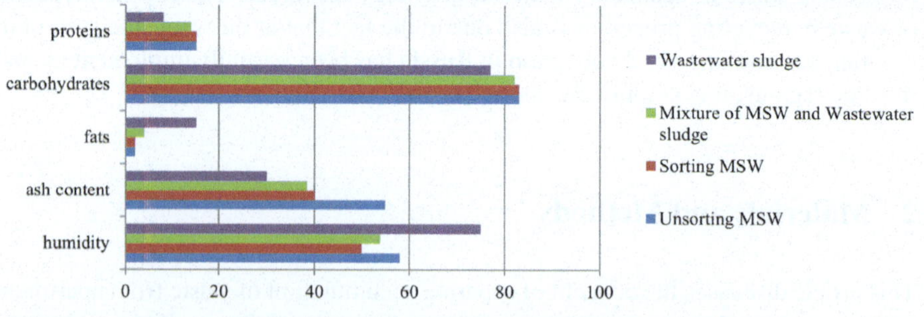

Fig. 4 Fermentable part of waste from residential areas (defined by the authors)

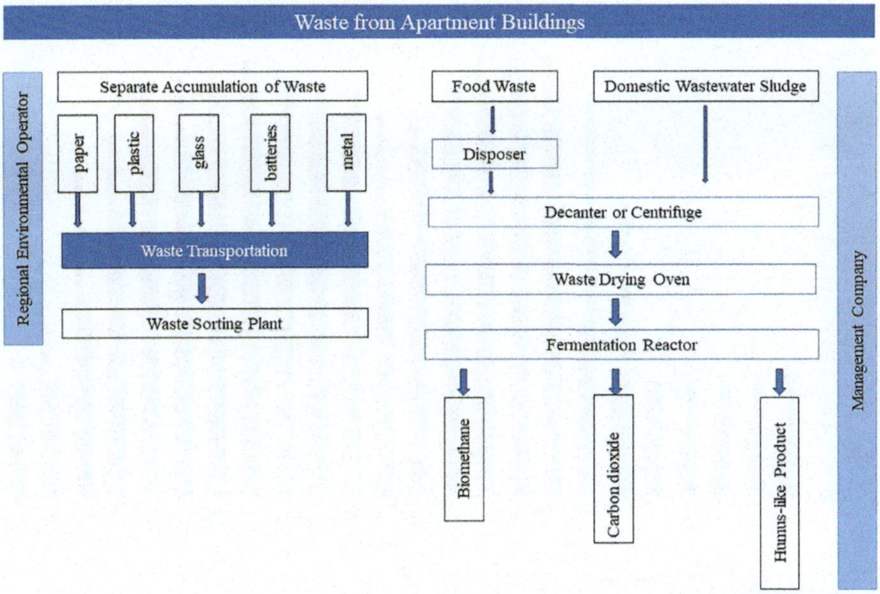

Fig. 5 Conceptual representation of food waste recycling scheme (developed by the authors)

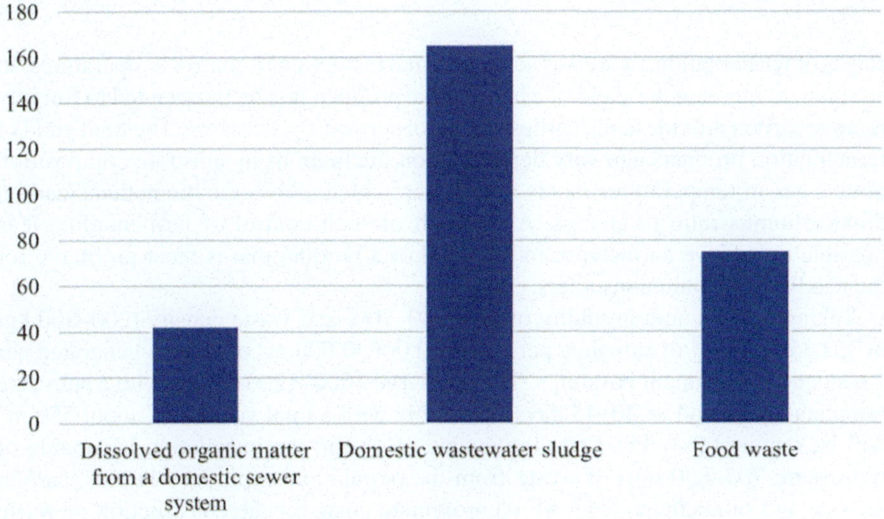

Fig. 6 Types of waste required for the production of the substrate, in kg per person per year

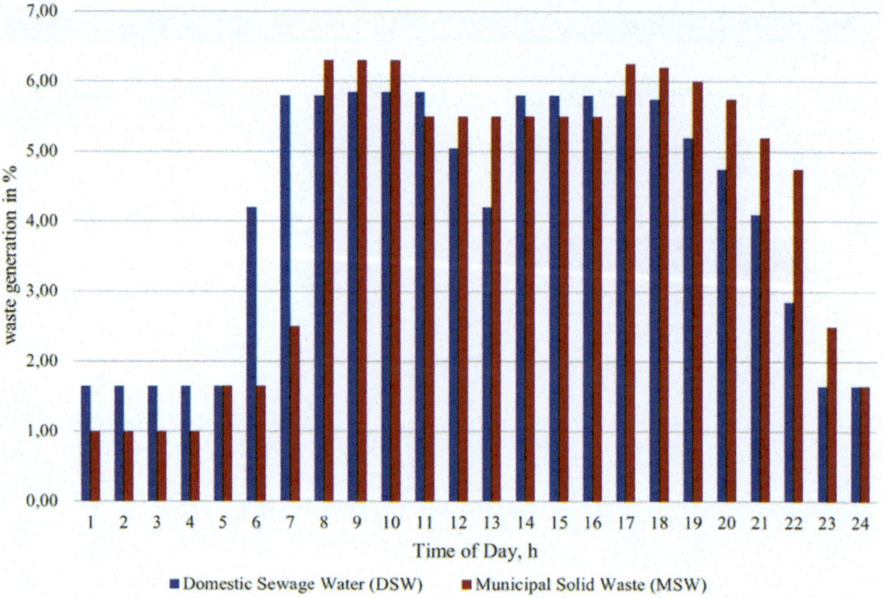

Fig. 7 Irregularity of waste and wastewater generation during the day, in% (using the example of the Tyumen city)

stages of fermentation, as well as active thermal insulation to control temperatures. In addition, to increase the yield of a humus-like product, it is recommended to forcibly remove carbon dioxide at the earliest stages of substrate processing. The final yield of fermentation products can vary depending on fluctuations in substrate composition or changes in temperatures inside the reactor, which allows the biomethane/carbon dioxide/humus ratio to change. As a result of such control of fermentation, it is possible to achieve an increase in the yield of a product that is most profitable for the needs of the main consumers.

Taking into account humidity of about 60–70% and waste density (200–300 kg/m^3), 2.35–3.53 m^3 of substrate per year or 0.006–0.009 m^3 per day is generated per 1 resident of apartment buildings. Thus, to serve 1000–1200 people with a substrate fermentation period of 30–45 days, 2 reactors with a total volume of about 550 m^3 will be required [14, 15]. Over the course of a year, the installation is capable of processing 705–890 tons of waste from the population to produce humus, carbon dioxide and biomethane (Fig. 8). Approximate costs for the construction of waste co-digestion bioreactors are determined in Table 1.

The total cost of the installation for the joint processing of waste and sewage sludge at the place of their generation is 6.37 million rubles [16]. The total operating costs for this installation, including electricity costs, labor costs, replacement and repair of equipment, amount to 2.78 million rubles per year. With the maximum possible methane output of 284 kg/t per year, the installation can produce 157–198 thousand m3 of biomethane and its cost will be 14–17 rubles/m^3. If the volume

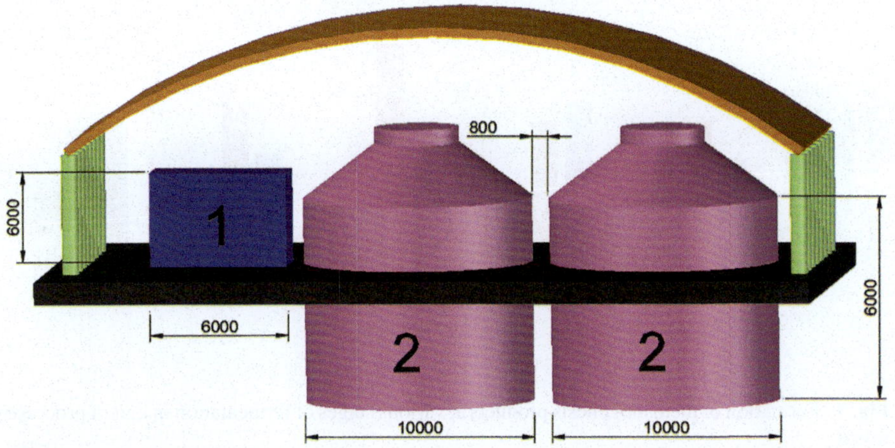

Fig. 8 Methanogenesis reactors for processing the fermentable part of waste at the point of generation (developed by the authors); 1—Waste collection and primary treatment pavilion; 2—Waste fermentation reactors

Table 1 Estimated cost of organic waste fermentation facilities

Cost type	Required amount, thousand rubles
Excavation	1100
Basic equipment	1500
Automation	1450
Technical pavilion and communications	450
Compliance	1500
Organization of additional separate collection	370

of exported waste decreases by 30%, taking into account the current tariff for the removal of MSW in the Tyumen region of 8907.46 rubles [17], the total monetary burden on management companies will decrease by 1 883.928–2373.749 thousand rubles per year or 1883.93 rubles per 1 resident.

The use of a waste recycling plant allows several times to reduce the area of land allocated for centralized disposal of household waste, as well as reduce emissions of natural fermentation products into the atmosphere. In addition, the energy value of biogas allows the production of both electricity and heat, which leads to resource savings. This aspect can be taken into account when constructing autonomous energy sources in the Arctic and the Far North. Biogas plants can be located in any climatic region, they do not require large investments and are quite easy to operate.

Unlike natural resources, biogas is a renewable energy source, and its use as a fuel leads to the solution of both environmental and energy problems. In addition, the resulting biomethane can serve as a raw material for organic synthesis reactions. The carbon dioxide produced as a result of fermentation can be used as a plant

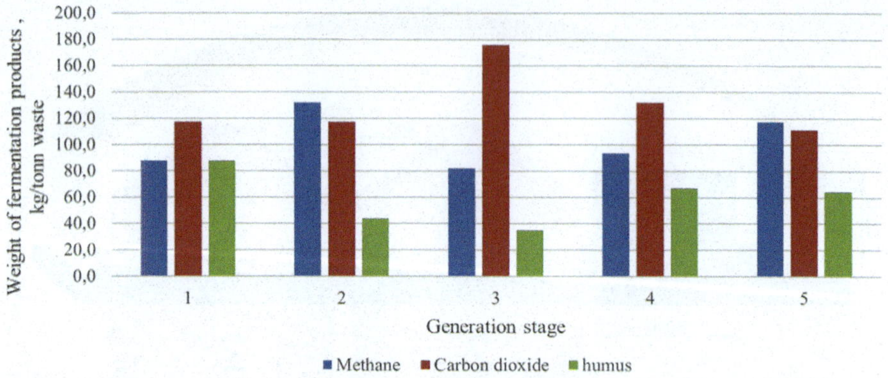

Fig. 9 Formation of methanogenesis products at various stages of fermentation in kg/t of processed waste

growth accelerator, increasing the efficiency of photosynthesis by 1.2–1.5 times. Humus obtained as a result of fermentation of solid waste is also a useful product for agriculture in restoring soil fertility or reclamation of technogenically disturbed lands (Fig. 9).

3 Conclusion

When using the proposed scheme, we obtain secondary raw materials that are not contaminated with food waste, including plastic, glass, metals, and paper. This makes it possible to improve the quality of waste sorting at waste sorting plants and their unhindered return to recycling processes. In addition, reducing the intensity of the flow of waste and sludge from domestic wastewater into sewer networks will reduce the number of blockages and accidents in both local and urban drainage systems. This will have a positive impact on reducing the costs of repairs and liquidation of the consequences of accidents in household sewerage networks. Products obtained from the digestion of organic waste can be used for household needs or collected and transported for subsequent use [18, 19]. Also, the synergistic effects of introducing this waste disposal scheme include a reduction in the area of MSW disposal facilities [20], not only due to a decrease in the amount of waste transported to landfills, but also due to an increase in the volume of waste sold by waste sorting complexes as secondary material resources.

References

1. Global Waste Index (2022) Sensoneo, Brastislava. https://sensoneo.com/global-waste-index. Accessed 1 Dec 2023
2. State (national) report on the state and use of land in the Russian Federation (2022) Federal Service for State Registration, Cadastre and Cartography, Moscow. https://rosreestr.gov.ru/upload/Doc/16-upr/doc_nation_report_2022.pdf. Accessed 1 Dec 2023
3. SanPiN 2.1.3684–21 "Sanitary and epidemiological requirements for the maintenance of territories of urban and rural settlements, for water bodies, drinking water and drinking water supply, atmospheric air, soils, residential premises, operation of industrial and public premises, organization and implementation of sanitary and anti-epidemic (preventive) measures". Available via Consultant Plus. https://www.consultant.ru/document/cons_doc_LAW_376166. Accessed 1 Dec 2023
4. SR 320.1325800.2017. Polygons for solid communal waste. Projecting, operation and reclamation. Available via Consultant Plus. https://www.consultant.ru/document/cons_doc_LAW_415846. Accessed 1 Dec 2023
5. Environmental protection in Russia (2022) Federal Service State Statistics, Moscow. https://rosstat.gov.ru/folder/210/document/13209. Accessed 1 Dec 2023
6. Public cadastral map of Russia (2023) Federal Service for State Registration, Cadastre and Cartography, Moscow. https://kadastrovayakarta.org. Accessed 1 Dec 2023
7. Statistical publications (2023) Federal Service State Statistics, Moscow. https://rosstat.gov.ru. Accessed 1 Dec 2023
8. Sharova OA (2014) Vliyanie poligona TBO na sostoyanie gruntovyh vod (na primere poligona R.P. Krasnyj Yar, Astrahanskaya oblast') (Influence of solid domestic waste landfill on the state of groundwater (for example, landfill R. P. Krasny Yar, Astrakhan Region). Vestnik Tambovskogo universiteta. Seriya: Estestvennye i tekhnicheskie nauki (Bulletin of the Tambov University. Series: Natural and Technical Sciences) 19(5):1741–1743
9. Zhukova NS, Samarskaya NS (2014) Ekologo-ekonomicheskie osobennosti sistemy potrebleniya tverdyh othodov (Environmental and economic peculiarities of the system for hardening waste consumption). Inzhenernyj vestnik Dona (Engineering Bulletin of the Don). https://ivdon.ru/ru/magazine/archive/n3y2014/2488. Accessed 1 Dec 2023
10. Baramzin KN et al (2017) Effektivnost' modernizacii poligonov TBO (Efficiency of modernization of MSW landfills). Investicionnaya ocenka (Investment valuation) 3(17):9–31
11. Afonin KV, Zhilina TS, Zagorskaya AA, Pavlova MN (2022) Ispol'zovanie metangeneriruyushchih ustanovok na baze TBO v maloetazhnom stroitel'stve (The use of methane-generating installations on the base municipal solid waste in low-rise construction). Stroitel'stvo i geotekhnika (Construction and geotechnics) 13(1):73–81. https://doi.org/10.15593/2224-9826/2022.1.06
12. Abu-Qdais HA, Al-Ghazawi Z, Awawdeh A (2023) Assessment of Greenhouse Gas Emissions and Energetic Potential from Solid Waste Landfills in Jordan: A Comparative Modelling Analysis. Water 15(1):155. https://doi.org/10.3390/w15010155
13. Xiang L, Li H, Wang Y, Qu L, Xiao D (2023) Energy Utilization Assessment of Municipal Sewage Sludge Based on SWOT-FAHP Analysis. Water 15(2):260. https://doi.org/10.3390/w15020260
14. Afonin AA, Zhilina TS, Zagorskaya AA (2019) The reduction the load on the MSW landfill by optimizing the waste management system of residential territories. IOP Conf. Series: Materials Science and Engineering 481:012006. https://doi.org/10.1088/1757-899X/481/1/012006
15. Eder B, Schulz H (1996) Biogas plants. Practical guide. Zorg Biogas, München
16. GOST 53790-2010 Unconventional technology. Energy biowaste. General technical requirements for biogas plants. Available via Information network "Techexpert". https://docs.cntd.ru/document/1200083240. Accessed 1 Dec 2023
17. Database of regional operators and housing and communal services management companies in Russia (2023) Moscow. https://regoperatory.ru/tarify-tko-tyumenskaya-oblast. Accessed 1 Dec 2023

18. Kulagin VA (2020) Perspektivy razvitiya mirovoj energetiki s uchetom vsego tekhnicheskogo progressa (Prospects for the development of global energy, taking into account all technological progress). Institut energeticheskih issledovanij RAN (Institute of Energy Research of the Russian Academy of Sciences), Moscow
19. Chernin SYa, Parubets YuS (2011) Rossijskij opyt vnedreniya biogazovyh tekhnologij dlya proizvodstva elektricheskoj i teplovoj energii (Russian experience in introducing biogas technologies for the production of electrical and thermal energy). Novosti teplosnabzheniya (Heat Supply News) 08(132), Moscow. https://www.rosteplo.ru/Tech_stat/stat_shablon.php?id=3922. Accessed 1 Dec 2023
20. Aziz HA, Ramli SF, Hung Y-T (2023) Physicochemical Technique in Municipal Solid Waste (MSW) Landfill Leachate Remediation: A Review Water 15(6):1249. https://doi.org/10.3390/w15061249

Mechanoactivated Mineral Compositions Based on Portland Cement Clinker Containing Blast-Furnace Dump Slag

S. A. Dergunov, A. B. Satyukov, S. V. Serikov, A. Yu. Spirina, and D. S. Odincova

Abstract The study reflects the goals of global sustainable economic development. The work presents the research results of the metallurgical rational waste use in composite cement production. Metallurgical waste use is considered for the production of concrete, mortars and dry building mixtures of various functional purposes in factories. The centrifugal impact grinding process determines the formation of surface nanostructuring and the production of the mechanoactivated composite. Alternative solutions are considered for composite cement production. There are both intergrinding components and their mixing in a crushed state. The optimal composite cement formulation was determined with the maximum use of slags. These composite cements meet the requirements of the current regulatory and technical documentation. Conclusions present information about the feasibility of using metallurgical waste. The diagrams below illustrate the kinetics of stone strength gain of composite cements and the compositional cement stone microstructure obtained by various methods.

Keywords Composite cement · Slag · Mechanoactivation · Mechanocomposites

1 Introduction

Environmental industrial policy is associated with industrialization, innovation and modernization of the economy today.

Composite cements development is a resource and energy saving trend in the modern construction industry. At the same time, technogenic waste using question is relevant in the development of resource-saving technologies.

S. A. Dergunov (✉) · A. B. Satyukov · S. V. Serikov · A. Yu. Spirina · D. S. Odincova
Orenburg State University, 13, Pobedy Ave., Orenburg 460018, Russia
e-mail: dergunow79@mail.ru

S. A. Dergunov · A. B. Satyukov
Testing Center "Orenburgstroyispitaniya", 13, Pobedy Ave., Orenburg 460018, Russia

L. V. Radionova and D. V. Ulrikh (eds.), *Advances in Ecology and Environmental Engineering*, Springer Proceedings in Earth and Environmental Sciences,
https://doi.org/10.1007/978-3-031-64423-8_34

The mineral systems mechanoactivation is one of the production areas. The efficiency is obvious for the implementation of cement-saving technologies. It is also useful for finding ways to improve the quality of binders and the manufacturability of their production.

Therefore, many enterprises are actively working to create new functional materials and products based on technogenic waste. Then enterprises introduce waste-free technologies at their production sites. This is reflected in the policies of organizations which are situated both in market relations and development strategies in the near and long term.

The study goal is to obtain an energy-efficient composite cement formulation with the addition of dump metallurgical slag. It is used for the production of concretes, mortars and dry building mixtures of various functional purposes [1–7].

2 Actual Material and Research Methods

Components were preparated according to the various modes in the laboratory grinding complex KI-0,36 ZAO "Ural-Omega", which includes a centrifugal impact mill MC-0,36.

The experimental studies results were processed by statistical and mathematical methods.

The cement stone structure was studied using microscopic observations, differential thermal analysis, dilatometry and electron microscopy.

The work uses data from geological surveys of the Orenburg region fields, research and development results of waste recycling from the region, studies of the slag dumps formulation and potential volumes of OOO "Akkermann Cement".

The research was carried out using the equipment of the Department roads and building materials, the Testing center "Orenburgstroyispitaniya", the Institute of Micro- and Nanotechnology of the Orenburg state university, "Ural-Omega," IC "Laboratoriya nanotekhnologij cementnyh sistem im. professorov A.F. Polaka i N.H. Karimova" of the Ufa State Petroleum Technical University.

3 Results and Their Analysis

Materials grinding cannot be described only as a mechanical process. Mechanical energy directly affects the molecular structure of the solid. It produces a catalytic effect on the physicochemical material conversion processes. Therefore, the result of grinding materials is not only a dispersion, but also a significant change in their structure and physical and mechanical properties [8, 9].

Two methods have been investigated for preparing the composite binder:

Table 1 The component formulation of composite cement samples

Formulations of composite cement samples, obtained by intergrinding, g			
Sample marking	clinker	slag	plaster
C0	20,00	0,00	1,30
C10	18,00	2,00	1,16
C20	16,00	4,00	1,03
C30	14,00	6,00	0,90
Formulations of composite cement samples, obtained by mixing, g			
P5	19,0	1,00	1,23
P10	18,00	2,00	1,16
P15	17,00	3,00	1,09
P20	16,00	4,00	1,03
P30	14,00	6,00	0,90
P40	12,00	8,00	0,78
P50	10,00	10,00	0,65

1. All components are intergrinded in a centrifugal impact mill to a specific surface area from 400 to 600 m^2/kg.
2. Mixing separately ground components of composite binder occurs in a high-speed blade mixer with a specific surface area from 400 to 600 m^2/kg.

Table 1 presents the component formulation of composite cement samples.

Samples dispergation was carried out in the laboratory grinding complex KI-0,36. It made it possible to obtain a homogeneous grain-size component formulation of composite cement. Results analyzing makes it possible to exclude the influence of the grain-size formulation on the properties of composite cement which is obtained by various methods [10].

Preliminary results showed that the introduction of up to 20% blast-furnace dump slag changes the rheological and physical–mechanical characteristics of stone significantly into composite cement formulation both intergrinding and mixing of pre-dispersed components.

A maximum extreme value of the stone strength is noted when the optimal content of the additive consists of 20%. Therefore, an increase in slag fraction improves the dynamics of strength gain in the early hardening composite cement periods.

This effect is explained by the formation of grains on the surface during grinding in a centrifugal impact mill. This happens in excess of active centers (mechanocomposites) which determine the hydration process of the entire system.

Table 2 presents the kinetics of stone strength gain of composite cements. The corresponding diagram is shown in Fig. 1.

The following scheme shows a hydrate formation process in composite cement which is obtained by components mixing:

Table 2 The kinetics of stone strength gain of composite cements, obtained by different methods

Marking	Sample age, day					
	1	2	3	7	14	28
C0	9,9	15,6	22,1	25,4	34,0	37,3
C20	12,5	21,3	28,6	32,3	36,4	39,9
P20	11,5	18,8	27,9	30,2	34	38,1

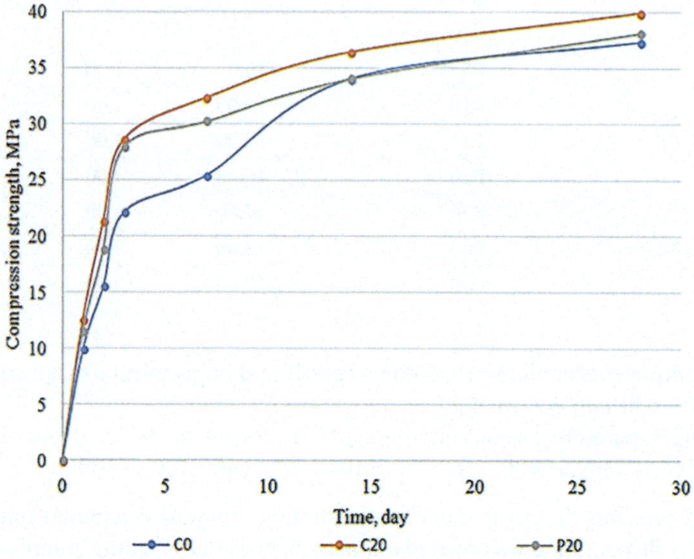

Fig. 1 The kinetics of stone strength gain of composite cements, obtained by different methods

$$\mathrm{Kl} + \mathrm{B} \rightarrow \mathrm{CH} + \mathrm{G1} \quad (1)$$

$$\mathrm{Md} + \mathrm{B} + \mathrm{CH} \rightarrow \mathrm{G1} + \mathrm{G2} \quad (2)$$

where Kl is a clinker component of composite cement; Md is a mineral component of composite cement; B is a water; CH is a calcium hydroxide; G1 is a hydrate neoplasm of clinker component (hydrosilicates, calcium hydroaluminates of various basicity, ettringite); G2 is a hydrate neoplasm of mineral component (hydrosilicates and calcium hydroaluminates).

The hydrate formation scheme has the form below for composite cement, obtained by intergrinding of components in a centrifugal impact mill. There are available mechanocomposites.

$$\mathrm{Kl} + \mathrm{B} \rightarrow \mathrm{CH} + \mathrm{G1} \quad (3)$$

$$Mk + B \rightarrow CH + G2 \tag{4}$$

$$Mk + B + CH \rightarrow G2 \tag{5}$$

$$Md + B + CH \rightarrow G1 + G2 \tag{6}$$

where Mk is a mechanocomposite in composite cement.

Mechanocomposites have a highly developed surface and a large concentration of active centers. Therefore, their interaction process is accompanied by the formation of nanostructured boundary layers with both hardening water and calcium hydroxide, which can be considered precursors for the subsequent formation of the cement stone structure.

According to the first scheme, hydration processes proceed sequentially in a composite cement components system when calcium hydroxide accumulates.

According to the second scheme, hydration processes proceed simultaneously. It contributes to the growth of the hydrate neoplasms volume in the hardening system. Ultimately, it should lead to the strength growth in the cement stone at the early strength.

The strength increase was noted in composite cements of both separate and intergrinding methods. It is explained by the presence of additional active centers. Nonetheless, strength and dynamics are more pronounced in the intergrinding method. The additional amount of mechanocomposites formation occurs by "spreading" one of the mixture components to another at the points of their friction contact.

Such layered formations are continuously crushed during machine processing. Then new formations are formed with a more complex internal structure. The mechanocomposites presence has an effect on hardening and properties in composite cement formulation. Therefore it is an important task to determine the size and concentration of these formations. Additional structural defects are created in composite cement components with grinding and the formation of mechanocomposites. There is excess energy that is accumulated in comparison with rough material [11].

The mechanocomposites volume fraction is 76% in the sampling composite cement. This introduces certain features into the process of its hardening. It is associated with the formation of nanosystems in the binder system.

The stone compression strength at the age of 28 days is equal with the strength of stone both from non-filler cement and a similar composite cement which is obtained by mixing separately crushed components. The early stone strength (at the age of 2 days) from this cement exceeds the strength of both non-filler cement and separate grinding cement.

The results correspond to the hydration scheme of composite cement containing mechanocomposites [12–15].

The main limiting factor is the low chemical activity of the silica-containing component in the composite cements hydrating process. Therefore, this component is more subject to mechanical activation. A domain slags structure is dominated by ion-covalent bonds –Si–O–Si– and a certain number of O–M + bonds with a high proportion of ion. It is necessary to break siloxane bonds Si–O–Si and ion-covalent –Si–O–M + for hydrate neoplasms formation [16, 17].

OH- anions play a determining role in the destruction of slag structures. They have a large polarizing effect, forming a donor–acceptor bond with Si4 + , Al3 + , Mg2 + atoms and breaking their bonds with oxygen. Therefore, the leading role belongs to reactions (3) and (4) in the process of composite cement hydration. So they ensure the formation of calcium hydroxide and an increase in the pH of the binder system.

The specificity of the mechanocomposites formation is the possibility of maintaining an intermediate activated non-equilibrium state in mechanocomposites for a long time. As a result, the reactivity of calcium hydroxide arising from reaction (4) is higher than reaction (3).

Therefore, water molecules form "bridges" at the interfacial boundaries of mechanocomposites. The "bridges" contribute to the hydration development process [11].

As a result of reactions (3), (4), (5) and (6), calcium hydrosilicates and hydroauminates are formed. They have a nanometer size. It leads to the development of dissipative self-organization in the binding system due to the exothermic hydration process, which provides energy inflow [18].

X-ray phase and differential thermal analyses showed that the main neoplasms were formed as a result of composite cements hydration. They are similar to the products of non-filler cement hydration. Nonetheless, the content of calcium hydroxide was reduced during composite cement hydration. It is actively involved in the above-mentioned reactions for the hydrate neoplasms synthesis.

It should be noted an increase in the content of X-ray amorphous hydrates, including calcium hydroaluminates of varying formulation and low-base C–S–H.

This formulation of neoplasms determines the non-classical mechanism of their crystallization when the structure of cement stone is formed.

Thus Fig. 2. shows the microstructure of the compositional cement stone, obtained by various methods at amplification $\times$ 600.

According to the classical theory of phase formation, a new phase nucleuses are formed as a result of sequential atoms attachment to each other. According to the two-step mechanism of crystal nucleation, the first step is substance clustering preceding the nucleus. The second step is the formation of nanoscale clusters or quatarons in crystal-forming media.

The crystal formation scheme is described below.

Firstly quatarones are formed in the crystal-forming medium. When they reach certain sizes, quatarones can transform into crystalline embryos.

Quatarones are clusters of the transition substance state. They have different from traditional aggregate states. This substance state is characteristic only for the nanoworld.

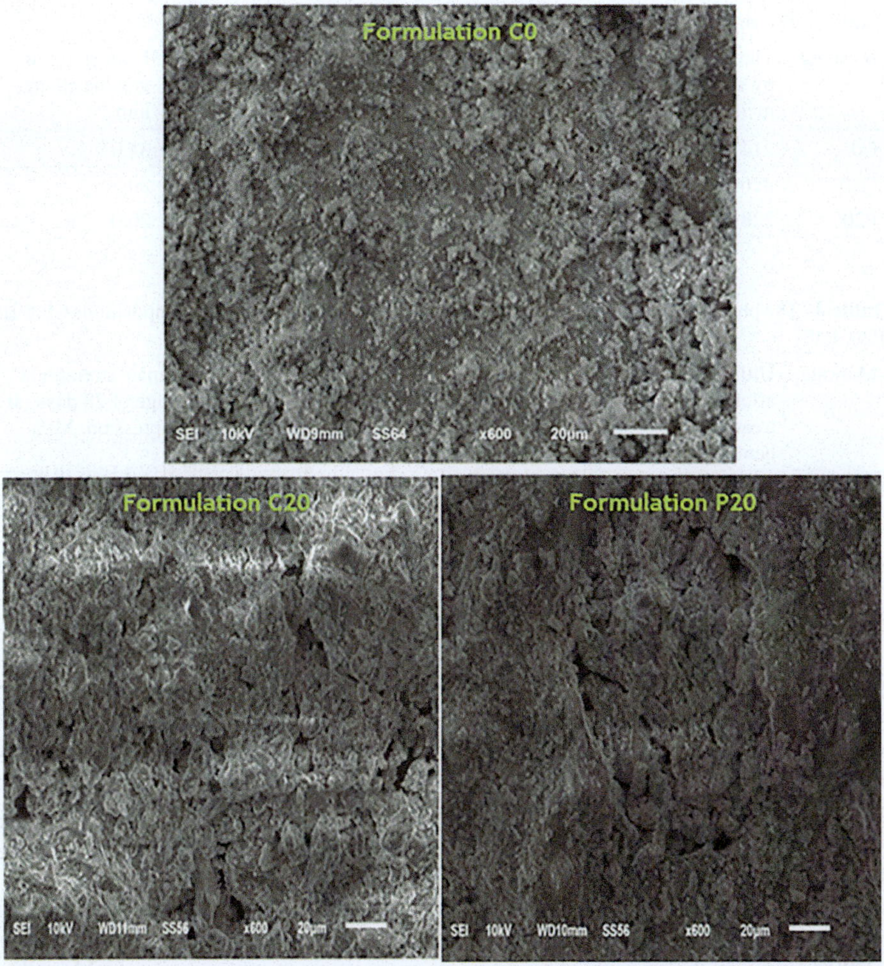

Fig. 2 The compositional cement stone microstructure obtained by various methods at amplification × 600

Quatarons ability to aggregate without fusion is a morphogenetically important property. As a result, they form structures in the form of chains, rings, tori, etc.

Subsequent crystallization leads to the formation of various morphological types of nanoscale crystalline formations. Many X-ray amorphous mineral formations also have a quataronic structure. Clusters with fractal structure are often formed in quataronic structure [19–22].

Tables 3 and 4 present a comparative analysis of the physical and mechanical characteristics of composite cements obtained by various methods using polyfraction sand.

Table 3 The rheological characteristics of composite cements (C0, C20, P20)

Marking	The fineness of grinding by specific surface area, m^2/kg	True density, g/sm^3	Normal density, %	Start of setting, min	End of setting, min	Uniformity of volume change, mm
C0	214	3,13	25,00	180	230	34,0
C20	226	3,10	25,50	180	220	20,5
P20	228	3,08	25,20	230	270	26,0

Table 4 The physical and mechanical characteristics of composite cements (compositions C0, P10, P20, P30)

Marking	Ultimate strength after heat treatment, at bending, MPa	Ultimate strength after heat treatment, under compression, MPa	Ultimate strength at the age of 28 days, at bending, MPa	Ultimate strength at the age of 28 days, at compression, MPa
C0	4,74	27,46	5,19	35,89
C20	5,57	29,06	6,51	42,00
P20	4,70	27,52	6,38	37,41

Table 3 shows the grain size distribution of composite cements with a slag content of 20%. It is obtained by various methods which are represented by a fairly narrow specific surface area range from 214 to 228 m^2/kg. Slag additive by various methods did not affect the change in the normal density. At the same time, the setting time increased in the composition obtained by mixing the components after separate grinding. The volume change with respect to the non-filler cement decreased in both uniformity methods. The volume change is minimal on composite cement obtained by joint grinding.

The physical and mechanical characteristics of composite cements are higher when cements are obtained by different methods. Moreover, slag additive consists of 20% in comparison with non-addition composition at compression and bending both at heat-and-moisture treatment and at normal hardening conditions. Composite cement obtained by components intergrinding will have the most significant strength gain (Table 5).

The data show that the intergrinding components is the most efficient way to obtain composite cement.

4 Conclusions

The practicability of using dump metallurgical slags includes economic and environmental components. The economic component consists of fees for renting land, which are occupied by dumps. That dumps have not found sales of dropouts and

Table 5 The strength gain of composite cements obtained by various methods for non-additive composition

Marking	Strength increase with respect to non-filler cement, %			
	Ultimate strength after heat treatment, at bending	Ultimate strength after heat treatment, when compressed	Ultimate strength at the age of 28 days, at bending, MPa	Tensile strength at the age of 28 days, when compressed
C0	4,74	27,46	5,19	35,89
C20	+ 17,5	+ 5,8	+ 25,4	+ 17,0
P20	0,0	+ 0,2	+ 22,9	+ 4,2

fines for emissions into the atmosphere. Economical effectiveness has maximum value capture of available resources.

The economic effect can be estimated of composite cements and heavy concrete's use in comparison with the heavy concretes of slag portland cement. The economic effect will be provided by the difference in the cost of these concretes. The cost will be the same for building the factory and purchasing equipment for concrete production.

Studies prove that the use of a centrifugal impact mill for crushing cements allows to reduce energy costs by 10,5–30%. The parts mortality is 7 times less in comparison with a traditional grinding system.

Acknowledgements The work was prepared as part of a grant in the form of a subsidy for the implementation of the Priority-2030 strategic academic leadership program within the framework of the federal project "The Development of Integration Processes in the Field of Science, Higher Education and Industry" in the direction of "Innovation in environmental design".

References

1. Dergunov SA, Garcavy MS, Satyukov AB, Serikov SV, Odincova DS (2022) The man-made mineral wastes in production of construction materials of various functional. In: international Russian Conference on Ecology and Environmental Engineering, RusEcoCon, Sochi, 16–20 May 2022
2. Fedorov AS (2021) "Clinker 732" of the factory "Akkermann cement". Cement and it's use 2:45–47
3. Morozov KM, Ilina IS, Shilovskii VA, Dergunov SA (2017) New cement production horizons. Industrial and civil engineering 11:77–80
4. Kerton F (2006) Slag Cements Market Outlook in Europe. Cement and it's use 5:12–17
5. Ivaschenko UG (2019) Resource and energy efficient technologies in the construction complex. In: The 7th international scientific and practical conference, Saratov State University, 2019
6. Zbigniew G (2019) Fly ash and slag. Cement and Concrete Research. 124:1058264
7. Serikov SV, Dergunov SA, Makaeva AA, Satukov AB (2022) Development of dry construction mixture composition. In: The 4th Baikal materials science forum. Materials of Russian scientific and practical conference Ulan-Ude, 1–7 July 2022

8. Polak AF, Babkov VV, Andreeva EP (1990) Hardening of mineral binders. Bashknigoizdat, Ufa
9. Rahimbaev SM, Pospelova EA, Gridchin AM (1998) Slag and ash qualimetry. In: Construction news from universities 7:41–45
10. Garkavi MS, Vorobiev VV, Kushka VN, Svitov VS (2003) Modern equipment for grinding and classification of materials. Meddenger BGTU 6:280–284
11. Lapshin OV, Boldireva EV, Boldirev VV (2012) Role of mixing and dispersion in mechanochemical synthesis (overview). Journal of inorganic chemistry 3:402–424
12. Garkavi MS, Dergunov SA, Makaeva AA, Serikov SV, Latipova AS (2023) Functional fillers and precursors for composite materials. In: Current problems of modern science, technology and education: abstracts of 81st international scientific and technical conference, Magnitogorsk State Technical University named after Nosov, 17–21 April 2023
13. Garkavi MS, Artamonov AV, Kolodezhnaya EV, Dergunov SA, Serikov SV (2023) Formation of nanosystems during hardening of centrifugal impact grinding composite cements. Constructiom materials 3:39–42
14. Garkavi MS, Artamonov AV, Stavceva AV, Kolodezhnaya EV, Dergunov SA, Serikov SV (2021) Modeling of structural transformations during grinding of composite cement. Constructiom materials 11:41–46
15. Garkavi MS, Dergunov SA, Serikov SV (2021) Formation of composite cement structure during grinding. Constructiom materials 10:65–68
16. Panfilov MI, Shkolnik YS, Orininskii NV (1987) Slag processing and waste-free technologies in metallurgy. Metallurgy, Moscow
17. Gorshkov VS, Aleksandrov SE, Ivaschenko SI, Gorshkova IV (1985) Complex processing and use of metallurgical slags in construction. Stroiizdat, Moscow
18. Garkavi MS (2013) Thermodynamics of hardening of binding systems. Theoretical Principles and Technological Applications. Berlin Palmarium Academic Publishing, Berlin
19. Ashabov AM (2011) Quataronic concept: basic ideas and some applications. News of the Komi scientific center 3:70–77
20. Ashabov AM (2019) About the properties of pre-natal (protomineral) clusters. Reports of the Academy of Sciences 5:47–50
21. Garkavi MS, Artamonov AV, Kolodezhnaya EV, Dergunov SA, Serikov SV (2023) Mechanochemical synthesis of composite anhydrite binders. Constructiom materials 10:52–57
22. Satukov AB, Dergunov SA, Orehov SA (2019) Nanomodified composite binder. Orenburg State University, Orenburg

Assessment of Natural Resource Potential of Territories for Solid Waste Management Facilities Construction

E. V. Gubar, O. A. Samarina, O. V. Tupicyna, and V. N. Pystin

Abstract Development of an integrated waste management system implies reconstruction of existing and construction of new facilities for treatment, utilization and neutralization of waste and disposal of non-utilizable fractions. It is impossible to ensure uninterrupted operation of solid municipal waste (SMW) management without construction of SMW management facilities. The purpose of this research is to study the construction of SMW management facilities. SMW landfills are considered to be complex natural and technical systems. The evaluation criteria of technological and ecological solutions on waste management measures realization are developed for objects at the investment stage of life cycle. The comparative assessment of geo-ecological, technical-economic and operational parameters of SMW disposal facility construction is carried out by developing four variants of technical and technological solutions for a reference landfill with disposal capacity up to 30 thousand tons per year with a storage area of 3 ha. The most significant geo-ecological parameters providing a comprehensive assessment of landfill design and construction were identified and validated: operation period; capacity; landfill capacity, landfill capacity for compacted waste.

Keywords Circular economy · Sustainable development · Environmental safety · Geo-ecological assessment · SMW management facilities

1 Introduction

Continuous economic growth and ongoing urbanization are the key reasons for increasing waste volumes every year. Besides, the increasing variety of food and non-food products, their assortment and packaging types causes an increase in the SMW formation volumes.

E. V. Gubar (✉) · O. A. Samarina · O. V. Tupicyna · V. N. Pystin
Samara State Technical University, 244, Molodogvardeyskaya St., Samara 443100, Russia
e-mail: e.v.gubar@mail.ru

L. V. Radionova and D. V. Ulrikh (eds.), *Advances in Ecology and Environmental Engineering*, Springer Proceedings in Earth and Environmental Sciences,
https://doi.org/10.1007/978-3-031-64423-8_35

Managing SMW facilities as complex nature and technical systems in order to ensure rational nature use and their sustainable development is a complicated problem of ecological safety of regions. Assessment (audit) of measures on waste management facilities construction should be carried out with the use of technical and economic indicators that measure the efficiency of construction and installation works, determine the limit of capital investments, compliance of technological and technical solutions with regulatory and legal acts, redundancy of design solutions.

The primary task to be solved when designing waste disposal facilities is the selection of the land site, taking into account the requirements of regulatory and technical documents (absence of sanitary protection zones of sources of household and drinking water supply, water protection zones, specially protected natural areas, mineral deposits; highway proximity, etc.) and environmental technology capacity of the territory. It is the choice of a land site as part of the implementation of measures for the SMW landfills construction that is the most responsible stage to ensure economic feasibility and capacity availability of the designed facilities in addition to the abovementioned [1].

To ensure the formation and development of an integrated system of SMW management it is necessary to consider the negative impact of these factors on the environment (e.g. exhaust emissions, fuel and lubricant leaks, noise pollution, violation of the integrity of soil cover, destruction of green spaces, etc.), but also their economic component [2–4].

To assess the effectiveness of technological solutions of SMW disposal facilities with limited land site area, a system of geo-ecological rationale for the location of SMW disposal facilities with the use of an assessment tool for design solutions is proposed. The quantitative parameters obtained as a result of the study provide criteria for the redundancy of design solutions for the SMW landfill construction. The proposed tool can be applied to prove the marginal cost of financial and economic models of such facilities construction.

2 Materials and Methods

SMW management facilities are known to have four main life cycle stages: investment, operation, reclamation and post-reclamation stages [5].

According to the purpose of this study, the criteria for assessing the economic, technological and environmental parameters of solutions for the landfill construction are developed in relation to the objects at the investment stage.

In compliance with the current legislation, the investment stage includes:

- Selection of the land site for the facility location in accordance with the requirements of the current regulatory legal acts of the Russian Federation;
- Engineering surveys for preparation of design and estimate documentation;
- Development of design and estimate documentation;

- State environmental and construction expertise of design and estimate documentation and engineering survey results;
- Construction of engineering and environmental protection facilities [6–9].

Preliminary assessment of the work cost structure and value shows that the earthworks cost constitutes 30% of the total cost of SMW landfill construction works. In addition, these works are the most labor-intensive and conditioned by natural, climatic and infrastructural peculiarities of the construction site. Excavation works include excavation of soil, its further transportation and/or relocation, as well as the construction of temporary or permanent earth structures, namely trenches for laying engineering communications and foundations.

Table 1 presents the initial data of the research object regarding which the development of technical and economic indicators was carried out under the conditions of different constructive schemes of storage sites organization.

This paper applies methods of system analysis of data presented in the regulatory legal acts in the field of environmental management.

Technical and technological solutions of the SMW disposal facilities are developed in accordance with the current regulatory and legal acts [6–9]. Calculation of the earth mass plan (cartogram) is carried out in the program complex GeoniCS 2015. GeoniCS 2015 software package allowed to automate design and survey works and create a topographic plan of the landfill at a scale of 1:1000.

The capacity of the SMW landfill is calculated for compacted waste. It is assumed that during disposal SMW is compacted by means of a compactor roller. As a result, the density of technogenic formation of the disposal site will be up to 1.3 t/m^3.

Prior to engineering-geodetic surveys, collection, analysis and processing of engineering survey materials of previous years, as well as visual inspection of the work site were carried out. Engineering topographic survey was performed by RTK satellite positioning method. This survey method was used due to the openness of the terrain,

Table 1 Main parameters of the research object

Indicator	Unit of measurement	Value
Capacity	Thousand tons/year	Up to 30
Land site area	ha	6.55
Disposal area	ha	3.00
Density of uncompacted waste	Tons/m3	0.60–0.65
Density of compacted waste	Tons/m3	1.10–1.30
Height of tiers of solid waste storage	m	up to 2.5
Thickness of sanitary backfill layer	m	0.15
Minimum layout slope	m	0.005
Seismic activity	Points	5
Groundwater level depth	m	5.0–7.0
Soils at the land site		Medium loamy soils

as well as higher productivity of this method of surveying. Taking into account the peculiarities of the object of engineering and geodetic survey, special attention was paid to compliance with the requirements of regulatory documents on environmental protection and fire safety. When performing engineering-geodetic works satellite geodetic systems "JAVADTriumph-1" 08,463, "JAVADTriumph-1" 09,138, "JAVADTriumph1" 25,296 were used.

The cost of works on landfills construction with different technological solutions of waste disposal sites (maps) is estimated according to unit price collections, being a part of the pricing system.

3 Results and Discussion

The SMW landfill is a complicated natural-technical system, which environmental facilities complex provides SMW storage and disposal. Engineering infrastructure of SMW disposal facilities should isolate waste from groundwater; ensure collection, drainage and treatment of surface runoff and filtration runoff from the surface of the isolation screen; provide for landfill degassing [10–13].

The technical parameter of the SMW landfill is a set of main and auxiliary facilities supporting the waste disposal site operation in accordance with its functional zoning:

1. Production zone including:

   - Waste disposal site;
   - Mineral soil storage area calculated on the basis of the monthly demand for technological isolation of work cards;

2. Auxiliary (household) zone including:

   - Auxiliary and service facilities (radiation and weight control checkpoint; weighing room; disinfection bath; clothing room; dining room; fire-fighting water storage tank; monitoring wells);
   - Energy facilities;
   - External networks and facilities of water supply, wastewater disposal (filtrate storage pond; surface wastewater storage pond; filtrate sewage pumping station (SPS); surface wastewater pumping station (SPS); filtrate treatment station; surface wastewater treatment facilities; treated wastewater storage tank; concentrate storage tank) [14, 15].

The projected SMW landfill is assumed to receive the screenings of SMW sorting and non-recyclable ballast fractions, that excludes landfilling of waste with resource potential. The volume of insulating materials used for layer-by-layer sanitary backfill is 7% of the disposed waste volume. This principle is implemented in order to reduce the negative impact of waste disposal facilities on the environment due to the extraction of waste fractions in the secondary raw material form.

Ballast and screenings are delivered by means of multilifts of the waste sorting station as the tanks are filled: ballast is delivered to the waste disposal map, screenings containing organic matter are delivered to the organic waste composting area after sorting.

Table 2 presents data on the SMW morphological composition. The average values correspond to the morphological composition of waste entering the landfill.

Waste is placed on the daily map by the "thrusting" method. Within the boundaries of daily maps adjacent to the forming berms, the waste is placed by the "bumping" method. Bulldozers push the waste onto the map, forming layers up to 0.5 m high with subsequent compaction to a 0.1–0.2 m thick layer with 4 multiple passes of the compactor along the trail. Due to 15–20 compacted layers a roll with a gentle slope of 2.35–2.40 m above the level of the unloading site of garbage trucks is formed. When compacting waste, its density increases from 0.6–0.65 to 1.1–1.3 t/m^3. As a rule, the compactor moves along the long side of the daily working map.

Formation of the map is accompanied by the creation of an end slope with a safety slope of 1:3, necessary for the subsequent laying of reclamation screens.

At the end of the shift, the compacted waste is sanitary backfilled within the boundaries of the daily map using a bulldozer. Technogenic soils of the composting site such as organic-mineral soil are used as backfill material. The thickness of the sanitary backfill layer is 0.15 m. Cascades of filled daily maps form successively placed tiers up to 2.5 m thick.

The degassing of the landfill body is envisaged to prevent uncontrolled emission of landfill gas into the environment in order to ensure environmental safety, fire and explosion safety of the landfill. Due to the small volume of SMW landfill and, thus, relatively low gas formation, the project employs a passive degassing method. Passive degassing of the landfill body is carried out by means of a network of gas-draining horizontal corrugated pipes connected by intermediate gas pipelines, through which the gas is gathered and released into the atmosphere through the degassing pipe. The network of gas-draining horizontal pipes is arranged in each separate waste layer.

Table 2 Morphological composition of SMW entering the landfill

Component	Content, % by weight	
	Range	Average value
Paper	10.3–62.5	38.6
Food waste	10.7–59.1	27.9
Wood	0.0–7.2	2.1
Textiles	0.3–17.5	1.5
Leather, rubber	0.0–9.5	1.2
Polymers	11.5–32.0	15.1
Metals	0.8–5.9	2.5
Glass	0.0–7.6	2.9
Other (screenings less than 15 mm)	7.3–19.9	8.2

Trenches of 1 m depth are made in the compacted SMW layer, at the bottom of which a corrugated pipe is placed, with the distance of 14.5 m between the pipes. The pipes are connected to each other in a common network, on which a vertical degassing pipe is arranged [16].

Garbage trucks enter the territory of the SMW landfill through a checkpoint, where automatic radiometric, visual and weight control of vehicles is performed. Before weighing the garbage truck passes the post of automatic stationary continuous radiation control of transport. In case of exceeding the radiation level of the vehicle, the post warns the weighing operator with an audible signal and information on the display about the exceeding and prohibition to enter the facility. If the radiation level does not exceed the threshold value, the car goes to the weighing station. At the moment of weight stabilization, the car is photofixed, its license plate number is recognized and photographed. Information about the weight of the car is saved in the database. The weighed vehicle is directed to the section corresponding to its load.

After the vehicle returns from the unloading site, it is weighed for a second time, the net weight is calculated and a weighing card is issued. If necessary, a weighing report is printed, indicating the vehicle, the driver and the weight.

At the exit from the landfill before the weighing station the garbage truck passes the control and disinfection reinforced concrete bath for wheel disinfection.

Assessment of measures for the SMW landfill construction should be carried out using technical and economic indicators measuring the efficiency of construction and installation work, specifying the limit of capital investment, compliance of technological and technical solutions to regulatory acts, redundancy of design solutions [17, 18].

In this research the design solutions for waste disposal site organization are divided into the following types (Figs. 1, 2, 3 and 4):

1. Quarry-type landfills (buried in the relief);
2. Bulk type landfills (flat, high-rise);
3. Landfills of combined type (quarry landfills with subsequent formation of bulk dumps).

Screening of the disposal area is carried out at the following levels of the layer section of the land fill (Fig. 5):

- at the base of the landfill—laying of waterproof base;

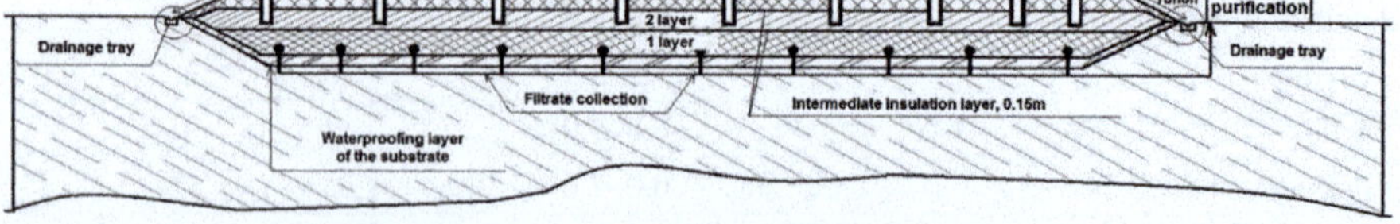

Fig. 1 Technological solutions for the design of the quarry-type landfill

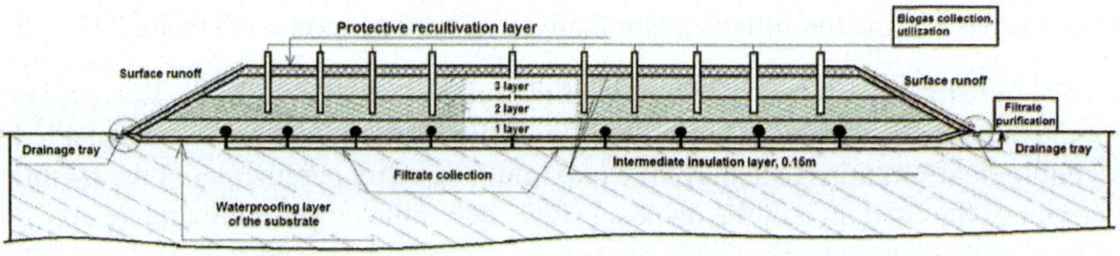

Fig. 2 Technological solutions for the design of the bulk-type landfill

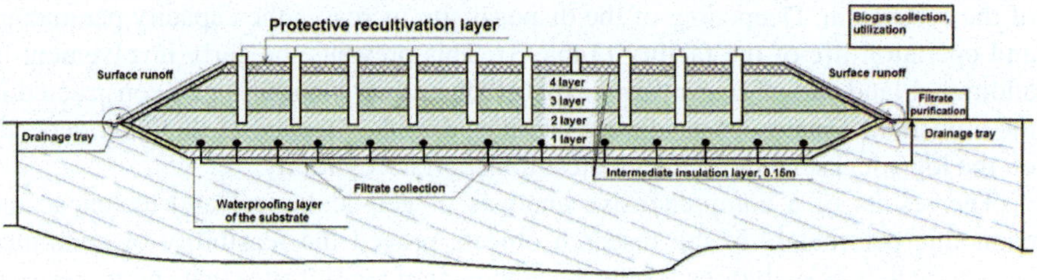

Fig. 3 Technological solutions for the design of the combined type landfill

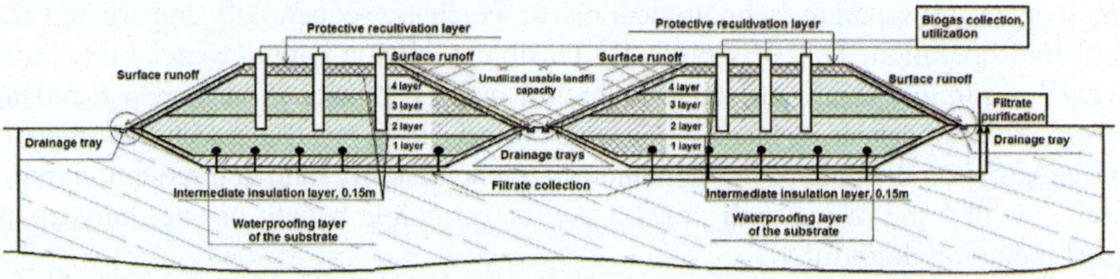

Fig. 4 Technological solution of landfill site design of combined type (quarry with bulk dump) divided into operation maps

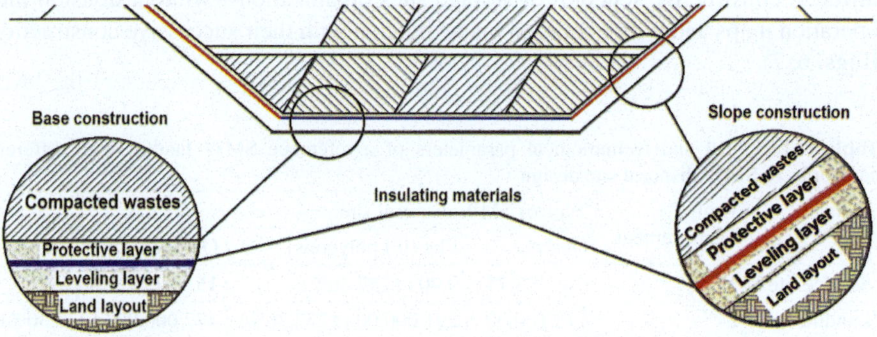

Fig. 5 Anti-filtration screen design

- at the base of the landfill excavation an anti-filtration screen is installed [19, 20].

The general operational parameters of the reference facility (Table 3), calculated for the three implementation options, differ significantly. Whether it is reasonable for technological solutions to achieve the maximum capacity parameters of the facilities per unit of the used area of the disposal site is determined by a comparative analysis of operational parameters.

The degree of impact of economic activity on the geological environment is a negative environmental factor due to the removal of geological resources disturbance of the ecosystem. Deepening of the disposal site increases the capacity parameters and operation life of the facility (Table 3). This prevents the early involvement of additional land sites for a similar, "exhausted capacity" landfill to be constructed in the immediate vicinity of the closed facility, since the location of the latter depends on the logistical scheme of the greatest transport accessibility.

The results of a comprehensive analysis of geo-ecological and technical and economic parameters of the research objects proved the feasibility of combined (quarry with bulk) solutions for the development of landfill sites that should serve as a basis for technological solutions of prospective landfill facilities.

Structural scheme of organization of storage sites may include a unified landfill map or several separate technological maps, which are sequentially constructed and put into operation. It is economically feasible to divide capital expenditures into years of commissioning of separate landfill maps with reduction of non-recurring costs in the first year of facility implementation. However, the forecast changes in price levels on the date of implementation of each facility map and reduction in the unit cost of 1 m3 of waste disposal make the economic benefit of one-time capital cost reduction doubtful.

A criterion assessment of the feasibility and validity of the choice in technological solutions of application / failure of landfill site mapping is required. For the combined type landfill the development of technical and economic indicators was carried out in different constructive schemes of storage sites organization—without division into operation maps and with operation maps division with their successive construction (Figs. 6, 7).

Table 3 Technical and technological parameters of a reference SMW landfill with different categories of SMW disposal site design

Indicator	Unit of measurement	Type of SMW disposal site		
		Quarry	Flat (fill polygons)	Quarry with bulk dump
Service life	Years	3.77	9.70 (↑257.30%)	15.92 (↑422.30%)
Landfill capacity	t	113 100.00	291 000.00 (↑257.30%)	477 000.00 (↑422.30%)
Landfill capacity for compacted waste	m^3	86 991.00	223 732.00(↑257.30%)	367 426.00(↑422.30%)

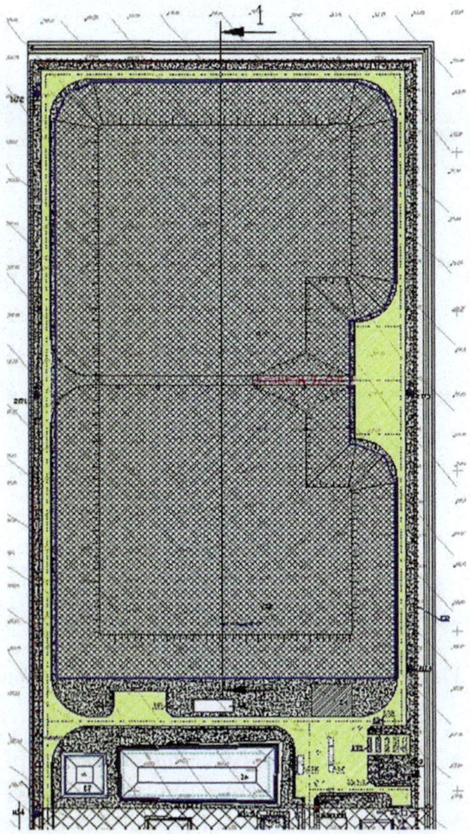

Fig. 6 General plan of the facility without operation maps division

In the examined objects the geo-ecological parameter of the economic activity impact degree on the geological environment is considered from the maximum possible deepening to the refusal from deepening of the SMW disposal site.

4 Conclusions

Based on the estimated marginal values of technical and economic indicators (Table 3), we made conclusions about the feasibility of construction of combined type landfills under other equal conditions. The developed criteria can be applied to assess the redundancy of technological solutions in the construction of SMW disposal facilities.

The comparative analysis of the main technical and economic indicators of putting into operation the stages of waste disposal sites, the cost indicators and facility capacity allowed to draw the following conclusions:

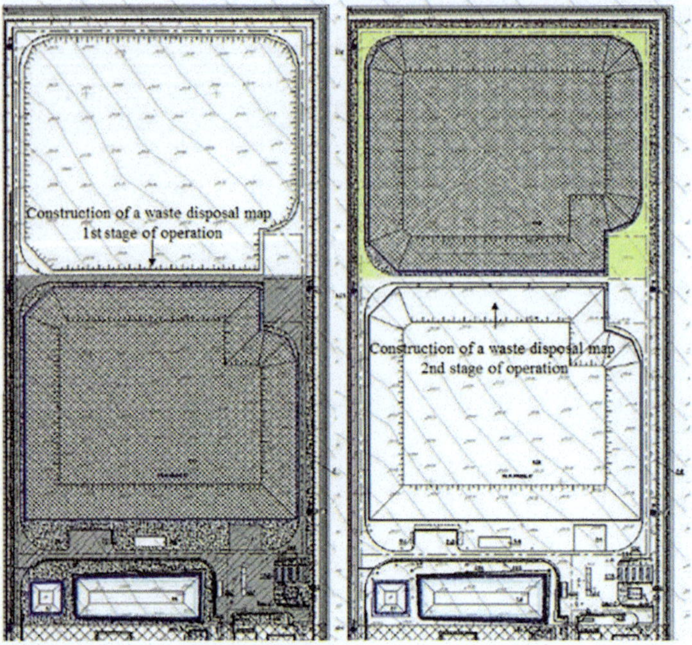

Fig. 7 General plan of the facility divided into operation maps with their successive construction

- the possibility of reducing the amount of financial capital expenditures at different constructive schemes of organization of storage sites is absent;
- one-time reduction of capital expenditures in the first year of facility construction by realizing the disposal site according to the maps leads to an increase in the unit cost of the unit volume of useful capacity of the landfill up to 45%;
- the increase in the cost per unit of useful capacity of the landfill results from a 1. 2 times reduction in the facility's capacity due to irrational use of the area of the allocated land site for the facility.

The obtained technical and economic indicators allow to justify the feasibility and assess the cost-effectiveness of investments for each constructive scheme of landfills construction, but also to identify the redundancy of technological solutions, which significantly increased financial and time costs.

The developed model allows to solve the following tasks:

- To form a set of main design objects and determine the marginal cost of their construction under different forecast scenarios;
- To assess technical and economic indicators of design solutions for SMW management facilities construction;
- To optimize financial and time costs of SMW management facilities construction;
- To assess suitability of land sites for SMW management facilities construction;

- To assess the impact of the construction site location on the marginal cost of financial and economic models of SMW treatment facilities;
- To form analogous objects for the development of the waste management industry;
- To assess the efficiency of technological solutions of SMW treatment facilities under the restriction of the land site area.

References

1. Elsakov VV, Lapteva EM, Vasilevich MI et al (2022) Justification of potential risks of placing a landfill of solid municipal waste. Theoretical and applied ecology 4:72–79
2. Voronkova TV, Viskov MV (2013) Analysis of changes in the capacity and service life of the landfill, taking into account the output of biogas. Bulletin of the Perm National Research Polytechnic University. Urbanistics 2(10):131–139
3. Gracheva EV, Voronkova TV (2013) Theoretical foundations of the formation of the water balance of the landfill of solid household waste. Modernization and scientific research in the transport sector 2:97–105
4. Kantor GYa, Syrchina NV, Ashikhmina TYa (2022) Carbon balance modeling of municipal solid waste landfills. Theoretical and applied ecology 1:198–204
5. Bykov DE, Tupitsyna OV, Chertes KL (2007) Reclamation of organo-mineral waste. Samara State Technical University, Samara
6. Federal Law "On Production and Consumption Waste" dated June 24, 998. No. 89-FZ (2023) ConsultantPlus Legal Reference System. https://www.consultant.ru/document/cons_doc_LAW_19109/?ysclid=lsn5cy4f1q865780718. Accessed 11 Sept 2023
7. Decree of the Government of the Russian Federation of October 12, 2020. No. 1657 "On Unified Requirements for Facilities for Processing, Recycling, Neutralization, and Disposal of Solid Municipal Waste" (2023) The Russian Government. http://static.government.ru/media/acts/files/1202010160038.pdf. Accessed 11 Sept 2023
8. A set of rules 320.1325800.2017 amended. 1 "Landfills for municipal solid waste. Design, operation and reclamation" (2023) Ministry of Construction, Housing and Utilities of the Russian Federation. https://minstroyrf.gov.ru/docs/182137/. Accessed 11 Sept 2023
9. Information and technical guide to the best available technologies 17-2021 "Disposal of production and consumption waste" (2023) Bureau of the Best Available Technologies. https://burondt.ru/NDT/NDTDocsDetail.php?UrlId=1669&etkstructure_id=1872&ysclid=lsn1t6rthy112982812. Accessed 11 Sept 2023
10. Sereda TG, Faizrakhmanov RA, Kostarev SN (2006) High technology in the design of artificial waste storage ecosystems. Perm State University, Perm
11. Zavizion YuV, Glushankova IS, Slyusar NN (2016) The use of synchronous thermal analysis to assess the stability of municipal solid waste buried in landfills. Ecology and Industry of Russia 6:43–49. https://doi.org/10.18412/1816-0395-2016-6-43-49
12. Levin EV, Sagitov RF, Gamm TAet al (2016) Ecological justification of the location of a solid waste landfill. Proceedings of the Orenburg State Agrarian University 3(59):182–184
13. Voronkova TB, Chudinov SYu (2013) Theoretical aspects of the water balance of landfills for solid household waste disposal with a filtrate recycling system. Theoretical and applied ecology 1:13–16
14. Konygin AA (2011) Complex technology of filtrate purification of landfills of solid household waste disposal. Academia. Architecture and construction 4:105–109
15. Vaisman YaI, Voronkova TV, Chudinov SYu (2013) Flow and quality management of filtration wastewater from landfills. Ecology and Industry of Russia 10:26–30
16. Sereda TG, Kostarev SN, Kochinov YuA et al (2023) Modeling of resource-saving treatment of solid municipal waste. Housing and communal infrastructure 1(24):95–104

17. Tupicina OV, Samarina OA, Gubar EV et al (2020) Ecological and economic regulation of the treatment of MSW in the Samara region. Safety, protection and protection of the natural environment: fundamental and applied research, pp 335–339
18. Gubar EV, Tupicina OV (2023) Development of financial and economic models of MSW treatment facilities on the territory of the Volga Federal District. Actual problems of subsoil use, pp 393–395
19. Voronkova TV, Chudinov SY (2013) The filtrate formation management system at the landfill. Solid household waste 8(86):36–41
20. Voronkova TV, Vaisman YaI, Chudinov SYu (2012) Recycling of filtrate at landfills of solid household waste. Bulletin of the Perm National Research Polytechnic University. Urbanistics 3(7):83–91

Automated Method of Sorting Municipal Solid Waste

M. A. Lemeshko and N. N. Semenov

Abstract The article is devoted to the problem of waste management, in particular solid municipal waste. One of the most important problems in waste management is the task of sorting. The effectiveness of the generalizing concept of waste management depends to a greater extent on the efficiency and cost of this technological operation. Assessed by such indicators as the productivity of the enterprise, the depth of sorting and recycling of waste, economic efficiency, safety and comfort of work. Among the most important problems of municipal solid waste recycling enterprises is the task of choosing a rational way to sort waste according to the degree of automation and technological equipment. The article discusses some trends in the creation of systems and methods for sorting municipal solid waste. The concept of using a human operator performing the functions of a system for recognizing and identifying sorted objects is described. Some variants of sorting lines are given with an assessment of sorting methods, according to the factor of their rational application. Options for improving automated sorting lines are proposed, taking into account the economic efficiency of a rational method for recognizing sorting objects. The expediency of using combined sorting methods using manual and machine sorting methods is substantiated. It is proposed to use robotic sorting lines with human participation in the recognition of sorting objects. The materials of the article are intended for specialists dealing with waste management problems.

Keywords Sorting · Waste · Automation tools · Recognition systems · Sensors · Manipulators

M. A. Lemeshko (✉) · N. N. Semenov
Institute of Service and Entrepreneurship (branch) of the DSTU in Shakhty, 147, Shevchenko St., Shakhty 346500, Russia
e-mail: Lem-mikhail@ya.ru

L. V. Radionova and D. V. Ulrikh (eds.), *Advances in Ecology and Environmental Engineering*, Springer Proceedings in Earth and Environmental Sciences,
https://doi.org/10.1007/978-3-031-64423-8_36

1 Introduction

Waste management, including sorting of solid municipal waste, is one of the important, modern tasks of safe life support for the population. In Russia, about 40 million tons of municipal solid waste (MSW) is generated annually, and at present, landfill disposal is mainly used. At the same time, the most negative impact of these wastes on the environment is manifested, including environmental pollution with toxic elements and compounds. Most of the volumes of these wastes should be recycled and used as secondary raw materials [1, 2]. However, in Russia this issue began to be resolved relatively recently. One of the most time-consuming and expensive stages in waste management is the task of sorting them, including the preliminary separation of some components, for example, oversized, sand, etc. For this, as a rule, separate machines are used that perform narrow technological tasks. After preliminary sorting of the waste, a detailed separation of useful components from the total mass of waste is then carried out. For this, both a manual, non-automated method and an automated sorting method, including robotic technologies, are used [3–5].

The software-controlled robotic manipulators used, equipped with well-known gripping devices of various designs, are designed to select individual recognized waste objects from the total mass of waste, moving them into storage containers. Robotic manipulators use grippers designed to capture and move various types of waste, such as: plastic of various configurations, film elements, cardboard, paper, textile objects and objects made of other materials of various shapes. Gripping devices move objects from a general waste conveyor to specialized conveyors or storage containers. For the gripping device to operate, its executive body must be oriented in relation to the object being grabbed. Orientation can be performed by the operator with a visual approach of the gripper to the object, or work in automatic mode according to the coordinates of the object.

Various systems and complexes for waste sorting are known. For example, a complex for sorting and processing solid household waste [6]. This complex uses the manual labor of sorters. The sorter performs the function of recognizing a waste object on a conveyor, grabbing it by hand or hands and moving it to specialized conveyors or other storage facilities according to the type and type of each waste object. Specialized machines for sorting waste, equipped with special gripping devices, are also known.

Each of the grippers is designed to grip specific objects on a waste conveyor. For example, the well-known design of a gripper for sorting waste [7]. This gripper is made in the form of rotating spikes with sharp tips, which grip wood, plastic, rubber, books, cardboard, thin-walled metal cans.

When sorting waste, the main task is to separate ingredients from the total mass of waste for the purpose of their further processing as secondary raw materials. Some waste is useful for reuse or to produce another useful product that can be used, such as fertilizer or fuel. The task of separating organic waste is also important [8].

Various ingredients in a waste mixture are characterized by varying degrees of separability from the total mass of waste, varying degrees of accuracy in their recognition by people or scanning electronic means at the stage of preparation for mechanized or manual movement into storage containers. And at the same time, the problem of payback of the equipment and technologies used is relevant, first of all, in the process of recognizing sorting objects [3–5].

It is known that the simplest and relatively cheapest method is manual waste sorting, which is currently used in Russia at domestic waste processing plants. However, this method allows for working conditions that are not safe for the health of sorters, and the dependence of the effective separation of waste ingredients on the fatigue of human sorters.

Obviously, it is necessary to rationally automate the process of waste sorting, so that the payback of the automated recognition system is ensured [9–11]. The rationality of the sorting process also implies solving environmental problems [12–15].

The effectiveness of manual sorting largely depends on the organization of work at the stage of waste collection and transportation. In automated sorting, individual auxiliary operations (feeding of material to the sorting conveyor, pre-screening by size, grinding) are mechanized, but recognition is carried out by a person. The disadvantages of the manual and automated sorting method are higher safety requirements and incomplete extraction of various solid waste fractions [16, 17].

When using manual sorting, it is necessary to take into account some physical limitations in human capabilities. Such as eye fatigue, impaired attention, physical fatigue, psycho-physical overstrain. It is relatively difficult for a person to isolate an object with geometric dimensions less than 50 mm from the waste stream on the conveyor.

Automatic sorting is certainly a promising area for improving technologies, equipment, and software in the field of waste management. On fully automatic material sorting lines, the entire waste sorting process (identification of selected materials and their separation from the general flow) takes place without the participation of personnel. As a rule, the technology of automatic sorting is based on the use of sensors for the optical determination of materials by irradiating the waste stream with radiation with a certain wavelength and subsequent spectral analysis of radiation reflected from the surface of the material [18].

With poor-quality sorting, the areas with buried waste in the ground (landfills) will continue to grow. The negative impact of landfills on the environment is caused by emissions of pollutants into the atmosphere, the discharge of toxicants into water sources and the formation and accumulation of significant amounts of waste, directly or indirectly affecting all natural environments [3]. Using the example of the Rostov region, the Russian Federation, let's consider information on the volumes of waste generated by sectors of the economy, see Fig. 1.

Analysis of the diagram shows that municipal solid waste (MSW) in terms of waste generation volumes ranks eighth in the list of industries in the Rostov region. And the largest volume of waste is generated by the following industries: Provision of electric energy, gas and steam; Mining As you can see, for the Rostov region waste sorting will have its own nuances. In general, across the country, the composition of waste will depend significantly on the region and on the structure of sources generating waste. However, for the sorting problem this does not matter much. The

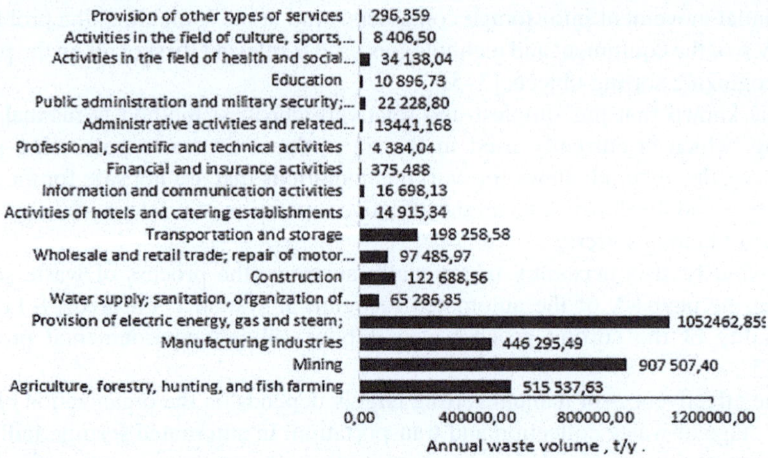

Fig.1 Information on waste by sectors of economic activity in Rostov region for 2022

waste sorting systems and methods discussed are equally applicable to construction waste, municipal solid waste, and industrial waste.

Considering the economic side of the issue, which is related to the environmental problem, two approaches to sorting quality should be contrasted. Let's call these approaches economic and environmental. The "economic" approach involves extracting the most "profitable" components from the total mass of waste. For example, materials with a high calorific value and/or components with a high sales value and low content of harmful substances. This approach is not environmentally beneficial—it will lead to an increase in the volume of waste to be disposed of in landfills. Another approach is ecological, which means sorting and separating mainly those components that are environmentally hazardous, but with low added value.

When using this approach, the volume of waste subject to disposal at landfills can be reduced, since other types of waste, which may contain increased amounts of harmful substances, will end up in the final product [19]. Of course, the ideal is a sorting option with the highest possible quality of separation of individual components, but it is obvious that such an option will require expensive equipment and software. The payback period for the sorting system in this case can be very long. However, in the future, the concept of "zero waste" is relevant, which many countries are discussing and implementing programs to achieve it [20].

2 Materials and Methods

Obviously, the most promising method of waste sorting is the automated and robotic sorting method [21–23]. In Russia there are no such domestic systems yet; in many developed countries such systems are being developed, but they are very expensive and not effective enough in the conditions of sorting municipal and urban waste.

Known fully automated sorting methods are very expensive and technologically relatively complex, primarily due to the high complexity and high cost of recognition systems for sorted objects. The solution to the problem is a method in which the disadvantages of manual and automated methods of waste sorting are minimized, at rational economic costs and with a high degree of recognition of sorting objects. In any case, at the current stage of improving the sorting process.

It is proposed, conceptually, to solve the problem of automated sorting through partial human participation in recognizing sorting objects. It is proposed to replace (or combine) the electronic recognition system with a cheaper method of recognizing waste components—by humans. In this case, it is possible to use relatively simple and cheap devices and systems for identifying sorting objects with the participation of a human operator, simultaneously with visual analysis and recognition of waste components.

Let's consider several options for well-known automated sorting systems. To partially automate the waste sorting process, rotating drums with holes are used to separate paper and film waste, as well as magnetic separation, which separates certain types of metal [3].

The technology of pre-sorting of solid waste is the key to industrial waste disposal and increasing the utilization rate of secondary resources, Table 1. Describes some well-known systems for sorting municipal solid waste.

Based on the data in the table, we can conclude that various sorting systems are currently known with varying degrees of completeness in recognizing the sorted objects.

In 2014, an automated waste sorting system ZenRobotics Recycler was developed [7]. The robot of this system collects waste from construction sites and construction sites and places it on a conveyor belt. At the same time, automated sorting of such fractions as wood, stone and metal was carried out on the conveyor. In this system, the objects to be sorted are picked up from the conveyor by robotic arms placed along the conveyor. The robot's gripper is equipped with tactile sensors that notify about the reliability of the grip of the sorted object. The robot is equipped with means for optical recognition of the waste object and is equipped with means for spectrometric evaluation of the material composition of these objects.

However, this well-known waste sorting system has a significant drawback; its performance is directly related to the number of robots, each of which must be equipped with an expensive object recognition system in the form of various scanners, memory units and software control, detectors and spectrometers. Obviously, this is a very expensive system and its payback period is very long, and for waste processing

Table 1 Sorting systems for municipal solid waste

№	Name	Short description
1	Waste sorting system from ZenRobotics	The capture of sorted items from the conveyor is used, placed along it by robots with manipulators, with gripping bodies in the form of a hand. The robot arm is equipped with tactile sensors that signal the reliability of the grip of the sorted item. The robot is equipped with a garbage object recognition system, consisting of conventional and laser scanners, spectrometers, 3D laser scanners controlled by the robot software. The recognition system determines: the size, shape and color of the object, its material composition, physical properties [7]
2	Qinetiq Waste Sorting System	Hyperspectral cameras are used to create "color" from the ultraviolet and infrared portions of the electromagnetic spectrum. One such camera distinguishes between plastic, metal, glass. A sorting system based on data on the orientation and location of objects gives a signal, for example, to manipulators to collect and sort garbage [3]
3	Automatic Waste Identification System by V. Sinram, J. Levasseur, N. Lorty	Includes conveyor belt and identification unit above the conveying surface for material recognition. The identification unit includes a lens receiving reflected light from a material and a first processing unit associated with the lens for spectral analysis of the reflected light captured by the lens to determine the nature of the material. The second processing unit associated with the first block for comparing the results of spectral analysis with data on various materials stored in the database of the processing unit [4]
4	Waste sorting method by Yu.A. Pack	Items are captured from the conveyor by means of manipulators controlled by object recognition systems; containing scanners, spectrometers, detectors and other sensors and recognition systems, software. In this case, the capture of the sorted objects by the manipulators is carried out using the code detectors of these marks. In addition, the conveyor is illuminated with ultraviolet lamps, and marks on objects are painted with fluorescent dyes. Also, the software block is equipped with images of 3-D models of parts of garbage items that are most resistant to deformation [5]
5	Screening plant by E. Williams and J. Bentil	The system is used to separate organic waste from inorganic. It uses a gas sensor that sends data to a microcontroller, and based on this data, the waste is differentiated and later transported to a waste processing plant [6]

enterprises with low productivity, or for conditions with a low share of recovery of useful ingredients, it will be unprofitable.

The next step in the development of automated sorting systems is a system created to minimize the cost of recognizing images of sorted objects [10]. In this system, one post for recognizing sorting objects is used, and the capture of sorting objects and their movement is performed by several manipulators equipped with mark recognition systems. At the same time, manipulators are not equipped with expensive recognition systems, multiple detectors and scanners of different operating principles.

At the same time, in the recognition system, ingredients are marked with encoded physical marks, for which inkjet printers, dyes and other means of applying marks to the surface of recognized objects are used. The authors of the development under consideration proposed to use the color of the marks and the shape of the icons applied with paint to recognized objects. In this method, manipulators that move recognized objects are equipped with mark detectors that recognize the color of the marks and their shape.

From the recognition system and devices for placing tags on sorted objects, information is transmitted to executive manipulators, which move garbage objects into individual containers or onto additional conveyors. Such conveyors are used in situations where the volume of individual fractions. separated from the total mass of garbage is relatively large.

The manipulators capture these objects using data from code detectors of these marks and those gripping devices that correspond to the type of object being moved. This method uses only one recognition post for sorting objects, with a maximum set of recognition tools, and manipulators moving sorting objects do not have such tools, which significantly reduces the cost of such sorting systems.

Indeed, in the known method of waste sorting under consideration, applying coded marks with dyes to sorted objects will reduce the number of object recognition systems, which will reduce the cost of the robotic sorting system. However, both the process of applying coded marks with dyes to sorted sorting objects and the process of identifying these marks is a relatively complex and unreliable way to identify a waste object on a conveyor. As the conveyor belt moves, the sorting object may tilt or tip over, making it difficult or impossible to scan some tags. Thus, the process of creating marks on recognized—identified objects is a complex and unreliable process, which is a disadvantage of the considered method of sorting garbage.

We have modernized this waste sorting method [11]. The essence of the modernization is that marks on objects are not applied with paint, but are made in the form of electronic identifiers of each recognized object. The technical result is a reduction in cost, an increase in the speed of the process of recognizing objects on a conveyor, and an increase in the reliability of information transfer from the system for identifying individual objects on a conveyor to robotic manipulators.

The effectiveness of the system we have developed, which includes processes for recognizing sorting objects, moving objects from the conveyor with robotic manipulators, is ensured by the fact that the recognition process is complemented by the process of determining and recording the coordinates of the center and boundaries of the identified sorted object on the conveyor, for which purpose along the length of the conveyor the canvas has support marks with an equal distance between them in the form of selected lines perpendicular to the conveyor belt. In this case, the recognition process control system contains a subroutine for determining and recording the coordinates of the center of the recognized object in relation to the edge of the conveyor and one of the reference marks on the conveyor.

Data about the location on the conveyor of each sorting object within the area of the conveyor belt from mark to mark, the coordinates of the center of each object, its boundaries, properties, characteristics of the object are transferred to the robot manipulator for each identified object or transferred to the control system for several manipulators, gripping and moving sorted objects.

In this case, the identification of each object is carried out using devices for scanning, spectrometry and detection of sorted objects, by comparing the data with images in computer sorting software.

The essence of the developed sorting method is illustrated in the figure, see Fig. 2. The following designations are used: 1—conveyor, 2—direction of movement of the conveyor; 3—object identification system, which has devices for scanning, spectrometry and detection of sorted objects, 4—computer, 5—information line, 6—robotic manipulator; 7—recognized objects, 8—reference marks on the conveyor; (a;b), (c;d)—coordinates of identified objects.

The developed method is implemented as follows. Solid household waste 7, freed from oversized and small fractions, is fed to conveyor 1 in the form of a thin layer of objects sufficient for their examination. The software performs scanning, detection and other actions to analyze each recognition object. The received information is compared with images of objects in the database, thereby identifying each object on the conveyor. In this case, a subroutine works for determining and recording the coordinates of the center of the identified object (a; b) and (or) (c; d) in relation to the edge of the conveyor and one of the reference marks 8 on conveyor 1. The identification program records the characteristics of objects 7 and transmits information via each recognized object to computer 4. Data from computer 4 for each recognized object 7 is transmitted to robotic manipulator 6 in the format: object number + object coordinates + object characteristics.

Unlike well-known analogues, the use of electronic information about recognized objects instead of paint marks will simplify the process of sorting a waste

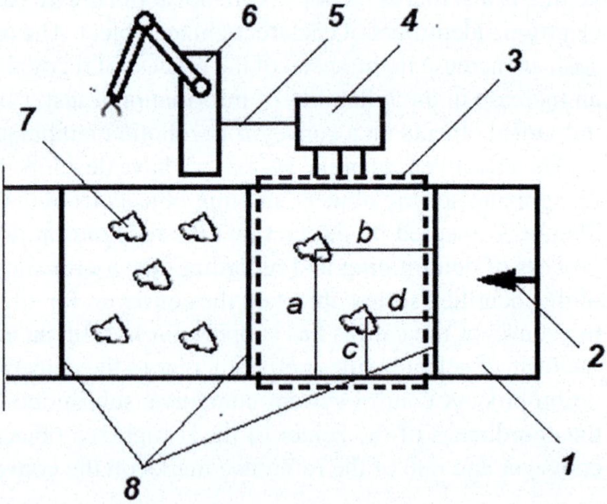

Fig. 2 Scheme of the method for creating electronic tags

object, increase the reliability of the recognition system and increase the percentage of recognition of waste objects on the conveyor. At the same time, the number of robotic manipulators is reduced, which reduces the cost of the waste sorting site. In addition, the claimed method will increase the throughput of the sorting conveyor, because the likelihood of inaccurate recognition of paint markers on sorting objects and errors when robot manipulators recognize paint marks will be reduced. Essentially, the paint marks in the prototype are replaced in the claimed method with electronic information about the coordinates of each recognized object and the type of this object. This information is electronically transmitted to manipulators, which, at specified coordinates, capture the recognized object and move it to the appropriate container.

The information received by the robot manipulator about the type of object that needs to be captured at the specified coordinates can be used to use various gripping devices, if such a robot manipulator has removable and replaceable gripping devices. For systems that use a group of robotic manipulators, it is possible for individual robotic manipulators to grip and move one or more types of sorted objects; such systems are used when the productivity of the sorting line is high.

We have also developed an automated system for recognizing sorting objects with human participation. To explain the features of this system, we present in tabular form the main advantages and disadvantages of the known automated sorting systems and manual sorting systems, see Table 2.

Table 2 Advantages and disadvantages of manual, automated and automatic sorting method

Type of sorting	Advantages	Disadvantages
Manual, manual with simple devices for identifying sorting objects; pre-manual sorting or secondary manual sorting	–Quick recognition of sorting objects: –Cheap labor for recognizing sorting objects -lack of expensive databases –Simplicity of equipment;	–High safety requirements; –Incomplete extraction of various solid waste fractions; –low selection speed, manual labor to move sorting objects
Automated (human participation in the recognition of sorting objects)	–Low cost of recognizing sorting objects; -comfortable working conditions of the sorter; -the function of training recognition systems, including neural networks	–The need for pre-sorting;—the use of human labor with its limitations;–limited resource for increasing sorting speed
Automatic	The ability to sort by many parameters (e.g. color, shape, density, etc.)	–High cost of equipment and recognition system; -the need for modern robotic systems

3 Results

Based on a review of existing solid waste sorting systems, it was established that it is advisable to use the method of recognizing sorting objects by humans, in conjunction with other automation tools.

Next, we will describe a solution to the problem of waste sorting in automated systems with human participation in the procedure for recognizing sorting objects.

The concept of the proposed automated sorting of waste, mainly solid municipal waste, extends to sorting systems, which include: the main transport conveyor with waste, waste transfer devices, storage containers and auxiliary conveyors for placing sorted objects in them. Such a system is equipped with a subsystem for recognizing sorting objects on a conveyor.

The function of recognizing a sorting object on a conveyor can be performed by a robot equipped with scanners, detectors, and other electronic means of identifying objects on a conveyor. At the same time, an important element of such a subsystem is software, including a very voluminous database. The recognition subsystem can be made up of a group of people standing at a moving conveyor wearing gas masks or with a local air supply system and being the actual sorters. There are known disadvantages of both the first option and the second. The essence of the developed concept is to use a person's ability to see well and recognize cheaply. Then the sorting system will include a human operator, a video camera (video cameras) aimed at the waste conveyor, a monitor, a keyboard, and a joystick. In this case, the human operator, monitor, keyboard and joystick are placed in a special room with a comfortable microclimate. For high-throughput sorting systems, there may be several such operators on the sorting line, whose function is to recognize sorting objects and identify them.

The operator recognizes sorting objects and marks these objects indicating their coordinates on the conveyor with electronic tags. These tags use manipulators to identify sorted objects with the coordinates for placing each recognized object on the conveyor.

A variant of a waste sorting system using a person in the recognition operation and using electronic tags of recognized objects on a conveyor is described in publication [24].

The marking subsystem is made in the form of a joystick or a computer mouse and keyboard, in the form of a special computer system consisting of a monitor and keyboard with symbols such as sorted waste and a manipulator for controlling a pointer on the screen (mouse or joystick).

The manipulators are equipped with gripping devices of various types so that each manipulator is focused on moving one type of waste and is equipped with special gripping devices for one type of waste, and the manipulator control subsystem is designed as a distributor of manipulator functions, while the manipulator control subsystem is connected to the marking and addresses each recognized object, its coordinates, and type of waste to the appropriate manipulator. Each manipulator

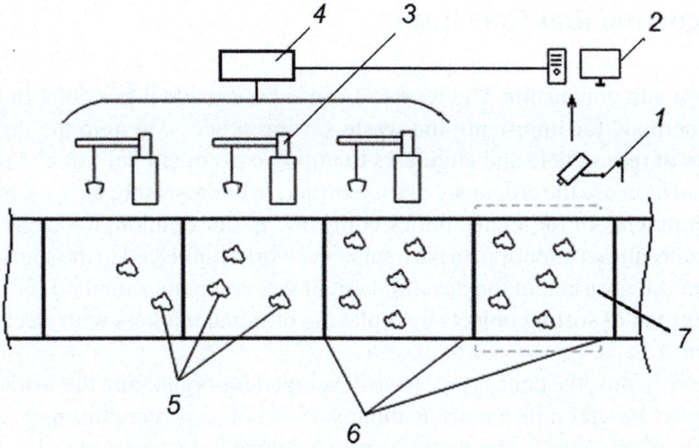

Fig. 3 Waste sorting system: 1—video camera, 2—monitor, 3—manipulators, 4—manipulator control system, 5—waste, 6—reference marks, 7—windows

is designed to move a certain type of object and is equipped with special gripping devices.

Figure 3 shows a diagram of a sorting system for implementing the proposed sorting method using a person in the recognition operation.

The waste sorting system includes a subsystem for recognizing objects on the conveyor, which is designed as a video system. The video system includes a video camera 1, which is aimed at the waste conveyor 5, and a monitor 2, which is located in the sorter operator's room. The sorting system includes 3 manipulators that capture the garbage objects to be sorted and move these objects to the appropriate drives. On the transport conveyor, support marks 6 are made at equal distances from each other along the length of the conveyor, forming windows 7 of processed objects with the edges of the linen. Reference marks 6 are a coordinate system for manipulators for each window.

The system operates as follows. Solid municipal waste 5, freed from oversized and small fractions, is fed to the conveyor in the form of a monolayer. The operator on the monitor, within the next window, recognizes the sorting objects, using the keyboard and joystick, records the coordinates and type of each recognized object, generates information for each window about the recognized objects and transmits the information to the control system of 4 manipulators 3, capturing and moving the sorted objects. objects 5. Individual robotic manipulators are specialized in one or more types of recognized objects that are most often found in the total mass of waste.

4 Discussion and Conclusion

Discussion and conclusion The research carried out made it possible, in the key of various methods for improving the waste sorting process, to note the directions of the efforts of researchers and engineers to improve occupational safety during waste sorting and increase the efficiency of this sorting. In our research, we have highlighted the developments of the ZenRobotics company. In our opinion, the technologies of this company attract attention, in particular, the works published in the sources [9]. We have proposed a variant of the development of this company's method for organizing the recognition of sorting objects by replacing physical markers with electronic ones (our patent).

Along with this, the concept we have developed for organizing the work of robotic manipulators based on information about sortable objects recognized by a human is briefly described. Due to the insufficiently advanced electronic systems for recognizing sorting objects, and their relatively high cost, we have proposed and developed a sorting method in which object recognition is performed by a human "sorter operator". At the same time, the sorter operator is located in a room with a comfortable microclimate, and performs monitoring, identification and identification of sorting objects remotely using a video camera, a high-quality monitor (screen) and a special electronic system for identifying sorting objects and electronic labeling. Several sorter operators can be used at the same time.

The technical result is achieved by replacing an expensive recognition system with a human. The described solution to the problem will significantly reduce the cost of the sorting system, specifically for recognition operations. At the same time, this sorting method will increase the reliability of the recognition system and increase the percentage of recognition of garbage objects on the conveyor, significantly improve sanitary working conditions. The main advantage of such a system is to reduce the cost of sorting equipment and increase the possibility of relatively rapid distribution of the developed method in Russia and in the World.

The above materials only partially reveal the authors' concept of using human capabilities in complex recognition systems. The authors emphasize that the proposals (ideas) described in the work relate to a temporary, transitional stage of improving the creation of highly efficient electronic computer recognition systems. Moreover, the method of using humans to recognize sortable objects in robotic systems can be used as a training system, a software improvement system, databases, for electronic computing systems for recognizing objects. The idea of using electronic tags to identify already recognized objects can be used in a robotic sorting system with human participation in the process of recognizing objects.

5 Conclusions

1. The analysis of various municipal solid waste sorting systems has been carried out in order to develop a rational option for automating the waste sorting process, without using modern "intelligent" recognition systems.
2. 2Taking into account the peculiarities of the state of waste management in Russia, it is proposed to use human capabilities to recognize sorting objects on an automated conveyor. This is like a temporary, quickly realized stage in the development of automatic waste sorting tools.
3. A sorter working behind a monitor and performing the function of recognizing sorting objects, in conjunction with other automation tools, solves tasks that are useful for trained electron-optical object recognition systems. The work of such an operator can be integrated into the work of a recognition system, for example, in convolutional neural networks, as a source for training an intelligent recognition system.
4. The proposed automation system, which is quickly implemented and low-cost, can be a basic system for solving other automation issues in sorting lines, such as transmitting information to a robot manipulator to move recognized garbage objects and other tasks of fully automating the garbage sorting process.

References

1. Grigoriev VN, Parshakova SV (2013) Optimization of the technological scheme for sorting solid household waste. Ecology and technological progress 1:39–45
2. Kolychev NA (2013) Optimization of the handling of solid household industrial waste close to them in composition in large and medium-sized settlements of Russia. Biosphere 4:391–417
3. Mullins J (2008) All-seeing garbage sorter, New Scientist 6:1–3. https://www.newscientist.com/article/dn14148-invention-allseeing-garbage-sorter.html
4. Sinram BW, Levasseur I, Lortie N (2014) System and method for identifying and sorting material. Patent Canada 13/787,339
5. Pak YuA (2017) Waste sorting method. Patent Russian Federation 2624288, IPC B09B 3/00, 04 May 2016
6. Razmyslov AN (2012) Complex for sorting and processing of solid household waste. Patent Russian Federation122918U1, IPC B09B 3/00, 20 December 2012
7. Park Yu A (2014) Waste sorting gripper. Patent Russian Federation141015, IPC B09B 3/00 27 May 2014
8. Williams EA, Bentil J (2016) Design and implementation of a microcontroller-based automatic waste management sorting unit for a recycling plant. American Journal of Engineering Research (AJER) 5(7):48–52
9. Lukka TJ, Tossavainen T, Kujala JV, Raiko T (2014) Zenrobotics recyclerrobotic sorting using machine learning. Proceedings of the International. Conference on Sensor-Based Sorting (SBS), p 1–8
10. Paulra SG. Hait S, Thakur A (2016) Automated municipal solid waste sorting for recycling using a mobile manipulator International Design Engineering Technical Conferences and Computers and Information in Engineering Conference. American Society of Mechanical Engineers, p 53–60

11. Karelin AE, Kozhemyachenko AV, Lemeshko MA (2021) Overview of municipal solid waste sorting systems. Engineering Bulletin of the Don 7(79):9–16
12. Bennear L, Coglianese C (2010) Measuring progress: program evaluation of environmental policies. Environment: Science and Policy for Sustainable Development 47(2):22–39
13. Côté R (1998) Designing eco-industrial parks: a synthesis of some experiences. Journal of Cleaner Production 6(3–4):181–188
14. Chalfan L (2001) Zero waste is the key to our future. Power point presentation. http://www.zerowaste.org/publications/06i_Case_for_ZW.pdf
15. Chertow MR (2007) Uncovering» industrial symbiosis. Journal of Industrial Ecology 11(1):11–30
16. Graedel T, Allenby B (2010) Industrial Ecology and Sustainable Engineering. Upper Saddle River. Pearson, NJ, p 352
17. Petrov VG (2005) Garbage sorting lines. Prospects for application, Institute of Applied Mechanics, Ural Branch of the Russian Academy of Sciences, Izhevsk, p 112
18. Slyusar NN, Borisov DL, Grigoriev VN (2011) Urbanism, Vestnik PNIPU 12:75–82
19. Lamzina IV, Zheltobryukhov VF, Shaikhiev IG (2015) Analysis of solid household waste sorting methods. Bulletin of the Technological University 18(5)
20. Paul SP, Terry T, Helen B et al (2009) Acritical review of a key. Waste Strategy Initiative in England: Zero Waste Places rojects 2008–2009. https://doi.org/10.1016/j.resconrec.2010.10.006
21. Anastas PT, Zimmerman JB (2003) Design through the Twelve Principles of Green Engineering. Env. Sci. and Tech. 37(5):94A–101A
22. Sokolov LI (2018) Waste management: Textbook. Infra-Engineering, Vologda, p 208
23. Antrekowitsch J, Steinlechner S (2011) The recycling of heavy-metalcontaining wastes: mass balances and economical estimations. JOM 63(1):68–72
24. Lemeshko MA, Bashnyak SE et al (2023) A rational way of sorting municipal solid waste, IOP Conference Series: Earth and Environmental Science 1154(1):012074

Use of Coal and Oil Mining Waste in the Production of Effective Silicate Materials

V. S. Romanyuk, V. D. Tkachenko, L. V. Klimova, E. A. Yatsenko, and A. V. Ryabova

Abstract Today, the issue of accumulation, burial and disposal of industrial waste is actively discussed in scientific circles, since industrial waste is a source of anthropogenic impact on the environment on a global scale and is formed as a by-product of coal and oil production, and also has a low resource utilization rate. Also, an important problem of our time is the reduction of reserves of non-renewable high-quality natural raw materials used in the production of silicate building materials. In this regard, the use of accumulated technogenic waste from coal and oil mining as secondary raw materials in the production of silicate materials makes the topic under study relevant and promising. The purpose of these studies is the need to develop resource-saving technology for the production of effective silicate materials using coal and oil mining waste in the raw material mixture.

Keywords Technogenic waste · Ash and slag waste · Drill cuttings · Silicate materials · Clay · Resource-saving technologies · Recycling

1 Introduction

Currently, in the Russian Federation, industrial enterprises annually generate more than 60 million tons of dust and gaseous waste, about 65 km^3 of wastewater and up to 30 billion tons of solid waste, the main part of which is ash and slag from thermal power plants (TPPs), mining waste and mineral processing, metallurgical waste, construction waste, as well as waste from the chemical and forestry industries (Fig. 1) [1, 2].

As a result of human activity, the degree of environmental pollution has reached catastrophic proportions; in particular, these issues are acute in the oil, gas and coal industries. The technology of drilling oil and gas wells causes the formation of

V. S. Romanyuk (✉) · V. D. Tkachenko · L. V. Klimova · E. A. Yatsenko · A. V. Ryabova
Platov South-Russian State Polytechnic University (NPI), 132, Prosveshcheniya St., Novocherkassk 346428, Russia
e-mail: lera_romanyuk_1999@mail.ru

L. V. Radionova and D. V. Ulrikh (eds.), *Advances in Ecology and Environmental Engineering*, Springer Proceedings in Earth and Environmental Sciences,
https://doi.org/10.1007/978-3-031-64423-8_37

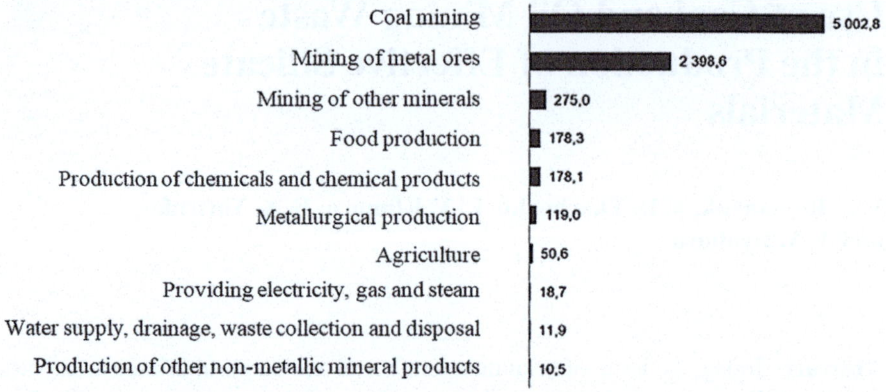

Fig. 1 Distribution of industrial waste by area of activity (million tons)

industrial waste, such as: drill cuttings, spent drilling fluid and drilling wastewater, and waste from the coal industry includes overburden and mine rocks, rock after gravity enrichment and flotation tailings of processing factories, ash and slag waste in the form of ash. entrainments and slags after combustion and gasification. These types of waste reduce the economic efficiency of enterprises due to the need to alienate the territory of enterprises for their storage facilities, increasing environmental payments for waste storage and emissions of pollutants.

Since 2009, the amount of waste generated that is not naturally recycled has increased tenfold. And since the amount of industrial waste significantly exceeds the amount of waste from housing and communal services, the development of new methods for their disposal is relevant. According to Rosprirodnadzor, from 2019 to 2023, the volume of recycled and neutralized industrial waste increased, but at the same time, its formation also increased in volume, which exceeds the volume of processed waste by 2.2 times.

The most common methods of handling waste from mining have become recycling, neutralization, and storage. Their number accounts for 45.6% and 40.4% of the total waste, respectively; landfill accounts for about 12% (Fig. 2) [2, 3].

In recent years, the dynamics of waste management from crude oil and gas production has been in a positive direction, expressed in an increase in the share of recycled and neutralized waste in the total amount of waste from 26.7% in 2018 to 40.7% in 2023 (Fig. 3).

One of the main reasons for the increase in the percentage of recycled waste in the industry is the state policy in the field of waste management; a secondary reason can be noted the increase in environmental awareness of enterprises. Making changes to the production technology of building materials, namely to the raw material composition of products, is one of the solutions. Since 2016, the Russian Federation has been implementing the Development Strategy for the Construction Materials Industry, which provides state-level support for manufacturers of construction materials involved in the recycling of industrial waste.

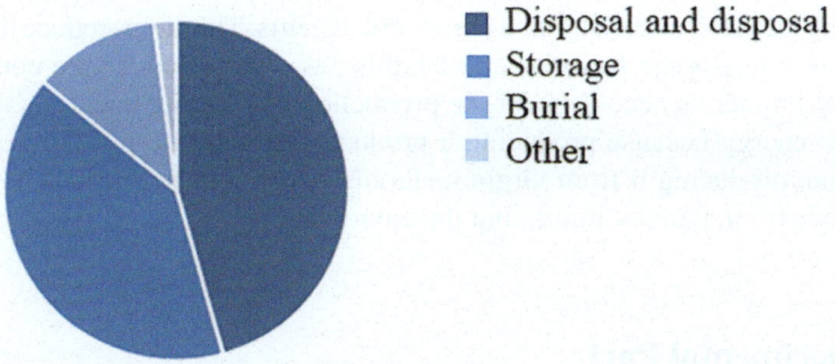

Fig. 2 Management of production and consumption waste from the mining industry

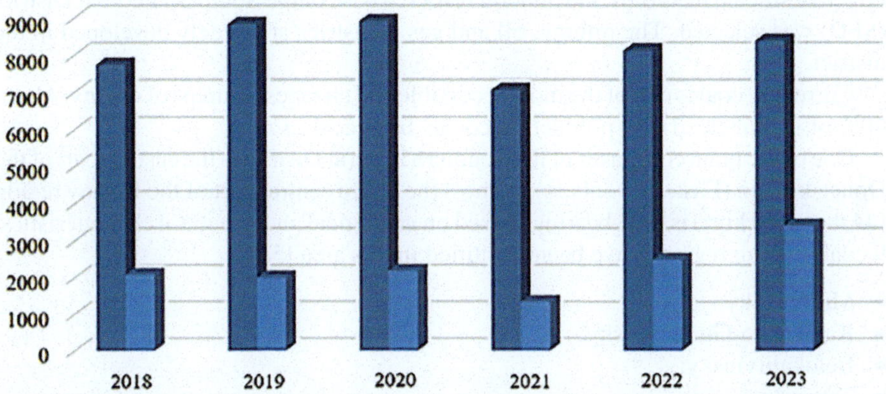

Fig. 3 Dynamics of waste management from oil and gas production and consumption

Currently, restoring the natural process of the ecological system of our planet is one of the most important tasks, requiring an immediate search for all possible solutions. Taking into account the important factor that the ecology of the environment suffers most from the massive pollution of various ecospheres, the launch of its restorative cleansing action cannot take place without the participation of humanity.

There are many ways to dispose of man-made waste, depending on its chemical composition and physical state. Each type of waste requires its own optimal processing method. Waste from galvanic, metallurgical and mining industries, such as slag and sludge, can be used as a filler for building materials, for example, in ceramics, cement and asphalt concrete mixtures. They can also be used for backfill when laying roads and leveling terrain. Waste glass is usually recycled into glass wool, but can also be used to produce foam glass, interior tiles and liquid glass. Rubber and plastic waste can be used for the production of polymer sand tiles, thermal insulation materials, as well as for obtaining fuel and energy resources using pyrolysis [3, 4].

Reusing recycled materials has a number of benefits. This helps reduce the amount of waste sent to storage facilities and landfills, as well as reduce the consumption of extracted minerals necessary for the production of building materials. Recycling also saves energy because producing a product from waste typically requires less energy than producing it from virgin materials. This, in turn, will help reduce fuel use in production, thereby improving the environment.

2 Experimental Part

To study the possibility of using coal and oil mining waste in the production of effective silicate materials, the mineral resource base of the Southern Federal District (SFD) was selected. The mining, oil and gas industries are widely developed in the district.

Currently, coal is one of the most accessible and cheapest sources of energy. About 40% of global electricity production comes from coal.

One of the main coal bases of the Southern Federal District is the eastern part of the Donetsk basin (Eastern Donbass). Almost the entire central part of the Rostov region (44 thousand km^2) is coal-bearing. Based on geological and structural characteristics, 9 coal-bearing regions have been identified in this area [5, 6]:

- Millerovsky;
- Kamensko-Gundorovsky;
- Belokalitvinsky;
- Tatsinsky;
- Krasnodonetsky;
- Gukovo-Zverevsky;
- Sulino-Sadkinsky;
- Shakhtinsko-Nesvetaevsky;
- Zadonsky.

The total coal resources of Eastern Donbass amount to 24.5 billion tons (0.4% of the total Russian total). The total amount of coal reserves in the Rostov region as of 01/01/2023 is about 13.5 billion tons, including explored (category $A + B + C_1$)—about 6.5 billion tons, preliminary estimated (category C_2)—about 3,2 billion tons, off-balance sheet—about 3,8 billion tons.

The licensed fund of coal subsoil plots in the Rostov region is represented by 18 deposits, of which 10 are fields of existing mines, 4 are areas for the construction of new mines, 4 are reserve explored areas for the construction of new coal enterprises [7, 8].

Exploration and production are carried out by 12 coal mining enterprises:

- JSC "Donskoy Anthracite" No. 410 (Dalnyaya mine);
- Obukhovsky Northern Reserve (section No. 412 of the Obukhovskaya mine);
- OJSC "Donugol" Obukhovskaya No. 1 (Sherlovskaya-Naklonnaya);

- LLC "Zamchalovskaya Mine" (Zamchalovsky Southern Reserve);
- OJSC "Sadkinskaya Severny Mine" (Sadkinskaya Severny site No. 1);
- LLC "Mine Rostov";
- JSC SHU "Obukhovskaya";
- LLC "Delta" Bystryansky 1–2;
- Sulinugol LLC (Novomikhailovsky reserve site, Mine No. 37);
- LLC "Oktyabrskaya Yuzhnaya Mine" (Mine named after M. Chikh);
- LLC "Mine Sadkinskaya-Vostochnaya" under construction (Sadkinsky Vostochny 2);
- LLC "TSGM" Likhovskoy South.

The coal industry is one of the industries that has a significant negative impact on the environment. Existing coal mining and beneficiation technologies at various stages of the process lead to the formation of solid waste, which leads to significant loss of land and deterioration of water resources. Coal industry enterprises also face high costs associated with waste management.

In 2023, the coal industry generated more than 4.5 billion tons of waste. About 3 billion tons of waste were used for various production purposes, and over 1.5 billion tons were placed in external rock dumps. The amount of waste generated increased by 7–8% compared to 2022; the specific indicator of waste generation per ton of coal production increased by 5%. The volume of waste used increased by 18%, at the same time, the amount of waste disposed in external rock dumps increased by 21.5% [9, 10].

To date, more than 15 billion tons of waste have been accumulated in dumps and sludge ponds on the territory of the Russian Federation as a result of both coal mining and their enrichment at factories. These wastes are mainly represented by overburden and host rocks, as well as products of mineral processing.

Disposal of waste from coal mining and processing is usually carried out at the very sites of mining and processing of mineral raw materials, where they can be used for backfilling mine workings or land reclamation. Such waste also has the potential to be used as secondary raw materials in construction. They can be used, for example, to create road bases or produce building materials such as Portland cement, aggregates, ceramic materials and others.

However, in the Russian Federation, the share of the use of coal industry waste as secondary raw materials does not correspond to the level observed in developed countries. Increasing the level of use of industrial waste is an important task at the state level.

The main types of solid fuel are hard and brown coals, the extraction and enrichment of which is accompanied by the formation of by-products: mine and overburden rocks, coal processing waste (mudstones, siltstones, sandstones, limestones, granites, etc.). In addition, mining waste also includes the remains of burning waste rocks accompanying coal deposits. These are dense rocks formed in coal deposits during underground fires (gliege, earthen slag, porcellanite) [11].

One of the main problems when using waste from coal mining and processing as secondary raw materials is the heterogeneity and instability of their composition. This makes it necessary to prepare waste before using it.

Waste from coal mining and processing as a raw material for the production of building materials can be used:

- as a component of the raw material mixture when producing cement;
- in the manufacture of ceramic (ceramic tiles and bricks, drainage pipes) materials and silicate (silicate brick, glass, etc.) materials;
- as a raw material in the production of porous aggregates (agloporite, azerite, expanded clay);
- as a concrete filler;
- as a replacement for earthen soil in the construction of hydraulic structures, foundations, and highways.
- as a raw material in the production of porous sand;
- as a raw material in the production of mineral wool.

Natural hydrocarbon raw materials play a significant role in the fuel and energy complex. Russia ranks third among the leaders in oil production, according to official data for 2021–2023. The amount of waste generated from oil production amounts to millions of tons annually. In the SFD, 226 fields are taken into account, including oil, gas and oil, oil and gas and oil and gas condensate fields. Most of these fields are in late stages of development, have already been developed and watered [12].

Activities for the development of hydrocarbon raw materials in the Southern Federal District are carried out by 49 enterprises, including OJSC Gazprom, OJSC NK Rosneft, OJSC LUKOIL and their subsidiaries, which are the largest players in this area.

In the Rostov region, 23 fields have been explored, of which 16 are gas, 4 are gas condensate, 1 are oil and gas condensate, 1 are gas and oil and 1 are oil. As of January 1, 2023, in the state balance of oil reserves, all fields are in the distributed subsoil fund:

- JSC "Leonovskoye"—Leonovskoye (gas-oil);
- LLC "Tishkinskoye"—Tishkinskoye (oil and gas condensate);
- LLC "YugGeo"—Romanovskoe (oil).

As of 01/01/2023, extracted oil reserves of the distributed subsoil fund amount to about 994 thousand tons in category $A + B_1$, about 446 thousand tons in category B_2 and about 5301 thousand tons in category $C_1 + C_2$ [13, 14].

When drilling oil and gas wells, man-made wastes are generated, such as drill cuttings, spent drilling fluid and drilling wastewater. One of the most challenging tasks is the disposal of waste drilling fluids and cuttings, in which the solid particles consist of various substances, such as fragments of broken rock and worn-out tools that are used to drill wells. The solid component of drill cuttings may include clay minerals and other materials.

It is drill cuttings that are a significant source of secondary raw materials for the production of construction materials. They contain various mineral components that can be used in the production of construction products such as cement, ceramics, asphalt and other materials. Recycling drill cuttings will reduce waste and reuse these materials, promoting sustainable resource use and reducing environmental impact.

An effective technology for the disposal of drilling waste is its solidification. The neutralized product produced in this way is used in the manufacture of materials for construction and road construction [15–17].

Among the methods for recycling coal and oil mining waste, the most promising and rapidly developing area of processing is their use in the production of building ceramics, in particular silicate materials.

According to modern conditions for the production of silicate materials, there is a need to use local loamy raw materials, given the depletion of reserves of high-quality clay raw materials and the high costs of their delivery. To increase the strength characteristics of ceramic materials made from loamy raw materials, an additional component is added—waste from the local oil or coal industry, which allows expanding and diversifying the sources of raw materials for the production of ceramic materials. This can reduce dependence on imported high-quality clay and reduce transportation costs.

The most important properties of coal and oil mining waste, which determine the scope of their further application, are the chemical composition and physical and mechanical characteristics. In addition, toxicological, fire and explosion characteristics may have an impact on the area of application.

The objects of the study were drilling cuttings from the Slavyansk oil field in the Krasnodar Territory and a sample of coal waste from the Khmelnitskaya mine in the Sverdlovsk region (Lugansk region). To determine the possibility of using the waste under study as a raw material component in the production of silicate materials, a comparative analysis of the chemical composition of the waste with clay mined in the Southern Federal District (Vladimirovskoe deposit, Rostov region) was carried out. All studies were carried out in the laboratory "Recycling of fuel energy waste" on the basis of the Platov South-Russian State Polytechnic University (NPI).

Studies of the chemical composition of the studied raw materials were carried out by X-ray spectral fluorescence analysis (XRF) on a sequential vacuum spectrometer of the PW2400 model by melting samples with a mixture of lithium borates. Mass loss during calcination was determined according to the methodology of the Scientific Council on Analytical Research Methods using secondary ion mass spectrometry 418-X at a temperature of 1050 °C [18]. When determining the chemical composition of drill cuttings, coal rock and clay, the following data were obtained, presented in Tables 1, 2 and 3.

Table 1 Chemical composition of drill cuttings from the Slavyanskoye oil field (Krasnodar region)

Material	Content, mass %											
	Na_2O	MgO	Al_2O_3	SiO_2	K_2O	CaO	TiO_2	Fe_2O_3	MnO	P_2O_5	$BaSO_4$	L.O.I
Drilling waste	1,47	0,79	5,41	22,72	0,91	4,44	0,26	3,19	0,11	0,11	50,59	10,01

Table 2 Chemical composition of coal waste heap from the Khmelnitskaya mine, Sverdlovsk region (Lugansk region)

Material	Content, mass %										
	Na_2O	MgO	Al_2O_3	SiO_2	K_2O	TiO_2	Fe_2O_3	C	O	SO_3	L.O.I
Coal rock	0,15	0,18	4,76	9,14	1,00	0,50	1,20	41,46	33,11	0,47	8,03

Table 3 Chemical composition of clay from the Vladimirovskoye deposit, Rostov region

Material	Content, mass %						
	SiO_2	Al_2O_3	Fe_2O_3	CaO	MgO	TiO_2	L.O.I
Clay	64,25	20,88	4,48	0,74	0,79	1,14	7,72

As can be seen from Tables 1, 2, chemical analysis confirms the presence in drill cuttings and coal rock of clay-forming elements, such as SiO_2 and Al_2O_3, which serve as scavengers. Silicon and aluminum oxides reduce the temperature coefficient of linear expansion (TCLE), at low values of which the heat resistance of the raw material increases.

A significant content of $BaSO_4$ was detected in the drill cuttings—50.59%, which is used as a weighting agent for drilling fluid in oil production. Fe_2O_3 in drill cuttings is 3.19%, which acts as a dye. The MgO content is 0.79%, and the CaO and TiO_2 content is no more than 5% [19].

A significant content of carbon and oxygen is observed in the coal rock sample, C—41.46% and O—33.11%. Carbon is the main element; the heat of combustion of coal depends on its quantity. And with a high oxygen content, the processes of oxidation and combustion of the material are accelerated. The presence of coloring oxide—Fe_2O_3 in an amount of 1.20% was also noticed.

From Table 3 one can notice a significant content of the main oxides SiO_2 (64.25%) and Al_2O_3 (20.88%). The clay of the Vladimirovskoe deposit consists mainly of kaolinite (37.8%) and a small amount (10.6%) of hydromica. Includes a package of montmorillonite (9.6%) with an admixture of sericite (7.7%) and quartz grains (26.4%). The clay under study has a cellular microstructure with a feldspar content of 5.9% [19].

Despite the fact that the chemical composition of drill cuttings and coal rock differs from the chemical composition of clay, the presence of clay-forming oxides in them makes it possible to use this type of technogenic waste as a raw material component in the production of effective silicate materials, with full or partial replacement of clay.

As can be seen from the chemical composition (Tables 1 and 2), drill cuttings and coal rock contain oxides of heavy metals, the content of which is within acceptable limits in accordance with SanPiN 2.6.1.2523 "Radiation Safety Standards (NRB-99/2009)" (no more than 2.0 wt.%) [20].

Table 4 Results of radiological studies of waste

Name of the material under study	Defined indicators	Research results; units of measurement; confidence probability	The value of the permissible level; units of measurement	ND on the research method
Drilling waste	Effective specific activity of natural radionuclides Ra-226, Th-232, K-40	(62 ± 10) Bk/kg; P = 0,95	No more than 370 Bk/kg (1st class)	GOST 30,108-94 "MI CMII GNMC "VNIIFTRI" dated 22.12.2003
Coal rock		(134 ± 10) Bk/kg; P = 0,95		

Sources of heavy metals in nature are natural (weathering of rocks and minerals, volcanic activity, etc.) and man-made (mining and processing of minerals, road transport, etc.). Drilling fluid contains natural radionuclides and harmful bacteria. In this regard, radiological and sanitary-epidemiological studies were carried out. Radiological studies of drilling fluids were carried out on the MULTIRAD spectrometric complex with PROGRESS software. The results of radiological studies of waste are presented in Table 4.

According to the obtained X-ray data from Table 4 shows that the waste under study belongs to class I of building materials depending on their specific effective activity, which indicates the possibility of their use in all types of construction, in particular in the production of effective silicate materials.

3 Conclusion

Industrial waste often has a composition and properties close to natural raw materials, and sometimes even advantages over them. The use of such waste in the production of building materials is one of the main ways to reduce material intensity in this industry. The use of new construction technologies using industrial waste as secondary raw materials makes it possible to reduce material consumption, reduce the cost of building materials and, ultimately, reduce construction costs in general.

The use of technogenic waste, in particular coal and oil mining waste, in a raw material mixture of effective silicate materials makes it possible to obtain an environmental effect as a result of waste disposal and saving the main clay raw materials used in the production technology of ceramic building materials.

During the tests, it was revealed that partial replacement of clay raw materials with waste from the oil and coal mining industries of the Southern Federal District in the production of effective silicate materials is acceptable. At the same time, there will be waste disposal and subsequently a reduction in the volume of man-made non-recyclable waste in nature, as well as the preservation of natural reserves of clay raw materials.

Acknowledgements The work was carried out as part of a project under an agreement on the provision of grants from the federal budget in the form of subsidies in accordance with paragraph 4 of Article 78.1 of the Budget Code of the Russian Federation "Carbon-neutral technologies for recycling large-scale fuel energy waste with the production of functional geopolymer materials" from "29" June 2022, No. 075-15-2022-1111.

References

1. Information on the state and prospects for using the mineral resource base of the Southern Federal District for 2021. https://www.rosnedra.gov.ru. Accessed 23 Nov 2023
2. Dynamics of industrial production in 2022–2023. https://rosstat.gov.ru. Accessed 23 Nov 2023
3. Kuvykin NA, Bubnov AG, Grinevich VI (2004) Hazardous industrial waste. Ivan. State chemical-technological un-t, Ivanovo, p 148
4. Pichugin EA (2013) Assessment of the impact of drilling waste on the environment. Young scientist 9:122–124
5. Tretyak AA, Yatsenko EA, Onofrienko SA, Karelskaya EV (2021) Identification of drilling wastes and their use. Bulletin of the Tomsk Polytechnic University. Georesource Engineering
6. Yatsenko EA, Tretyak AA, Chumakov AA, Golovko DA (2021) Prospects for the use of drilling fluids for the synthesis of aluminosilicate proppants. Materials today: Materials 38:1886–1888
7. Yatsenko EA, Goltsman BM, Chumakov AA, Vilbitskaya NA, Wensheng Li (2021) Research on the synthesis of proppants used for oil production by hydraulic lining. Material Science Forum 1037:181–188
8. Goltsman BM, Yatsenko EA, Komunzhieva NYu, Yatsenko LA, Gerashchenko VS, Smoliy VA (2020) Influence of fluxes on the synthesis of porous materials based on native silicate raw materials. Glass and Ceramics (English translation of Glass and Ceramics) 77(5–6):240–244
9. Federal Agency for Technical Regulation and Metrology. Production of ceramic products (2015) BAT Bureau, Moscow, pp 66–78
10. Wendlandt W (1978) Thermal methods of analysis. Mir, Moscow, p 527
11. Shestak Ya (1987) Theory of thermal analysis: Physical and chemical properties of solid inorganic substances. Mir, Moscow, p 456
12. Landfill for the disposal and processing of drilling waste and oil production: Fundamental technological solutions. (1996) Book. 3. Development of fundamental solutions for the neutralization and disposal of drilling waste and oil-contaminated sands. NGDU, Surgut, p 101
13. Deneko Yu (2014) About the problem of drilling waste processing. Oil and gas of Siberia 1(14):29–30
14. Aminova AS, Gaibullaev SA, Juraev KA (2015) Tech. The use of oil sludge is a rational way of their disposal. Young scientist 2:124–126
15. Pashchenko AA (1986) Physical chemistry of silicates: a textbook for universities. Higher school, Moscow, p 367
16. Gorshkov BS, Saveliev VG, Fedorov NF (1988) Physical chemistry of silicates and other refractory compounds: textbook. Higher School, Moscow, p 400
17. Strelov KK (1985) Teoreticheskie osnovy tekhnologii refractory materials: a tutorial. Metallurgy, Moscow, p 480
18. Al'myashev VI, Gusarov VV (1999) Thermal methods of analysis: textbook. Allowance. SPbGTU (LETI), St. Petersburg, p 40
19. Romanyuk VS, Klimova LV, KurdashovVM, Izvarin AI, Yatsenko VS (2022) Prospects for the Use of Painted Ceramic Facing Materials Using Man-Made Waste. Lecture Notes in Civil Engineering

20. SanPiN 2.6.1.2523 "Radiation Safety Standards (NRB-99/2009)" (2010) Standard inform, Moscow, p 222

Hemp Waste (Husk) as a Sorbent for Treatment of Oily Wastewater

O. V. Tyukalova, S. S. Timofeeva, and D. V. Ilkov

Abstract With the growth of production and processing of industrial hemp seeds as a source of many valuable products, the amount of waste, including the husk that remains after the seeds are hulled, is increasing. Hemp husk can be used as a promising sorbent for the treatment of wastewater from petroleum products. The aim of this present work was to investigate the possibility of using native and modified hemp husk as a sorbent for the treatment of wastewater contaminated with oil products. The native hemp husk can extract up to 65% of the petroleum products dissolved in water. The sorption capacity of the husk increases with increasing concentration of dissolved petroleum products. Treatment of the husk with liquid glass or foaming agent PO-6-RZ-1 increases the sorption capacity of the husk by 1.5 times.

Keywords Hemp husk · Petroleum products in water · Adsorption capacity · Sorbent modification

1 Introduction

In Russia there is a steady tendency to expand cultivation of industrial hemp and production of food products made from hemp seeds. Food production generates hemp husk, which is the flakes left over after the hulling of hemp seeds. This waste is sold for use as mulch, animal bedding or incinerated. One of the promising uses of hemp husk could be as a sorbent.

At present, pollution by oil and petroleum products is one of the most widespread and dangerous ecological burdens on the environment. This is primarily due to the development of the oil industry and increased consumption of petroleum products. Extraction, transportation, processing of hydrocarbon raw materials and use of

O. V. Tyukalova · S. S. Timofeeva (✉)
Irkutsk National Research Technical University, 83, Lermontova Str., Irkutsk 664074, Russia
e-mail: sstimofeeva@mail.ru

D. V. Ilkov
Medal LLC, Office 119, 1, Kuznetsova, Chelyabinsk 54028, Russia

L. V. Radionova and D. V. Ulrikh (eds.), *Advances in Ecology and Environmental Engineering*, Springer Proceedings in Earth and Environmental Sciences,
https://doi.org/10.1007/978-3-031-64423-8_38

petroleum products are often accompanied by emission of harmful substances into the atmosphere and spills that pollute soil and water. All components of ecosystems are affected: soil, water, atmosphere, flora and fauna.

Oil pollution of water and soil is often caused by improper compliance with environmental standards by companies involved in oil extraction and refining, storage of oil and petroleum products, their transportation by rail and water, and gas stations. The main causes of oil pollution of water are wear and tear of oil pipelines, safety violations and accidents [1].

Pollution of surface waters with petroleum products in wastewater continues to be a major problem. Petroleum products enter surface waters with untreated wastewater from many industries, as well as with ballast water from ships and in other ways.

Analysis of data on the discharge of pollutants in wastewater into natural surface waters of the Russian Federation shows that in the last 10 years (2013–2022) there was an increase in the discharge of a number of pollutants in wastewater into natural surface waters; in particular, the discharge of oil and petroleum products increased by 15.9% to 2657.2 tons in 2022 [2].

Petroleum products form a thin film on the surface of the water, and in the water column they are in emulsified or dissolved form, heavy fractions settle to the bottom—this causes great damage to inhabitants of the hydrosphere and biosphere. Petroleum products entering lakes, rivers, seas cause death of most of them. Effluents greatly deteriorate the consumer properties of water, making it unsuitable for drinking, domestic and industrial needs [3, 4].

Despite the existence of a large number of methods and approaches to cleaning surface and wastewater from oil and petroleum products, the problem remains relevant. To date, there are various methods to eliminate water pollution by petroleum products: mechanical, biological and physicochemical. One of the most widespread and effective methods of additional water treatment from petroleum products is adsorption. Various mineral and organic materials are used as adsorbents [5]. Currently, the search for effective but relatively inexpensive sorption materials for the removal of various contaminants from water is actively continuing.

A universal adsorbent with high adsorption capacity and widely used in production is activated carbon. However, due to the high cost of activated carbon and the need to regenerate it, it is necessary to search for alternative sorption materials that have high sorption capacity and low cost. As such sorption materials, wastes from various industries, especially cellulose-containing wastes, can be used. The disadvantage of this type of sorbent is high water absorption, which reduces the buoyancy and sorption activity of the sorbent. To increase the sorption properties, the surface of the materials is usually modified by treatment with chemical reagents that increase the hydrophobicity of the sorbents [6].

Wood waste (sawdust), which can be used as a sorbent for petroleum products in wastewater treatment technologies, is of great interest. The sorption activity of sawdust increases significantly (1.4–4 times) under physical and physicochemical influences on the sorbent (heating, microwave irradiation, ozonization) [7].

The use of such natural materials as leaves of various trees [1, 6, 8, 9], fibers of date palm trunk [10], leaves and stems of reed [11], cobs of cattail [12], coconut and walnut shells [3] is proposed as sorbents of petroleum products. The sorption activity of materials can be significantly increased by various modifications.

It is well known that a large amount of waste is generated during the production of plant products. For example, such wastes as rice and buckwheat hulls account for 14–22% and 16–20%, respectively, of the mass of husked products [13]. As the gross collection of edible plant raw materials increases, so does the amount of waste generated during harvesting and processing. In addition, valuable plant raw materials are constantly renewed and are often not used. The use of plant wastes of the agro-industrial complex is limited either to processing in the hydrolysis industry in insignificant amounts, or to burning as a low-calorie fuel, which leads to atmospheric pollution by combustion products, often these wastes go to landfills. Recently, these wastes have been considered as promising raw materials for the production of various materials, including sorbents for neutralization of oil-contaminated soils and water treatment from petroleum products [14–16].

Plant wastes are mainly composed of cellulose, hemicellulose and lignin [17]. They often have a unique porous structure. With the help of special treatment (various methods of chemical and physicochemical activation, pyrolysis, carbonization) it is possible to obtain highly active sorbents with specified properties. Used sorbents can be used as fuel briquettes with increased calorific value. Petroleum sorbents from agroindustrial wastes have the ability to biodegrade under the action of indigenous soil or artificially introduced microorganisms [18].

There are reports on the use of such plant wastes as peanut shells [19, 20], sunflower seed hulls [3], rice husks [13, 21], rice straw, ground corn cob kernels [22], buckwheat husks [13, 23], flax bark [24], etc. as sorbents for neutralization of oil-contaminated soils and for treatment of water from petroleum products.

Chemical, physical and physicochemical modification methods are used to increase the sorption activity of sorbents based on plant wastes. These include treatment with acids, alkalis, silicones, and organic solvents [17, 18]. A highly effective sorbent for the removal of oil and petroleum products from liquid media has been obtained by heat treatment of rice hulls [21] or co-thermolysis of rice hulls and straw with oil sludge [25].

In recent years the production of industrial hemp has been revived and developed in the Russian Federation. The sown areas of hemp in 2022 reached 13.8 thousand hectares [26]. In connection with the revival of hemp production in Russia, the volume of secondary raw materials is increasing, including the husk remaining after the seed hulling (the husk makes up to 35% of the seed mass [27]). The use of these wastes is currently unsustainable, as they are usually burned in open areas under poorly controlled conditions, causing environmental damage due to the formation of heavy smog [28]. However, hemp husk contains high amounts of cellulose (40.8%) and lignin (28.7%), which may be a prerequisite for its versatile use as a feedstock in many industries [29].

Studies conducted with hemp husk allow us to consider it as a promising sorbent for the treatment of wastewater from heavy metals and other pollutants [30, 31].

The aim of this study was to investigate the possibility of using hemp (*Cannabis sativa L.*) husk as a sorbent for the treatment of oily wastewater.

2 Study Objects and Methods

Studies of sorption characteristics of industrial hemp husk and its modifications were conducted on model wastewater obtained by dissolving Ichedinskaya oil in water under ultrasonic treatment with the operating frequency of ultrasound 22 kHz for 1 h.

As sorbents of petroleum products were used native dried husk, as well as husk modified by different methods: mechanical grinding (fraction 0.4–1.0 mm), washing with hot water, alcohol, treatment with surfactants: diluted liquid glass, silicone resin solution, foaming agents PO-6-RZ-1 and PO-6-RZ-2.

The native husk is a mixture of hemispherical, slightly elongated, greenish-brown, dull hard shells and residues of hemp seeds.

Studies of sorption capacity of husk and modified materials were conducted under static conditions. The content of petroleum products in water was controlled by the fluorometric method [32] on the device FLUORAT-2.

3 Results and Discussion

Hemp husk is a by-product of hemp seed hulling (Fig. 1). Table 1 shows the characteristics of hemp husk obtained during food production in Medal LLC, Chelyabinsk.

It has been experimentally found that the native husk is characterized by higher sorption capacity than that treated with hot water and alcohol (Fig. 2).

Fig. 1 Appearance of hemp husk

Table 1 Characteristics of native hemp husk

Indicator	Value
Ash content,%	4.2 ± 0.2
Humidity,%	6.3 ± 0.2
Cellulose content,%	41.1 ± 2
Lignin content,%	27.8 ± 1.5
Kernel residue,%	7.5 ± 0.3
Crude fat,%	9.2 ± 0.6

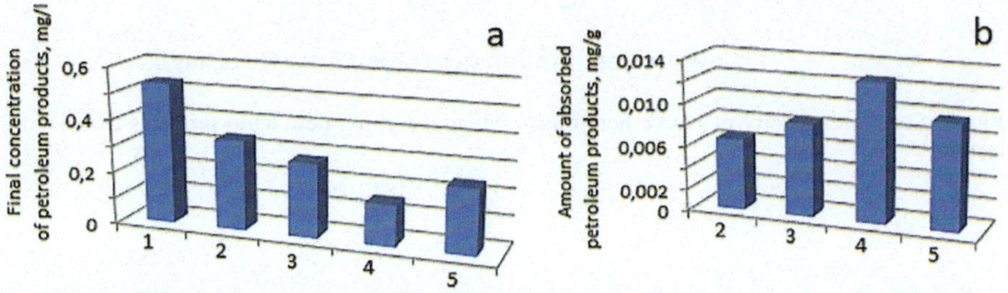

Fig. 2 Sorption of petroleum products from aqueous solutions by hemp husk: **a** residual concentration of petroleum products; **b** sorption capacity of husk; 1—initial concentration of petroleum products; 2—husk washed with hot water; 3—husk washed with alcohol and water (1:1); 4—untreated husk; 5—crushed husk

According to known studies [33], when using sorbents for water purification from oil contamination, in which pores in size are inaccessible for molecules of the removed substance, oil sorption occurs due to adhesion on the surface of sorbent particles. In this case, the amount of absorbed petroleum products is determined by the surface properties (hydrophobicity and oleophilicity), as well as the value of the specific surface area of the material.

Obviously, unwashed husk contains surface substances (residues of hemp oil) with affinity to petroleum products. Treatment with hot water and alcohol removes these substances from the surface of the husk and reduces its sorption capacity. Washing the husk with hot water removes from its surface such ballast extractive substances as monosaccharides, glycosides, dyes, as well as proteins, amino acids, pectin substances, mono-oligo- and polysaccharides [18]. Washing with alcohol solution removes organo-soluble substances.

In the work we found a linear dependence of sorption capacity of native husk on the content of petroleum products in water (Fig. 3), described by the linear equation $y = 0{,}014x + 0{,}0081$ with correlation coefficient $r = 0{,}99{,}975$. This is most likely due to the fact that as the layer of petroleum products accumulates on the surface of the sorbent, its adhesion increases.

The process of sorption of petroleum products by native hemp husk is described by the first order equation (Fig. 4) with the first order constant $k = 0.54\ h^{-1}$.

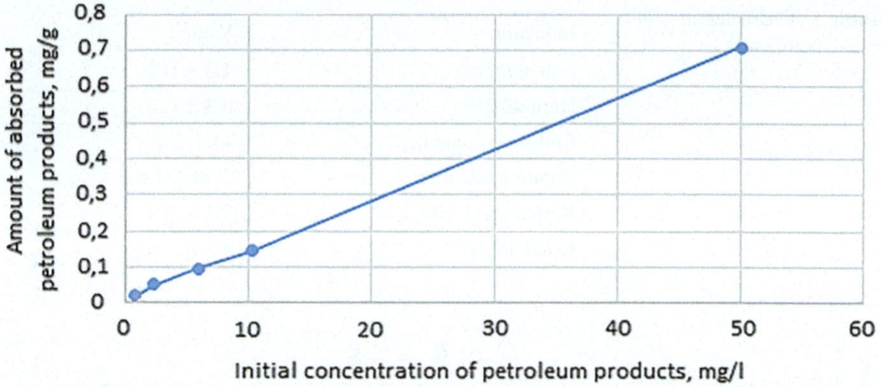

Fig. 3 Sorption capacity of native hemp husk when absorbing petroleum products dissolved in water

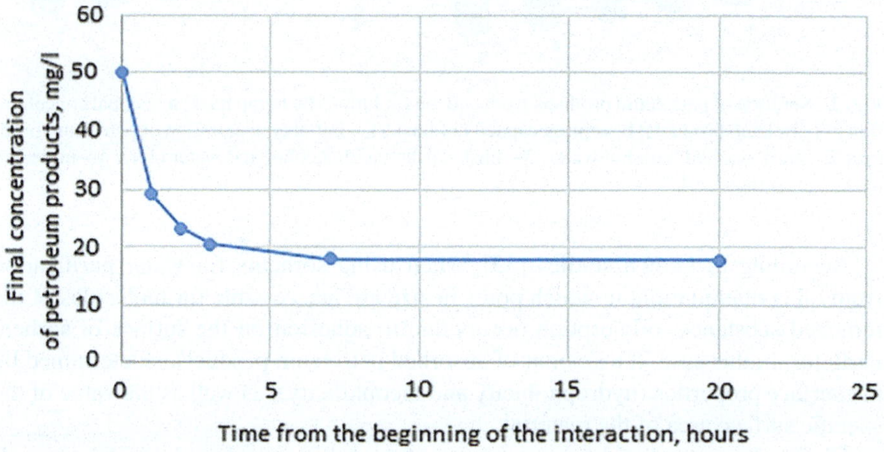

Fig. 4 Kinetics of sorption of petroleum products (50 mg/L) by native hemp husk from water

When native hemp husk was treated with liquid glass diluted with water in the ratio of 1:2, silicone resin solution and propellants PO-6-RZ-1 and PO-6-RZ-2 to increase the hydrophobic properties of the husk and to increase its affinity for petroleum products, it was found that a significant increase in sorption capacity was observed during modification with liquid glass and propellant PO-6-RZ- (Table 2).

It should be noted that no treatment improves the buoyancy of the husk. The buoyancy of the husk after treatment with diluted liquid glass and silicone resin solution does not change compared to the dried untreated husk, and after treatment with foaming compositions PO-6-RZ-1 and PO-6-RZ-2 it is significantly reduced—the husk sinks. These data are important for the use of hemp husk and its modified samples as sorbents for the removal of oil films from the water surface.

Table 2 Efficiency of treatment of model oily wastewater under static conditions with surfactant-modified husk (Cin = 6 mg/L)

Modified sorbents	Treatment efficiency, %
Native husk	48.5
Husk modified with liquid glass	79.2
With silicone resin solution	43.0
With foaming agent PO-6-RZ-1	77.2
With foaming agent PO-6-RZ-2	32.3

4 Conclusion

Indicators of the process of sorption of dissolved petroleum products by hemp husk have been determined. It is shown that the sorption capacity of native husk increases with increasing concentration of dissolved petroleum products in water. Comparison of sorption properties of native and modified hemp husk for extraction of petroleum products from wastewater showed that treatment with hot water, alcohol, silicone resin solution, and foaming agent PO-6-RZ-2 does not increase the treatment efficiency.

The best modifiers of hemp husk are liquid glass and foaming agent PO-6-RZ, widely used in fire extinguishing. Treatment with diluted liquid glass and foaming agent PO-6-RZ increases the sorption capacity of the husk by 1.5 times. Thus, the mass waste of crop production—native and modified husk of industrial hemp—can be used as a sorbent in the treatment of wastewater from petroleum products.

References

1. Badmaeva SV, Dashinamzhilova EC, Khanhasaeva SC (2018) Application of sorbents derived from plant waste for the absorption of petroleum products. Bulletin of Buryat State University. Chemistry. Physics 4:30–35. https://doi.org/10.18101/2306-2363-2018-4-30-35
2. On the state and protection of the environment of the Russian Federation in 2022. State report. Ministry of Natural Resources of Russia; Moscow State University, Moscow, 2023, p 686
3. Yamansarova ET, Gromyko NV (2015) Perspective types of plant sorbents for removal of petroleum products from natural waters. Education and Science in the Modern World. Innovations 1:180–185
4. Korshunova TYu, Loginov ON (2019) Oil pollution of aquatic environment: peculiarities, influence on various objects of hydrosphere, main methods of treatment. Ecobiotech. 2(2):157–174. https://doi.org/10.31163/2618-964X-2019-2-2-157-174
5. Veprikova EV, Tereshchenko EA, Chesnokova NV, Shchipko IL, Kuznetsova BN (2010) Features of water treatment from petroleum products using oil sorbents, filtering materials and active carbons. Journal of the Siberian Federal University. Chemictry 3:283–304
6. Prokhorova SV, Stepanova SV (2017) Treatment of oil-contaminated water by modified sorption material based on waste of plant origin. The Samara Bend: problems of regional and global ecology 26(3):221–223

7. Malyshkina ES, Vyalkova EI, Osipova EYu (2019) Use of natural sorbents in the process of water treatment from petroleum products. Bulletin of Tomsk State University of Architecture and Civil Engineering 21(1):188–200. https://doi.org/10.31675/1607-1859-2019-21-1-188-200
8. Sverguzova SV, Bomba IV, Voronina YuS (2018) Purification of oily emulsions with leaf litter of cherry and rowan. Chemical Bulletin 1(4):4–10
9. Alexeeva AA, Stepanova SV (2015) Study of physicochemical bases of the process of oil film sorption from water surface by mixed leaf fall. Water: Chemistry and Ecology 4:87–90
10. Boleydei Hamid, Razavi Zahra (2017) Utilization of Date Palm Trunk Fibers (DPTFs) Waste as a Sorbent for the Removal of Crude Oil from Aqueous Media: Equilibrium Isotherms and Kinetic Studies. Iranian Journal of Health, Safety & Environment 5(4):1099–1106
11. Sokolova NA, Khlobzheva IN, Vasenev II, Kostin VE, Kochetkov VG, Gamaga BB (2022) Prospects for the use of leafy biomass of the common reed for water treatment. AgroEcoInfo: Electronic Scientific and Production Journal 4. https://doi.org/10.51419/202124433
12. Gorovykh OG, Sayevich KF, Alzhanov BA (2019) Use of fibers of cattail cob perianth for treatment of water bodies from oil pollution. Proceedings of BSTU Series 2. 2:183–192
13. Abdullin ISh, Israfilov IKh, Shayekhov MF (2002) Activation of sorbents based on rice husk and buckwheat husk using high-frequency discharge of reduced pressure. Proceedings of the 3rd International Symposium on Theoretical and Applied Plasma Chemistry. Ivanovo 2:302
14. Tsombueva BV, Goryashkieva ZV, Shcherbakova LF (2017) Method of soil treatment from oil pollution using natural sorbents. Bulletin of Volgograd State University 7(2):19–25. https://doi.org/10.15688/jvolsu11.2017.2.3
15. Yakubovsky SF, Bulavka YuA, Mayorova EI (2017) Obtaining sorbent for collection of spilled oil and petroleum products by utilization of agro-industrial waste. Bulletin of Polotsk State University. Series B. Industry. Applied Sciences. 11:84–89
16. Liu L, Wang L, Song W, Yang L, Yin L, Xia Sh, Wang H, Strong PJ, Song Zh (2018) Crude oil removal from aqueous solution using raw and carbonized Xanthoceras sorbifolia shells. Environmental Science and Pollution Research 25(29):29325–29334. https://doi.org/10.1007/s11356-018-2895-0
17. Levchuk AA, Rashid ID (2015) Modification of properties of lignocellulosic wastes of plant production. Electronic online Polythematic Journal Scientific Works of KubSU 5:175–198
18. Bulavka YuA, Yakubovsky SF, Mayorova EI (2017) Utilization of agro-industrial wastes for the production of oil sorbents. 21 century. Technosphere Safety 2(4):38–47
19. Hunadi L, Shaykhiev IG, Svyatchenko AV, Sapronova ZhA, Bomba IV (2018) Utilization of peanut processing waste as a sorption material. Chemical Bulletin 1(4):40–49
20. Karabaeva MI, Mirsalimova SR, Salikhanova DS, Mamadalieva SV, Ortikova SS (2022) Main directions of utilization of waste vegetable raw materials (peanut shells) as adsorbents (review). Chemistry of Plant Raw Materials 1:53–69. https://doi.org/10.14258/jcprm.2022019956
21. Gafarov IG, Kuznetsov AI, Rastorguev YuI, Timofeev VS, Tyomkin ON, Hoang Kim Bong (2005) Sorbent for removal of oil and petroleum products from liquid media and method of its preparation from rice husk. Patent of the Russian Federation RU2259875C2
22. Ivanova MA, Zenitova LA (2021) Treatment of water bodies from oil spills with sorbent based on polyurethane foam, vegetable and polymer wastes. Innovative Scientific Research: Online Journal 2–3(4):186–191. https://doi.org/10.5281/zenodo.4604865
23. Yamansarova ET, Gromyko NV, Abdullin MI, Kukovinets OS, Zvorygina OB (2015) Study of sorption properties of materials based on plant raw materials in relation to oil water pollution. Bulletin of Bashkir University 20(4):1209–1211
24. Shaykhiev IG, Nizamov RKh, Stepanova SV, Friedland SV (2010) Flax processing wastes as sorbents of petroleum products. Determination of oil capacity. Bulletin of Bashkir University 15(2):304–306
25. Appazov NO, Baharbayev BM, Akylbekov NI, Zhapparbergenov RU, Kanzhar SA, Diyarova BM (2020) Obtaining a quality sorbent from rice wastes and oil sludge. Oil and Gas 3–4(117–118):169–179

26. National report "On the progress and results of implementation of the State Program for the Development of Agriculture and Regulation of Markets of Agricultural Products, Raw Materials and Food in 2022". Moscow (2023), p 120
27. Zverev SV, Zubtsov VA, Roslyakov YuF, Efremov DP, Yanova MA (2020) Physical and technological properties of hemp seeds. Bulletin of Krasnoyarsk State Agrarian University 11:240–247. https://doi.org/10.36718/1819-4036-2020-11-240-247
28. Zaitseva LA, Volkova AV, Minevich IE (2023) Preparation of cellulosic product from hemp husk. Polzunov Bulletin 2:174–183. https://doi.org/10.25712/ASTU.2072-8921.2023.02.023
29. Minevich IE, Nechiporenko AP, Volkova AV, Zaitseva LA (2023) On the prospects of processing hemp husk into cellulosic product. Chemistry of Plant Raw Materials 2:345–354. https://doi.org/10.14258/jcprm.20230211359
30. Samodolova OA, Samodolov AP, Ulrich DV (202) Hemp husk as a promising sorbent for wastewater treatment of different origin. Proceedings of Higher Educational Institutions. Construction 12:58 PM-65. https://doi.org/10.32683/0536-1052-2021-756-12-58-65
31. Samodolova OA, Samodolov AP, Ulrich DV, Lonsinger TM (2023) Use of plant wastes in the treatment of wastewater contaminated with heavy metals. Bulletin of Moscow State Construction University 18(5):747–756. https://doi.org/10.22227/1997-0935.2023.5.747-756
32. PND F 14.1:2:4.128-98 Methodology for measuring the mass concentration of petroleum products in samples of natural water, drinking water, and wastewater by fluorimetric method on liquid analyzer Fluorat-02. Moscow (2012)
33. Gurvich LM, Sherstnev NM (2011) Multifunctional compositions of surfactants in technological operations of oil production. All-Russian Scientific Research Institute of Management Organization of Oil and Gas Industry Economy Management, Moscow

Energy Science, Green Technology and Eco-innovation

The Use of Renewable Energy of Sea Waves for Desalination of Seawater

V. V. Mironov, Yu. A. Ivanyushin, D. A. Suglobov, and D. V. Mironov

Abstract This article discusses an innovative technology for desalination of seawater using renewable energy from ocean waves. This technology can replace traditional energy-intensive equipment for desalination of water by reverse osmosis and has prospects for use in the coastal areas of the Black and Japanese Seas. The main advantage of this technology is the use of renewable energy from ocean waves. This makes it possible to significantly reduce the cost of producing fresh water and reduce the environmental burden. A technical solution has been developed for the use of low-potential renewable energy of sea waves for desalination of seawater. The purpose of this work is a feasibility study of the application of innovative technology for desalination of seawater using renewable energy from sea waves to provide clean fresh water for non-residential facilities (beach glamping). To achieve this goal, mathematical analysis methods were used using statistical data and long-term observations from domestic and foreign sources with analysis and comparison of the data contained. The technology of desalination of seawater using renewable energy from ocean waves pays off within two years, which makes it economically profitable. In addition, it can be used in various climatic conditions and does not require high maintenance costs.

Keywords Reverse osmosis · Desalination of seawater · Wave energy

1 Introduction

Currently, there is a growing shortage of clean fresh water in many regions of the world, due to climate changes on the planet and man-made activities of mankind. One of the ways to solve the growing shortage of clean fresh water is the desalination

---

V. V. Mironov · Yu. A. Ivanyushin · D. A. Suglobov (✉)
Industrial University of Tyumen, 38, Volodarskogo Str., Tyumen 625000, Russia
e-mail: suglobovda@gmail.com

D. V. Mironov
LLC "ELECTRORAM", Przhevalskogo Str., 33, Corp. 1, Ap. 79, Tyumen 625023, Russia

L. V. Radionova and D. V. Ulrikh (eds.), *Advances in Ecology and Environmental Engineering*, Springer Proceedings in Earth and Environmental Sciences,
https://doi.org/10.1007/978-3-031-64423-8_39

of seawater by reverse osmosis. It is possible to reduce energy consumption for water purification from salt and other contaminants by reverse osmosis using renewable energy from the sun, wind and fluctuations in the surface of water areas. Unlike the renewable energy of the sun and wind, the volume of wave energy is truly limitless. The wave energy density exceeds similar indicators for solar and wind energy by several times. The greatest interest for practical use in water purification plants from salt and other contaminants is the energy of small fluctuations in the surface of water areas. The availability and repeatability of low-potential energy of sea waves for most regions of the world ocean is 80–90%. We can say that low-potential energy exists almost always and everywhere. The question is how to effectively take it and turn it into the power supply of high-pressure pumps of standard reverse osmotic desalination plants. If we take into account that to obtain one cubic meter of fresh water in reverse osmotic installations, it takes from 4 to 8 kWh of electricity, then using the converted low-potential wave energy with a capacity of only 1 kW/m, it is possible to obtain from 3 to 6 cubic meters of fresh water per day from one meter of the width of the wavefront. At the same time, you can save about 24 kWh of electricity per day. There are known studies on the use of the energy of sea waves for the operation of reverse osmotic desalination plants. However, most installations of this kind convert the energy of sea waves first into electrical energy, and then into the mechanical operation of the drive of high-pressure pumps, which reduces the efficiency of energy conversion [1–4]. There are also known studies on the direct conversion of the energy of sea waves into useful power, but the scope of application of such technologies is limited. They cannot use the low-potential energy of sea waves with a wave height of less than a meter. In addition, the known energy conversion units cannot create a pressure of more than 12 excess atmospheres, which is insufficient for the operation of reverse osmotic desalination plants with medium and high salinity seawater [5, 6].

2 Materials and Methods

The purpose of the work is a feasibility study of the application of innovative technology of desalination of seawater for the life support of non–capital construction facilities (beach glamping) based on the use of renewable low-potential energy of sea waves.

Technical solution. Developed by a team of authors (Industrial University of Tyumen and LLC "ELECTRORAM"), the technical solution (Fig. 1) allows you to remove the low-potential energy of sea waves with a height of 20 cm and convert it into high-pressure hydraulic power required for the operation of standard reverse osmotic desalination plants. The developed technical solution makes it possible to eliminate environmental pollution by greenhouse gases.

The technical solution works as follows. A buoy with positive buoyancy (1) is placed on the surface of the sea area. The buoy (1) is connected to a fixed vertical

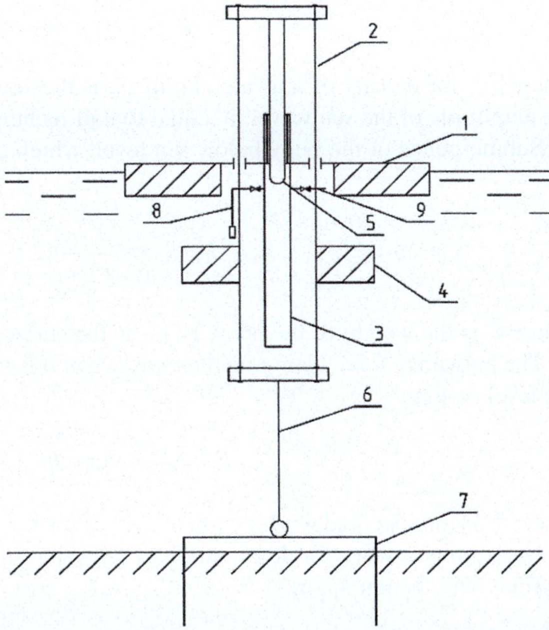

Fig. 1 Diagram of a marine wave energy puller with a hydraulic power generator: 1—a movable buoy with positive buoyancy; 2—a vertical fixed frame; 3—a high-pressure pump housing rigidly connected to a movable buoy; 4—a float rigidly connected to a fixed frame; 5—a high-pressure pump plunger rigidly connected to a fixed frame; 6—cable; 7—bottom anchor

frame (2) with the possibility of relative movement of the buoy (1) and the frame (2). The height of the vertical frame is selected depending on the amplitude of vibrations of the surface of the water area. The height of the frame (2) can reach up to 10 m. The buoy (1) is rigidly connected to the high-pressure pump housing (3). The frame (2) is provided with a fixed float rigidly connected to it (4) and a high-pressure pump plunger rigidly connected to the frame (5). The frame (2) is connected to the bottom anchor (7) by a cable (6). The bottom anchor (7) is made in the form of a heavy load or a vacuum pile. The housing of the high-pressure pump (3) is supplied with suction (8) and discharge (9) pipelines with check valves. Marine or polluted fresh water under high pressure sufficient for the operation of reverse osmotic desalination membranes, under wave action on a mobile buoy (1), is fed from the pump housing (3) through the discharge pipeline (9) to the collector, and then to the reverse osmotic module to obtain purified water.

When the surface of the water area is agitated, the buoy (1) performs reciprocating movements up and down. When the buoy (1) moves down due to its own weight, the surrounding water is sucked into the pump housing (3). When the buoy (1) moves up, water is injected due to the pushing force of seawater acting on the buoy (1) and the movement of the plunger (5) inside the pump housing (3).

Energy The seawater desalination plant is a modular design. The number of modules is selected depending on the need for fresh water.

Calculated dependencies. Consider the operation of a power station consisting of floating modules, operating on low-potential energy of sea waves of high security and repeatability. And the excess pressure p, Pa, at the depth H_p in the wave trough:

$$p = \rho \cdot g \cdot (H_p - a), \tag{1}$$

where p is the density of seawater, kg/m^3; g is the acceleration of gravity, m/s^2; a is the amplitude of the sea wave, m, equal to half its height H.

Submergence of the buoy below sea level, which compensates for its weight h_1, m:

$$h_1 = \frac{G}{\rho \cdot g \cdot \omega_1}, \tag{2}$$

where G is the weight of the buoy, N; ω_1 is the surface area of the buoy, m^2.

The buoyancy force acting on the buoy when it is submerged under the waters of the level of F_2:

$$F = \rho \cdot g \cdot \omega_1 \cdot h_b, \tag{3}$$

where h_b is the specified working immersion of the buoy, which provides the pressure in the pump necessary for desalination of seawater, m.

Total buoy height h_p, m:

$$h_p = h_1 + h_b + h_3, \tag{4}$$

where h_b is the height of the reserve, not less than 0,05 m.

Maximum buoyancy force F_3:

$$F_3 = \rho \cdot g \cdot \omega_1 \cdot h_p. \tag{5}$$

Pressure p_1 developed by the plunger in the pump chamber, Pa:

$$p_1 = \frac{F - G}{\omega_2}, \tag{6}$$

where ω_2 is the cross-sectional area of the plunger, m^2.

Plunger stroke h_n, depending on the wave height, m:

$$h_n = H - h_b \tag{7}$$

Bellows generator water capacity:

- per hour of sea water, m^3/h:

$$q_{hr}^{sea} = \omega_2 \cdot h_n \cdot \frac{3600}{T}, \tag{8}$$

where T is the sea wave period, s;

- per hour of fresh water, m^3/h:

$$q_{hr}^{fr} = k \cdot q_{hr}^{sea}, \tag{9}$$

where k is the proportion of fresh water generated from sea water when water is supplied to the reverse osmosis unit [7–9];

- average daily fresh water productivity, m^3/day:

$$Q_t^{fr} = t \cdot k \cdot q_{hr}^{sea}, \tag{10}$$

where t is the duration of operation of the bellows generator and reverse osmosis unit during the day, h;

- average annual fresh water productivity, m^3/year:

$$Q_{year}^{fr} = n_y \cdot Q_t^{fr}, \tag{11}$$

where $n_y = 365$ days is the duration of a calendar year.

Source data. The main initial data for calculating the characteristics of the power plant, including the performance of the technology for fresh water, are the height of the seawave H, m, and its period T, s. The data of long-term observations of the values of H and T in the Black Sea and the Sea of Japan are presented in Fig. 2, made according to the data of [7, 10]. Similar data is available for other maritime territories of the Russian Federation.

The Black Sea and the Sea of Japan are dominated by waves up to two meters high, the total annual frequency of which is 84.8%–82.6%, respectively. Waves with a period of 2–6 s are most common in water areas. At the same time, if in the Black Sea waves with a period of 2–4 and 4–6 s occur in almost equal proportions (46.2–48.2% of the total number of waves), then in the Sea of Japan waves of a small period predominate, up to 4 s—64.3% of the total number of waves.

In addition, in the Black Sea area, there are significant seasonal changes in the average wave heights, which reach their maximum values in the winter period of the year (a twofold increase) [11–14], this confirms the data on the average wave heights in the coastal zone (Fig. 3). In this regard, the seasonal wave regime in the coastal zone of the Sea of Japan, which is much more uniform (stable) [15], based on the average wave height, is of particular interest for practical application of the technology (Fig. 3).

However, the magnitude of wind waves is not constant and varies in a fairly large range during the month and individual days (Fig. 4).

Please note that the sea partially freezes in winter. The ice cover reaches its maximum development in mid-February. On average, 52% of the area of the Tatar Strait and 56% of the Peter the Great Bay are covered with ice [16] with a duration of up to 4–5 months annually. In this regard, when determining the annual productivity of the technology in formula (11), it is necessary to enter a correction factor that takes into account the projected duration of the presence of ice cover in the projected area.

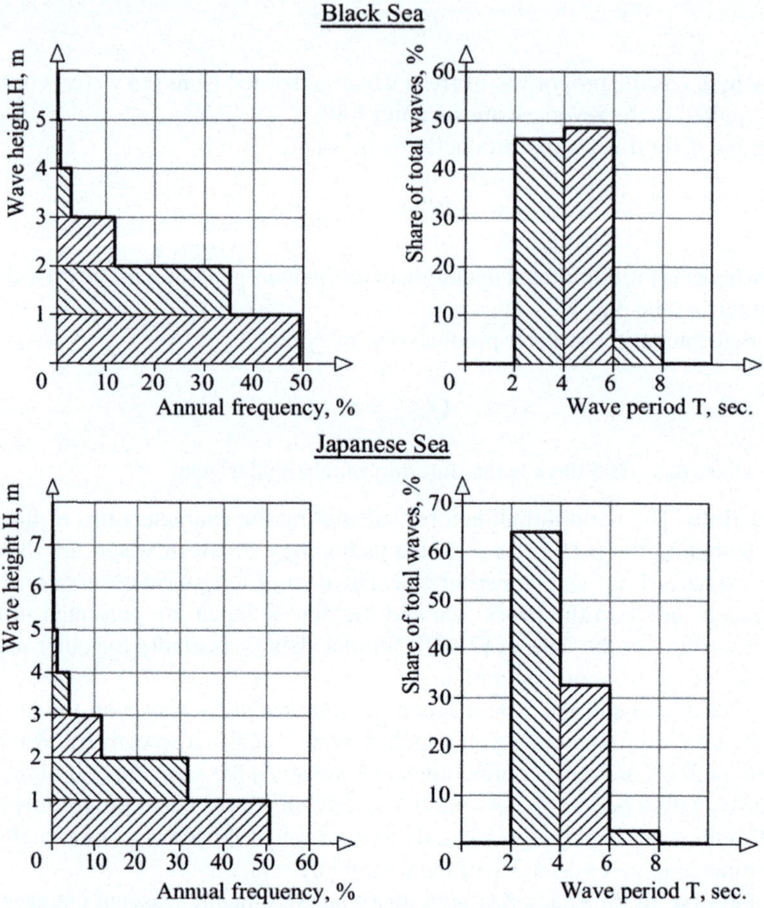

Fig. 2 Data of long-term observations of height and period of waves in the waters of the black sea and sea of Japan

Taking into account the above, formula (11) for determining the annual productivity of fresh water takes the form:

$$Q^{fr}_{year} = t \cdot k \cdot k_1 \cdot \omega_2 \cdot \sum\nolimits_{1}^{i=12} h_{ni} \cdot n_i \cdot \frac{3600}{T_i}, \quad (12)$$

where T_i ti is the average wave period of the i-th month, s; n_i ni is the duration of the i-th month, day; h_{pi} is the plunger stroke, depending on the average wave height H_i during the i-th month, m; k_I is the coefficient that takes into account the duration of the presence of ice cover in the location.

Taking into account the percentage ratio of wave periods (according to the data of [17, 18], Fig. 2) and assuming that this ratio does not change during the year, formula

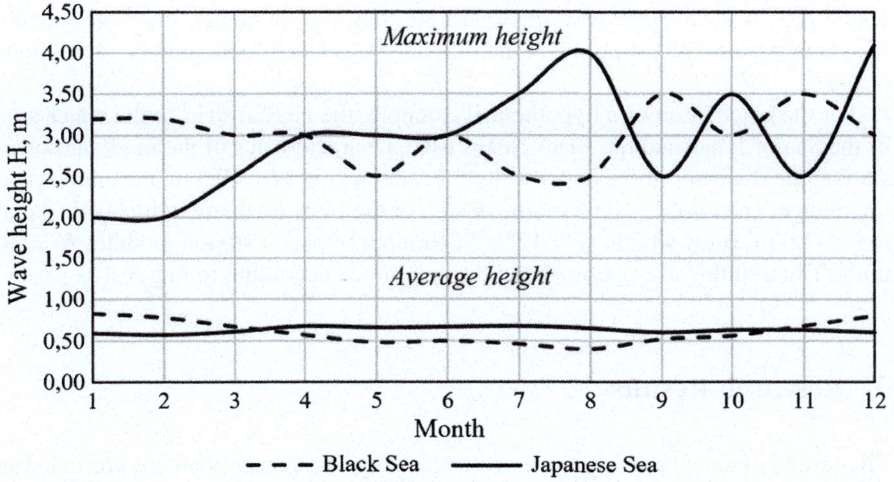

Fig. 3 Average and maximum wave heights in the coastal strips of the Black (from Anapa to Pitsunda) and Japanese (from Posyet to Nakhodka) seas (average monthly long-term data)

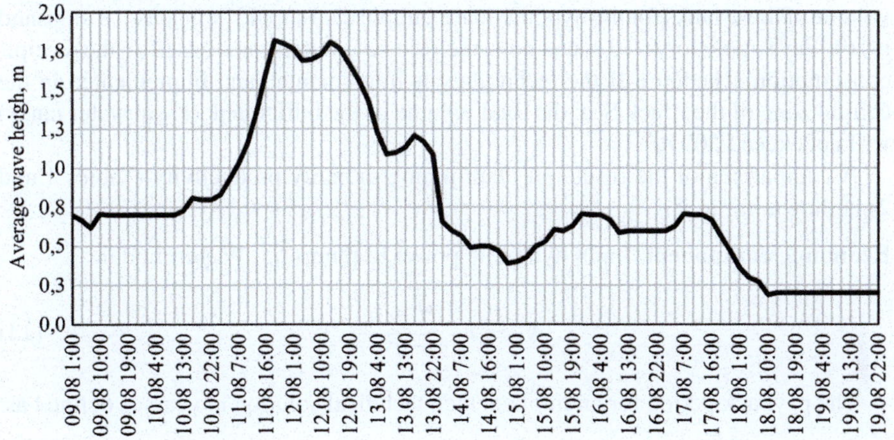

Fig. 4 Average wave height in the sea of Japan near Rynda Bay (Russian Island) in the period from 08/09/2023 to 08/19/2023

(12) will take the form:

$$Q_{year}^{fr} = t \cdot k \cdot k_1 \cdot \omega_2 \cdot \sum_{1}^{i=12} h_{ni} \cdot n_i \cdot \frac{3600}{\sum\left(\frac{m_j}{T_j^{mid}}\right)}, \tag{13}$$

where T_j^{mid} is the average wave period in the selected *j*-th range;*j* mj is the specific weight of waves with the period of the *j*-th range from their total number, in fractions of one.

For the calculation of a hypothetical example, the coastal strip of the water area of the Sea of Japan and the Black Sea is used. Characteristics of the modular power station: the diameter of the upper wall of the air chamber $D = 0.3$ m, the diameter of the plunger (piston) $d = 0.05$ m, the design of the buoy is square, with a side equal to $b = 3.6$ m. Buoy weight $G = 1250$ N. Number of power station modules $N = 30$ units. The monthly average wave height is assumed according to Fig. 3.

3 Research Results

The results of calculating the fresh water capacity of a power plant are presented in Tables 1 and 2.

The disadvantage of implementing the technology in the coastal zone of the Sea of Japan is the presence of ice cover during the three winter months. In this case, it is necessary to provide buffer tanks (clean water tanks) for the water supply. The total volume of these tanks for the hypothetical example data should be at least 18 thousand m^3 (90 day supply for 1030 consumers, the volume is determined based on the receipt-consumption schedule). If the technology is implemented and implemented in the coastal zone of the Black Sea, the total volume of the buffer tank (clean water tanks) will not exceed 300 m^3.

A comparison of the results of determining monthly productivity (Tables 1 and 2) is shown in Fig. 5.

Economic component. Input hydraulic power to pump N_1, W, per module

$$N_1 = F \cdot \frac{H - h_b}{T}, \tag{14}$$

The payback period for an implemented technical solution in years is defined as:

$$T_{pb} = \frac{C_1}{E_{year}}, \tag{15}$$

where C_1—the cost of one module does not exceed $C_1 \leq 300$ thousand rubles; E_{year}—annual savings on electricity due to the introduction of a modular power station, thousand rubles:

$$E_{year} = c_{kW-h} \cdot F \cdot \sum_{1}^{i=12} \frac{H_i - h_b}{\sum\left(\frac{m_j}{T_j^{mid}}\right)} \cdot t \cdot k_1 \cdot n_i, \tag{16}$$

Table 1 Technology performance in the case of implementation in the western part of the coastal strip of the sea of Japan

Wave periods T, sec		0–2	2–4	4–6	6–8	≥ 8	Average hourly (over the course of a month) fresh water productivity, m^3/h	Monthly volume of fresh water produced by one module, m^3
Accepted wave period in the selected range T^{mid}, sec		1	3	5	7	10		
Month	Specific weight m of the selected range of periods	0.002	0.6413	0.3270	0.0272	0.0018		
	Average wave height H, m	Average hourly productivity for fresh water, l/h, in the selected range of wave periods from one module						
1	0.59	2.50	267.32	81.78	4.86	0.23	0.357	265.37
2	0.57	2.42	258.25	79.01	4.69	0.22	0.345	231.57
3	0.61	2.59	276.38	84.56	5.02	0.23	0.369	274.37
4	0.68	2.88	308.09	94.26	5.60	0.26	0.411	295.99
5	0.66	2.80	299.03	91.49	5.44	0.25	0.399	296.86
6	0.67	2.84	303.56	92.87	5.52	0.26	0.405	291.63
7	0.68	2.88	308.09	94.26	5.60	0.26	0.411	305.85
8	0.65	2.76	294.50	90.10	5.35	0.25	0.393	292.36
9	0.60	2.54	271.85	83.17	4.94	0.23	0.363	261.17
10	0.63	2.67	285.44	87.33	5.19	0.24	0.381	283.36
11	0.63	2.67	285.44	87.33	5.19	0.24	0.381	274.22
12	0.60	2.54	271.85	83.17	4.94	0.23	0.363	269.87
In just a year from one module, m^3								**3342.63**
In total for a year from N = 30 modules, thousand m^3								**100.279**
including taking into account the presence of ice cover during 3 winter months ($k_1 = 0.75$), thousand m^3								**75.209**
Average daily (per year) productivity from N = 30 modules, m^3/day								**206.053**
Equivalent number of consumers (at the accepted specific water consumption—200 l/day per person), person								**1030**

where $c_{\text{kW-h}}$ is the cost of electricity at the location of the technology, rub/kWh; H_i is the average wave height in the i-th month, m.

A graphical representation of formula (16) is shown in Fig. 6. The horizontal line is the numerator of the formula, and the oblique lines are the denominator. The intersection of lines is a payback point. The payback period only for the production of fresh water (in comparison with the use of standard centrifugal pumps for feeding

Table 2 Technology performance in the case of implementation in the coastal strip of the Black Sea

Wave periods T, sec		0–2	2–4	4–6	6–8	≥ 8	Average hourly (over the course of a month) fresh water productivity, m^3/h	Monthly volume of fresh water produced by one module, m^3
Accepted wave period in the selected range T^{mid}, sec		1	3	5	7	10		
Month	Specific weight m of the selected range of periods	0.0006	0.4623	0.4859	0.0050	0.0016		
	Average wave height H, m	Average hourly productivity for fresh water, l/h, in the selected range of wave periods from one module						
1	0.83	1.06	271.09	170.96	12.57	0.28	0.456	339.23
2	0.78	0.99	254.76	160.66	11.81	0.26	0.428	287.94
3	0.67	0.85	218.83	138.00	10.14	0.23	0.368	273.83
4	0.57	0.72	186.17	117.40	8.63	0.19	0.313	225.45
5	0.48	0.61	156.78	98.87	7.27	0.16	0.264	196.18
6	0.50	0.64	163.31	102.99	7.57	0.17	0.275	197.76
7	0.46	0.58	150.24	94.75	6.96	0.16	0.253	188.01
8	0.40	0.51	130.65	82.39	6.06	0.14	0.220	163.48
9	0.52	0.66	169.84	107.11	7.87	0.18	0.286	205.67
10	0.56	0.71	182.90	115.34	8.48	0.19	0.308	228.88
11	0.68	0.86	222.10	140.06	10.29	0.23	0.374	268.96
12	0.80	1.02	261.29	164.78	12.11	0.27	0.439	326.97
In just a year from one module, m^3								**2902.35**
In total for a year from N = 30 modules, thousand m^3								**87.071**
There is no ice cover on the surface of the water area, thousand m^3								–
Average daily (per year) productivity from N = 30 modules, m^3/day								**238.549**
Equivalent number of consumers (at the accepted specific water consumption—200 l/day per person), person								**1190**

to reverse osmotic membranes) is from 11 months to 2 years, depending on the cost of electricity—from 5 to 10 rubles per 1 kWh.

The lower the cost of floating modules and the higher the price of electricity, taking into account logistics (for example, when generating from an autonomous source—a diesel generator), the more efficient the use of renewable energy of sea

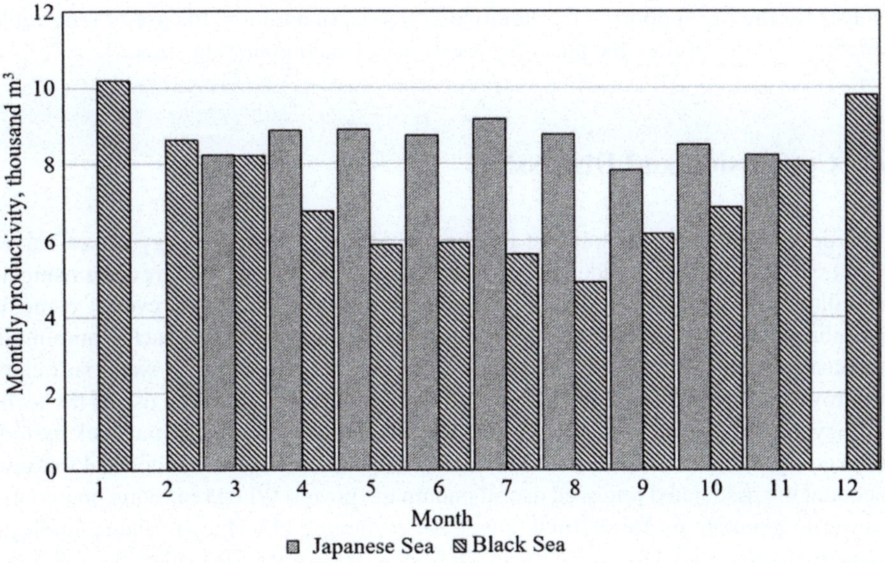

Fig. 5 Monthly productivity of the technology during implementation and implementation in the coastal strip of the Japan and Black Seas

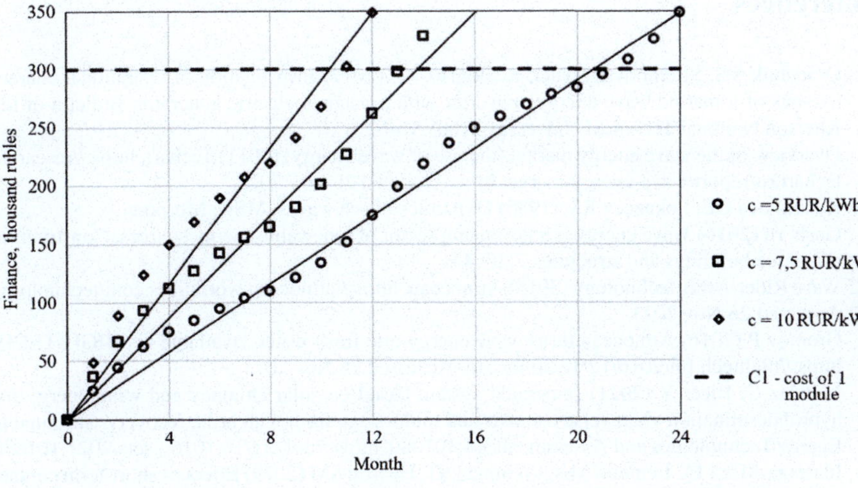

Fig. 6 Cost-effective use of sea wave energy to provide fresh water for beach modular hotels in the coastal strip of the black sea

waves for the life support of beach modular hotels. In addition, the use of renewable energy does not pollute the air in the location of beach glamping sites.

4 Conclusion and Discussion

The developed technical solution for the use of low-potential energy of sea waves makes it possible to exclude energy—consuming equipment from reverse osmosis installations—high-pressure pumps for supplying seawater to reverse osmosis membranes (Fig. 1). The proposed technology for life support of beach glamping in the coastal zone of marine areas allows obtaining volumes of fresh water sufficient to provide a significant number of consumers (Tables 1 and 2). The use of the technology in freezing seas is justified for non-year-round operation. The payback period, depending on the cost of electricity, does not exceed two years without taking into account the associated removed useful pneumatic power. When removing pneumatic power to generate useful thermal power, the economic efficiency of the technology increases.

References

1. Osadchuk VS, Strembovsky YuG, Kostogryz AP (2017) Investigation of operational characteristics of a marine wave energy converter with a magnetoelectric generator. Bulletin of the Kherson National Technical University 1(60):51–62
2. Overview of the wave energy market (the use of wave energy) (2011) Techart. https://research.techart.ru/report/wave-energy-market.htm. Accessed 26 Nov 2023
3. Kononkova GE, Pokazeev KV (1985) Dynamics of sea waves. MSU, Moscow
4. Gush Yu (2016) Wave energy is a promising sector of renewable energy sources. Construction of unique buildings and structures 2:30–44
5. Wave Rider Array technology (2015) Atmocean. https://atmocean.wordpress.com/technology. Accessed 26 Nov 2023
6. Gromov P (2016) Atmocean turns wave energy into fresh water. Available via HIGH-TECH. https://hightech.fm/2016/12/16/atmocean. Accessed 26 Nov 2023
7. Méndez C, Bicer Y (2021) Integrated system based on solar chimney and wind energy for hybrid desalination via reverse osmosis and multi-stage flash with brine recovery. Sustainable Energy Technologies and Assessments 44:101080. https://doi.org/10.1016/j.seta.2021.101080
8. Idarraga-Mora JA, Lemelin MA, Weinman ST, Husson SM (2019) Effect of short-term contact with C1–C4 monohydric alcohols on the water permeance of MPD-TMC Thin-Film composite reverse osmosis membranes. Membranes 9(8):92. https://doi.org/10.3390/membranes9080092
9. Solovieva ZA (2019) Vodnyy sektor arabskikh stran: defitsit vody i zelenyye tekhnologii (The water sector of the Arab countries: water deficit and green technologies). Trudy Instituta vostokovedeniya RAN (Proceedings of the Institute of Oriental Studies of the Russian Academy of Sciences) 22:263–285
10. Lopatukhin LI, Bukhanovsky AV, Chernyshova ES (2009) Spravochnyye dannyye po rezhimu vetra i volneniya Yaponskogo i Karskogo morey (Reference Data on the Wind and Wave Regime of the Japan and Kara Seas). Russian Maritime Register of Shipping, St. Petersburg

11. Arkhipkin VS, Gippius FN, Koltermann KP, Surkova GV (2014) Wind waves in the Black Sea: results of a hindcast study. Natural Hazards and Earth System Sciences 14:2883–2897. https://doi.org/10.5194/nhess-14-2883-2014
12. Gippius FN, Arkhipkin VS, Surkova GV (2016) Rezhim vetrovogo volneniya na Chernom more po dannym retrospektivnykh raschotov (Wind wave regime in the Black Sea according to retrospective calculations). Kompleksnyye issledovaniya morey Rossii: operativnaya okeanografiya i ekspeditsionnyye issledovaniya (Comprehensive studies of the seas of Russia: Proceedings of the youth scientific conference). Sevastopol, pp 265–270. https://doi.org/10.13140/RG.2.1.1185.1128
13. Rusu L (2019) The wave and wind power potential in the western Black Sea. Renewable Energy 139:1146–1158. https://doi.org/10.1016/j.renene.2019.03.017
14. Bingölbali B, Jafali H, Akpınar A, Bekiroğlu S (2020) Wave energy potential and variability for the south west coasts of the Black Sea: The WEB-based wave energy atlas. Renewable Energy 154:136–150. https://doi.org/10.1016/j.renene.2020.03.014
15. Kamranzad B, Takara K (2020) A climate-dependent sustainability index for wave energy resources in Northeast Asia. Energy 209:118466. https://doi.org/10.1016/j.energy.2020.118466
16. Plotnikov VV, Chetyrbotskii AN, Gordeychuk TV (2010) Otsenka sostoyaniya ledyanogo pokrova Yaponskogo morya (Assessment of the state of the ice cover in the Sea of Japan). Meteorologiya i gidrologiya (Meteorology and hydrology) 3:46–55
17. Khan MZA, Khan HA, Aziz M (2022) Harvesting energy from ocean: technologies and perspectives. Energies 15(9):3456. https://doi.org/10.3390/en15093456
18. Safari M, Sohani A, Sayyaadi H (2021) A higher performance optimum design for a tri-generation system by taking the advantage of water-energy nexus. Journal of Cleaner Production 284:124704. https://doi.org/10.1016/j.jclepro.2020.124704

The Effect of Reactive Oxygen Species and Antioxidants on the Survival of Yeast Cells Saccharomyces Cerevisiae and Saccharomyces Carlsbergensis Under the Action of Ionizing Radiation

Phyo Myint Oo, V. I. Panfilov, A. E. Kuznetsov, I. G. Antropova, and S. V. Kalenov

Abstract *Saccharomyces yeasts* are widely used microorganisms for studying various effects on eukaryotic cells, their radiosensitivity. It is known that microorganisms, including yeast, often have a cross-reaction to various types of stress, including oxidative. The presence of free radicals in the body contributes to its aging. Thus, ionol can also be used to inhibit this pathological process. In this study it was established that the introduction of ionol led to an increase in the survival rate of yeast cells by 2–3 times. The survival of *Saccharomyces yeasts* in the presence of ionol under the effect of ionizing radiation was increased by participation of ionol in free radical reactions. The strain of yeast *Saccharomyces cerevisiae* T-985 and *Saccharomyces carlsbergensis* are used to study the oxidative effects of reactive oxygen species, antioxidative effect of ionol, X-rays (ionizing radiation) on the growth characteristics and fermentation activity of yeast under the combined effects of reactive oxygen species and anti-stress factors (low-intensity visible light). Conditions have been found under which such effects improve the yield of bio-synthesis of *Saccharomyces yeasts.*

Keywords Yeasts · Cells · X-ray · Antioxidant · Ionol · ROS · Free radicals

1 Introduction

It is known that yeast Saccharomyces cerevisiae is one of the most studied model organisms. All types of ionizing radiation lead to a wide range of biological changes in the body. Despite the fact that the body exhibits various protective reactions and

---

P. M. Oo (✉) · V. I. Panfilov · A. E. Kuznetsov · I. G. Antropova · S. V. Kalenov
Mendeleev University of Chemical Technology of Russia, 20, Geroyev Panfilovsev Str., Moscow 125480, Russia
e-mail: phyomyintoo2502@gmail.com

L. V. Radionova and D. V. Ulrikh (eds.), *Advances in Ecology and Environmental Engineering*, Springer Proceedings in Earth and Environmental Sciences,
https://doi.org/10.1007/978-3-031-64423-8_40

some resistance to ionizing radiation, ionizing rays have a negative effect on living organisms. In many X-ray experiments, microorganisms are suspended in an aqueous solution during irradiation; therefore, cell concentration may be an important variable. Studies under these conditions may also be complicated by changes in the suspending medium during irradiation as a result of metabolic activity, the nature and magnitude of which will vary depending on the species under study. In general, such changes will be proportional to the number of cells. In the cell, reactive oxygen species (ROS) arise as a result of various redox reactions occurring in it. ROS include H_2O_2, O_3, O, OH, $^1O_2$, $O_2^{\bullet-}$, $HO_2^{\bullet}$, HO_2^-, NO, some organic free radicals: alkyl $R^{\bullet}$, alkyl peroxide $RO_2^{\bullet}$, alkoxyl $RO^{\bullet}$, anion radicals of a reducing nature $D^{\bullet-}$ and some others, and are formed as a result of reactions catalytic initiation, dissolution of active gases from the atmosphere, radiation, cavitation exposure, photo-radiation, biological emission. ROS are extremely toxic metabolites for the cell and therefore there are many systems in the cell to combat them. Exposure to ROS causes changes in natural biocenoses and their components, which are usually characterized as stressful, and stress–resisting factors as anti-stress. Thus, the abiogenic decomposition processes of hydrogen peroxide, catalyzed by micro quantities of metal ions of variable valence, can participate in the leveling of stress caused by exposure to hydrogen peroxide. There are various ROS protection systems in the human body, including radioprotectors, antioxidant systems, and recovery processes.

It is known that the role of free radicals in many painful conditions is associated with the formation of oxygen forms during various biochemical reactions [1]. Free radicals can damage important biomolecules and are the main cause of pathological changes in the human body [2]. The harmful effects can be reduced by regular consumption of antioxidants with a high reactivity to free radicals. Nowadays, substances with antioxidant potential that inhibit radicals and reduce toxicity are widely studied. There is evidence that antioxidants stop the chain reaction, and inactive compounds are formed. More effective synthetic antioxidants are used to stabilize fuels, lubricating oils, plastics, lacquers and paints. However, in the food industry [3], as in perfumery and medicine [4], only a few of them can be used, since they should not be toxic and affect the aroma and taste of products. One of such non–toxic highly effective antioxidants is ionol (2,6-di-tert-butyl-4-methylphenol) [5]. Ionol is a lipophilic organic substance, an agent of the phenol class, which are widely used in the chemical industry as an antioxidant and radioprotector. It is also a starting compound for the synthesis of various derivatives of spatially hindered phenols, many of which have biological activity or are used in industry. It is known about the use of ionol(dibunol) for the treatment of certain types of cancer, radiation and trophic lesions of the skin and mucous membranes, etc. There is also an opinion that ionol has the strongest antiviral effect if combined with vitamin C or L-lysine. Phenols, a hydroxyl group with tert-butyl substituents, exhibit unique antioxidant properties, but such compounds are practically not found in nature. Therefore, synthetic spatially hindered analogues of tocopherols or flavonoids are of interest not only as compounds with high antioxidant activity, but also as a tool for studying free radical processes. The possibility to study the free radical reactions of ethanol using radiation-chemical modeling of redox reactions will reveal the effect of ionizing radiation, radiation dose. This substance is able to fight even the herpes virus [6]. It is known that the introduction of an antioxidant, for example, ionol, into the yeast culture medium allowed to increase the yield of biomass, the specific growth rate, and improve the quality of biomass. However, at higher concentrations, antioxidants have antimicrobial activity [7]. The aim of this work is to obtain dose-effect curves when exposed to ionizing radiation on yeast cells Saccharomyces cerevisiae and Saccharomyces carlsbergensis and to study the effect of visible light on the growth and viability of cells after irradiation, as well as the effect of additives (reactive oxygen species–H_2O_2, antioxidant–ionol) on the viability of yeast cells in the post-radiation period.

2 Materials and Methods

The following solvents were used in the work: distilled water, ethyl alcohol. Ionol 2,6-Di-tert-butyl-4-methylphenol, 99.8% from ACROS ORGANICS, 70% ethanol solution containing 5×10^{-3}–5×10^{-4} M of ionol, hydrogen peroxide, yeast S. cerevisiae strain T-985 race Mayen and S. carlsbergensis was used as the study samples.

The optical absorption spectra of the ionol solution were measured using a SF-2000 spectrophotometer. The chosen solvent, ethanol, served as the reference standard, the length of the optical path of the quartz cuvette is 1 and 10 mm. The ionol chromatogram was recorded on a Waters chromatograph in isocratic mode (UPLCBEH column, stationary phase C18 1.7, size μm 21×10 mm). Mobile phase acetonitrile: water 90:10, flow 600 μl/min, volume of the injected sample 6 μl. The work uses a spectrophotometric detector to determine the absorption spectra of ionol radiolysis products. X-ray irradiation was used as ionizing radiation. To study the effect of ionizing radiation on yeast cells, the latter were grown in flasks on a thermostatically controlled rocking chair, followed by the selection of part of the suspension of growing yeast cells from the flasks together with the nutrient medium. At the same time, adaptive changes in the resistance of cells to irradiation were monitored during replanting and sequential passivation of yeast cells to the introduction of sublethal doses of H_2O_2 into the medium with grown yeast with illumination of the fermentation medium with visible light. For this purpose, in the course of replanting, a part of the suspension (5–10% Vol.) was selected from the flasks, which was sown on a nutrient medium, followed by the determination of growth indicators in lighting conditions.

In the course of passivation and adaptation to H_2O_2 and ionol, two lines of yeast were conducted, 2 flasks in each line. line 1—control without the introduction of H_2O_2, ionol and line 2—with the introduction of H_2O_2 and ionol for 24–36 h of cultivation in the amount of 0.6 g/l H_2O_2 and 1 ml of ionol with a concentration of 5×10^{-3}–5×10^{-4} M on all passages. At the end of cultivation, aliquots of yeast suspension were taken from the flasks and transferred to glass tubes with a volume of 1 ml, then they after 24 h were irradiated on the Model-KALAN 4 X-ray machine at the Institute of Materials of Modern Energy and Nanotechnology, Mendeleev

University of chemical technology at the absorbed dose rate of 3 Gy/s according to the fricke dosimeter [8, 9].

The cultivation of yeast S. cerevisiae T-985 and S. carlsbergensis was carried out under aerobic conditions with stirring of the medium in flasks on a thermostatically controlled rocking chair at 28 °C, 150 rpm. The volume of the flasks was 100 ml and the volume of the sterilized medium in the flasks was 50 ml. The seed material was, depending on the conditions of the experiment, from 5 to 10% volume. For comparative evaluation, cultivation variants with yeast consistently adapted to the introduction of H_2O_2 and ionol in light with a light illumination level on the surface of the medium of 750 Lux were used. Composition of the medium for cultivating yeast S. cerevisiae and S. carlsbergensis, g/l: $(NH_4)_2SO_4$—5.0; KH_2PO_4—1.0; KCl—0.15; $MgSO_4 \cdot 7H_2O$—0.2; yeast extract—0.5; sucrose—30; tap water, pH 5.8. Adaptation of yeast to the introduction of hydrogen peroxide and ionol in the light was used as a standard procedure for obtaining microorganisms and sensitive to visible light in a state of oxidative stress and antioxidative effects [10, 11]. The change in the viability of yeast was observed by the standard method by introducing a methylene blue dye into the system with further calculation of the proportion of dead cells under a microscope.

3 Results and Discussions

Figure 1 shows the optical absorption spectra of 1 mM ionol in ethanol. An increase in the intensities of the absorption bands at wavelengths of 250–320 nm has been recorded, this is due to the formation of new products of the radiation-chemical transformation of ionol. The main product of radiolysis in the presence and absence of trace elements is a product with a retention time of 0.51 min. During ionol radiolysis after irradiation at a dose of 1 kGy, the presence of two products with retention times of 0.63–0.51 min was recorded. Their optical absorption spectra recorded using a spectrophotometric diode-matrix detector (Fig. 2).

The radiation-chemical yields of ionol equal to 0.36 ± 0.02 molecules/100 eV were calculated by the Eq. (1).

$$G_{(-BHT)} = \frac{9.65 \cdot 10^9 \cdot C_{BHT}}{\rho \cdot D} \tag{1}$$

where C_{BTH} - concentration of BTH, mol/l; ρ–density of the solvent, g/l; D–absorbed dose, Gy.

Their antioxidant activity is shown by intercepting free radicals and reducing the radiation-chemical yield of BHT.

The mechanism of the anti-radical action of ionol, as well as other spatially hindered phenols, is well described by reactions (2–4) [12, 13];

$$\text{ArOH} + \text{RO}_2^{\bullet} \rightarrow \text{ArO}^{\bullet} + \text{ROOH} \tag{2}$$

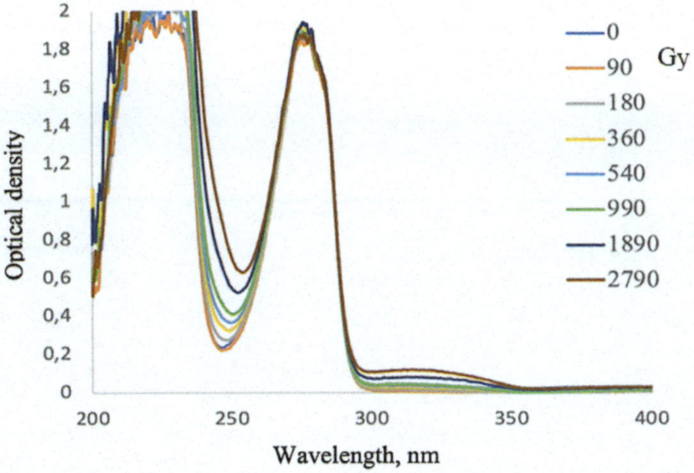

Fig. 1 Changes in the optical absorption spectra of 1 mM of ionol in alcohol from the dose

$$ArO^{\bullet} + RO_2^{\bullet} \rightarrow \text{Molecular products} \quad (3)$$

$$ArO^{\bullet} + ArO^{\bullet} \rightarrow \text{Molecular products} + ArOH \quad (4)$$

It is known that the introduction of synthetic antioxidants into the cultivation medium can increase the yield of biomass, the specific growth rate, and improve the quality of biomass. However, at higher concentrations, antioxidants have antimicrobial activity. With the periodic method of yeast cultivation on a laboratory fermenter, the introduction of optimal ionol concentrations of 5×10^{-3}–5×10^{-4} M without irradiation of X-ray radiation allowed to increase the main indicators of yeast quality on average relative to the control. To determine the effect of ionol, at the beginning, tests were carried out on yeast with the addition of yeast cells to the suspension with the addition of (5×10^{-3} M) and (5×10^{-4} M) and without the addition of ionol. It can be seen (Fig. 3) that the accumulation of S. cerevisiae yeast biomass at 800 Gy increases in comparison with doses of 400 Gy. And the growth of yeast during the irradiation period increases than when acting after 3 h of irradiation. The accumulation of S. carlsbergensis yeast biomass after 3 h at 400 Gy and 800 Gy decrease than the control and after 24 h at 400 Gy and 800 Gy specifically doesn't change with the comparison of control in Fig. 3.

As Fig. 4 is showed, the change in the proportion of dead yeast cells that are not adapted to hydrogen peroxide, depending on the dose of X-ray irradiation. At an irradiation dose from 0 to 400 Gy, the proportions of dead cells in both strains are similar in all exposure variants, however, at higher doses, *S. carlsbergensis* cells are more sensitive to radiation compared to *S. cerevisiae* cells. The introduction of hydrogen peroxide leads to a sharp decrease in yeast resistance to stress and cell

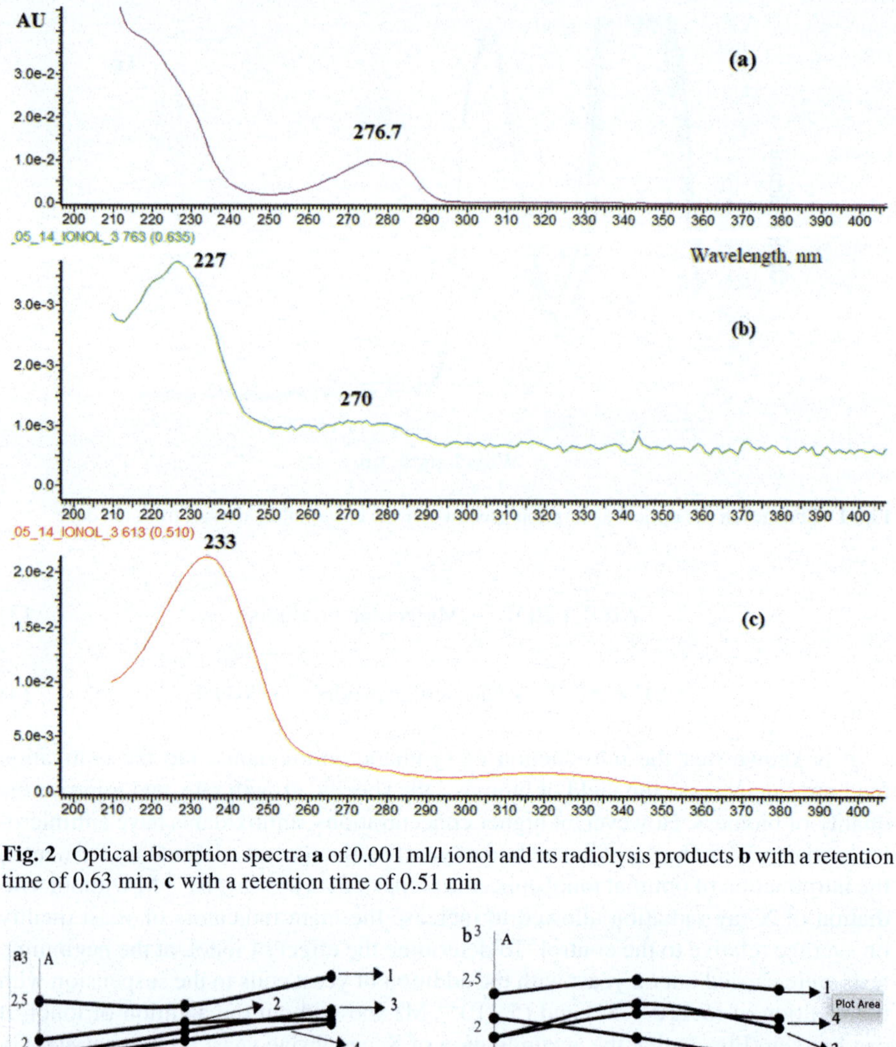

Fig. 2 Optical absorption spectra **a** of 0.001 ml/l ionol and its radiolysis products **b** with a retention time of 0.63 min, **c** with a retention time of 0.51 min

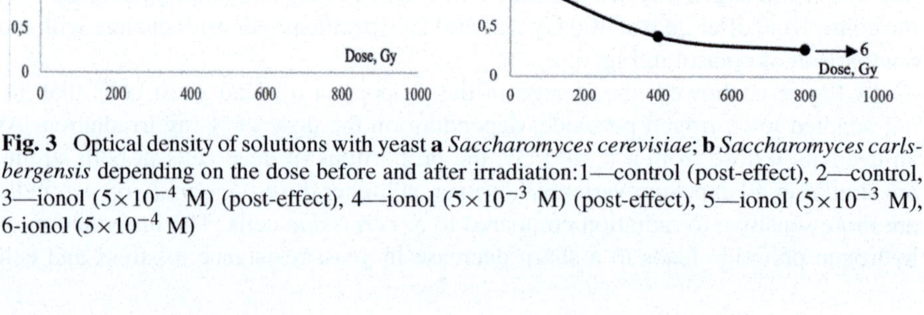

Fig. 3 Optical density of solutions with yeast **a** *Saccharomyces cerevisiae*; **b** *Saccharomyces carlsbergensis* depending on the dose before and after irradiation:1—control (post-effect), 2—control, 3—ionol (5×10^{-4} M) (post-effect), 4—ionol (5×10^{-3} M) (post-effect), 5—ionol (5×10^{-3} M), 6-ionol (5×10^{-4} M)

death in the post-radiation period [14]. On the other hand, the introduction of ionol with the concentration of (5×10^{-3} M) and (5×10^{-4} M) against the background of the introduction of hydrogen peroxide in the post-radiation period sharply reduces the proportion of dead cells, i.e., the resistance of yeast to ionizing and oxidizing effects increases sharply. Thus, the introduction of ionol into the nutrient medium contributes to the development of yeast resistance to the oxidative effect of hydrogen peroxide and ionizing X-rays [15–17].

Figure 5 shows the change in the proportion of dead yeast cells of *S. cerevisiae* and *S. carlsbergensis* before and after irradiation with a dose of 400–800 Gy. The change in the proportion of dead cells decreases after 24 h irradiation. With the addition of ionol (5×10^{-3} M and 5×10^{-4} M), the number of dead *S. cerevisiae* yeast cells in the post-radiation period [18], equal to 24 h after irradiation, is 3 and 4 times less than in the control sample.

To determine the effect of ionol, yeast tests were performed at the beginning with the addition of yeast cells to the suspension with the addition of (5 mM) and (0.5 mM) and without the addition of ionol. It was found that irradiation affects the survival of

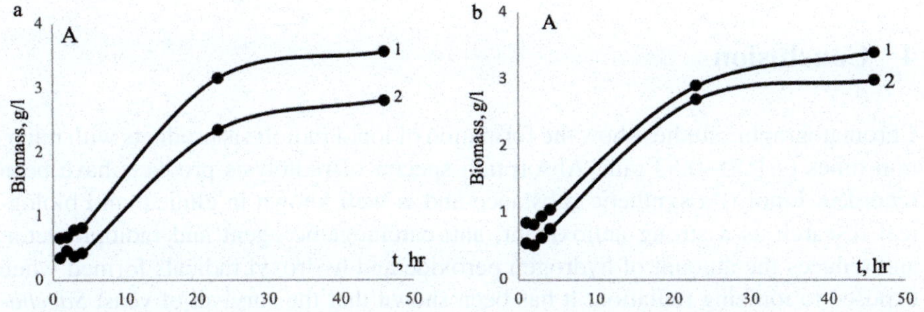

Fig. 4 A **a** Dynamics of growth of yeast *S. cerevisiae*; **b** Dynamics of growth of yeast *S. carlsbergensis* from the time of cultivation: 1—control, 2—in addition of H_2O_2

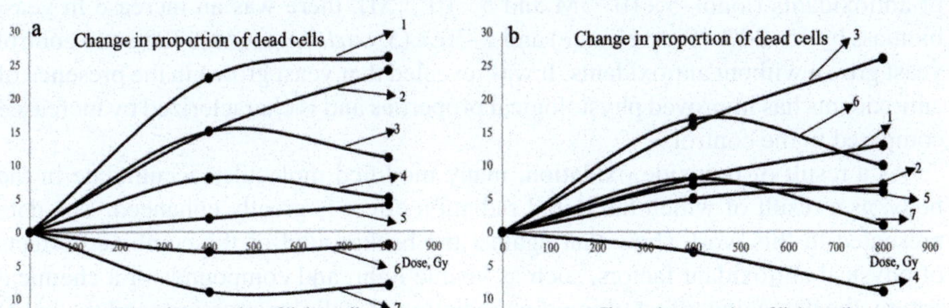

Fig. 5 Change in the proportion of dead and living **a** *S. cerevisiae*;**b** *S. carlsbergensis* yeast cells from the radiation dose during subsequent maintenance of the yeast suspension: 1—control, 2—ionol (5×10^{-3} M), 3—H_2O_2, 4—control (post- effect), 5—ionol (5×10^{-4} M), 6—ionol (5×10^{-3} M) post- effect, 7—ionol (5×10^{-4} M) post-effect

S. cerevisiae yeast cells: the proportion of dead *S. cerevisiae* yeast cells is 17% before irradiation, 40% at a dose of 400 Gy and 43% at a dose of 800 Gy. It was shown that the introduction of the well-known antioxidant ethanol-ionol solution with a concentration of 5–0.5 mM led to an increase in yeast survival in the post-radiation period, after 3 h the viability of yeast cells increased almost 3 times, after 24 h it decreased, but cell viability was 2 times greater than in the control group. It has been shown that the accumulation of biomass of both of *S. cerevisiae* and *S. carlsbergensis* yeast at 800 Gy increases in comparison with doses of 400 Gy. And yeast growth in the post-radiation period increases compared with cell survival data obtained after 3 h of irradiation. The proportion of dead yeast cells of *S. carlsbergensis* is 18% before irradiation, 26% at a dose of 400 Gy and 31% at a dose of 800 Gy. The introduction of antioxidant ethanol ionol solution with a concentration of 5 mM and 0.5 mM led to an increase in yeast survival in the post-radiation period, after 3 h the viability of yeast cells increases by 3–10%, after 24 h it will increase further, and cell viability is up to 10% higher than in the control group. And yeast growth in the post-radiation period increases compared to cell survival data obtained after 3 h of irradiation.

4 Conclusion

Chromatographic studies show the formation of ionol radiolysis products with retention times of 0.51–0.63 min. Absorption spectra of radiolysis products have been recorded. Ionol is a synthetic substance and is well known in clinical and biological research as a strong antioxidant, anti-carcinogenic agent and radioprotector, and reduces the amount of hydrogen peroxide and hydroxyl radicals formed when exposed to ionizing radiation. It has been shown that the survival of yeast *Saccharomyces cerevisiae* is influenced by exposure to ionizing radiation and hydrogen peroxide [19, 20]. The introduction of ionol led to an increase in the survival rate of yeast cells by 2–3 times. It was found that in the presence of optimal concentrations of antioxidants (ionol–5×10^{-3} M and 5×10^{-4} M), there was an increase in yeast biomass by 15–20% (*S. cerevisiae*) and 5–30% (*S. carlsbergensis*) relative to control yeast grown without antioxidants. It was revealed that yeast grown in the presence of antioxidants has improved physiological properties and is characterized by increased compared to the control.

As a result of peroxide oxidation, many modified molecules accumulate in the body, as a result of which the initial radiation effect is greatly enhanced. The data presented in this work show that against the background of the combined effects of physical antioxidant factors, such as visible light, and compounds of a chemical nature, the damaging effect of ionizing radiation on cellular structures and the lethal effect on living cells of microorganisms is significantly leveled.

Acknowledgements The work was supported by Mendeleev University of Chemical Technology of Russia.

References

1. Harman D (1992) Free radical theory of aging. Mutat Res 275(3–6):257–266
2. Menshchikova EB, Zenkov NK, Lankin VZ, Bondar IA, Trufakin VA (2008) Okislytenii stress: Pathalogichzskye sostoenye i zabolevaanya (Oxidative stress: Pathologolical conditions and diseases), Novosibirsk edn "ARTA", p 284
3. Yashin YaI, Ryzhnev VYu, Yashin AYa, Chernousova NI (2009) Prerodnye antioksidaanti (Naturalaantioxidants), Sodezhanye v pishevik produktak i ik vleyaanye na zdorove i starinye cheloveka (The content in food products and their impact on human health and aging) M Edt "TransLit", p 212
4. Alexandrova VA, Snigireva GP (2015) Sozdanye makromoleculyanik antioksidantov dlya zashiti genoma ot ioniziruyushii radiasii (Creation of macromolecular antioxidants to protect the genome from ionizing radiation), Bioantioksidanti (Bioantioxidants), abstracts of the dok IX International Conference, Moscow, p 8
5. Ziyatdinova GK, Gilmetdinova DM, Officerov EN (2006) Elektrochimichezckoe opredelenye proizvodnik digidrokvertsetina i ionola (determination of dihydroquercetin and ionol derivatives), Scientific Notes of Kazan State University, Volume 148, Book 2. https://foodandhealth.ru/dobavki/butilgidroksitoluol-e321/
6. Resshetnik OA, Pobedimsky DG, Mingaleeva ZSh (1989) Biotechnology 3:358–362
7. Phyo Myint Oo (2018) Reactivity of Ledum palustre L. (L. palustre), Melilotus officinalis (M. officinalis), Murraya paniculata and some coumarins in their composition. Dissertation, D. I. Mendeleev Russian State Technical University
8. Alexandrova VA, Snigireva GP (2015) Creation of macromolecular antioxidants to protect the genome from ionizing radiation. Bioantioxidant: abstracts of the dokl. IX International Conference. Moscow, p 8
9. Kalenov SV (2007) Cultivation of yeast and halobacteria under controlled oxidative stress. Diss. on the job. Learned. Degrees of Candidate of Technical Sciences, Mendeleev University of chemical Technology, Moscow, p 197
10. Kalenov SV, Kuznetsov AE (2010) Method of yeast cultivation for alcohol production. Patent RF No. 2394098. Byul No. 19
11. Menshchikova EB, Lankin VZ, Zenkov NK, Bondar IA, Krugovykh NF, Trufakin VA (2006) Okislytenii stress, prooksidanti i aantioksidanti (Oxidative streass: prooxidantsa and antioxidants). Firm "Slovo", Moscow, p 556
12. Askerova UF, Askerov FB (2011) Aktivnye formi kisloroda b zhibik sistemak (Reactive oxygen species in living systems), II nternational Scientific Conference dedicated to the 80th anniversary of Academician of the Russian Academy of Sciences, Astrakhan, November 2–3, p 41
13. Phyo Myint Oo, Panfilov VI, Kuznetsov AE, Antropova IG, Kalenov SV (2023) Influence of the reactive oxygen species and ionizing radiation on yeast cells Saccharomyces cerevisiae and Saccharomyces carlsbergensis under illumination by visible light. Journal of Agriculture and Environment 1(29). https://doi.org/10.23649/jae.2023.1.39.006
14. Lisanevich MS, Galimzyanova RYu (2021) Vlianye gamma- izluchenya na svoistva litevoi marki polipropilina (The effect of gamma radiation on the properties of the injection molding grade of polypropylene). Butlerov Communications 68(11):150–155. https://doi.org/10.37952/ROI-jbc-01/21-68-11-150
15. Lisanevich MS, Galimzyanova RYu (2021) Effect of radiation sterilization on air permeability and structure of Holofiber® nonwoven fabric. Butlerov Communications C, vol.2. no.4. id.3. 10.37952/ ROI-jbc-C/21-2-4-3
16. Myint Oo Phyo, Panfilov VI, Kuznetsov AE, Antropova IG, Kalyonov SV (2023) Evaluation of the survival of the yeast Saccharomyces cerevisiae and Saccharomyces carlsbergensis under conditions of illumination of yeast cells with visible light in the presence of hydrogen peroxide and naphthalene in the post-radiation period. Butlerov Communications C, vol.5. no.1. id.4. https://doi.org/10.37952/ROI-jbc-C/23-5-1-4

17. Usmanov SM (2001) Radiation: Reference materials. Humanit. Ed. Center VLADOS, Moscow, p 176
18. Faizullina LKh, Khalilova YuA, Galimova YuS, Nurlanova SN, Abzalilov TA, Mochalov KS, Valeev FA (2021) Evaluation of antioxidant activity of levoglucosenone and some of its derivatives. Butlerov Communications C 68(11):108–113. https://doi.org/10.37952/ROI-jbc-01/21-68-11-108
19. Faizullina LKh, Khalilova YuA, Galimova YuS, Nurlanova SN, Abzalilov TA, Mochalov KS, Valeev FA (2021) Evaluation of antioxidant activity of levoglucosenone and some of its derivatives. Butlerov Communications C, vol.2. no.3. id.18. https://doi.org/10.37952/ROI-jbc-C/21-2-3-18
20. Kuznetsov AE (2021) Highly efficient environmentally friendly combined systems of microbiological synthesis and wastewater treatment with oxidative stress effect. Dissertation, D.I. Mendeleev Russian State Technical University

Chemical Activation as Promising Method of Converting Fossil Coal from the Tigyit Deposit into Activated Carbons

Zaw Ye Naing, Saw Win Myint, A. V. Nistratov, A. A. Kurilkin, and V. N. Klushin

Abstract The article provides the results of studying the chemical activation processes using potassium carbonate of fossil coal from the Tigyit deposit (Tigyit, Myanmar) to obtain granular activated carbons. The results of the petrographic analysis of raw materials in the form of a maceral composition and an arbitrary vitrinite reflection index are characterized, indicating the potential possibility of its use to solve the problem stated above. Based on the data of thermographic tests performed in a protective atmosphere, the rational limits of thermal effects on raw materials during chemical activation are estimated. In the form of a number of technical parameters of the target products, the expedient conditions for the implementation of the process and the results of chemical activation are provided; they indicate the possibility of obtaining active carbons based on these raw materials with structural and adsorption properties acceptable for practical use. The authors point to the necessity of improving the developed technology to optimize.

Keywords Fossil coal · K_2CO_3 · Active coals · Technical parameters

1 Introduction

Fossil coals, the significant deposits of which are located in various regions of the Republic of the Union of Myanmar, are one of the most important state natural resources. Fossil coal open development has a significant impact on the development of both individual industries and the country's economy as a whole. Alongside with this, numerous sources of scientific and technical information indicate the fundamental possibility of obtaining rare and valuable products in the form of activated carbons on the basis of a significant number of grades (varieties) of fossil coals. There

Z. Y. Naing (✉) · S. W. Myint · A. V. Nistratov · A. A. Kurilkin · V. N. Klushin
Mendeleev University of Chemical Technology of Russia, 9, Miusskaya Pl., Moscow 125047, Russia
e-mail: zawye7@mail.ru

L. V. Radionova and D. V. Ulrikh (eds.), *Advances in Ecology and Environmental Engineering*, Springer Proceedings in Earth and Environmental Sciences,
https://doi.org/10.1007/978-3-031-64423-8_41

is currently no production of active carbons on the basis of local raw materials in the Republic. The organization of the latter in the country is very relevant, in particular, due to the urgent need to solve numerous environmental problems caused by the functioning of various industrial companies.

The economy of the Republic of the Union of Myanmar is currently developing rapidly, which is largely associated with the increasing pace of industrial production development. Nevertheless, the country virtually does not have its own facilities for the production of activated carbon, the use of which is critical for the effective solution of a large number of technological and environmental problems. The availability of these adsorbents is significantly limited by their significant prices on the global market. This fact largely determines the expediency and necessity of using cheap own resources, in particular, in the form of fossil coals and a wide range of large-tonnage plant waste from woodworking, agricultural and food industries for the production of these products.

In the context of the growing production of currently combustible natural gases and oil, fossil coals remain a very important raw material, which is almost universally used, both for energy purposes and for the range of other purposes [1, 2]. Their importance is also significant as a very cheap raw material in the industrial production of some brands of activated carbons and other carbon adsorbents, as well as a wide range of products based on them used in various economic fields. In particular, these porous absorbers are increasingly used in solving the progressive problems of pollution of biosphere objects by anthropogenic emissions and discharges.

The climatic conditions of the Republic of the Union of Myanmar, which belongs to the tropical belt countries, and its rich natural resources provide a reliable guarantee of the fundamental possibility of organizing and implementing the production of carbon adsorbents using widely available, but still not effectively used, cheap plant raw materials, represented, in particular, by large-tonnage waste from the cutting and mechanical processing of unique wood species in most cases as well as residues of cultivation and processing of food (fodder) and industrial crops [3]. The expediency and effectiveness of their transformation of this orientation are proved and substantiated by many sources of scientific and technical information [4–7]. Also, such a focus in the country has not been evaluated in relation to cheap raw materials in the form of fossil coals from Myanmar's domestic deposits, 16 of which have been explored with total reserves of 258 million tons. A special attraction from the point of view of the cost of coal mining is the fact that the deposits exploited among the explored ones are developed as open pits. These include the Tigyit deposit, the operation of which is very significant for the state economy.

The field of application of carbon adsorbents, including activated carbons, is currently very wide and diverse, due to the spectrum of their unique physico-chemical properties, combining, in particular, hydrophobicity, high and stable absorption capacity at low concentrations of target components in processed media (streams), minimal catalytic effect on such objects, the possibility of regeneration of saturated absorbers by various methods, sufficient strength and a number of other practically important indicators. Their consumption is most widespread in solving the tasks of drinking water treatment, environmental protection (purification and neutralization of emissions and discharges), respiratory protection. A variety of carbon-containing materials can serve as raw materials for the production of adsorbents of this type: wood, peat, fossil coals of varying degrees of metamorphism (from brown to anthracite ones), liquid and gaseous hydrocarbons, many products and wastes of industrial and agricultural production [8, 9].

The Republic of the Union of Myanmar has a number of coal deposits [10], among the most developed of which is one near the city of Tigyit. The identification of opportunities, efficiency and, consequently, prospects for the non-fuel use of its reserves is very important for the country's economy. The authors of this research work conducted a set of studies aimed at processing a representative sample of fossil coal from this deposit into active coals by the chemical activation of this raw material, the optimal conditions of which were established using the data from the research work [11]. In this regard, the authors assessed the expediency and effectiveness of achieving this purpose by chemical activation of the named raw materials using potassium carbonate as an activating agent—a relatively cheap, large-tonnage and almost universally available product [12–20].

The first steps in this direction have been taken in this research work in relation to representative samples of fossil coal at the Tigyit deposits.

Table 1 presents a number of characteristics of the fossil coals of the specified deposits of the Republic of the Union of Myanmar.

It follows from the data in Table 1 that both coal types are characterized by a sufficiently high content of volatile substances, which necessitates their preliminary volatilization during processing into activated carbons [2–4], and a very low sulfur concentration, which significantly simplifies the problem of gas purification in such production.

It should be emphasized that the Internet contains commercial information (generally in the form of corresponding price lists of many trading companies) about possible supplies of various brands of activated carbons produced from a number of types of coal raw materials, which is also indicated by the available catalogs of these products (in particular, [21]) and a significant number of patent information sources. The composition of fossil coals from Russian deposits recommended [22]

Table 1 Parameters of the fossil coal composition, the Tigyit deposits

Sample	Component	
	Name	Content, % by weight
Tigyit	Carbon	84.0
	Oxygen	4.3
	Nitrogen	2.6
	Sulfur	0.03
	Ash	11.4
	Volatile substances	48.4
	Moisture	17.9

Table 2 Fossil coal parameters

Coal grade	Volatile output, %	Content, %					Ash content, %
		C	H	O	S	N	
T	13	90–95	3.4–4.4	>1.6	<4.5	1.2	8–12
SS	20–32	74–90	4–5	5.3–8.2	0.8	2.1	8–45
D	34–42	70–86	5.0–6.0	>10	0.5–1.0	1.8	24–30
G	40	78–89	4.5–5.5	>6.8	<16.0	1.7	4–16
Tigyit	48.4	84	5.4	4.3	n.t	2.6	11.4

n.t. not detected

for processing to obtain adsorbents is provided in Table 2 according to [23–26], where it is compared with the composition of fossil coal from the Tigyit deposit according to the analysis conducted in the laboratory of physical chemistry of coals of the Mining Institute of NUST MISIS (Moscow Moscow).

As can be seen from the tabular data, the sample of fossil coal from the Tygyit deposit is distinguished by an increased content of volatile substances, hydrogen and nitrogen relative to the characterized coals of Russian deposits (especially, the grades T and CC) with a virtual sulfur absence. Otherwise, the composition parameters of the compared coals are close, which indicates the likely prospects of using the studied raw materials for this purpose. Thus, the information provided gives rise to hope for the success of studying the fossil coal of the mentioned Myanmar deposit in this regard as the country does not have its own coal-based production of activated carbon. The demand for cheap carbon adsorbents is very significant and is progressively increasing due to the simultaneously developing national production in particular, and the problems of protecting the environment from toxic emissions and discharges of its enterprises. The latter do not have deep purification, which is primarily due to the virtual absence of their own production of carbon adsorbents and their high cost on world markets. Thus, the assessment of the possibility and feasibility of using fossil coal from the Tigyit deposit as a raw material for producing activated carbons seems to be quite relevant against other organizational and technical state measures.

2 Experimental

For the purpose of such an assessment, the authors predetermined the limits of the appropriate temperature effect on the mentioned raw materials during pyrolysis by heating its powder fraction of less than 200 microns with an intensity of 9° C/min to ~900° C in the current of bottle nitrogen in the furnace of the Q-1200 derivatograph (MOM, Hungary), while the above-mentioned laboratory of physical chemistry of coals studied some of its petrographic properties which make it possible to indirectly judge the suitability of the studied fossil coal for solving the problem set. The study of these petrographic properties consisted in determining the maceral composition and an arbitrary reflection index of the vitrinite of the rough fossil coal of the Tigyit deposit, containing 84.0% of carbon. The results obtained reflect the data in Table 3 in the form of percentages of the volume fractions of macerals of the vitrinite (*Vt*), semivitrinite (*Sv*), inertinite (*I*) 4.1 and liptinite (*L*) groups, the content of fusenized components on pure coal ($\sum OK$) and an arbitrary index of vitrinite reflection in oil ($R_{o,r}$), as well as the values of the mean square deviation (MSD max.).

As can be seen, the value of an arbitrary vitrinite reflection index of the characterized fossil coal corresponds to the lower limit of the interval of this indicator, which is 0.40–0.79% [27] for long-flame coals (E) used in Russia on an industrial scale for the production of activated carbons [21]. Thus, this circumstance can serve as an additional argument in favor of the expediency of conducting the above-mentioned assessment.

Micrographs of samples obtained using the SIAMS 620 complex are shown in Figs. 1 and 2.

According to Figs. 1 and 2, the macerals of the vitrinite group are mainly represented by collinite in the form of separate sections, or as part of suberinite. Inertinite is mainly represented by large areas with an intact cell wall structure, or in the form of individual grains. Liptinites are mainly represented by cutinite, as well as in the form of suberinite, a substance with a well–preserved cellular structure, the cell walls of which are formed by liptinite, and the intracellular space is filled with vitrinite. Pyrite inclusions ranging in size from 1 to 10 microns were found in the coal sample. Thus, the use of standard-compliant petrographic methods for the study of solid combustible minerals makes it possible to determine the coal quality from

Table 3 Petrographic composition and reflection index of the showcase of the excavated coal of the Tigyit deposit

Indicator	*Vt*	*Sv*	*I*	*L*	$\sum OK$	$R_{o,r}$	MSD max
Value	89.5	3.8	4.4	2.2	7	0.425	0.056

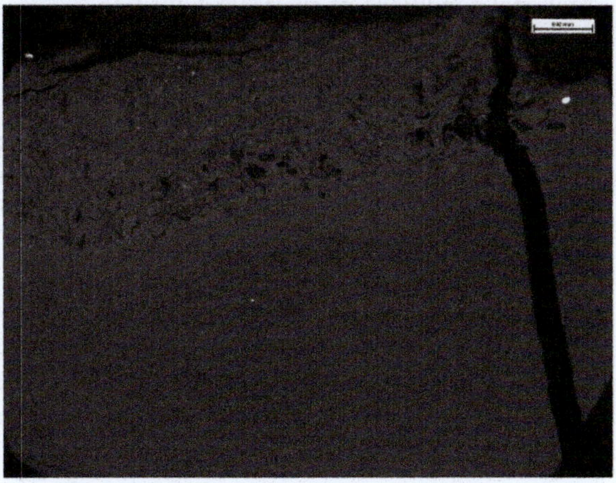

Fig. 1 Vitrinite (collinite) and suberinite

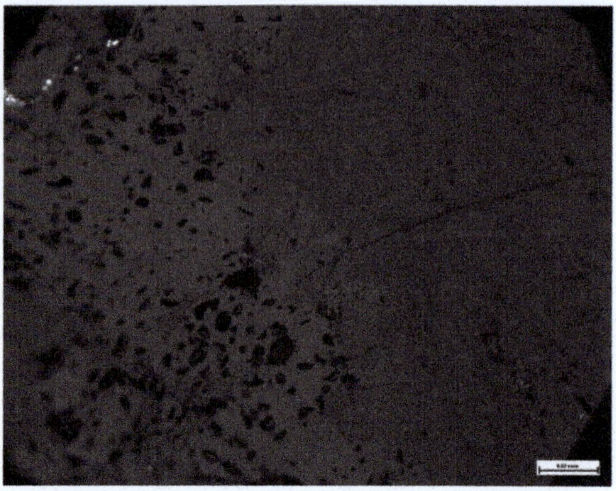

Fig. 2 Inertinite and vitrinite

its oxidized samples, since neither the quantitative petrographic composition nor the signs of metamorphism of microcomponents change. The conducted research makes it possible to judge the high prospects of carrying out research on the use of fossil coal from the Tigyit deposit for the production of activated carbons.

The main body of research was conducted using a laboratory-scale pyrolytic unit placed under thrust, the core of which was a steel cylindrical reactor equipped with fittings and containing a suspension of the test sample, located in a tubular electric furnace with controlled heating equipped with a nichrome winding and thermal insulation. To control the temperature in the reactor working area, a junction of a thermocouple connected to a potentiometer was placed between its outer wall and the inner wall of the electric furnace. Raw compositions dried to a constant mass at ~ 105° C were subject to thermal processing, obtained also by impregnating coal grains of 1–2 mm fraction dried to a constant mass with a solution of potassium carbonate providing for mass ratios "K_2CO_3: fossil coal" equal to 1:1 and 2:1. The assessment of rational conditions of chemical activation (values of heating intensity, final activation temperature and duration of isothermal exposure of the target product) was based on a combination of values of the yield indicators of the product specified and its structural and adsorption characteristics: total porosity in water V_{Σ}, volumes of sorbing pores Vs in water vapor, carbon tetrachloride and benzene, iodine absorption values (J_2) and methylene blue dye (MG) from their solutions [28, 29].

The information of a similar plan regarding the results of chemical activation of the raw material composition K_2CO_3: raw materials = 2:1 partially characterize the data in Fig. 3.

A total of the data in Fig. 3 indicates the obvious expediency of providing a maximum temperature of 900° C for the studies of chemical activation.

The impact of the heating intensity of the composition "K_2 CO_3: raw materials = 2:1" on similar results of processes implemented at a set maximum temperature is characterized by the data in Table 4.

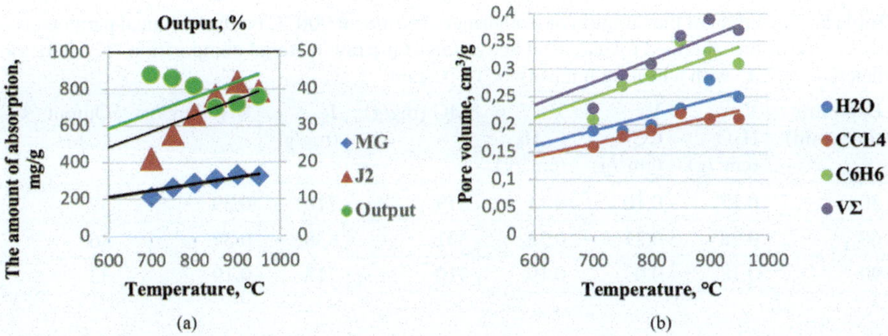

Fig. 3 The trends in the impact of the maximum activation temperature on the structural parameters (**a**), yield and absorption capacity (**b**) of activated carbons obtained by heating the composition "K_2CO_3: raw materials = 2:1" with an intensity of 10° C/min and an isothermal exposure time of 60 min

Table 4 The effect of the heating intensity of the composition "K_2 CO_3: raw materials = 2:1" to the final chemical activation temperature of 900° C on the technical parameters of activated carbons (60 min of isothermal exposure)

Heating rate (°C/min)	Vs H_2O (cm^3/g)	Vs CCl_4 (cm^3/g)	Vs C_6H_6 (cm^3/g)	MG (mg/g)	J_2 (mg/g)	V $_\Sigma$ (cm^3/g)	Output (%)
5	0.19	0.07	0.19	309	711	0.20	35
10	0.28	0.21	0.33	331	838	0.39	36
15	0.17	0.07	0.20	304	714	0.23	30
20	0.13	0.03	0.16	301	589	0.24	39

The information in Table 4 indicates the feasibility of using the heating intensity of the raw material composition of 10° C/min under the characterized conditions of activation.

The effect of the isothermal exposure duration on the same parameters of activated carbons obtained by the activation of the same raw material composition under conditions of its heating to 900 °C with an intensity of 10 °C / min is demonstrated by the information provided in Table 5.

The information given in Table 5 indicates the need to use an isothermal exposure of the target product for 1 h.

Thus, the rational conditions for the activation of the raw material composition to "K_2 CO_3: raw materials = 2:1" express the heating intensity of 10 °C / min, the maximum temperature of 900 °C and the 60 min exposure of the target product.

Table 6 describes similar activation results under identical conditions of a raw material composition with a mass ratio of fossil coal: K_2 CO_3 = 1:1.

The comparison of the data in Tables 5 and 6 indicates the positive effect of an increase in the proportion of K_2 CO_3 in the raw material composition for the chemical

Table 5 The effect of the duration of isothermal exposure at 900° C on the technical parameters of activated carbons obtained by chemical activation of the raw material composition "K_2 CO_3: raw materials = 2:1" with a heating intensity of 10 °C/min

Exposure time (min)	Vs H_2O (cm^3/g)	Vs CCL_4 (cm^3/g)	Vs C_6H_6 (cm^3/g)	MG (mg/g)	J_2 (mg/g)	V $_\Sigma$ (cm^3/g)	Output (%)
30	0.17	0.10	0.11	319	710	0.39	39
60	0.28	0.21	0.33	331	838	0.39	36
90	0.18	0.09	0.19	310	712	0.19	33

Table 6 Technical characteristics of active carbons of chemical activation K_2CO_3 at a heating rate of 10 °C/min and an exposure time of 60 min at a temperature of 900° C

Ratio raw materials: K_2CO_3	Vs H_2O (cm^3/g)	Vs CCL_4 (cm^3/g)	Vs C_6H_6 (cm^3/g)	MG (mg/g)	J_2 (mg/g)	V $_\Sigma$ (cm^3/g)	Output (%)
1:1	0.17	0.08	0.10	290	610	0.18	43

activation and the obvious rationality of using the mass ratio of salt impregnation of raw materials equal to 1:2.

The target product obtained by its activation under these rational conditions is characterized by a gravimetric density of 0.52 kg /dm^3, grain abrasion strength of 68.8%, ash content of 8.9% and moisture content of 17.4%. Its iodine adsorption activity is higher than that of commercial activated carbon based on the coal grade AG-3 (650–670 mg/g), the bulk density is slightly higher (0.4–0.5 kg/dm^3), while the abrasion strength is slightly lower (70–75%) and the ash content is lower (12–15%) [30].

3 Conclusion

Thus, the characterized results indicate the fundamental possibility of obtaining activated carbons on the basis of fossil coal from the Tigyit deposit characterized by a sufficiently high quality. The improvement of the optimization task for the composition of raw materials to obtain these products from the mentioned raw materials by its chemical activation with K_2CO_3 should be associated with the study of heat processing of such with close 1:2 impregnation ratios under conditions of providing the heating intensity of 10 °C/min to 900 °C and the 60 min duration of isothermal exposure under such an intensity.

Acknowledgements The work was supported by Mendeleev University of Chemical Technology of Russia

References

1. Shifeng Dai, Finkelman RB (2018) Coal as a promising source of critical elements: progress and future prospects. Int. J. Coal Geol., 186:155–164
2. Shifeng Dai, Finkelman RB (2019) The importance of minerals in coal as the hosts of chemical elements. Int. J. Coal Geol. 212:103–251
3. The raw materials complex of foreign countries. Myanmar. http://www.mineral.ru/facts/world/116/145/index.html. Accessed 15 Feb 2019
4. Hameed BH, Ahmad AL, Latiff KNA (2011) Adsorption of basic dye (methylene blue) onto activated carbon prepared from rattan sawdust. Chemical Engineering Journal 172(1):326–334
5. Akpen GD, Nwaogazie IL, Leton TG (2011) Optimum condition of color removal from waste water by mango seed shell based activated carbon. Indian Journal of Sience and Technology 4(8):890–894
6. Omri Abdessalem, Benzina Mourad (2012) Characterization of activated carbon prepared from a new raw lignocellulosic material: ziziphusspuna-christiseeds. Journal de la Société Chimique de Tunisie 1-4:175–183
7. Pandharipade SL, Moharkar Yogesh, Thakur Raj (2012) Synthesis of Adsorbents from waste materials such as Ziziphus Jujube Seed & Mango Kernel. International Journal of Engineering Research and Applications (IJERA) 2(4):1337–1341
8. Kinle H, Bader E (1984) Activated carbons and their industrial application. Trans. from German. Chemistry, Leningrad, p 216
9. Mukhin VM, Tarasov AV, Klushin VN (2000) Active coals of Russia. Under the general editorship of Prof. Doctor of Technical Sciences A.V.Tarasov. Metallurgy, Moscow, p 352
10. Fossil coals. http://www.mining-enc.ru/u/ugli-iskopaemye. Accessed 25 Oct 2017
11. Zaw Ye Naing, Klushin VN (2017) Features of thermal and thermo-oxidative decomposition of fossil coal from the Teychik deposit. Journal Successes in Chemistry and Chemical Technology 31(9):34–36
12. Xia HY, Peng JH, Zhang LB (2015) Preparation of high surface area activated carbon from *Eupatorium adenophorum* using K_2CO_3 activation by microwave heating. *Green Processing and Synthesis* 4(4):299–305. https://doi.org/10.1515/gps-2015-0025
13. Ahmed MJ, Theydan SK (2012) Physical and chemical characteristics of activated carbon prepared by pyrolysis of chemically treated date stones and its ability to adsorb organics. Powder Technol 229:237–245. https://doi.org/10.1016/j.powtec.2012.06.043
14. Altenor S, Carene B, Emmanuel E et al (2009) Adsorption studies of methylene blue and phenol onto vetiver roots activated carbon prepared by chemical activation. J Hazard Mater 165:1029–1039. https://doi.org/10.1016/j.jhazmat.2008.10.133
15. Balajii M, Niju S (2019) Biochar-derived heterogeneous catalysts for biodiesel production. Environ Chem Lett. https://doi.org/10.1007/s10311-019-00885-x
16. Foo K, Hameed B (2012) Preparation, characterization and evaluation of adsorptive properties of orange peel based activated carbon via microwave induced K_2CO_3 activation. Bioresour Technol 104:679–686. https://doi.org/10.1016/j.biortech.2011.10.005
17. Galhetas M, Mestre AS, Pinto ML et al (2014b) Chars from gasification of coal and pine activated with K_2CO_3: acetaminophen and caffeine adsorption from aqueous solutions. J Colloid Interface Sci 433:94–103. https://doi.org/10.1016/j.jcis.2014.06.043
18. Mestre AS, Bexiga AS, Proença M et al (2011) Activated carbons from sisal waste by chemical activation with K_2CO_3: kinetics of paracetamol and ibuprofen removal from aqueous solution. Bioresour Technol 102:8253–8260. https://doi.org/10.1016/j.biortech.2011.06.024
19. Yahya MA, Al-Qodah Z, Ngah CZ (2015) Agricultural bio-waste materials as potential sustainable precursors used for activated carbon production: a review. Renew Sustain Energy Rev 46:218–235. https://doi.org/10.1016/j.rser.2015.02.051
20. Chingombe P, Saha B, Wakeman RJ (2005) Surface modification and characterisation of a coal-based activated carbon. Carbon 43 (15):3132–43.https://doi.org/10.1016/j.carbon.2005.06.021

21. Mukhin VM (ed) (2003) Active coals. Elastic sorbents, catalysts, desiccants and chemical absorbers based on them. Nomenclature catalog. Publishing house "Ore and metals", Moscow, p 280
22. Mukhin VM, Klushin VN (2018) Production and application of carbon adsorbents. LAMBERT Academic Publishing, p 308
23. Technical characteristics of fuel. http://котлы-кв.рф./fuel-specifications.html. Accessed 26 Oct 2017
24. Weakly glowing coals. http:ecogoroshek.ru/slabospekayushchiesya-ugli.html. Accessed 26 Oct 2017
25. Fossil coals. http://www.mining-enc.ru/u/ugli-iskopaemye. Accessed 26 Oct 2017
26. The new handbook of chemist and technologist. http://chtmanalytica.com/book/novyy-spravochnik_khimika_i_tekhnologa/05_syre_i_ Accessed 25 Oct 2017
27. Kolokoltsev SN (2017) Natural energy carriers and carbon materials: Composition and structure. Modern classification. Production technologies and mining. LENAND, Moscow, p 224
28. Keltsev NV (1976) Fundamentals of adsorption technology. Chemistry, Moscow, p 511
29. Kolyshkin DA, Mikhailova KK (1972) Active carbons: properties and test methods. Chemistry, Moscow, p 56
30. Mukhin VM, Kurilkin AA, Voropaeva NL, Leksyukova KV, Uchanov PV (2016) The place of activated carbons in ecology and economics, new technologies of their production. Sorption and chromatographic processes 16(3):346–353

Resource-Saving Environmentally Friendly Technology for Small-Scale Production of Lead-Free Solder

L. V. Radionova and S. R. Faizov

Abstract In the electrical industry, alloys containing from 10 to 90% lead are used for soldering. Lead, according to a large number of studies, has a negative impact on human health and the environment. In 2006, a European Union regulation came into force limiting the use of lead-containing solders in the electrical industry. In this regard, most leading corporations strive to use lead-free solders. This article presents research on the development of a resource-saving technology for producing lead-free solder wire and rod from 52In-48Sn alloy by direct extrusion. Using a mathematical model of direct extrusion and a computer program developed on its basis, the deformation and energy-force parameters for the manufacture of Ø 8.0 and 15.0 mm rods and Ø 2.0 mm wire were determined. According to the calculation results, the maximum extrusion force in the manufacture of the specified range is 106 kN, which was taken into account when selecting equipment and designing tools. Simulation in the QForm package determined the temperature and speed conditions for the extrusion of rods and wire. Using the developed technology, a pilot batch of 100 kg of Ø 2.0 mm solder wire wound on coils and 150 kg of Ø 8.0 mm rod was manufactured. The use of lead-free solder in the manufacture of electronics will reduce the harmful effects of lead on the environment and human health. Technology that eliminates many intermediate operations and the use of a large amount of equipment will also have a positive impact on the environment.

Keywords Extrusion · 52In-48Sn alloy · Lead-free solder · Rod · Wire

1 Introduction

The environmental agenda has become one of the strategic guidelines for the development of the economy and society in the world [1, 2]. It also affected the electrical industry [3]. Alloys of the Sn–Pb system are widely used for low-temperature

L. V. Radionova (✉) · S. R. Faizov
Moscow Polytechnic University, 38, Bolshaya Semyonovskaya Str, Moscow 107023, Russia
e-mail: radionovalv@rambler.ru

L. V. Radionova and D. V. Ulrikh (eds.), *Advances in Ecology and Environmental Engineering*, Springer Proceedings in Earth and Environmental Sciences,
https://doi.org/10.1007/978-3-031-64423-8_42

soldering [4]. The industry uses solders that can contain from 10 to 90% lead. Lead, according to a large number of studies, has a negative impact on human health [5, 6] and the environment [7].

In 2006, the European Union regulation (EU-WEEE, Waste from Electrical and Electronic Equipment) came into force limiting the use of lead-containing solders in the electrical industry [8]. Similar rules apply in the USA [9]. According to environmentalists, the removal of lead from the production of household, industrial, telecommunications and other equipment should help improve the environment. In their opinion, the placement of end-of-life radio-electronic industry products containing lead solder in landfills worsens the environmental situation. After disposal of computers, monitors, televisions, radios, toys, etc., lead enters the soil and subsequently enters the food chain [10].

In this regard, most leading corporations place lead elimination at the top of their priority lists [11]. Lead-free soldering undoubtedly has commercial prospects. Manufacturers of "green products" can claim increased sales of their products.

Tin-based alloys with the addition of Ag, Cu, Bi, Zn, In and other metals are considered lead-free solders [12–14]. Solders of the In-Sn system are promising and have already proven themselves quite well for use as connecting materials in electrical engineering, and especially in aerospace engineering. Indium-based solders are considered more reliable under long thermal cycles as they have been proven to maintain good ductility even under cryogenic conditions.

In Russia, indium is mined together with zinc and is obtained in its pure form from waste and intermediate products of zinc and lead production [15]. Chelyabinsk Zinc Plant JSC, being one of the main producers of zinc in Russia, manufactures indium as a by-product and then sells it on the market in the form of raw materials. Production of solder from indium and its sale on the market in the form of finished products will allow the plant to receive additional profit.

Traditional solder production technology consists of the following stages: preparation of the alloy, casting of ingots (Ø 100–200 mm), pressing, and drawing to the required wire diameter. In some cases, rolling of the workpiece may be used between pressing and drawing. This technology for producing solder requires a large amount of complex equipment and is accompanied by a negative impact on the environment.

In view of the above, this study develops a resource-saving technology for producing lead-free solder based on the direct extrusion method.

2 Determination of Energy-Power Parameters of Direct Extrusion of Rods and Wires

To make a reasonable choice of equipment and design tools for direct extrusion, it is necessary to calculate the energy and power parameters of the process. For these purposes, we have developed a mathematical model that allows us to carry out calculations taking into account the characteristics of the analyzed process [16]. The

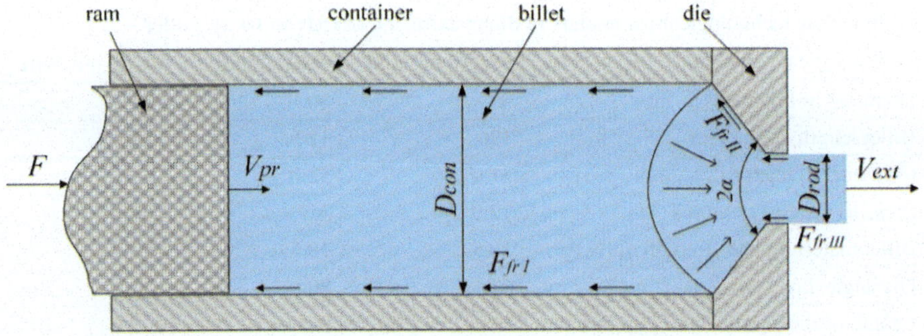

Fig. 1 Schematic of deformation zone direct extrusion

extrusion process is characterized by a deformation zone (Fig. 1) and a force F that consists of the force to deform the material F_{def} and the force to overcome friction at the contact between the billet and the tool F_{fr}

$$F = F_{def} + F_{fr}. \tag{1}$$

The force to overcome friction is determined by the geometry of the deformation zone (Fig. 1) and consists of three components—the effort aimed at overcoming the processed metal sliding friction against the container $F_{fr\,I}$, the surface of the die $F_{fr\,II}$, and the parallel land of the die $F_{fr\,III}$.

The deformation force F_{def} depends on the physical and mechanical properties of the material, the extrusion ratio and the geometric parameters of the deformation zone. The melting point of the eutectic alloy 52In-48Sn is 120 °C. Its tensile strength is 10–11 MPa and elongation is 32–34%.

Based on the mathematical model of direct extrusion, a PC program called "Calculation of power, temperature and speed parameters of low-temperature solder direct extrusion" was developed. Using this program, technological parameters were determined for the extrusion of Ø 15.0 and 8.0 mm rods, as well as Ø 2.0 mm wire (Table 1).

From the calculations, it follows that the maximum force during extrusion is 106 kN, which must be taken into account when choosing equipment and designing tools.

3 Study of Temperature and Speed Conditions for Direct Extrusion of Rods and Wires

During cold plastic deformation, the metal heats up due to contact friction and deformation. Considering the high extrusion ratio and low melting point of the 52In-48Sn alloy, it is necessary to evaluate the temperature-speed conditions and their impact on the stability of the process and the quality of the finished product. To do this, it

Table 1 The technological parameters of direct extrusion of rods and wire solder

Technological parameters	Ø 15.0 mm rod	Ø 8.0 mm rod	Ø 2.0 mm wire
Diameter of billet D_{billet}, mm	30	30	20
Billet length L_{billet}, mm	120	120	120
Container diameter D_{con}, mm	31.0	32.0	21.8
Pressing speed V_{pr}, mm/s	12.0	3.5	0.5
Billet material yield strength σ_s, MPa	10.5	10.5	10.5
Die angle 2 α, °	67	80	37
Length of die parallel land l_{dpl}, mm	5	5	5
Friction coefficient f	0.5	0.5	0.5
Extrusion ratio μ	4.3	16.0	118.8
Extrusion speed V_{ext}, mm/s	51	56	59
Force at the start of pressing F_{start}, kH	92.8	106.0	105.0
Force at the end of pressing F_{end}, kH	41.0	51.3	71.9

is advisable to use simulation in the QForm package. This will allow you to evaluate not only the average temperature but also the temperature distribution over the cross-section of the wire or rod. The methodology for conducting numerical experiments to study temperature during extrusion is shown in detail in [17]. An example of temperature simulation during extrusion of a Ø 8.0 mm rod at a pressing speed of 3 mm/s is shown in Fig. 2. The maximum temperatures in Ø 8.0 and 15.0 mm rods and Ø 2.0 mm wire during single extrusion are given in Table 2. Semi-continuous extrusion leads to heating of the tool (Fig. 3) and, as a consequence, some additional increase in the temperature of the wire and rod.

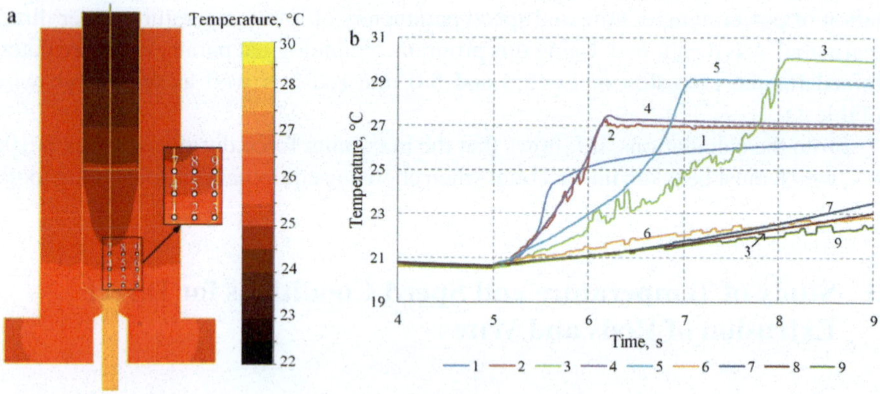

Fig. 2 The temperature field simulation (**a**) in the QForm program and graphs of changes in metal temperature over time at points 1–8 (**b**) the Ø 8.0 mm solder rod during extrusion

Table 2 Maximum temperatures in the cross-section of the solder rod (wire) during extrusion

Rod (wire) diameter D_{rod}, mm	Pressing speed V_{pr}, mm/s		
	0.5	1.0	3.0
15.0	21.1	22.3	25.5
8.0	22.5	24.7	30.0
2.0	33.4	44.0	75.8

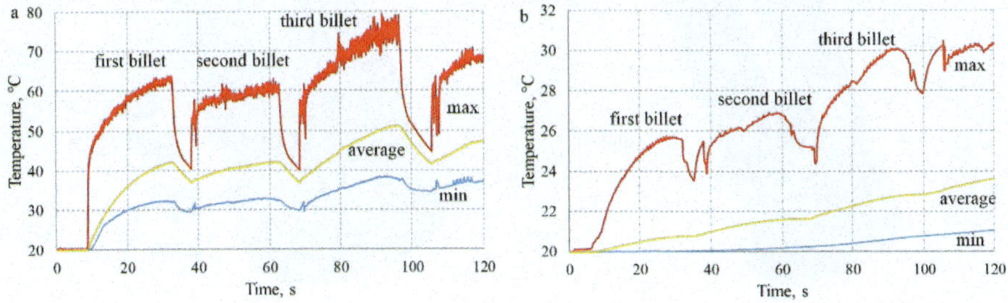

Fig. 3 The die (**a**) and container (**b**) temperatures at semi-continuous extrusion of Ø 2.0 mm wire at a pressing speed of 3 mm/s

Analysis of the results of numerical experiments revealed that at pressing speeds that can technically be implemented on a vertical hydraulic press (at a pressing speed of 3 mm/s, the flow rate of Ø 2.0 mm wire from the die will reach 356.4 mm/s) after the third billet, the temperature of the die and the container is stabilized. The temperature of the die does not exceed 80 °C, and that of the container—31 °C. The conducted research allows us to assert that the extrusion process in the analyzed modes is technologically feasible for all standard sizes of products. There is no heating of the alloy above the melting point, there are no reasons for loss of process stability and a decrease in the quality of the finished product.

4 Development of Technological Equipment for Direct Extrusion of Rods and Wires

Two devices were designed as tools mounted on a hydraulic press (Fig. 4). In the manufacture of wire, to ensure the versatility of the tool, monolithic drawing dies of the required diameter are used as industrial dies. This ensures high wear resistance and low cost of consumable tools.

The strength analysis of the designed tooling was performed in the SOLIDWORKS Simulation program and was described in detail by us earlier [18]. The results of the analysis showed that the container should be made of steel with a tensile strength greater than 460 MPa. This will provide a structural safety margin of at least 2, which is sufficient for metallurgical tools and equipment. The press

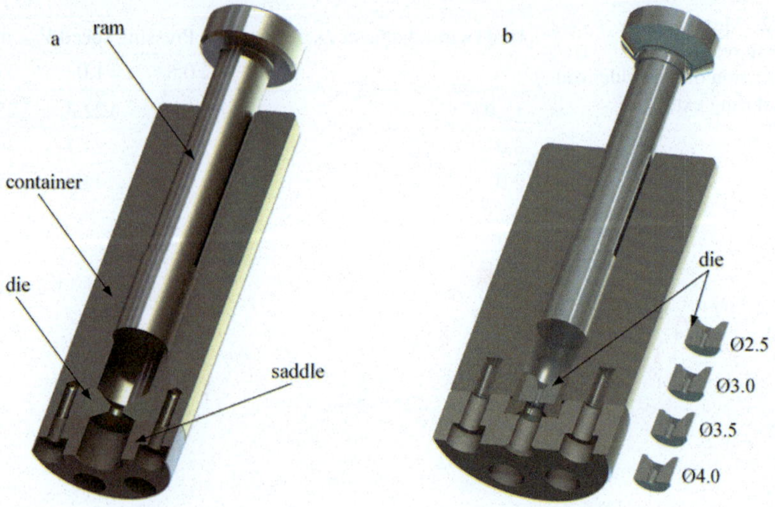

Fig. 4 Extrusion tooling for rod (**a**) and wire (**b**)

stamp is made of steel grade C 80 W2 (1.1625) and has a safety factor of 2.1. The design decision to use a monolithic die as an extrusion die turned out to be correct. A die made of HG20 alloy in a case made of C45 (1.1191) steel is not an element that reduces the overall strength of the tool. The safety factor of the die will be 5.7.

5 Production of Pilot Batch Rods and Wires from 52In-48Sn Alloy

As equipment for organizing small-scale production, a D2428 hydraulic press was used, equipped with a data collection system that allows measuring the extrusion force (Fig. 5). This made it possible to verify the developed mathematical model [17] and experimentally determine the friction coefficients for various technological lubricants [19]. Studies of the coefficient of friction during extrusion have shown the promise of using lubricants. In the course of experimental and analytical studies, it was established [19] that the use of lubricants, namely graphite (C) and molybdenum (MoS_2) lubricants, allows reducing the friction coefficient by more than 10 times. The results of experimental measurements showed high convergence with the results of mathematical modeling and simulation in the QForm program.

Microstructural analysis and energy dispersive X-ray spectral (EDS) mapping of the 52In-48Sn alloy after extrusion confirmed that it is not hardened by plastic deformation at room temperature. In the structure, both before and after deformation, we observe a eutectic mixture of two phases β-In_3Sn and γ-$InSn_4$ (Fig. 6).

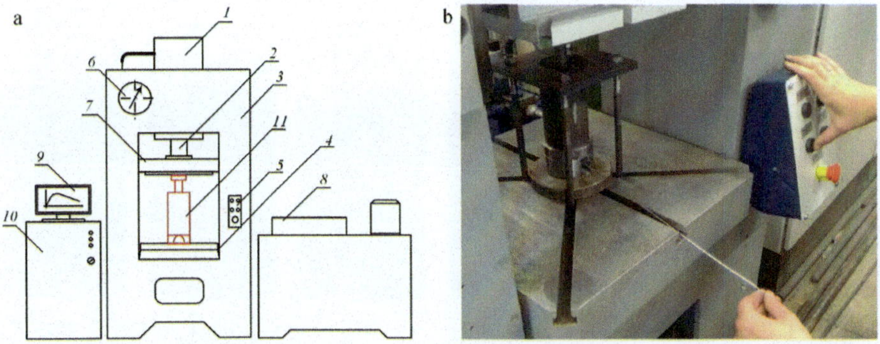

Fig. 5 Scheme (**a**) and photo (**b**) of laboratory hydraulic press D2428: 1—main cylinder; 2—stroke; 3—housing; 4—bench; 5—control desk; 6—manometer; 7—slide; 8—oil pump; 9—monitor; 10—block of control and data collecting; 11—tooling of rods or wire extrusion

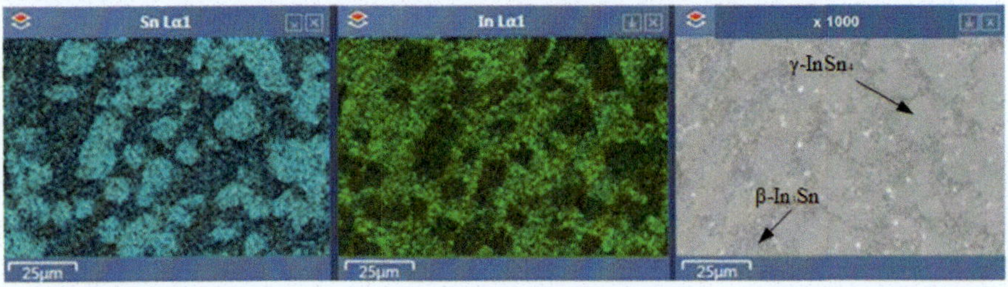

Fig. 6 Energy-dispersive X-ray spectroscopy (EDS) mapping of the 52In-48Sn alloy after extrusion

The production of Ø 15.0 mm rods from a Ø 30.0 mm billet in a Ø 31.0 mm container, as well as Ø 2.0 mm wire from a Ø 20.0 mm billet in a Ø 21.8 mm container, was not accompanied by the formation of defects and passed stable. When producing Ø 8.0 mm rods from a Ø 30.0 mm billet in a Ø 32.0 mm container, a surface defect "air bubbles" was observed. Carrying out additional research using computer modeling in the QForm package made it possible to establish that the reason for the formation of a bulge on the surface of the finished product filled with air is an excessive gap between the container and the billet.

The simulation was carried out on a Ø 30 mm billet, which corresponds to the technological parameters of experimental extrusion, as well as to better visualize the behavior of the billet in a Ø 25.0 mm container. The length of the billet during the study was 120; 60 and 40 mm. The simulation results are shown in Figs. 7 and 8.

The contact zones of the alloy with the container during simulation are painted in blue, and in the zones where there is air between the container and the billet, we see it in gray (Figs. 7 and 8). In the designation "40/1" the number "40" indicates the length of the billet, and the number "1" indicates the number of billets in the container at the same time. According to the results of the simulation, the pressing of a Ø 30.0 mm billet in a Ø 32.0 mm container, it can be noted that the deformation

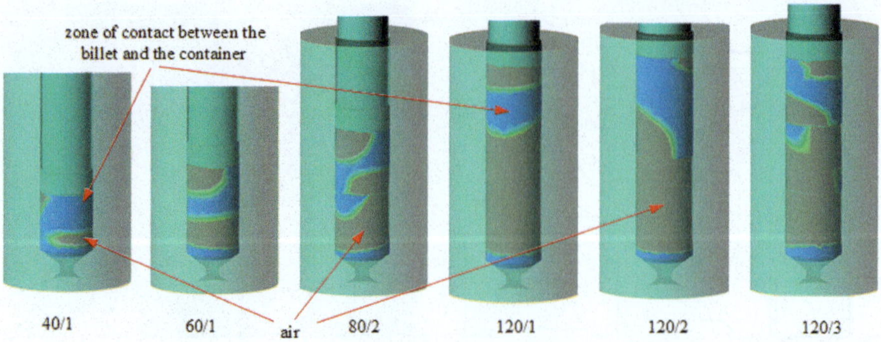

Fig. 7 Simulation of pressing a Ø 30.0 mm billets in a Ø 32.0 mm container

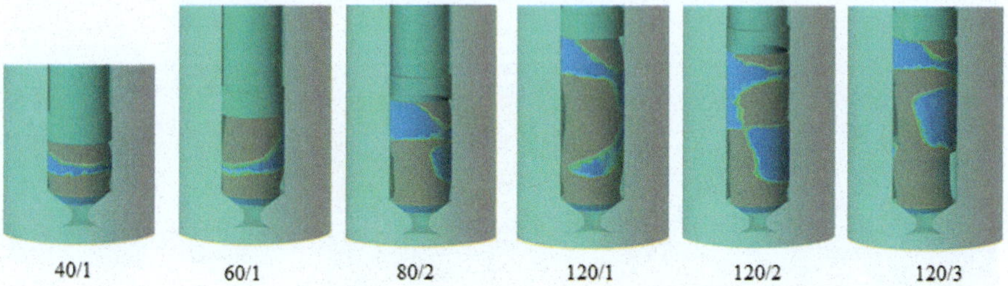

Fig. 8 Simulation of pressing a Ø 25.0 mm billets in a Ø 32.0 mm container

of the billet begins from its lower and ends in the upper region of the container. This leads to the closing of the middle part of the container with a deformable alloy in which air remains. Visually, the bend of the Ø 30.0 mm billet is not visible. The behavior during pressing of a billet of smaller diameter is visualized in Fig. 8. The simulation result with the designation "120/1" visualizes well the bending of the billet during pressing if the ratio is not fulfilled $L_{billet}/D_{billet} \leq 2.5$.

Based on the simulation results, the air volume in the container was calculated for various options for filling the container with a billet (Fig. 9). The diameter of the billet has a significant impact on the air volume locked in the container. The most indicative variant of simulation is one Ø 40 mm billet in containers. With a Ø 30 mm billet, the air volume locked in the container is 68 mm^3, and with a Ø 25 mm billet, this volume increases to 747 mm^3. The billet length also has an effect on the air volume locked in the container. The air volume locked in the container when pressing one Ø 40 mm billet with a length of 120 mm ("120/1") is 3548 mm^3 and with a length of 120 mm consisting of three billets ("120/3") the volume of trapped air will decrease to 2998 mm^3.

The results of the simulation and calculation of the air volume showed that the main reason for obtaining a surface defect "air bubbles" is the diameter and length of the billet during extrusion. The gap between the billet and the container should be as small as possible. Also, to reduce the air volume in the container, you can use

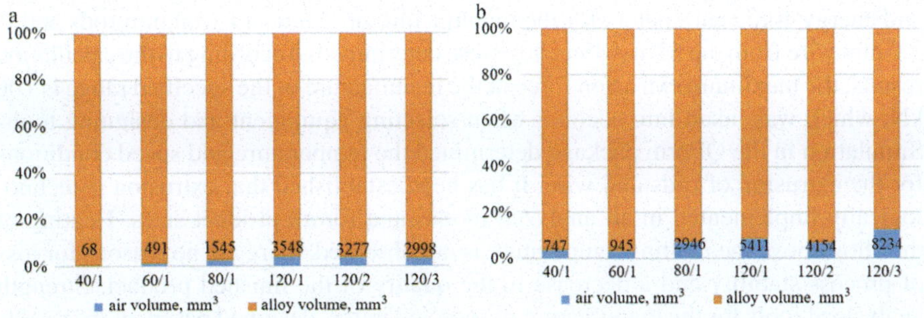

Fig. 9 The results of calculating the air volume in the container during the extrusion of a Ø 8 mm rod from the billets: **a** Ø 30.0 mm; **b** Ø 25.0 mm

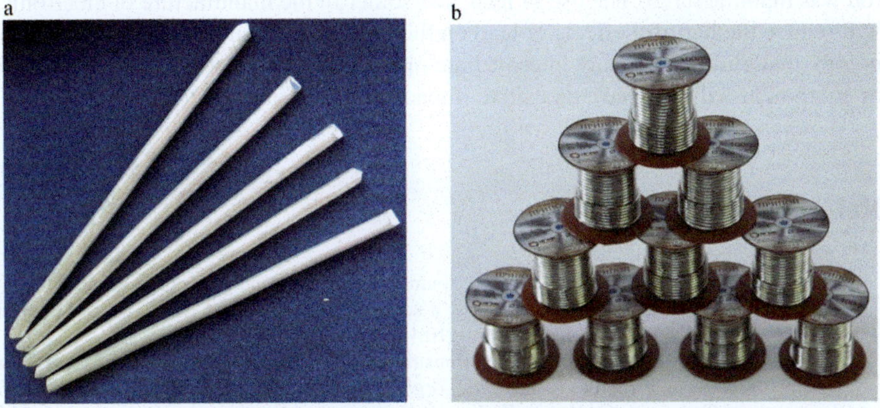

Fig. 10 Ø 8.0 mm solder rods (**b**) and Ø 2.0 mm solder wire (**a**)

billets in the amount necessary to fulfil the container. According to the results of the research, for extruding rods to Ø 8.0 mm the diameter of the billet was adjusted. For Ø 8.0 mm rods, the cast billet diameter was 31.0 mm. This eliminated the formation of a surface defect "air bubbles" in extruded rods (Fig. 10a). Using the developed technology, a batch of 100 kg of Ø 2.0 mm solder wire was produced, wound on coils with a solder mass of 100 $\mp$ 5 g, and 150 kg of Ø 8.0 mm rod for JSC Chelyabinsk Zinc Plant. The finished products are shown in Fig. 10.

6 Conclusions

A resource-saving technology for the production of lead-free solder rods and wires using the direct extrusion method has been developed. Using a mathematical model of direct extrusion and a computer program developed on its basis, the deformation and energy-force parameters for the manufacture of Ø 8.0 and 15.0 mm rods and Ø 2.0 mm wire from the 52In-48Sn alloy were determined. According to the calculation results, the maximum extrusion force in the manufacture of the specified range is 106 kN, which was taken into account when selecting equipment and designing tools. Simulation in the QForm package determined the temperature and speed conditions for the extrusion of rods and wire. It has been established that extrusion is technologically implemented in all analyzed modes and for all product sizes. Heating of the alloy above the melting temperature is not observed, there are no reasons for loss of process stability and a decrease in the quality of the finished product. Strength analysis of tools for the manufacture of rods and wires, designed based on the results of calculations, showed a safety margin of more than 2.1, which is considered sufficient for metallurgical equipment and tools. Using the developed technology, a pilot batch of 100 kg of Ø 2.0 mm solder wire wound on coils and 150 kg of Ø 8.0 mm rod was manufactured. The use of lead-free solder in the manufacture of electronics will reduce the harmful effects of lead on the environment and human health. Technology that eliminates many intermediate operations and the use of a large amount of equipment will also have a positive impact on the environment.

References

1. Ahmad H, Yaqub M, Lee SH (2024) Environmental-, social-, and governance-related factors for business investment and sustainability: a scientometric review of global trends. Environ Dev Sustain 26:2965–2987. https://doi.org/10.1007/s10668-023-02921-x
2. Söderholm P (2020) The green economy transition: the challenges of technological change for sustainability. Sustain Earth 3:6. https://doi.org/10.1186/s42055-020-00029-y
3. Twagirayezu G, Irumva O, Huang K, Xia H, Uwimana A, Nizeyimana JC, Itangishaka AC (2022) Environmental Effects of Electrical and Electronic Waste on Water and Soil: A Review. Polish Journal of Environmental Studies 31(3):2507–2525. https://doi.org/10.15244/pjoes/144194
4. Shunfeng Cheng, Chien-Ming Huang, Michael Pecht (2017) A review of lead-free solders for electronics applications, Microelectronics Reliability 75:77–95. https://doi.org/10.1016/j.microrel.2017.06.016
5. Wani AL, Ara A, Usmani JA (2015) Lead toxicity: a review. Interdiscip Toxicol 8(2):55–64. https://doi.org/10.1515/intox-2015-0009
6. Olufemi AC, Mji A, Mukhola MS (2022) Potential Health Risks of Lead Exposure from Early Life through Later Life: Implications for Public Health Education. Int J Environ Res Public Health 19(23):16006. https://doi.org/10.3390/ijerph192316006
7. Yang J, Li X, Xiong Z, Wang M, Liu Q (2020) Environmental Pollution Effect Analysis of Lead Compounds in China Based on Life Cycle. Int J Environ Res Public Health 17(7):2184. https://doi.org/10.3390/ijerph17072184
8. Stevels ALN (2012) The present recast of the WEEE Directive, Editor(s): Vannessa Goodship, Ab Stevels, In Woodhead Publishing Series in Electronic and Optical Materials, Waste Electrical and Electronic Equipment (WEEE) Handbook, Woodhead Publishing, pp 53–65. https://doi.org/10.1533/9780857096333.1.53
9. Label PI, Label P, oHS Compliance Engineer R (2005) Directive 2002/95/EC of the European Parliament and of the Council of 27 January 2003 on the restriction of the use of certain hazardous substances in electrical and electronic equipment. Artesyn Technologies: Boca Raton, FL, USA

10. Kumar A, Kumar A, M M S CP, Chaturvedi AK, Shabnam AA, Subrahmanyam G, Mondal R, Gupta DK, Malyan SK, S Kumar S, A Khan S, Yadav KK Lead (2020) Toxicity: Health Hazards, Influence on Food Chain, and Sustainable Remediation Approaches. Int J Environ Res Public Health 17(7):2179. https://doi.org/10.3390/ijerph17072179
11. Jayaram V, Gupte O, Bhangaonkar K, Nair C (2023) A Review of Low-Temperature Solders in Microelectronics Packaging. IEEE Transactions on Components, Packaging and Manufacturing Technology 13(4):570–579. https://doi.org/10.1109/TCPMT.2023.3271269
12. Zhou J, Tan XF, McDonald SD, Nogita K (2022) Mechanical Properties and Microstruc-ture of Binary In-Sn Alloys for Flexible Low Temperature Electronic Joints. Materials 15:8321. https://doi.org/10.3390/ma15238321
13. Kang H, Rajendran SH, Jung JP (2021) Low Melting Temperature Sn-Bi Solder: Effect of Alloying and Nanoparticle Addition on the Microstructural, Thermal, Interfacial Bonding, and Mechanical Characteristics. Metals 11:364. https://doi.org/10.3390/met11020364
14. Gao Y, Bian X, Qiu X, Jia Y, Yi J, Wang G (2023) Investigation of Microstructure and Mechanical Properties of SAC105 Solders with Sb, In, Ni, and Bi Additions. Materials 16:4059. https://doi.org/10.3390/ma16114059
15. Grigorieva NA, Fleitlikh IYu, Tikhonov AYa, Mamatyuk VI, Karpova EV, Logutenko OA (2022) Recovery of indium from sulfate solutions with D2EHPA in the presence of organic proton-donor additives. Hydrometallurgy 213:105925. https://doi.org/10.1016/j.hydromet.2022.105925
16. Radionova LV, Faizov SR, Sarafanov AE (2020) Mathematical modelling of low temperature solder direct extrusion. IOP Conf. Ser. Mater. Sci. Eng. 969(1):012107. https://doi.org/10.1088/1757-899X/969/1/012107
17. Faizov S, Sarafanov A, Erdakov I, Gromov D, Svistun A, Glebov L, Bykov V, Bryk A, Radionova L (2021) On the Direct Extrusion of Solder Wire from 52In-48Sn Alloy. Machines 9:93. https://doi.org/10.3390/machines9050093
18. Radionova LV, Safonov EV, Gromov DV, Lisovskiy RA, Faizov SR (2023) Strength Analysis and Modeling of Direct Extrusion Tooling for Fusible Solder. In: Radionov AA, Gasiyarov VR (eds). Proceedings of the 8th International Conference on Industrial Engineering. ICIE 2022. Lecture Notes in Mechanical Engineering. Springer, Cham. https://doi.org/10.1007/978-3-031-14125-6_27
19. Radionova LV, Faizov SR, Gromov DV, Erdakov IN, Siverin OO (2021) Friction Coefficient in the Process of Direct Extrusion of Wire and Bar Solder from 52In–48Sn Alloy. Bulletin of the South Ural State University. Ser. Metallurgy 21(3):56–66. (in Russ.). https://doi.org/10.14529/met210307

Technology Potential for Environmental Management 4.0

Accident Reduction Technology at Oil Storage Enterprises

O. V. Udartseva

Abstract The article considers the most probable causes of explosive and fire-hazardous situations in the tank farms of oil storage companies. The article analyzes the specifics of the combustible medium and technological processes associated with the possible formation of explosive mixtures. The geometric volume of the explosive zone within the radius of the tank location has been confirmed by calculations.

The research proposes the approaches to reduce the probability of fire hazard in the process of storing and pumping oil. The necessity of the safety system upgrade at a hazardous facility is justified. The authors suggest a technical solution for equipping the tank farm with a gas equalization system with additionally installed concentration monitoring sensors based on the use of an information and software complex. The adoption of digital technology will allow for the automatic monitoring, processing, visualization and transmission of information on the concentration of explosive mixtures in the area of the tank location. The function of this system is the storage and transmission of a large array of data providing for a high computing speed which is necessary to make decisions on the technological process safety provision.

Keywords Tank · Explosive and flammable mixtures · Oil storage

1 Introduction

The study of statistical data on the accident causes at oil and gas facilities highlights the technological process of storing petroleum products in a tank farm as the main object of potential explosion and fire hazard [1], Fig. 1.

The analysis of the causes and nature of accidents in the oil and gas industry showed that 45% of all cases comprise liquefied hydrocarbons exploding in oil storage systems.

O. V. Udartseva (✉)
Industrial University of Tyumen, 38, Volodarskogo Street, Tyumen 625000, Russia
e-mail: udartsevaov@tyuiu.ru

L. V. Radionova and D. V. Ulrikh (eds.), *Advances in Ecology and Environmental Engineering*, Springer Proceedings in Earth and Environmental Sciences,
https://doi.org/10.1007/978-3-031-64423-8_43

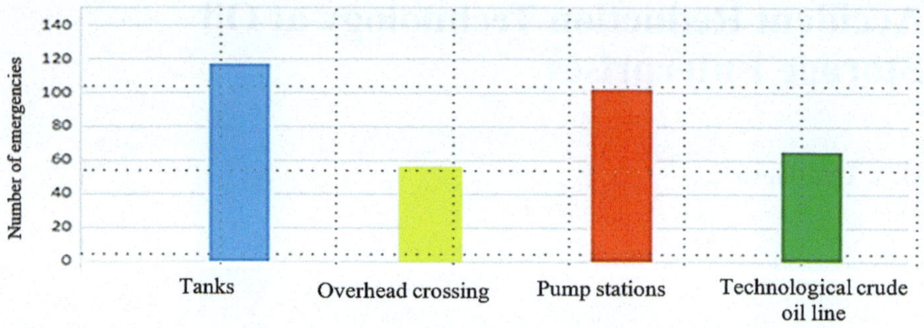

Fig. 1 Statistics of emergency situations at oil and gas facilities

2 Investigation of the Emergency Causes

There is a need to investigate the emergency causes and search for the solutions aimed at improving the safety of the petroleum product storage process.

The analysis of the main causes and factors of emergency development allows differentiating them as follows:

- mixtures of hydrogen sulfide corrosion products and organic compounds capable of spontaneous combustion resulting from leakage (42.9%);
- fire and repair work in the area of the location of the vertical stock tanks (VST) (23.5%);
- sparks of electrical units (14.7);
- uncontrolled emergencies (9.2%);
- static electricity discharges (9.7%), [2–4]

Statistics show that most accidents at oil and gas industry enterprises are caused by depressurization of technological equipment with subsequent emissions and spills of explosive substances. According to the analysis of the causes, 42.9 of accidents in the oil and gas industry occur as a result of explosions of liquefied hydrocarbons in oil storage systems.

The main reasons for the formation of an explosive situation in hydrocarbon storage areas include:

- increase in pressure in the system;
- “large” and “small” breathing tanks;
- non-compliance with technical conditions for the operation of tank farms.
- An increase in pressure in the system occurs as a result of the process of storing petroleum products, which is accompanied by the formation of a sufficiently large volume of hydrocarbons that create pressure several times higher than atmospheric pressure. For the tank to function normally, the pressure in the system must be distributed evenly. Therefore, the permissible pressure indicator in a closed tank must be calculated at the stage of its design, taking into account calculations, the requirements of building regulations and technical documentation.

- "Little" breathing of the reservoir is a consequence of daily temperature fluctuations, which contribute to a decrease or increase in pressure in the system. The highest danger is posed by the inversion state of the atmosphere at minimum wind speeds at night and in the morning.
- "Big" tank breathing is caused by the tank emptying or filling. When emptying, air enters the tank, which contributes to the formation of an explosive mixture. When the tank is filled, the internal gas space of the tank enters the atmosphere through the breathing fittings, resulting in the formation of zones of fire and explosion hazard concentrations. The fire-explosive mixture with "large" breathing is significantly superior to the mixture with "small" breathing.

According to an analysis of the causes of accidents, at oil and gas enterprises explosions pose a greater danger than fires, since they have greater destructive power with maximum damage caused by the failure of automatic fire extinguishing systems.

At the moment, various systems for preventing and minimizing the accident aftermath are used to protect tank farms. However, all systems operate separately and perform different functions.

To facilitate constant monitoring of the air environment state in the tank farm, it seems necessary to adopt a comprehensive information security system based on an existing control system.

Information support is a system of classification and coding of received information in the form of documentation systems and arrays of information. The main task of information support is to structure information into information arrays, which are tables corresponding to each parameter. A set of information arrays forms a database containing data of a temporary and permanent nature, which represent normative reference information and analyzed indicators. Information from the installed sensors enters the system automatically with the required frequency.

The expediency of using information technologies is justified by the efficiency of decision-making if a risk of a fire and explosive situation occurs [5–10], Fig. 2.

Currently, to reduce the fire hazard risk, oil products storage facilities are equipped with gas equalization systems.

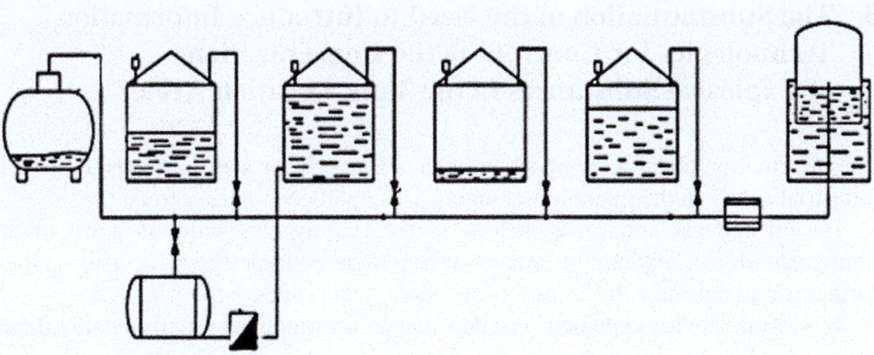

Fig. 2 Gas equalization system

At the moment, various systems for preventing and minimizing the consequences of accidents are used to protect tank farms. However, all systems operate separately and perform different functions. To ensure comprehensive protection and constant monitoring of the condition of the tank farm, the task arises of developing an information security system based on the already existing gas equalization system.

Sensors are the main element of information collection in the system, which are stationary gas analyzers, with the help of which, through digital and analog connectors, the parameters of the air environment and the concentration of gases contained in it enter the system. The use of sensors in an information system consists only of collecting information about the concentration of elements of the steam–air mixture.

Stationary gas analyzers based on the principle of absorption of infrared radiation by gases can be used as sensors.

The action of the gas analyzer is based on the property of gases to absorb infrared rays of a certain spectrum. The gas analyzer includes an infrared light sensor with a sensor between the optical filter and the measuring cell. The gas enters the measuring cell and absorbs infrared light, at which time the sensor detects the decrease in the intensity of the incoming infrared light based on the calibration curve, generating an output signal. The main advantages of this type of gas analyzer include:

- stability of use in open areas;
- frequent repeatability of measurements;
- resistance to poisoning and aging.

The concentration measured by the gas analyzer is converted into an analog output current signal and a digital output signal.

To store and process information in the information complex, it is proposed to use a database management system (DBMS) based on Microsoft Access. A DBMS based on Microsoft Access allows you to create interconnected tables in which all information received from sensors and entered into reference tables will be stored.

3 The Substantiation of the Need to Introduce Information Technologies for Controlling the Concentrations of Explosive Substances in the Tank Location Area

The information technology epoch stipulates a search for new solutions to provide industrial safety in the operation of enterprises in all economic sectors.

The oil and gas enterprises belong to the leading companies in terms of the implementation of modern information technologies, which allows focusing on their further use to optimize the control of technological processes.

To substantiate the expediency of introducing information systems into the safety control process, the geometric volume of the explosive zone in the tank location area was calculated [10, 11].

The group of oil tanks is located in the banking of S—15,872 m^2, V–10,000 l, d–34.2 m, h—17.88 m, ε—0.8.

The estimated volume of combustible vapors formed during the "large" breathing of the tank is:

$$\tau = K \cdot \frac{V}{q} \cdot ln\frac{\varphi_{start}}{\varphi_{fin}}, \text{kg/cycle} \tag{1}$$

where K is the volatility coefficient of the liquid—10; V_1–V_2 is the volume of the liquid, at $\varepsilon = 0.8$;

$$V_1 - V_2 = \varepsilon \cdot V_{sol} = 0,8 \cdot 10000\ \text{m}^3 = 8000\ \text{m}^3$$

φ_s is the volume fraction of saturated vapors at T_r:

$$\varphi_s = \frac{P_s}{P_p} \tag{2}$$

where P_s is the pressure of saturated oil vapor at $T_p = 292$ K; P_p is the operating pressure of the system,

$P_p = P_{bar} = 1$ bar $= 100{,}000$ Pa.

P_s is determined by the Antoine equation:

$$P_s = 133,332 \times 10^{A - \frac{B}{t_p + Ca}}, \text{Pa} \tag{3}$$

The values of the constants are determined by gasoline AI-72: $A = 5.07020$; $B =$ 682,876; $C = 222{,}066$.

$$P_s = 133,332 \cdot 10^{5,07020 - \frac{682,876}{19+222,066}} = 23,035, \text{Pa}$$

Define φ_s:

$$\varphi_s = \frac{23035\ \text{Pa}}{100000\ \text{Pa}} = 0.2304.$$

The molecular weight of oil is 90 kg/kmol.

The amount of combustible vapors during filling is as follows:

$$G_b = 8000 \cdot \frac{1 \cdot 10^5}{293} \cdot 0.2304 \cdot \frac{90}{8314.31} = 6809.59\ \text{kg/cycle}$$

The volume of combustible oil vapors of explosive concentration in the radius of the tank is the following:

$$V_{exp} = \frac{G_b}{\varphi_{lel}} \tag{4}$$

At the value of—$G_b = 6809.59$ kg/cycle.
Index φ of the "lel":

$$\varphi_{lel} = \frac{M_{\varphi lel}}{V_t},\ \text{kg/m}^3 \tag{5}$$

where V_t is defined as:

$$V_t = V_0 \cdot \frac{T_p}{T_0} \cdot \frac{P_0}{P_{gen}},\ \text{kg/m}^3 \tag{6}$$

where $V_0 = 22.41$ m^3/kmol—molar volume of oil vapor under normal conditions; $T_r = 292$ K is the operating temperature of the oil in the tank; $T_0 = 273$ K is the temperature under normal physical conditions; $P_0 = R_{work}$, hence, $P_0/R_{work\cdot} = 1$.

$$V_t = 22.41 \times \frac{292\text{K}}{273\text{K}} = 23.96,\ \text{m}^3/\text{kmol}$$

Hence the index φ_{lel}:

$$\varphi_{lel} = \frac{90 * 0.08}{23.96} = 0.3\,\text{kg/m}^3$$

The performed calculation makes it possible to determine the geometric volume of the explosive zone:

$$V_{ex} = \frac{6809.59\frac{kg}{cycle}}{0.3\frac{kg}{m^3}} = 22,700\ \text{m}^3$$

Thus, in the absence of wind, the estimated volume of the explosive mixture is V—22,700 m^3, which poses an emergency threat.

4 Proposal for Optimizing the Fire and Explosion Safety Control System of the Tank Farm

Based on the calculation conducted, there is a need to upgrade the control system for fire and explosive concentrations of vapor-air mixtures not only inside the tank system, but also in the environment. Modernization involves the use of an information system that allows automatic monitoring, processing, storing, visualizing and transmitting information about the concentration of vapor-air mixtures in the tank location area [12–17].

This information system contains a block for collecting data on the concentration of explosive substances in the tank area location, information processing and visualization systems.

The information system diagram, Fig. 3.

The algorithm of interaction of all elements of the information control system consists in the algorithm of operation shown in Fig. 4.

The introduction of an information system for monitoring the technological process will further allow storing and transmitting large data amounts in the system providing for a high computing speed necessary to provide the technological process safety.

To control the fire and explosive concentrations, additional equipment is required for the gas equalization system with monitoring sensors [14, 18, 19].

The gas equalization system ensures the movement of an air mixture of oil vapors in a closed loop, thereby reducing the release of explosive oil vapors into the atmosphere. However, with "large" and "small" respirations, emissions into the atmosphere remain at a sufficiently high level that can lead to an accident. High concentrations of oil vapors in the atmosphere pose a danger to technological processes,

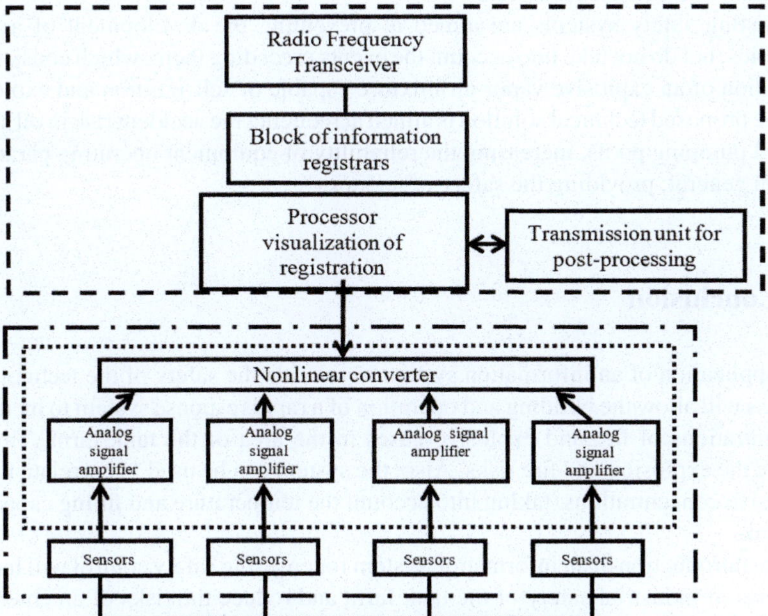

Fig. 3 The sequence of stages of the information system for monitoring the parameters of the oil storage process

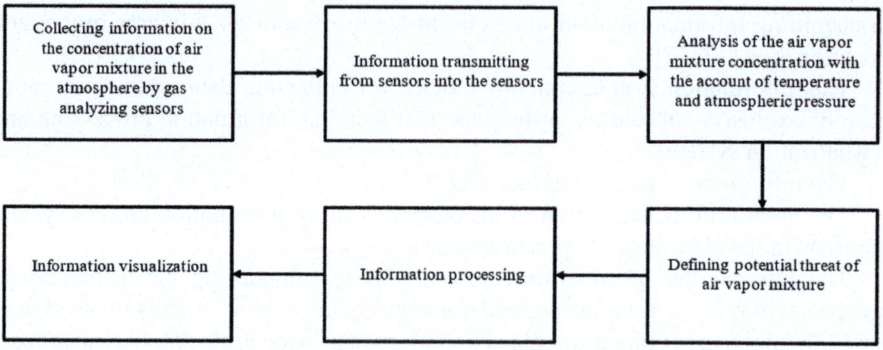

Fig. 4 Algorithm of operation of the information system for monitoring the technological process of oil storage

creating not only man-made risks, but also economic and reputational ones. Moreover, in the event of a fire in the gas equalization system, there is a risk of flame propagation. This is due, in particular, to the fact that the pipelines and breathing pipes of the tanks are equipped with fuses of the same type, which do not always fulfill their functions, since the operating modes of the breathing pipes and the gas equalization system are different.

Existing safety systems are aimed at preventing the development of accident scenarios, but do not take into account the events preceding them, which consist in the formation of an explosive vapor-air mixture capable of self-ignition and explosion.

The proposed technical solution is aimed at reducing the accident risk in oil storage and oil pumping pools, increasing the reliability of equipment operating parameters and, in general, providing the safety of territories.

5 Conclusion

The application of an information system to provide the safety of the technological process will allow the building and operation of a rapid response system to increasing concentrations of fire and explosive gases in the area of the tank farm. This will reduce the explosion and fire risks. Also, the system can be used to correlate fire and explosive concentrations, taking into account the temperature and filling capacity of the tank.

The introduction of an information system for ensuring safety control will increase the level of industrial safety of the tank farm and reduce the risks of an emergency as a result of the formation of a fire and explosion hazard zone on its territory.

References

1. Newsletter of the Federal Service for Environmental, Technological and Nuclear Supervision. Moscow, No. 5(122), 2022
2. Saraeva AA (2019) Emergency oil spills and the environment. High-tech research as the basis of innovative development of society, No.1. https://ami.im/sbornik/MNPK-265_pdf. Accessed 04 Feb 2023
3. Plisko NN (2017) Safe storage of oil and petroleum products in tank farms. Natural and technical sciences 6:10
4. Aralov ES, Pomogalov VN, Kartashov NA (2020) Analysis of modern problems of ensuring the reliability of hydrocarbon storage facilities. Urban planning. Infrastructure. Communications 4:6–11
5. Koval AV (2021) Analysis of oil spill features at stationary oil and petroleum products storage facilities. Science and youth: new ideas and solutions: proceedings of the XVth International scientific and practical conference of young researchers, Volgograd, pp 77–78
6. Kireev IR, Tukbayeva RB, Yenikeeva TM (2017) Modern technologies for ensuring the safety of storage of petroleum products. Modern trends in the development of science and technology: a collection of scientific papers based on the materials of the XXVth International Scientific and Practical Conference 6:11–19
7. Klimantov AA, Marchenko MA, Priymak VV et al (2017) Assessment and forecasting of reliability and level of industrial safety of petroleum product storage tanks: monograph. St. Petersburg University of the State Fire Service of the EMERCOM of Russia, St. Petersburg, p 176
8. Dorosh IV, Nafikov EV (2020) Reduction of oil losses in tank farms and measures to provide safety during their operation. Problems of security provision 13:259–262
9. Budykina TA, Budykina KYu (2019) Modern methods and means of safety provision during the liquidation of emergency oil spills in tank farms. Progressive technologies and processes 17:41–47
10. Udartseva OV (2022) Reducing the probability of formation of a fire-explosive mixture in the area of oil storage tank farms: scientific article. Problems of risk management in the technosphere
11. Strehlow RA, Ricker RE (2018) The blast wave from a bursting sphere. Loss Prevention
12. Held M, Jager EH, Stolz D (2018) TNT-equivalence for uncontrolled explosion of gas pressure vessels. Paper at Euro-Mech
13. Roth D, Hubmann N, Havel C, Herkner H, Schreiber W, Laggner A (2021) Victim of carbon monoxide poisoning identified by carbon monoxide oximetry (English). The Journal of Emergency Medicine 40(6):51–58
14. Sekerin VD, Gaisina LM, Shutov NV, Abdrakhmanov NKh, Valitova NE (2018) Improving the Quality of Competence-Oriented Training of Personnel at Industrial Enterprises. Quality—Access to Success 19(165):68–73
15. Khalilova DI, Elizarieva EN (2016) Economic efficiency of methods for eliminating soil oil pollution. Forum of Young Scientists, No. 4. https://docplayer.com/52863947-Effektivnost-metodov-likvidacii-neftyanyh-zagryazneniy-pochvy.html. Accessed 24 Nov 2023
16. Shevtsov SA, Kargashilov DV, Bykov IA, Usachev DK, Belomytceva EG (2018) Fire-explosion-proof and environmentally safe technology for storing petroleum products in tanks. Environmental protection in the oil and gas complex 3:8–11
17. Suchkov VP, Kolpin NG, Derevyanko NS, Lyakhov GA (1991) Protection of tanks with pontoons from fires. NTIS "Transportation and storage of petroleum products" 4:14–18
18. Kluban VS, Yuryev VI (2019) Assessment of the explosion and fire hazard of vertical steel VST tanks with a pontoon and an aluminum roof. Fires and emergencies: prevention, elimination 2:55–60
19. Roth D, Hubmann N, Havel C, Herkner H, Schreiber W, Laggner A (2011) Victim of carbon monoxide poisoning identified by carbon monoxide oximetry (English). The Journal of Emergency Medicine 40 (6):51–58

Low-Waste Construction Technologies Based on the Use of Formwork Systems with Composite Formwork Facing

S. G. Abramyan, S. V. Klyuev, and I. L. Kiselev

Abstract The paper views sustainable urban development in the context of ensuring environmental safety of natural and industrial ambience. Because nowadays in-situ construction proves one of the most rapidly advancing construction technologies, the study focuses on improving formwork systems, which, as compared to reinforcement and concrete works, involve higher installation and stripping labor costs. The authors' approach to improving formworks' environmental performance relies on enhancing their operational reliability by upgrading such characteristics as turnaround, technological adaptability, maintainability, which helps reduce the construction time and extend the construction systems' lifespan. The experience of the global formworks producers specializing in composite facing formworks was explored. Thee long-term use of such formworks has proven these systems highly reliable. This research compares formwork facing made of composite armatilen™ board developed by the Russian company LLC MATIL and laminated plywood globally widespread in formwork fabrication. Armatilen™ board is experimentally confirmed to have better moisture resistance, maintainability, serviceability, concrete surface quality etc. Considering these advantages, the board is strongly advised for application in formwork systems used for shop-fabrication of three-dimensional modular building blocks. Some facing characteristics are provided in tabular form for comparing the formwork facing options. Armatilen™ formwork facing is also found to be undecayable and disposable without causing significant damage to the natural and industrial environment. Armatilen™ composite board is multifunctional and can be used outside the construction industry. The use of armatilen™ board in place of natural wood and layered wood-based materials ensures the ecological balance of natural systems.

S. G. Abramyan (✉)
Volgograd State Technical University, 28, Lenin Avenue, Volgograd 400005, Russia
e-mail: susannagrant@mail.ru

S. V. Klyuev
Belgorod State Technological University Named After V.G. Shukhov, 46, Kostyukova St, Belgorod 308012, Russia

I. L. Kiselev
LLC "MATIL", 11-104 Izmailovsky Proezd, Moscow 105037, Russia

L. V. Radionova and D. V. Ulrikh (eds.), *Advances in Ecology and Environmental Engineering*, Springer Proceedings in Earth and Environmental Sciences,
https://doi.org/10.1007/978-3-031-64423-8_44

Keywords Construction System · Erection · Cast-in-place Concrete · Formwork · Facing · Turnaround · Technological Adaptability

1 Introduction

The analysis of a number of foreign [1–3] and Russian [4–6] sources shows that a sustainable development of modern cities is impossible outside the environmental safety context. Construction is one of the economic sectors that cause a profound damage to nearly all geosphere shells of the planet [7–9], which necessitates appropriate consideration of the environmental safety issues with regard to all lifecycle stages of a building or facility. Despite the availability of individual recommendations, the contemporary stage of development of the construction industry is sadly lacking the approved regulatory documents that would thoroughly tackle the issues of environmental upgrading of various lifecycle stages of a construction system.

This paper discusses the stages of in-situ building construction with cast-in-place concrete and, partially, modular construction with 3D modular building blocks. It is known that the installation and dismantling of formwork systems account for more than 45% of the overall labor intensity of construction per cubic meter of cast-in-place and reinforced concrete structures. The cost of building construction and its environmental safety depend to a large extent on the turnaround of formwork systems. The higher the formwork turnaround rate (multiple usage), the lower the price per cubic meter of a resulting structure. Enhanced formwork turnaround helps cut costs involved in the formwork facing handling and replacement operations and also reduces the labor intensity of the process flows and, as a result, the overall construction time. This is also the reason behind the minimized impact on the environmental balance within the natural and industrial system.

2 Material and Methods

The current advancement of in-situ construction alongside modular construction is explained by the fact that these technologies allow constructing buildings and facilities resistant to man-induced impacts and the effects of various hazardous environmental factors. Furthermore, construction systems made with the use of in-situ or modular construction technologies provide enhanced seismic stability stemming from the rigidity and strength of individual structural elements. No less important is that site-cast buildings and facilities have the greatest potential in terms of architectural expression. The degree of customization and the quality of final structural elements as well as the construction and installation works depend on the characteristics of the formwork systems employed, and specifically, the formwork facing, which directly contacts the surface of the future cast-in-place concrete or reinforced concrete structural element. The current offering of formwork systems is quite extensive [10–12].

The primary exploration methods featured in this paper are analysis, benchmarking and comparison of the existing types of formwork facing used in making vertical, horizontal and inclined concrete and reinforced concrete structural elements. The paper also draws upon the experience of the Chinese company HolyCore Composite Materials Co., Ltd. manufacturing sandwich panels (HolyPan) with glass-fiber reinforced hot-melt cellular polypropylene filler used as formwork facing [13] and the Swiss company Quadrant Plastic Composites also specializing in making multifunctional sandwich panels, in particular with formwork facing made by applying "a very light polypropylene (PP) and glass-fiber composite between the two coating layers of high-performance GMTex™ glass-fiber reinforced plastic (GMT)" [14]. The manufactures point out that the turnaround rate of formwork systems using the suggested facing reaches 200 cycles and "the facing easily separates from the concrete structure after setting, thus creating a nearly perfect concrete structure finish" [14].

The purpose of the paper is to identify the specifics and key differences in formwork turnaround as applicable to the two composite materials in question: laminated plywood (which has become extremely widespread on the global formwork market) and armatilen™ composite board developed by LLC MATIL using modern high-strength polymer binders combined with reinforcing materials.

The form-building parts of armatilen™ board are made of reinforced solid wood. The wood reinforcement is accomplished with structural glass-fiber threads and polymer binder. Depending on the operating conditions and product specifications, a variety of reinforcing layers can be used. The reinforcing layer of the external coating of armatilen™ board consists of a polymer matrix and continuous structural threads binding the form-building layer and the external layers. It is worth noting that the armatilen™ board reinforcing technologies are similar to the technologies used in the aerospace industry. The main functional purpose of the outer laminating layer is protection. This layer is made of modified polyolefin-based polymer matrices and used to cover all the product surfaces. As experimentally proven, it guarantees facing panel integrity by protecting it against various aggressive chemicals and moisture impacts.

The experiment designed to determine the moisture resistance of the facing panel involved wetting the surfaces of the two samples used for comparison and their abrasive treatment by a grinding machine (Fig. 1).

The cyclic workflow (five-fold wetting and subsequent abrasive treatment) has demonstrated the definitive advantage of armatilen™. board. For instance, Fig. 1 features the original samples (on the left—armatilen™ board, on the right—laminated plywood) before the abrasive treatment, Fig. 1b—the surface after the initial machine grinding with an abrasive brush head, Fig. 1c—the surface after the final machine grinding, and Fig. 2d—the final result. As can be seen in Fig. 1d, the

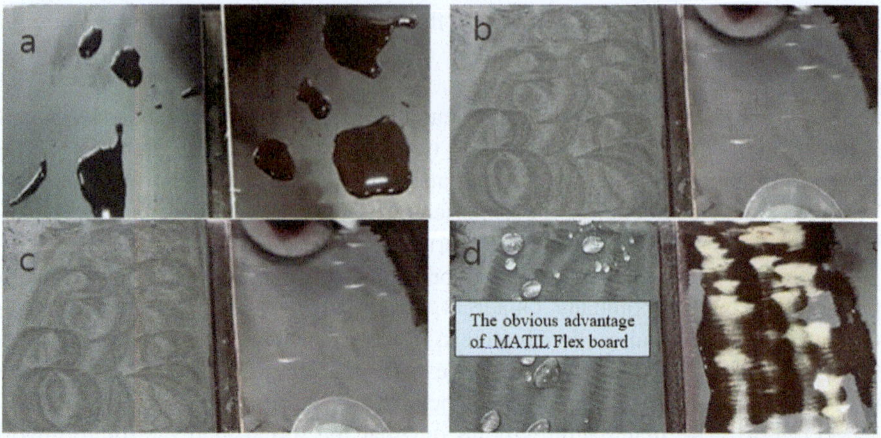

Fig. 1 Formwork facing moisture resistance testing process: **a** wetting, **b** initial machine grinding with a rigid brush head, **c** final machine grinding with a rigid brush head, **d** final wetting

laminating layer is removed by abrasion and the plywood is wetted. This way, laminated plywood, if subjected to even a short-term contact with water, can swell and disintegrate.

One of the key reliability metrics of any product is its reparability; with this in view the experiments were conducted to assess the reparability of armatilen™ board. For this, the surface of the board was dented with a locksmith hammer. The rectification process was accomplished using the hot-dip soldering method (Fig. 2).

Formwork panels often get damaged during their transportation from one construction site to another or handling while on the construction site. The board surface restoration (repair) processes shown in Fig. 2 can well be applied on a construction site. Certainly, after several turnaround cycles or damages, laminated plywood facing also can be repaired. However, note should be taken that, while the conditions of making concrete works may be the same, armatilen™ board facing will serve a larger number of turnaround cycles before requiring a repair. This is due to the fact that the laminating layer of plywood facing is made of a 0.15mm melamine coating and not designed to bear significant operating loads. There two solutions to this problem: complete facing replacement or partial repair. In the first case, this entails extra costs to purchase and deliver laminated plywood, replace the facing and so on, which makes the work more expensive; this is why the second repair option proves more popular, though it, in its turn, drastically deteriorates the quality of the surface of a concrete structure.

For comparison, Fig. 3 shows the facing surfaces after repair. As can be seen in Fig. 3b, after the repair and further use of laminated plywood facing, the joint between the two parts is heaved, the patches let water and air seep from the concrete, and as a result, the concrete surface is covered with cavities.

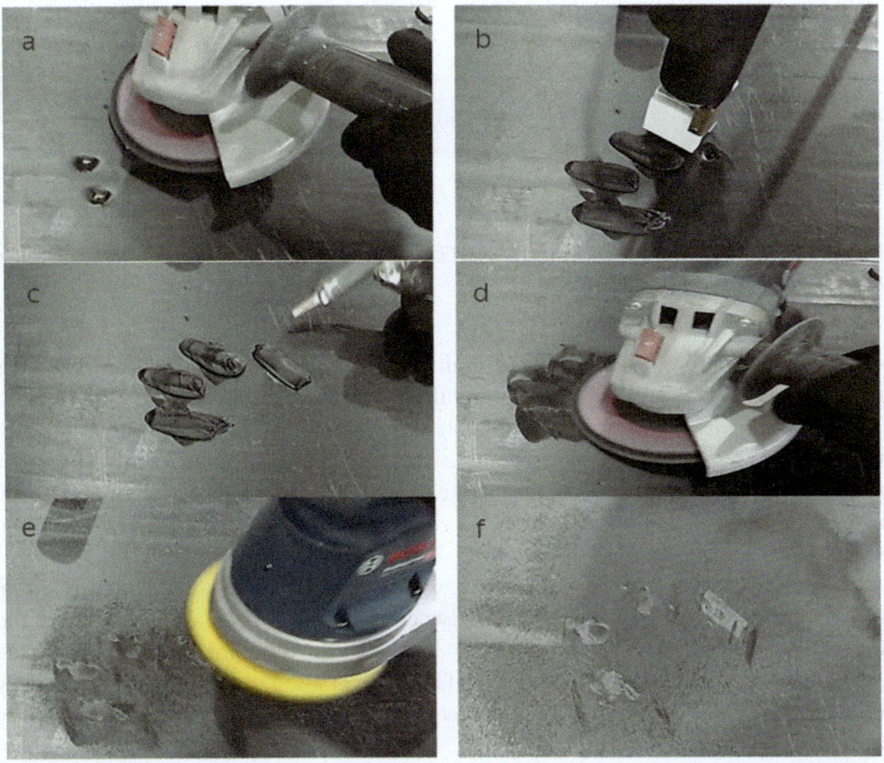

Fig. 2 Dented armatilen™ board surface rectification process: **a** grinding side scales off the dented spots; **b** applying the soldering material to the dents; **c** forced cooling of the applied soldering material with a hot air gun; **d** surface levelling by grinding after applying the soldering material; **e** final grinding of the surface with a nylon brush head; **f** finished surface after the repair

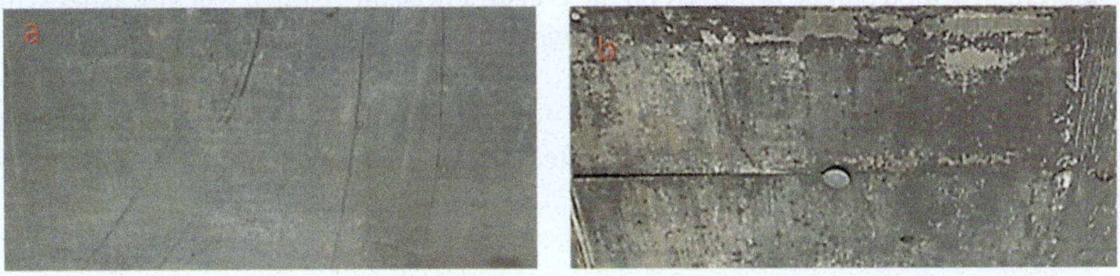

Fig. 3 Surface of formwork facing after rectification: **a** with armatilen™ board; **b** with laminated plywood

Meanwhile, the surface of the formwork facing made of armatilen™ board (Fig. 3a) only bears scratches and retains its integrity as well as air or water tightness. Practical observations show that laminated plywood provides A3 class surface

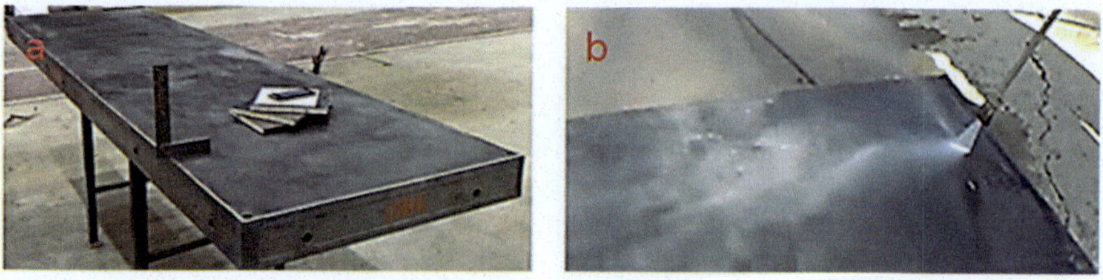

Fig. 4 Integrated formwork panel with DOKA Framax welded frame and armatilen™ board facing: **a** overall view, **b** on-site cleaning process

of concrete structures solely at the initial stages of operation (20 turnaround cycles instead of 60 cycles according to the regulatory documents).

Because armatilen™ board facing can be made in any sizes and configurations, it is easily integrated with steel or aluminum reinforcing ribs of formwork panels regardless of their manufacturer. Figure 4a shows a formwork panel where the facing is made of armatilen™ board and lateral rigidity is achieved using a DOKA Framax formwork grouting profile.

The formwork facing is levelled flush, which ensures a high surface quality of the resulting concrete structure. The gap between the DOKA Framax board frame and the side surface of the facing is closed by welding with a flexible polymer rod made of the same material as the working surface of armatilen™ board and finalized by grinding, which allows smoothing out any uneven spots and obtaining a high-quality formwork panel surface.

Another important advantage is the improved maintainability of formwork panels. The laminating polymer layer of the formwork facing prevents the concrete mix from sticking to the working surface of the facing; however, the height of the formwork panel is always a few centimeters greater than the height of a target vertical structural element and it is this area of the facing surface that accrues concrete crust during concreting, which can be removed on the construction site using a wheel washer with a high-pressure water jet (Fig. 4b). Formwork facing made of laminated plywood is not designed to bear such loads.

Formwork facing made of armatilen™ board also retains its high-strength and wear resistance properties when operated under extreme sub-zero temperature conditions.

3 Results and Discussions

In the course of building construction, a high quality of the surface of concrete and reinforced concrete structures helps reduce labor costs associated with the finishing works. The construction works performed with the use of formwork panels with armatilen™ board facing provide A3 class surfaces (face surfaces of walls and columns and lower surfaces of floors that are subject to special requirements, as well as surfaces that allow so called improved painting without preliminary puttying) practically over the entire formwork operation stage. Figure 5 presents a comparison of the concrete structures obtained with the use of the two types of formwork facing.

It should be noted that the photograph was taken after the first cycle of formwork use and the steel frame of the facing experimentally accommodated two facing halves, one made of armatilen™ board and the other of laminated plywood. The quality of the concrete surface is ensured by the moisture resistance and airtightness of the armatilen™ board facing, which therefore, does not affect the concrete hydration processes.

This circumstance enables the use of formwork systems with armatilen™ board facing in modular construction (considered one of the high-technology modern methods of arranging construction operations [15–17]) not only in case of new building construction, but also in case of reconstruction, in particular, by way of conversion [18, 19]. Due to its improved strength performance, this formwork facing can be used for fabricating buttress formworks when making foundations for high-hazard facilities, especially in earthquake prone areas [20].

It is known that the cost of concrete works significantly depends on the formwork turnaround factor: the higher the turnaround rate, the lower the cost of the work and, consequently, the total cost of the real property project. The diagram below (Fig. 6) shows the numbers of cycles characterizing the formwork turnaround for the facing options compared herein in making horizontal (floor slabs) structures and various type vertical structures.

A cycle is understood as the time required to install and dismantle a formwork system and the time taken while the concrete setting reaches at least 70% design strength, which, in turn, depends on a whole slew of factors [21–23]. When using a formwork with laminated plywood facing, the facing is lubricated with anti-adhesion compounds before each cycle and the facing surface is cleaned after the end of each cycle. Formwork facing made of armatilen™ board does not require either

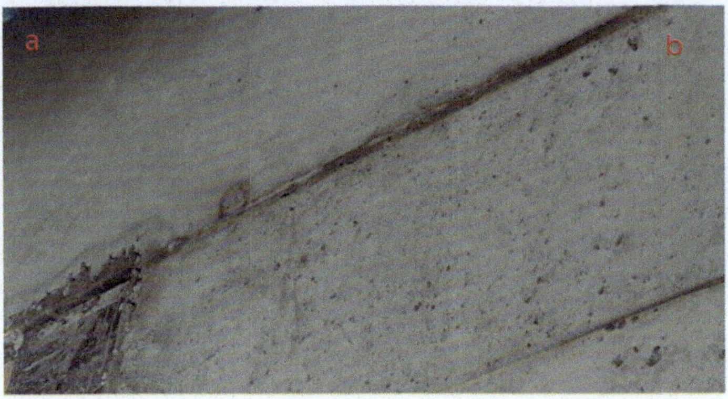

Fig. 5 Concrete structure surface after formwork stripping: **a** with armatilen™ board formwork facing; **b** with laminated plywood formwork facing

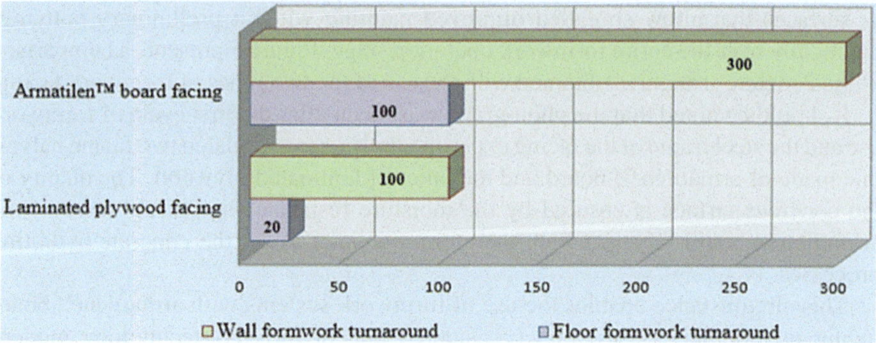

Fig. 6 Turnaround of formworks with armatilen™ board facing and laminated plywood facing

lubrication or cleaning. As the experience shows, concrete crust only emerges after thirty facing turnaround cycles and, as noted above, it only forms on the upper part of the formwork facing that does not directly contact the surface of the future concrete (reinforced concrete) structure. However, during maintenance (removal of concrete crust), formwork panels are cleaned so as to completely remove dust and various types of dirt that may occur immediately on the construction site.

Figure 7 shows the labor costs involved in formwork maintenance after 30 turnaround cycles during wall concreting for each 100 m^2 of formwork panels made of the facing types under comparison.

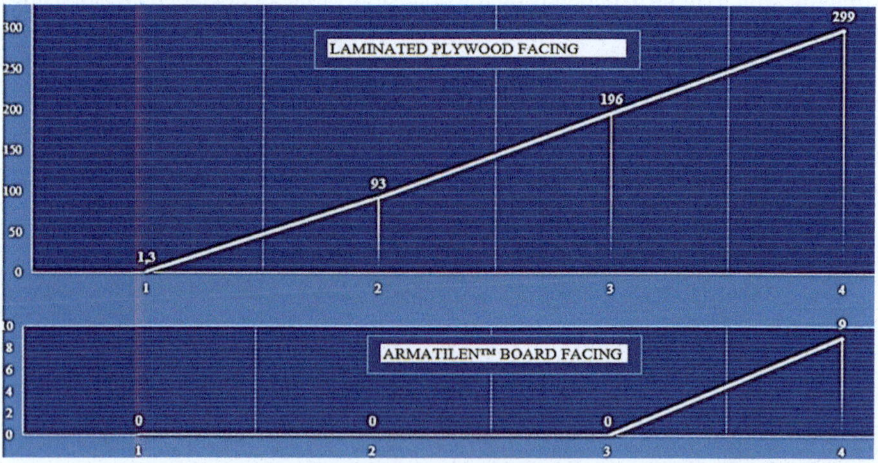

Fig. 7 Labor costs involved in formwork maintenance (lubricating and cleaning) per 100 m^2 of formwork facing made of the materials under comparison; man-hour

The chart data are referred to a turnaround rate of 10 cycles. Due to the fact that, before concreting, the laminated plywood facing must be lubricated with anti-adhesion compounds, the labor costs at the zero turnaround rate are equal to 1.3 man-hours given the mechanized application of a lubricating material.

When testing the armatilen™ board facing, some of the leading Russian construction organizations specializing in in-situ and modular construction operations revealed that, in terms of the turnaround rate, this facing is closer to metallic facing. However, in terms of maintenance, metallic facing is less efficient than composite facing, which reduces the technological adaptability of both formworks and concrete works. Metallic facing panels are heavy and their maintenance, as well as in case of laminated plywood facing, contributes to atmospheric pollution. One of the key indicators of improved technological adaptability is a reduction of labor costs, which predetermines a trend to use various light-weight items and tools in erecting construction systems.

Table 1, based on the armatilen™ manufacturer's data, sets out certain characteristics of the formwork facing types compared in this paper.

Table 1 Some key indicators characterizing the formwork facing under comparison

Indicator	Laminated plywood facing	Armatilen™ board facing
Area of application	Almost any formwork systems subject to modifications (milling, edge painting etc.)	Any formwork systems without modifications
Facing lubrication	Continuously until the start of concreting	Not required
Facing surface cleaning	Gradual deterioration of the working surface due to cleaning	Cleaning by abrasive treatment with a high-pressure hot or cold water jet
Drilling and nailing	Easy to drill and nail; the holes reduce the moisture resistance, speed up the destruction process, and leave traces on the concrete surface	Easy to drill and nail, the availability of holes does not affect the panel strength or the quality of the concrete surface. The holes are removed by soldering the panel's outer layer
Volumetric weight, kg/m^3	650	600
Moisture resistance	Swelling and disintegration on contact with water	No changes after a long-term contact with water
Resistance to aggressive chemicals	Disintegration on contact with aggressive media	Resistance to alkalis, acids and salts
Concrete surface quality	Good quality of the concrete surface during the initial cycles with subsequent progressive deterioration	Exceptional quality of the concrete surface after each cycles of formwork use

We would like to underscore that all the working surfaces of armatilen™ board facing are fitted out with chips, which store the useful product information and the panel's unique serial number.

The use of composite materials in order to create new formwork systems is not a recent trend; the renowned companies specializing in the development and introduction of innovative products have a long-term experience of using this approach (e.g., PERI DUO panel formworks, DOKA—X-LIFE working surfaces etc.).

The issues pertaining to the creation of new composite formwork systems that can be used for building construction and reconstruction are discussed in are previous publications [24–26].

In Russia and in many other countries where various advanced construction technologies are in use, contractors most often confine themselves to the tried-and-tested as well as inexpensive technologies, which not only artificially reduces the overall project cost, but also affects the quality of the completed works (in part, because some objects are commissioned without finishing). Yet, any technologies, goods, products etc. are only relatively expensive during the roll-out phase, while the initial costs are paid off during subsequent phases.

4 Conclusions

When talking about low-waste formwork systems where facing is made of armatilen™ board, attention should be paid to the following effectiveness indicators: a five-fold facing turnaround rate for making floor slabs and three-fold facing turnaround rate for concreting vertical structures, 33 times lower labor costs involved in formwork facing maintenance with a turnaround rate of 30 cycles, and the absence of need for purchasing lubricating materials. Formwork facing made of armatilen™ board is undecayable and reliable in operation as it is extremely resistant to damages and easily reparable. Furthermore, it can be disposed without causing significant damage to the natural and industrial environment.

The armatilen™ board is multifunctional, its foreign analogues are ALKUS® and SONOBOARD plastic panels (working surfaces), which are successfully used in the aerospace and aviation industries, in machine engineering and other sectors. In construction, the application area of armatilen™ board can be expanded with the adoption of new materials (parquet boards, facing panels for vertical outer and inner surfaces etc.) and making prefabricated buildings for construction camps etc.

The use of various types of equipment and auxiliary devices (including those made of armatilen™ boards) in place of natural wood ensures the ecological balance of natural systems due to their multiple usability during the operation phase and service life extension beyond recycling.

The efficiency of erecting construction systems made of cast-in-place concrete and reinforced concrete is ensured by a whole number of factors: longevity of structures; reduced duration of construction works; lower building weight (as compared to facilities made of other materials, including brick masonry), possibility of performing the construction works under various weather conditions. in a variety of climatic regions, as well as in congested spaces, customization of construction (architectural expression, variable functional purpose). Overall, the introduction of modern, well-proven technologies is an important stage in the holistic development of in-situ construction, which, in particular, fosters the development of new types of formwork systems.

Acknowledgements The work was performed using the equipment of the High Technology Center BSTU named after V.G. Shukhov.

References

1. De Iulio R, De Martino M, Isidori E (2022) Environmental education and its contribution to sustainable cities. Science for Education Today 12(2):136–150. https://doi.org/10.15293/2658-6762.2202.07
2. Riffat S, Powell R, Aydin D (2016) Future Cities and Environment. Future Cities and Environment 2(1):23. https://doi.org/10.1186/s40984-016-0014-2
3. Lin JY, Zhao LCh, Zheng YY, Du QK, Xiong L, Zhu BW, Tzeng GH (2023) Using a Hybrid Multiattribute Decision-Making Model for Evaluating the Sustainable Development Potential of Characteristic Towns and Exploring Development Planning Strategies. J of Environmental and Public Health 5:1–14. https://doi.org/10.1155/2023/9065729
4. Burmatova OP (2021) Koncepcija ustojchivogo razvitija umnogo goroda: jekologicheskij aspekt (The Conceptual Framework of Sustainable Development for a 'Smart' City: Ecological Aspect). ECO 6:139–160. https://doi.org/10.30680/ECO0131-7652-2021-6-139-160
5. Naumenkov AV, Shelomentseva MV (2018) Ekologicheskiy faktor v ustoychivom razvitii regionov (Ecological factor in sustainable development of regions). Russian J of Resources, Conservation and Recycling 1(5). https://doi.org/10.15862/06ECOR118. https://resources.today/PDF/06ECOR118.pdf. Accessed 19 Nov 2023
6. Klyuev SV, Kashapov NF, Radaykin OV, Sabitov LS, Klyuev AV, Shchekina NA (2022) Reliability coefficient for fibreconcrete material. Construction Materials and Products 5(2):51–58. https://doi.org/10.58224/2618-7183-2022-5-2-51-58
7. Abramyan SG (2017) Construction industry and sustainable development concept. Internet magazine "Naukovedenie" 9(5). https://naukovedenie.ru/PDF/64EVN517.pdf. Accessed 19 Nov 2023
8. Klyuev AV, Kashapov NF, Klyuev SV, Lesovik RV, Ageeva MS, Fomina EV, Ayubov NA (2023) Development of alkali-activated binders based on technogenic fibrous materials. Construction Materials and Products 6(1):60–73. https://doi.org/10.58224/2618-7183-2023-6-1-60-73
9. Abramyan SG, Burlachenko OV, Oganesyan OV (2017) Modeling of Construction Processes for a Specific Object Based on Environmental Parameters. Vestnik MGSU 7(106):797–803. https://doi.org/10.22227/1997-0935.2017.7.797-803
10. Ponomareva YuA, Lyakhov VM, Zhilnikova TN (2021) Raznovidnosti opalubochnyh sistem dlja karkasno- monolitnogo stroitel'stva (Types of formwork systems for frame and monolithic construction) Ingineering J. of Don 3:9. http://www.ivdon.ru/ru/magazine/archive/n3y2021/6862. Accessed 21 Nov 2023
11. Pogodin DA, Spiridonov NN, Khalidov AA (2019) Sovershenstvovanie sovremennyh tehnologij i vozvedenie mnogojetazhnyh zhilyh zdanij za schet optimizacii opalubochnyh rabot (Improving modern technology and construction of multistorey residential buildings by optimizing shuttering works). Russian J of transport engineering 2(6). https://doi.org/10.15862/07SATS219. https://t-s.today/PDF/07SATS219.pdf. Accessed 21 Nov 2023

12. Li W, Lin X, Bao DW, Xie YM (2022) A review of formwork systems for modern concrete construction. Structures 38:52–63. https://doi.org/10.1016/j.istruc.2022.01.089
13. Composite Sandwich Panels for Concrete Formwork. https://www.holycorepanel.com/sandwich-panel/composite-sandwich-panel. Accessed 21 Nov 2023
14. Composite sandwich panels prove ideal for building formwork systems. https://www.pressreleasefinder.com/Quadrant_Plastic_Composites/QPCPR004/en. Accessed 21 Nov 2023
15. Klyuev S, Fediuk R, Ageeva M, Fomina E, Klyuev A, Shorstova E, Sabitov L, Radaykin O, Anciferov S, Kikalishvili D, de Azevedo Afonso RG, Vatin N (2022) Technogenic fiber wastes for optimizing concrete. Materials 15(14):5058
16. Amran M, Fediuk R, Klyuev S, Qader DN (2022) Sustainable development of basalt fiber-reinforced high-strength eco-friendly concrete with a modified composite binder. Case Studies in Construction Materials 17:e01550
17. Thai H, Ngo T, Uy B (2020) A review on modular construction for high-rise buildings. Structures 28:1265–1290. https://doi.org/10.1016/j.istruc.2020.09.070
18. Abramyan SG, Klyuev SV, Polyakov VG, Sabitova TA, Akopyan GO, Guseynov KM (2023) Specifics of information model development for functional conversion of offshore oil platforms. Construction Materials and Products 6(4):42–57. https://doi.org/10.58224/2618-7183-2023-6-4-42-57
19. Abramyan SG, Klyuev SV, Emelyanova OE, Oganesyan OV, Chereshnev LI, Akopyan GO, Petrosian RO (2023) Improving reinforced concrete column strengthening techniques for reconstruction projects using composite jacketing formworks. Construction Materials and Products 6(5). https://doi.org/10.58224/2618-7183-2023-6-5-1
20. Abramyan SG, Burlachenko OV, Oganesyan OV, Burlachenko AO, Archakov IB, Pleshakov VV (2022) Technological solutions ensuring reliable operation of steel vertical reservoirs in seismic areas. Construction Materials and Products 5(5):5–16. https://doi.org/10.58224/2618-7183-2022-5-5-5-16
21. Makul N, Fediuk R, Amran HMM, Zeyad Abdullah M, Azevedo A, Klyuev S, Vatin N, Karelina M (2021) Capacity to develop recycled aggregate concrete in south east asia. Buildings 11(6):234
22. Fediuk R, Amran M, Klyuev S, Klyuev A (2021) Increasing the performance of a fiber-reinforced concrete for protective facilities. Fibers 9(11):64
23. Klyuev S, Fediuk R, Ageeva M, Fomina E, Klyuev A, Shorstova E, Zolotareva S, Shchekina N, Shapovalova A, Sabitov L (2022) Phase formation of mortar using technogenic fibrous materials. Case Studies in Construction Materials 16:e01099
24. Abramyan SG, Oganesyan OV (2020) Development of lining formwork for column expansion during reconstruction of building and structures. IOP Conference Series: Materials Science and Engineering. International Conference on Construction, Architecture and Technosphere Safety (ICCATS 2020) (6–12 September 2020, Sochi, Russia), Vol. 962, 2:7. https://doi.org/10.1088/1757-899X/962/2/022087/pdf
25. Abramyan SG, Burlachenko OV, Oganesyan OV (2018) Multi-Purpose Forming System for Strengthening Structural Elements of Buildings and Works. Materials and Technologies in Construction and Architecture. International Conference on Construction and Architecture: Theory and Practice of Industry Development (CATPID-2018) 931:405–410. https://doi.org/10.4028/www.scientific.net/MSF.931.4052018. https://www.scientific.net/MSF.931.405. Accessed 23 Nov 2023
26. Abramyan S, Polyakov V, Oganesyan O (2017) Pneumatic formwork used in strengthening of structural elements during reconstruction of buildings and structures. MATEC Web of Conferences. International Conference on Modern Trends in Manufacturing Technologies and Equipment (ICMTMTE 2017) (Sevastopol, Russia, September 11–15, 2017) 129:05001. https://doi.org/10.1051/matecconf/201712905001. https://www.matec-conferences.org/articles/matecconf/pdf/2017/43. Accessed 23 Nov 2023

Energy-Entropy Approach in the Study of a Class of Natural and Technical Systems in the Use of Water Resources

E. D. KHetsuriani, V. L. Bondarenko, A.-B. I. Ilyasov, M. I. Shtavdaker, and M. S. Khokhrin

Abstract The purpose of the study is to assess the relationship between the concepts of Entropy and Energy in in the class of natural engineering systems under consideration. The inherent relationship of the term Entropy in the concept of Energy is figuratively conditioned by the fact that if the queen of the world—Energy was freed from her shadow—entropy, the world would supposedly avoid "heat death", but another misfortune would inevitably overtake it—it would turn into a mechanism doomed to eternal repetition of the same movements, and each the day of life would be an exact repetition of the previous day. Entropy saves us from such a fate. Such is its twofold essence and as dialectically contradictory as the whole surrounding world, in which death is an essential condition for the renewal of life. The presence of entropy, on the one hand, threatens heat death, and on the other hand, is a source of the birth of a new one and stimulates the processes of biological activity at all hierarchical levels of the biosphere of our planet, the concept of entropy causes not only chaotic movements of heated water molecules, gas, it also includes gene mutations that create new biological species among plants and animals. If these processes are deprived of the entropy contained in them, they will not generate anything unexpected, new. This important conclusion can be extended to the whole surrounding world and, in particular, to river basin geosystems as elementary parts of the Earth's biosphere (W biosph = 10 10 km^3).

E. D. KHetsuriani (✉) · M. S. Khokhrin
Platov South-Russian State Polytechnic University (NPI), 132, St. Prosvescheniya, Novocherkassk 346428, Russia
e-mail: goodga@mail.ru

E. D. KHetsuriani
Don State Technical University (DSTU), 1, Gagarin Sqr, Rostov-On-Don 344003, Russia

V. L. Bondarenko · M. I. Shtavdaker
Novocherkassk Engineering and Land Reclamation Institute Named After A.K. Kortunov of Don State Agrarian University, 111, Pushkinskaya Str, Novocherkassk 346428, Russia

A.-B. I. Ilyasov
3L. Branch (USA) in Asia, 7A Novatorov Str, Moscow 119421, Russia

L. V. Radionova and D. V. Ulrikh (eds.), *Advances in Ecology and Environmental Engineering*, Springer Proceedings in Earth and Environmental Sciences,
https://doi.org/10.1007/978-3-031-64423-8_45

Keywords System · Energy · Entropy · River basin Geosystem · Natural system · Technical system

1 Introduction

For more than three billion years, the Earth's biosphere (W bush = 1010 km^3) has been manifesting itself as a system characterized by self-sufficiency and self-regulation and includes many biological elements interacting with inanimate elements as part of the "Natural Environment" component. The entire life activity of mankind is inextricably linked with the use of water resources, and they, in turn, are formed at the global level of the Earth's biosphere (577 km3) under the influence of energy flows (35.6 TW) coming from the primary source—the Sun. The water resources used in the branches of economic and other activities are formed within the spatial limits of river basin geosystems, as part of the Earth's biosphere, where almost all types of economic and other activities are conducted. The formation and use of a renewable water resource in various technological processes is interconnected with the concept of Energy (which causes the movement and interaction of all forms of matter), which remains constant in quantitative terms and in a systematic sense (the first law of thermodynamics), but in the process of use changes its shape, and its peculiar shadow is Entropy.

The formation of water runoff (surface, underground) in basin geosystem, as part of the global moisture turnover (577 thousand km^3), within the Earth's biosphere (W biosph = 10 × 10 km^3), is carried out due to the incoming continuous streams of solar radiation to the Earth during the year equivalent to 178,000 TW, of which 20% is spent on the processes of global hydrological moisture turnover and only about 1% on the processes of photosynthesis (journal "In the World of Science" No. 11, November 1990 pp. 7–15) [1].

The river basin geosystem, as an elementary part of the Earth's biosphere, where the processes of formation of water resources occur due to solar radiation, figuratively taken as a small universe. For a small world, the idea of entropy comes from the laws of thermodynamics. In any world, entropy always goes up with natural changes, and down with unnatural changes. That's just how it works. This understanding comes from Clausius' second law of thermodynamics, where he says that natural processes always lead to an increase in entropy and unnatural processes lead to a decrease. In other words, natural things happen naturally, and unnatural things happen unnatually. The idea of entropy was first proposed by Clausius back in 1865—"I propose to call the value S entropy from the Greek (entropē)—transformation, I specially selected the word entropy so that it would be consonant with the word energy, since these two quantities are so similar in their physical meaning that the consonance of their names seems useful to me" [2].

The relationship between the concepts of energy and entropy in the utilization of water resources within the technological processes associated with economic and other activities, when considered within the spatial boundaries of the river basin, determines the order of the processes of water flow formation and regulation through the use of different types of hydraulic structures.

The utilization of water resources for economic and other purposes is a distinct type of activity, as it is without this renewable natural resource that industrial, agricultural, and other forms of production would be unable to function, as well as the daily life of the population residing in the catchment area. The practical utilization of water resources takes place through the creation of natural-technical systems, or NTS, of the type "NE-OA-P". These systems comprise natural ("NE"), technological ("OA"), and social ("P") elements, which are interconnected, interact, and relate to each other (III) within the spatial and temporal framework of the river basin geosystem. This defines the NTS classification "NE-OA-P" [3].

2 Materials and Methods

The current and newly created NTs (natural terrestrial systems) "NE-OA-P" are all located within the boundaries of river basins, where natural water runoff—both surface and underground—is formed through hydrological processes on the global scale of the Earth. From a systemic perspective, NTs "NE-OA-P" sit at the bottom of the hierarchy of natural systems, and they're located between systems of higher levels of hierarchy, like the basins of the Black and Azov Seas, for example [4].

Class "NE-OA-P" NTs fall under the category of complex, open systems, where the natural interactions between the component "NE", which is made up of natural elements, dominate over the technological effects of "OA", which is part of the overall system. River basin geosystems as elementary parts of the Earth's biosphere, whose boundaries on the Earth's surface are determined by the watershed line of the catchment area in the classification hierarchical structure are divided by the size of the catchment area (F bg-km^2) and spatial volumes (W bg-km^3) within which the zones of influence of "OA" are formed" (Table 1) [5, 6].

Based on the fact that natural processes generate an increase of entropy due to the transformation of energy forms, the concept of entropy within the considered NTS "NE-OA-P" class determines the interactions between natural, human-made, and social components, which result in the transformation of energy during movement, interaction, and the combination of movement and interaction. This is accompanied by the increase of a low-grade, bound energy (Ebind) that is unable to do work in the overall energy balance (E full) upon entering the system.

3 Results

The entropy of a system is directly linked to its energy state, and changes in the system's state are defined by the amount of energy it has. The amount of energy in a system is determined by how much entropy it forms. To understand entropy in the

Table 1 Hierarchical system of quantitative indicators of the size of basin geosystems, within which the zones of influence of "OA" are located

No.	Quantitative indicators of zones of influence	
	Catchment area, km^2	Volumetric size, km^3
1	F zones of influence ≤ 10	W zones of influence ≤ 103
2	10 < F zones of influence ≤ 500	103 < W zones of influence ≤ 6500
3	500 < F zones of influence ≤ 1000	6500 < W zones of influence ≤ 1.3 × 10 3
4	1000 < F zones of influence ≤ 10 × 10 3	1.3 × 10 3 < W zones of influence ≤ 1.3 × 10 4
5	10 × 10 3 < F zones of influence ≤ 10 × 10 4	1.3 × 10 4 < W zones of influence ≤ 1.3 × 10 5
6	10 × 10 4 < F zones of influence ≤ 5 × 10 5	1.3 × 10 5 < W zones of influence ≤ 6.5 × 10 5
7	5 × 10 5 < F zones of influence ≤ 10 × 10 5	6.5 × 10 5 < W zones of influence ≤ 1.3 × 10 6
8	< F zones of influence ≤ 5 × 105	1.3 × 10 6 < W zones of influence ≤ 6.5 × 10 6
9	10 × 10 5 5 × 10 5 < F zones of influence ≤ 40 × 10 5	6.5 × 10 6 < W zones of influence ≤ 5.2 × 10 6

NTS course "NE-OA-P", there's a bit of a challenge because the areas of influence of OA cover almost the whole space of the river geosystem.

This includes the surface layer of the atmosphere where rain and snow form, the river hydrography network in the basin, and the upper layers of the lithosphere where groundwater flows out and enters the river system and the soil with its underlying rocks.

The entropy (S) function was introduced by R. Clausius in 1854 and published in the journal "Physical Yearbook" [Sedov], which reflects the ratio of the amount of heat (Q) reported to the system to the temperature (T) that the system in question has:

$$ds = \frac{dQ}{dT} \tag{1}$$

For example, it can be noted that in one liter of hot water the entropy value is higher than in one liter of cold water.

For the considered NTS class "NE-OA-P", the state of the system is characterized by an energy change, and the energy change of the system is determined by entropy [7, 8]. Therefore, when studying the processes of III between the components and the elements included in them in the composition of the considered NTS "NE-OA-P" it is advisable to proceed from the concept of energy, reflecting the quantitative indicator of the movement and interaction of matter, energy and information (MEI) in the components of the system, and the formed level of entropy characterizes the conditions of accumulation and storage of energy in the system. At the operating NTS "NE-OA-P", as evidenced by the system integrated monitoring studies (SIMS), the increase in entropy is a characteristic sign of the dominance of natural processes of energy storage in natural and man-made components with elements included in them (flora, soil cover, fauna on land and in the aquatic environment, for example, a reservoir, etc.). The change in entropy at the operating NTS "NE-OA-P" reflects the past, present, and future, and a typical example is the Tsimlyansk reservoir on the Don River, built in 1952, which provides many branches of economic and other activities in the lower reaches of the Don River and is highly productive in the production of biological resources [9, 10].

In the NTS (Natural Technological System) class "NE-OA-P", entropy determines the direction of energy transformation processes during operation. These processes are irreversible, according to the "heat cannot flow from a cold to a hot object" principle. By understanding entropy in this NTS class, and using it as a tool for cognition, we can see things in a new light and rethink how things work between the components and elements within the system and at higher levels in the Earth's biosphere. Entropy is a measure of how much energy is locked up and can't do any work. It's a quantitative way of saying that energy can't be transformed into work.

Entropy also measures how ordered or disordered a system is. If there are more ordered things in a system, it's considered more ordered. If there's more disorder, it's less ordered.

In other words, entropy is like a measure of chaos. It tells us how much chaos there is in a system.

Basically, all complex systems that include the "NE-OA-P" class behave the same and function according to the second law of thermodynamics. But the idea that the "NE-OA-P" just degrades can be refuted with examples of real-world "NE-OA-P"-type systems like in the rivers of the Lower Don and Kuban, as well as the Terek River, which have all been functioning for years due to self-organizing processes and the growth of internal order. This is supported by theoretical studies from Lars Onsager and Ilya Prigogine.

Miguel Ruby, from a journal called "In the World of Science," wrote in February 2009 that "the appearance of order from chaos doesn't contradict the second law and goes hand in hand with the increase of order and complexity. "So, the second law of thermo is universal and therefore applies to the NTS class "NE-OA-P". In this class, we introduce the concept of entropy as a measure of how much something can't be reversed. The second law is the best-known one, and if the first says that it's impossible to get anything from nothing, the second says that you can't get anything from something. [Miguel] In nature, everything tends to get more disordered, but the quality of energy also gets worse. This is something we need to take into account when we're looking at the impact of the NTS (NE-OA-P) on the natural and social world (see Fig. 1).

According to the Second Law of Thermodynamics, energy always tends to spread out and turn into chaos. That's why the energy reserves in a system increase in entropy (which means disorder). In nature, that means that things become less organized and more chaotic.

Fig. 1 Spatial limits of the basin geosystem of the Upper Kuban

High-quality energy, like in a coal mine or uranium mine, is concentrated energy. It's not as good as low-entropy energy, but it's still better than a lot of other forms of energy we use [11].

Energy in the ordered movement of atoms, molecules, electrons, etc., for example, the flow of water entering Hydroelectric power plants have a high quality. From a systemic perspective, the existing and newly created reservoir-based hydroelectric units can concentrate significant reserves of potential energy within the water mass, as well as store solar energy.

In the context of the NTS, "NE-OA-P", entropy characterizes the state of structure formations in both natural and man-made systems and their constituent components (biotic, abiotic, and artificial). The level of variability in the state of these formations is due to the fact that both natural and human-made systems consist of numerous structure formations such as flora and fauna, the soil cover, underlying rocks, the river hydrographic network, the atmospheric air environment, the geologic environment of the upper lithospheric layers, and a complex of hydraulics structures with elements such as fish protection devices, reservoirs, and other structures. The interactions between components and the elements included in these components, as part of the NTS' "NE-OA-P" influence zones, form an energy-entropy state that exists in space

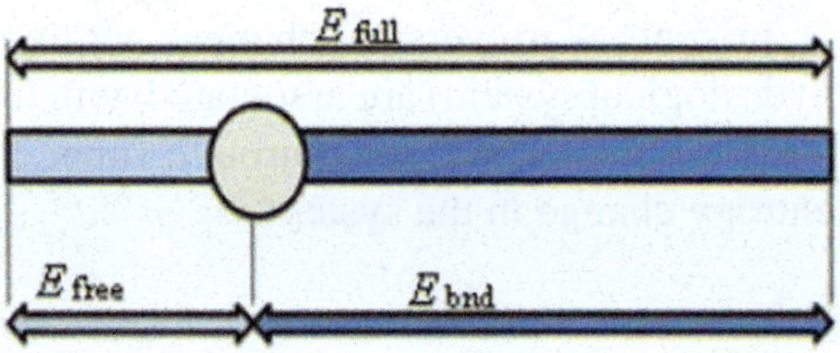

Fig. 2 Schemes of the energy balance in the NTS "NE-OA-P"

and time. This energy-entropy state causes variability in the III interactions between the elements within the natural "NE" and the social components [12, 13].

The state of this NTS is determined by the entropy level (S) within the balance of free (Efree) and bound (Ebnd) energy parts (see Fig. 2). Transitions between these energy components (Ebnd to Efree) are caused by dissipative processes. This means that high-quality energy (Efree) transfers to lower-quality energy (Ebnd), which cannot be used for any type of work.

Entropy describes the direction of ongoing processes within a system, which can be categorized as either reversible or irreversible. Systems can be considered to be either in equilibrium or out of equilibrium. Out of equilibrium systems, including the NTS (NE-OA-P), once the free energy (Efree) has been expended, it is no longer capable of performing work. However, it is still possible to import the necessary amount of Efree from external sources, such as natural resources or the economic activities of a maintenance service, in order to perform certain types of work.

The change in entropy of a nonequilibrium system, such as the NEO AP, is determined by the well-established equation proposed by I. Prigogine. [14]:

$$\frac{dS}{dt} = \frac{deS}{dt} + \frac{diS}{dt} \tag{2}$$

where dS is the total entropy change in the system over a given period of time, dt; diS is the entropy change due to irreversible processes occurring within the system, or entropy production; and deS is the amount of entropy that has been imported from the surroundings.

According to the Second Law of Thermodynamics, deS is always positive. It can take on both positive and negative values (Fig. 3).

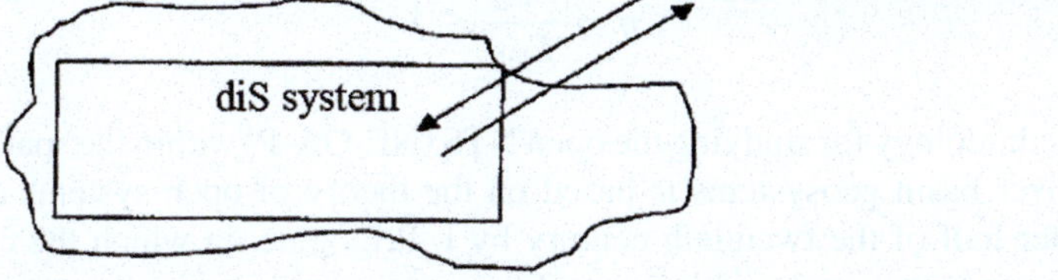

Fig. 3 The scheme of interaction of the NTS "NE-OA-P" with the external environment

In general, irreversible changes within the influence areas of the "OA" (or other hydrological system) are associated with the flow of energy or matter (such as water, solid particles, ions, etc.) through time. These changes can be represented by the entropy change in the system.:

$$diS = F \cdot dx \tag{3}$$

where F is the generalised (thermodynamic) force expressed as a function of variables such as temperature, relative humidity, and concentration of substances in bottom sediments, suspended sediments, and ion runoff. It also depends on the applied pressure Hm on structures, among others.

In the class of NTSs "NE-OA-P" under consideration, the total irreversible processes can be expressed as the sum of all changes resulting from flows dX, which can be determined through monitoring studies during the EIA process at the design, construction, and operation stages of "OA". These flows can be represented mathematically by the given expression:

$$diS = \sum F \cdot dx \geq diS0 \text{ or } \frac{diS}{dt} = \sum \frac{dx}{dt} \cdot F \geq 0 \tag{4}$$

This, in a generalized sense, reflects the Second Law of Thermodynamics, which states that the entropy of a system in each irreversible process is proportional to the product of the force applied and the flow rate:

$$J = \frac{dx}{dt} \tag{5}$$

The qualitative characteristics of states, both for individual structural formations within the composition of the NTS "NE-OA-P", and for the system as a whole, are defined by the level of entropy (S) expressed in terms of the balance between free (Efree) and bound (Ebnd) energy components (Fig. 2).

If the connected (Ebnd) component of energy predominates, then a larger number of structural units within the system will focus their functioning on reducing their operational efficiency and, correspondingly, functional effectiveness, which is measured by the efficiency coefficient (η) that is determined by a specific ratio:

$$\eta = \frac{E_{free}}{E_{full}} \leq 1 \tag{6}$$

The methodology for studying the open NTS (NE-OA-P) within the spatial boundaries of river basin geosystems is based on the theory of open systems developed in the latter half of the twentieth century by I. Prigogine, in which the concept of a closed system was replaced by that of an open system, which has the ability to exchange matter, energy, and information with its environment (MEI). Based on the analysis of existing natural systems within the global framework of the Earth's biome, it can be observed that the hierarchical organization of nature aims to reduce the total entropy of the universe. However, the dynamics between natural (biotic and abiotic) components within natural environments are aimed towards the natural increase in entropy [15, 16].

The interaction between these two principles determines the evolutionary progression of the physical universe, which can be observed in living organisms and natural environments. For example, the spatial boundaries of river basin geosystems illustrate this.

Within living organisms, reproductive, mutational, and recombinational processes tend to increase entropy. Conversely, aggregation of molecules within cells, cells within organisms, and organisms within communities aim to reduce entropy levels. In the river basin geosystem, as an elementary part of the Earth's biosphere, there is a similar trend for biotic and abiotic components in the natural environment to cause an increase in entropy in the form of increased order. For example, this can be seen in the river hydrography of the catchment area. This is accompanied by an increase in entropy as a result of the production of energy in the form of a bounded part (Ebnd) and its subsequent transformation into a stable state, such as heat.

The introduction of "OA" (or other measures) into the natural environment is associated with intra-basin water management and the selection of specific parts for economic use, such as in various technological processes. These activities may lead to a reduction in the rate of increase in entropy, as a result of changes in the concentration of free energy (Efree) in various structures, buildings, and technological processes. The open natural thermal system (NTS) "NE-OA-P" functions within the spatial boundaries of the river basin geosystem, an elementary component of the Earth's biosphere. The NTS consumes MEI (measured energy input) from the surrounding environment and at the same time releases the used substance and spent energy (as a bound form of energy (Ebnd)), causing entropy [17].

Therefore, it can be concluded that economic and other activities related to the use of water resources are intrinsically linked to the transformation of energy and matter from the external environment within the spatial limitations of the river basin geosystem. These activities result in the withdrawal of low-grade energy flows (Ebnd) from the system and eventual transformation into a more stable form of energy, heat. Therefore, it can be concluded that the establishment of the NTS "NE-OA-P" in the river basin geosystem results in a certain order among structural formations in the natural environment and, to some extent, the social environment, by controlling natural processes such as water runoff formation (surface and underground) within the catchment area. These processes include riverbed formation, flood protection, and flooding prevention of settlements. Additionally, the vital activity of the river ichthyofauna is also impacted. Transformation of solar energy within reservoirs does not contradict the Second Law of Thermodynamics. The second law of thermodynamics does not imply monotonic disorder and an increase in the bounded part of energy (Ebnd), and it is quite compatible with the emergence of order and the complication of system-forming relations between components and their constituent parts in the class of non-equilibrium open systems (NE-OA-P) under consideration [18, 19].

It should also be noted that, in systems at any hierarchical level, from a living cell to a powerful ship engine, similar processes occur in non-equilibrium open-system (NE-OA) structures, where the initial energy level (Ef) is converted into motion with an efficiency (η) that is less than 1. For example, in the process of converting the potential energy of water flow into electrical energy, the efficiency ranges from 0.68 to 0.75, and this value is the highest compared to other energy conversion technologies.

4 Conclusions

The energy-entropy approach to the creation and operation of the NTS Class "NE-OA-P" in relation to the use of water resources for technological processes in economic and other activities, allows us to objectively reflect on the transformation of energy forms in the process of interconnection, interaction, and interdependency between natural ("NE"), technological ("OA"), and social ("P") components within the NTS system "NE-OA-P".

In this context, entropy represents a quantitative measure of bound energy (E_ and) according to its order within the system, which determines the directionality of the system's processes [20].

References

1. Asatryan A, Khetsuriani E, Bondarenko V (2023) Basic irrigation methods and techniques woody and shrubby vegetation and flower crops on the objects recreational value. Water purification technologies TECHNOVOD-2023, Novocherkassk, p 101–105
2. Khetsuriani E, Asatryan A, Khetsuriani T, Byrin DS (2023) Monitoring of territories recreational facilities in an urban environment to draw up a reconstruction project. Water purification technologies TECHNOVOD-2023, Novocherkassk, p 116–122
3. Bondarenko V, Ylyasov A, Khetsuriani E, Semenova E, Larin D, Khetsuriani T (2020) Environmental safety conditions and factors in zones of influence of water facilities. Materials Science and Engineering, Novocherkassk, p 217–230
4. Badalyan L, Kurdyukov V, Ovcharenko A (2014) The current state of research in the field of integrated assessment of the safety of the ecosystem of the city under the influence of motor transport. The journal Rational Use and Conservation of Aquatic Biological Resources, Moscow, p 378–420
5. Bondarenko V (2010) Environmental management: territories of basin geosystems: study guide. Moscow, p 527–538
6. Bondarenko V, Klimenko O, Semenova E, Nikolaenko D (2016) Environmental safety in construction. Engineering and environmental surveys in the complex of surveys for the construction of water facilities. Monograph of Bondarenko V.L., Novocherkassk, p 309–315
7. Bondarenko V, Privalenko V, Kuvalkin A (2021) Solving environmental problems in the design of hydraulic structures. Monograph of the Publishing House of the SSC RAS, Moscow, p 306–310
8. Bondarenko V, Skibin G, Azarov V, Semenova E, Privalenko V (2016) Environmental safety in environmental management, water use and construction: assessment of the ecological state of

basin geosystems. Monograph of the South Russian State Polytechnic University (NPI) named after M. I. Platov, Novocherkassk, p 419–431

9. Budyko MI, Drozdov OA (1950) About moisture turnover in a limited land area. Moscow, p 248–256
10. Ged RD (1990) Energy for planet Earth. The journal “In the World of Science”, Moscow, p 7–16
11. Denisov OV, Zimenko VA, Kohanov YuB (2014) Assessment of the level of ecosystem safety by forecasting bioecological parameters based on an integral safety indicator. The journal “Rational Use and Conservation of Aquatic Biological Resources”, Rostov-on-Don, p 147–149
12. Kuzansky N (1979) Philosophical Heritage book series. Moscow, p 301–309
13. Nikolis G, Prigozhin I (1990) Cognition of the complex. Moscow, p 425–433
14. Nikolis G, Prigozhin I (1979) Self-organization in nonequilibrium systems. Moscow, p 440–456
15. Odum Yu (1987) Fundamentals of ecology/Trans. from English, Moscow, p 40–60
16. Prigozhin I, Stengers I (1986) Order from chaos. Moscow, p 89–97
17. Chernyaeva AM (1999) Russia: River basins. Yekaterinburg, p 201–210
18. Chernyaeva AM (1998) Russia: Water resource potential. Yekaterinburg, p 108–115
19. Tyurina TA, Basilaia MA (2014) Technology and its role in the environmental crisis. The journal “Rational Use and Conservation of Aquatic Biological Resources”, Rostov-on-Don, p 341–347
20. Atkins P (1987) Order and disorder in nature. Moscow, p 268–275

Analysis of Methods to Improve Occupational Safety

M. A. Lemeshko, A. G. Iliev, and N. I. Stuzhenko

Abstract The article analyses methods for reducing occupational injuries by examining defects in the functioning of the occupational safety management system in coal mining enterprises. It proposes improvements to the occupational safety management system at industrial enterprises by providing a comprehensive analysis of risk formation conditions. The text presents an algorithm for implementing the acceptable risk concept and analyses the distribution of accidents by work experience at coal mining enterprises. It identifies the main types of industrial conflicts and the factors that affect the risk of injury. The results can contribute to the development of effective risk management strategies and measures to improve labour safety at coal mining enterprises. Research indicates that to manage the likelihood of negative events, it is crucial to forecast their occurrence and identify the factors that influence their probability. Measures should be taken to intentionally reduce the likelihood of such events by modifying the factors that affect them.

Keywords Risk assessment · Injury · Risk management · System · Safety

1 Introduction

Statistical data analysis shows a significant decrease in the absolute rate of injuries in the Russian Federation over the last decade. However, the specific values of these indicators have remained virtually unchanged. Additionally, the total number of persons employed in coal mining and operating enterprises has decreased sharply during the period under review.

The decrease in employee numbers and mine closures were primarily due to the liquidation of inefficient and unsafe companies. The measures taken aimed to reduce the risk of potential injuries by improving overall safety in mining operations and

M. A. Lemeshko (✉) · A. G. Iliev · N. I. Stuzhenko
Institute of Service and Business (Branch) of DSTU in Shakhty, 147, Shevchenko Street, Shakhty 346500, Russia
e-mail: lem-mikhail@ya.ru

L. V. Radionova and D. V. Ulrikh (eds.), *Advances in Ecology and Environmental Engineering*, Springer Proceedings in Earth and Environmental Sciences,
https://doi.org/10.1007/978-3-031-64423-8_46

reducing the number of workers at high risk of injury. In coal mining operations, more than 2/3 of accidents are caused by organizational factors. The analysis of the causes of miners' injuries has revealed issues with the industrial safety system, which has led to a conflict between efficiency and safety objectives during the implementation of the industrial management process.

These issues reflect the contradictions between health and safety requirements and the actual conditions under which production functions are carried out. The presence of these links can lead to conflict situations, both overt and covert, which may cause miners to resort to hazardous working methods. These scenarios are commonly referred to as 'industrial conflicts'. Conflicts can result in safety violations that may lead to injuries. The situation will persist until the conflict in the work process is resolved.

2 Setting the Task

The analysis of factors that cause injuries in coal mines is crucial for occupational health and safety. This study is important for ensuring safe working conditions, protecting workers' health, and promoting economic interests. Occupational accidents can have physical and psychological consequences for workers and cause significant economic losses for companies [1, 2]. Identifying the organizational causes of injuries and broken links in the occupational health and safety management system are crucial for developing and implementing effective measures to prevent injuries in coal mining companies. Improving occupational safety and resolving industrial disputes can reduce the risk of injury, improve production efficiency, and lower the economic costs associated with occupational accidents. This study aims to comprehensively analyze and enhance the occupational safety management system in coal mining companies by developing an algorithm for implementation stages that considers the concept of acceptable risks. To achieve this goal, the following tasks must be addressed:

1. Explore methods to decrease workplace accidents.
2. Evaluate deficiencies in the occupational health and safety management system.
3. Develop and describe a step-by-step algorithm based on the concept of acceptable risks.
4. Describe the algorithm of stages based on the concept of acceptable risks.
5. Investigate the distribution of accidents in coal mining companies according to work experience.
6. Classify types of labour conflicts in the labour protection system.
7. Develop a risk management process for coal mining enterprises.
8. Develop measures to optimize the management system to improve safety at work and reduce accidents.

To determine specific (potential) threats caused by dangerous factors, their identification is required first. This process can be carried out by means of using specialized assessment of working conditions and possible traumatic situations. The effective way of reducing a total number of hazards is to identify those that do not get a real threat. The next step is to use qualitative and quantitative risk assessment methods to identify possible emergencies:

- Conducting the analysis of accidents and assessing their connection with disability.
- Application of the “failure tree” analysis method.
- Application of the “event tree” analysis method.

For a more detailed analysis of such events done by means of appropriate methods, see Fig. 1.

The results of the risk analysis are the formulation of objectives and the determination of strategic directions in the field of occupational safety within the organization and the development of action programmes aimed at improving working conditions. A fundamental feature of this approach is the identification of objectives and tasks based on the analysis of the identified risks. These goals and objectives become an integral part of the occupational safety and health programme and include the following aspects.

- Assigning responsibility for achieving the goals and objectives.

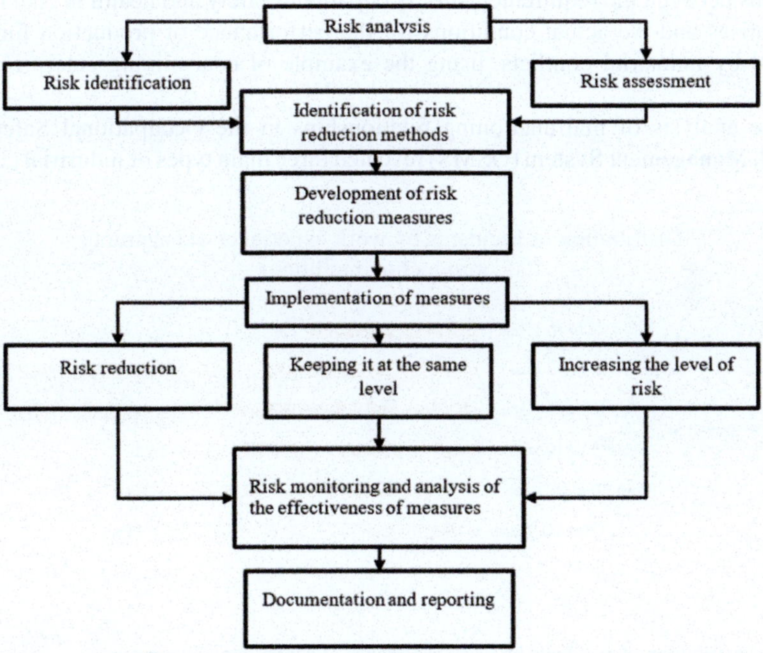

Fig. 1 General outline of the risk management process

- Assessment and mobilization of resources, including human, financial, technological and others•
- Setting a period for the tasks realization.
- Taking into account the latest scientific findings and strategic planning of different types of activities [3, 4].

The application of a systematic method makes it possible to eliminate potential sources of injury, thus creating the conditions for a gradual reduction in the rate of occupational accidents until their complete elimination. When applying innovative approaches to occupational safety and health management, the organization uses a fundamentally new methodology for assessing the functional role and activities of top and middle managers, as well as supervisors and workers.

The analysis of the causes of occupational accidents, taking into account the experience of the employees, shows that the highest number of accidents occurs among employees with 5 to 10 years of experience and among employees with more than 10 years of experience. Probably, such cases among experienced workers can be explained both by the loss of acute risk perception and incorrect assessment of the degree of occupational hazard, and by the use of unsafe working methods in order to "successfully" perform planned tasks [5–7].

The choice of unsafe working methods by employees is caused by "defective" links in the existing production safety management system, which include contradictions between the requirements for occupational safety and health in coal mining enterprises and the actual conditions for the performance of production functions caused by industrial conflicts, using the example of coal mining enterprises, see Fig. 2.

The analysis of malfunctioning relationships in the Occupational Safety and Health Management System (OSMS) revealed three main types of industrial conflict:

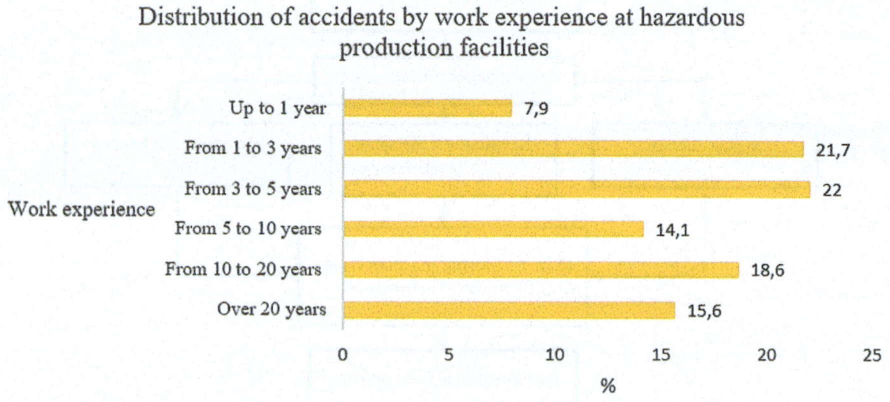

Fig. 2 Distribution of accidents by work experience at coal mining enterprises

- Technological conflict is characterized by the impossibility of using safe working methods when performing a certain operation due to the technological peculiarities of this operation.
- Technical conflict is related to unsafe equipment.
- Economic conflict is related to the increase of costs that do not contribute to the improvement of industrial safety.

Organizational conflict is characterized by the inability to ensure the safety of the production function in the conditions of the current organization of the technological process, according to Table 1.

The manifestations of an industrial dispute that lead to workers being forced to choose dangerous working methods do not depend on the nature of the conflict. Organizations operating a risky production facility determine the level of acceptable risk based on the concept of "tolerable" risk. It should be noted that the values for individual risks are usually set at lower levels than the actual values, but with the understanding that these levels can be achieved in the production facility. Thus, the value of acceptable risk changes to the value of forced risk, which may require significant resources to achieve [8–10]. The question of what level of risk is acceptable or unacceptable is therefore relevant. Answering this question requires management decisions, which can vary considerably from one situation to another. They are influenced by many factors, which are summarized below in the form of a diagram, see Fig. 3.

Next, it is necessary to consider the thresholds that may represent acceptable levels of risk, "Eq. 1".

$$R \geq R_{\text{acceptable}} \tag{1}$$

where $R = QW$ is the mathematical expectation of the amount of damage caused in the process of practical implementation of the project for the period under study; acceptable is an acceptable value adopted in the organization; Q is the risk of a negative event during the period of practical implementation of the project or during the period under study; $W = M(W)$ is the average amount of damage if a negative scenario is realized in practice. The risk of a negative event occurring in practice is

Table1 Types of industrial conflicts

Type of production conflict	Communication "defect" in OSM causing production conflict
Technological	Impossibility of meeting labor protection and industrial safety (OSM) requirements due to technological features of specific work
Technical	Impossibility of meeting health and safety requirements due to the design features of the process equipment
Economic	Compliance with health and safety requirements is not economically profitable
Organizational	Fulfillment of the HSE requirements is impossible (difficult) with the given organization of the technological process

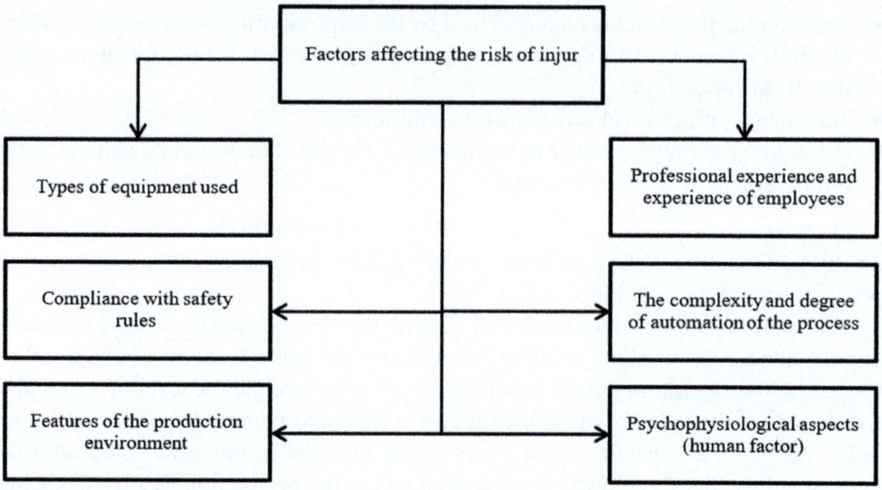

Fig. 3 Factors affecting the risk of injury

also calculated based on the "Eq. 2":

$$Q \geq Q_{\text{acceptable}} \tag{2}$$

where $Q_{\text{acceptable}}$ is the optimal value of the probability of a negative event during the implementation of a project or a certain period of time;

The probability of certain damage is determined by the condition "Eq. 3".

$$Q_{\text{catastrophic}} \geq Q_{\text{catast.acceptable}} \tag{3}$$

where $Q_{\text{catastrophic}} = QP(W \geq IS)$ is the risk of catastrophic losses, $(W \geq IS)$ is the criterion of catastrophic damage to the organization, I is the probability of an accident; S is the damage from an accident, (IS) is the probable amount of damage in an accident, $Q_{\text{catast. acceptable}}$ is the level of acceptable catastrophic damage [11, 12]. By analogy, it is possible to determine an acceptable value of the average damage W, "Eq. 4".

$$W \geq W_{\text{acceptable}} \tag{4}$$

where $W_{\text{acceptable}}$ is the level of an appropriate average damage value. The same range of criterion values can be used for the maximum possible loss value and the most probable damage, "Eq. 5".

$$W_{max} \leq W_{\text{acceptable}} \tag{5}$$

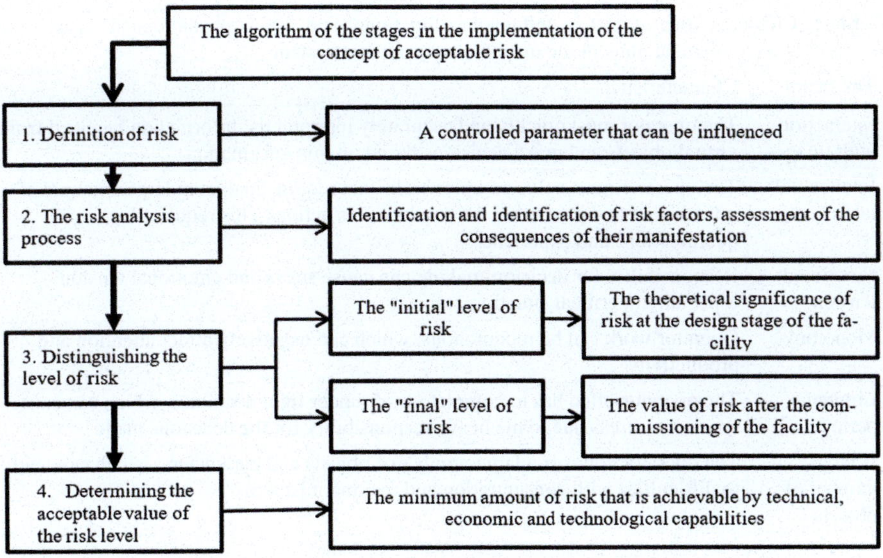

Fig. 4 The algorithm of the stages in the implementation of the concept of acceptable risk

In the context of this study, the normative value of the risk of injury can be considered as the level of probability of occurrence of a negative event that the organization's subjects are willing to accept in order to satisfy their socio-economic needs in the current conditions of the enterprise. This level is determined by the socio-economic situation of the industry in a given region [13–15].

Thus, the concept of "acceptable" risk directs the manager of the production system towards a conscious, rational attitude to risk and an awareness of the facts, which are structured in the form of a table (see Fig. 4).

It can therefore be concluded that the construction and use of kindergartens and preschool institutions is not allowed in the risk management zone in the range of values equal to 10^{-5} [16–18].

In addition, when assessing the risk of accidents at a hazardous production facility, it is necessary to take into account the fact that the operations of the operator in modern production have a number of specific characteristics.

These activities can negatively affect the functional and mental health of a person and, as a result, significantly influence the reliability and safety of their professional activities. This helps to increase the requirements for individual psycho-physiological characteristics of an operator in the "man–machine" system. The most important characteristics are presented in Table 2 [19–21].

Therefore, when assessing the risk of an accident, it is important to consider the diverse nature of the operator's activities. The psychophysiological state of a human operator can vary depending on many environmental factors. The comfort of working conditions, emotional events of everyday life, and in some cases family relations can

Table 2 Characteristics of factors influencing functional and psychophysiological states of an operator of technological units in the machine-environment system

The factor	Characteristic
Interaction with images	The operator must quickly and accurately interpret the information received from virtual objects and make decisions based on this information
Intellectual tasks	The operator must solve complex tasks that require analyzing large amounts of information and predicting the results of his actions This requires high qualifications and experience
Emotional tension	Responsibility for decision-making can cause stress and emotional tension, especially in critical situations
Monotony	Operator work can be monotonous, which can negatively affect attention and productivity
Limited period	The operator often has to complete work under tight deadlines, which can cause emotional stress due to the heavy responsibility for the decisions made
A large number of objects	The operator controls a large number of objects and parameters, which can lead to difficulties with concentration and making mistakes

all have an impact on a person's well-being. It is important to note that this statement is not based on objective evidence and should be treated as a subjective evaluation.

3 Recommendations

The results of the study can be used as a basis for industry regulations on the safety of hazardous production facilities. The recommendations developed will help improve the effectiveness of measures to prevent accidents and injuries.

When managing the risk of negative events, it is important to forecast the possible occurrence of these events. This includes identifying certain factors that influence the likelihood of a negative outcome. In order to reduce the risk, measures can be taken by purposefully altering these events and identifying factors. They can increase the effectiveness of these measures.

The results of the study may be of scientific and research interest in determining the influence of professional choices that ensure the safety of operators of industrial facilities, as part of the analysis of the characteristics of the socio-psychological climate of operators of technological units of hazardous production facilities, for example, in the coal industry. The process of managing the probability of negative events at potentially hazardous facilities consists of risk analysis, monitoring and collecting information on risks. One of the forms of obtaining information on the probability of negative events is the declaration of industrial safety of hazardous production facilities. As a result, we can conclude that the issue of choosing methods to improve the efficiency of management processes is in demand in the coal mining industry.

This production sector is very dangerous for personnel compared to other industries; the domestic coal mining industry is facing the task of urgently solving significant problems in ensuring labour protection and industrial safety. Without solving these problems, it is impossible to ensure the progressive development of the industry [22, 23]. During the study of new methods of managing occupational safety and health in the organization, the method of a systematic approach to risk assessment and conflict resolution in the production environment was used.

The fundamental novelty of this approach is the integration of risk assessment methods for setting OSH goals and objectives. Scientific analysis of industrial conflicts in the management system has shown that technological, technical, economic and organizational aspects play an important role in the formation of contradictions between occupational safety requirements and actual working conditions.

These conflicts can affect the adoption of safe working practices, the operation of equipment, the economic benefits of safety and the organization of processes. A detailed study of the distribution of accidents according to work experience confirms that even experienced workers are at risk of injury, not only because of reduced risk perception, but also because of the use of unsafe work practices to meet production targets.

Overall, the proposed approach provides a framework for systematic and effective OSH risk management that can contribute to a safer working environment in the organization.

4 Conclusions

1. The theoretical basis for the concept of "acceptable risk" with regard to the objects in question is clarified, as well as the factors that influence the thresholds for acceptable risk.
2. An algorithm for implementing the concept of accident-related acceptable risks is proposed, allowing the development of reasonable criteria for acceptable levels of hazard.
3. Quantitative indicators of social risk characterizing an unacceptable level of threat to the population in the event of accidents have been established.
4. Significant factors in the work process that influence the reliability of operators' actions in emergencies have been revealed.
5. The results obtained can be used to enhance the safety of industrial facilities in the coal mining sector. The general approach and proposed algorithm can also be applied to similar problems to determine acceptable risks.
6. The theoretical basis of the concept of "acceptable risk" in relation to the objects under consideration is clarified, and the factors influencing the thresholds of acceptable risk are identified.
7. The research results can improve safety management systems in other industries.

8. The data obtained can be used as a basis for further research on assessing and managing accident risks at industrial facilities.

References

1. Lemeshko MA, Molev MD, Iliev AG (2020) Analysis of Adaptive Machines with Two Process-Connected Working Movements. Lecture Notes in Electrical Engineering 641:120–127. https://doi.org/10.1007/978-3-030-39225-3_14
2. Kulkov EP, Panfilov AV, Bondarenko BI (2015) Organizational measures for bringing the object of industrial safety expertise into compliance with industrial safety requirements. Innovative Science 10–1:80–82
3. Stolyarenko KG, Yenikeeva TM, Zakirova ZA (2018) Analysis of common causes of emergency situations in the op-eration of hydro-treatment unit of diesel and kerosene fractions. Oil refining and petrochemistry, Scientific and technical achieve ments and best practices 1:40–44
4. Smorodova OF, Baykov IP, Kitaev SV, Sergeeva KV (2018) Assessment of industrial safety of oil and gas industry facilities by making a decision on the relative superiority of criteria. Scientific Journal Oil and Gas Business 3:199–214. https://doi.org/10.17122/ogbus-2018-3-199-214
5. Krizhanovsky MS (2019) Energy and resource conservation in thermal power engineering and the social sphere. Proceedings of the International Scientific and Technical Conference of Students, Postgraduates. Scientists 7(1):103–105
6. Derehuz I, Kulbovskyi I, Melnichenko O (2022) Modeling processes to improve. https://doi.org/10.33744/0365-8171-2022-111-234-240
7. Abdrakhmanov NH, Kireev IR et al (2018) Ensuring safety in gas mine development Ufa. Ufa State Petroleum Technical University, p 130
8. Dubrovin PV, Temerbayeva MV, Dubrovin PV et al (2023) Development of regulatory and technical documentation for the examination of equipment operating under pressure. Bulletin of the Innovative Eurasian University 3(91):125–133. https://doi.org/10.37788/2023-3/125-133
9. Abdrakhmanov NKh, Fedosov AV, Shaibakov RA (2019) Organization of safe management of fire operations on gas pipelines. Bulletin of the National Academy of Sciences of the Republic of Kazakhstan 6(382):272–279
10. Kuzyakina MV, Gura DA et al (2019) Assessment of Potential Forest Biomass Resource on the Basis of Data of Air Laser Scanning. International Scientific Conference "Energy Management of Municipal Facilities and Sustainable Energy Technologies" EMMFT 2018 2:403–416. https://doi.org/10.1007/978-3-030-19868-8_41v
11. Pietowska-Laska R (2016) Innovation in the management of the health and safety of workers in a company. International Multidisciplinary Scientific Conference on Social Science and Arts, SGEM 2016, Albena, Bulgaria, August 24–30 2016, Conference Proceedings 4(2):363–370. https://doi.org/10.5593/SGEMSOCIAL2016/B24/S07.048
12. Pokorski J (2015) Building the potential for innovative companies in Poland—the results of the Innovative Economy Operational Programme. In: Zadura-Lichota P. (ed.) Innovative entrepreneurship in Poland. Uncovered and hidden potential for Polishinnovation, the Report of the Polish Agency for Company Development, Warsaw, Poland, pp 73–88
13. Laska A (2012) Safety culture in companies functioning exemplified by big construction companies. Scientific notebooks of Rzeszow University of Technology, Management and Marketing 19(2):71–80
14. Krizhenovsky MS, Builushkina LN (2019) Innovative processes in science and technology of the XXI century: Proceedings of the XVII International Scientific and Practical Conference of Students, postgraduates, scientists, educators and practitioners, pp 220–224

15. Abdrahimov YR, Fedosov AV, Makulova DR et al (2014) The analysis of the changes in the procedure of the working conditions' assessment at the workplaces. The electronic scientific journal Oil and Gas Business 2:383–391
16. Fedosov AV, Vadulina NV, RyamovaAV et al (2015) Measuring a level of a harmful factor. Monograph, Publishing house of USPTU, Ufa, p 333
17. Onishchenko GG and Zaitseva NV (2013) Current trends in the development of social and hygienic monitoring and analysis of health risks, materials of the All-Russian Scientific and Practical Conference (May 15–17, 2013). Federal Research Center for Medical and Preventive Technologies for Risk Management of Healthcare Workers, Perm
18. Sokolova MM (2013) Reducing the risk of the occupational diseases among the workers of the petrochemical complex: problems and expertise of RT enterprises. Herald of the Kazan State Technological University 22(16):1–5
19. Verma S, Chaudhari S (2016) Highlights from the literature on risk assessment techniques adopted in the mining industry: A review of past contributions, recent developments and future scope. International Journal of Mining Science and Technology 26(4):691–702
20. Cheberyachko S (2012) Analysis of test methods of determining antidust respirator quality. Geomechanical Processes During Underground Mining—Proceedings of the School of Underground Mining, Dnipropetrovsk, Yalta, 24–28 sept., pp 123–126. https://doi.org/10.1201/b13157-22
21. Sribna Yu, Trokhimets O, Nosatov I et al (2019) Globalization of the global coal market—Contradictions and trends. E3S Web Conference 123:01044. https://doi.org/10.1051/e3sconf/201912301044
22. Zhang Yu, Jing L, Bai K et al (2018) A systematic approach to extremely large coal mine accidents in China from 1997 to 2011: applying the HFACS approach. International Journal of Occupational Safety and Ergonomics 25(2):181–193
23. Dreval Yu, Sharovatova OP, Bryhada O (2020) Fundamental principles of activity of international labour organization in occupational safety and hygiene. Scientific Bulletin of the National Mining University 6:89–95. https://doi.org/10.33271/NVNGU/2020-6/089

Study of the Effectiveness of the Use of Aerogel Thermal Insulation to Reduce Heat Loss

T. Abdullin, A. Kondratiev, and Yu. Vankov

Abstract Energy conservation is an important component of sustainable development and environmental safety. Saving energy helps to reduce emissions of carbon dioxide and other harmful substances into the atmosphere, thereby preventing environmental problems. Energy-saving technologies and solutions contribute to improving the efficiency of energy resources and improving the quality of life. The efficiency and expediency of using aerogel thermal insulation at industrial power facilities is being investigated at a thermal power plant and compared with traditional basalt fiber thermal insulation. Calculations and simulations are also carried out using the COMSOL Multiphysics software package, the required thickness of thermal insulation required in accordance with the operating standards at the thermal power plant is estimated. The use of aerogel as an alternative to traditional thermal insulation is justified under certain conditions due to the excellent thermal insulation properties and high environmental friendliness of the material. It was possible to obtain the same temperatures on the insulation surface with a 14 times lower aerogel thickness compared to basalt insulation, when the cost of aerogel is only 1.5 times more than the cost of basalt insulation.

Keywords Aerogel · COMSOL multiphysics · Energy saving · Industrial thermal power engineering · Basalt insulation · Thermal covers · Shut-off valves

1 Introduction

Thermal insulation of energy communications plays an important role in ensuring the efficient operation of heating, cooling and power supply systems. It helps to reduce heat or cold losses, as well as prevent overheating or hypothermia of equipment. This, in turn, leads to a reduction in energy costs, an increase in the service life of equipment and an improvement in the overall energy efficiency of the system. In addition,

T. Abdullin (✉) · A. Kondratiev · Yu. Vankov
Kazan State Power Engineering University, 51, Krasnoselskaya Street, Kazan 420066, Russia
e-mail: abdullin.t.r_25@kgeu.ru

L. V. Radionova and D. V. Ulrikh (eds.), *Advances in Ecology and Environmental Engineering*, Springer Proceedings in Earth and Environmental Sciences,
https://doi.org/10.1007/978-3-031-64423-8_47

thermal insulation prevents condensation from forming on pipes and fittings, which can lead to corrosion and other problems. In general, thermal insulation of energy communications is a key element to ensure reliable and economical operation of power systems [1].

An important element of energy communications is an adjustable shut-off valve.

The most common method of thermal insulation of shut-off valves and pipelines at industrial facilities is the use of insulation mats based on basalt fiber and its derivatives, wrapped with wire or metal ties [2].

Shut-off valves, such as valves, due to the abundance of their constituent elements, in industrial enterprises are subject to more significant risks of breakage, malfunction or leakage, compared with pipelines, while losses of steam, working substance or coolant throughout the enterprise amount to tens of millions of rubles. High-quality thermal insulation of valves, with the possibility of visual monitoring of its condition and, if necessary, rapid dismantling of thermal insulation, is an urgent need at most large modern energy enterprises [3].

At the same time, there is a steadily increasing interest in the introduction and use of innovative composite thermal insulation materials to improve the energy efficiency of the enterprise and reduce losses, and consequently fuel costs. Among the most common ones on our market are expanded polystyrene [4], fibrous [5], clay-based composites [6–8], from recycled materials [9–12], composite with phase changes [13], vacuum composites [14, 15] and aerogels [16, 17].

Thus, the purpose of this study is to determine how effectively aerogel thermal insulation in the form of fiber can be used for industrial energy facilities.

2 Preliminary Calculations and Modeling

As an object of research of the qualities and characteristics of aerogel thermal insulation at the Kazan CHPP-3 (Combined Heat and Power Plant) branch of JSC "TGK-16", one valve 7PC-1 with a nominal diameter of 100 mm and a pipeline section with a nominal diameter of 133 mm, 1 linear meter long, located adjacent to the recirculation line on one boiler unit TPE-429 were identified Article No. 7.

As part of the research, it is necessary to develop and manufacture quick-release thermal covers with an aerogel composite layer as thermal insulation. According to the authors-developers, the use of quick-release thermal covers helps to eliminate heat losses currently available when using traditional basalt fiber thermal insulation mats, while increasing the useful life of the thermal insulation layer and ease of installation by reducing the overall dimensions of the thermal insulation product, since the thickness of aerogel thermal insulation is much less than the thickness of basalt fiber with similar thermal characteristics [18].

Currently, there are not enough scientific publications and studies on the advantages and disadvantages of using thermal covers with aerogel composite materials in the composition, as well as in general, the use of aerogel thermal insulation for enterprises in the territory of the Russian Federation.

Mathematical modeling of thermal insulation based on aerogel material has been carried out as theoretical research.

To determine the required thickness of the thermal insulation layer of the product, ISO 12241:2022 "Thermal insulation for building equipment and industrial installations" was used as a guide [19]. Since one of the main requirements for thermal insulation in enterprises is to protect personnel from burns in case of possible contact with pipelines or heat fittings, a method was used to calculate the insulation thickness according to a given temperature of its outer surface. The set temperature of the outer surface is 40 °C. The thermal conductivity coefficient of the aerogel fiber used in the work is 0.025 W/m K at 200 °C (according to the manufacturer's technical data sheet).

Using (1)–(3) and repeatedly refining the final result by the method of successive approximations, the required thickness of the aerogel layer was obtained—10 mm.

To verify the calculations and analysis of temperatures in the layers of the finished product, the COMSOL Multiphysics software package developed a model and calculated the temperature on the outer surface of the thermal cover according to a given thickness of the aerogel layer obtained using mathematical calculations in (1)–(3). Figure 1 shows the temperature distribution in the layer of the thermal insulation product, where the surface temperature the pipeline temperature is 220 °C—the operating temperature of the coolant in this section, and the resulting temperature of the outer surface of the thermal insulation layer is 44 °C.

When comparing the results of mathematical calculations and modeling of a thermal insulation product, the difference was 4 °C, which can be explained by another method of calculating the model in COMSOL, in comparison with the mathematical calculation according to the set of rules.

The calculation of the thermal insulation thickness was performed for single-layer cylindrical surfaces according to (1):

$$\ln B = \ln \frac{d_{out}^{w} + 2\delta_{is}}{d_{out}^{w}} = 2\pi \lambda_{is} R_{out}^{L} \frac{t_{in} - t_{surf}}{t_{surf} - t_{out}}, \tag{1}$$

where B is a certain dimensionless quantity, d_{out}^{w} is the outer diameter of the wall of the insulated object, [m], δ_{is} is the wall thickness of thermal insulation, [m], λ_{is} is the thermal conductivity coefficient of thermal insulation, [W/(m °C)], R_{out}^{L} is the linear thermal resistance to heat transfer of external insulation, [m °C/W], t_{in} is the temperature of the environment inside the insulated object, taken into account when calculating the surface temperature of the object, [°C], t_{surf} is the temperature of the outer surface of the thermal insulation layer, [°C], t_{out} is the ambient temperature, [°C].

Then, according to (2), the value B is found and the required insulation thickness is determined:

$$\delta_{is} = \frac{d_{out}^{w}(B - 1)}{2}. \tag{2}$$

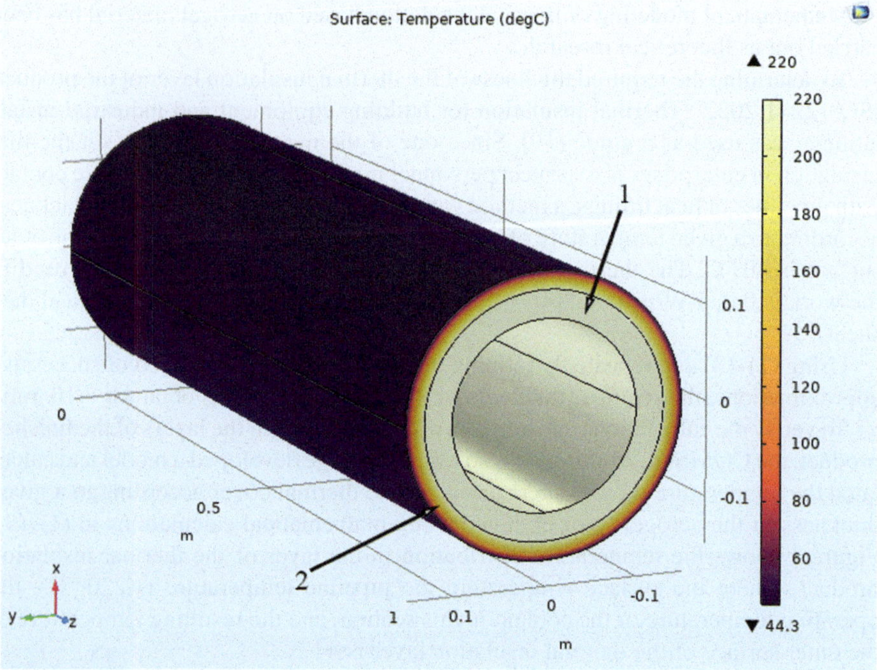

Fig. 1 Temperature distribution in the thermal cover layer. 1—a layer of pipeline, 2—a layer of thermal insulation of the product

The considered method is approximate. More accurate results can be obtained by the method of successive approximations.

Next, the calculation is performed by (3):

$$\left(\frac{t_{in} - t_{surf}}{t_{surf} - t_{out}}\right) = \frac{\ln \frac{d_{out}^{w} + 2\delta_0 i}{d_{out}^{w}} \alpha_{out} (d_{out}^{w} + 2\delta_0 i)}{2\lambda_{is}}, \tag{3}$$

where i is the iteration number in the calculation; α_{out} is the heat transfer coefficient on the surface of thermal insulation, [W/(m^2 °C)], δ_0 is the specified initial value of the insulation thickness, [m], determined by the required calculation accuracy, for example, 0.001 m.

In successive steps 1, 2, 3, …, i, for insulation thicknesses: $\delta_1 = 1\delta_0$; $\delta_2 = 2\delta_0$; $\delta_i = i\delta_0$, the values are calculated according to (3).

At each step of calculations i, (4) is compared with a given value of (5) until two adjacent values are comparatively equal.

$$\left(\frac{t_{in} - t_{surf}}{t_{surf} - t_{out}}\right)_i \tag{4}$$

$$\left(\frac{t_{in} - t_{surf}}{t_{surf} - t_{out}}\right)_p \quad (5)$$

In (5), p is the index of the previous iteration.

According to the results of the calculations and simulations, the thickness of the aerogel thermal insulation for the thermal cover is assumed 10 mm, the layers of fiberglass used in the manufacture of covers in the simulation do not bring the estimated temperature change, and therefore they were not taken into account in the preliminary calculations.

3 The Experimental Part of the Study

Experimental studies were conducted in accordance with the requirements of the international standard ISO 5725-1:2023 "Accuracy (trueness and precision) of measurement methods and results Part 1: General principles and definitions" [20].

According to the preliminary measurements made, a product design project was developed at the research facility and patterns were made. All materials used in the product are selected taking into account the requirements of technical safety and comply with the international class A1 incombustibility (in accordance with the European classification standard EN 13501-1 [21]), which is necessary in the operating conditions of an industrial enterprise [22, 23]. According to the patterns, 2 fiberglass covers were made, one thermal cover for the valve, the second for the pipeline section. Aerogel fiber with a thickness of 10 mm is used as the main thermal insulator. Fiberglass has a one-sided polymer coating, for moisture protection of the inner material and to counteract splashes of high-temperature coolant in case of possible leakage or breakage of fittings. The thermal covers are stitched with para-aramid refractory threads.

The installation of thermal insulation covers for temperature studies on the valve and the adjacent section of the pipeline was carried out at the stopped boiler unit TPE-429 №7 of the Kazan CHPP-3 branch of JSC "TGK-16" during the repair period.

The layered representation of the layers in the product is shown in Fig. 2, where 1 and 3 are the outer layers of fiberglass directed by a layer of polymer coating outside the thermal cover, 2 is a layer of aerogel fiber, 4 is the surface of an insulated element, for example, a pipeline or shut–off valves.

The manufactured and installed experimental thermal insulation is shown in Figs. 3 and 4.

For the next 4 weeks, during the operation of the boiler unit at operating parameters, temperature measurements of the insulation surfaces were carried out 3 times a week using a CEM DT-8869H pyrometer on the valves and pipeline section under study, in order to study changes in thermal losses in operating conditions and to study possible defects in the design of thermal insulation covers.

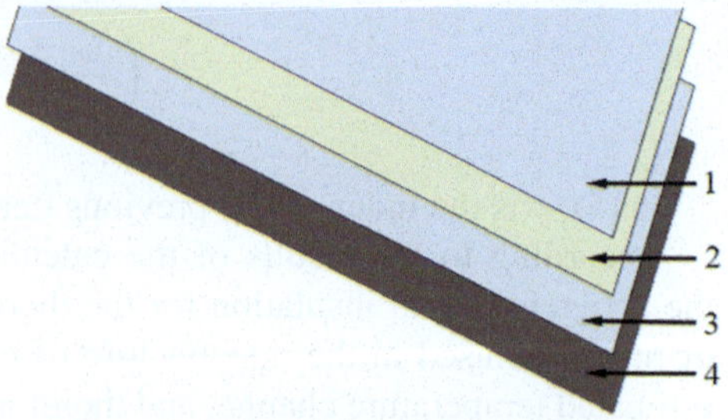

Fig. 2 Layered representation of thermal cover layers

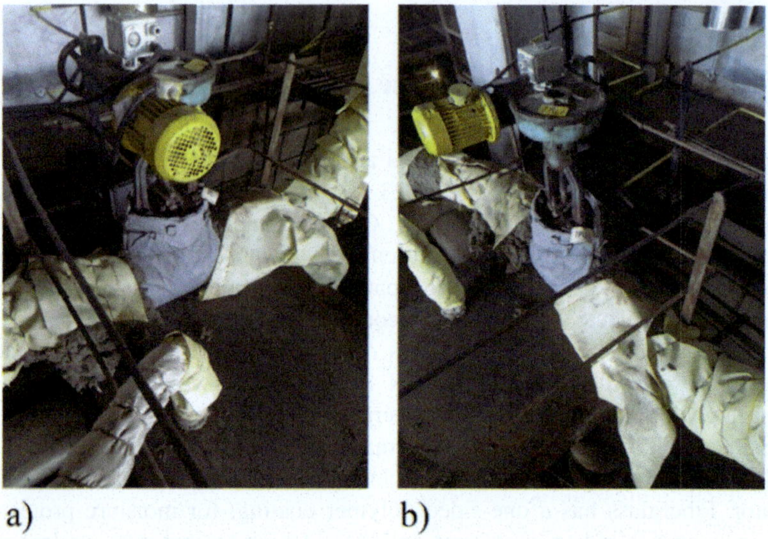

Fig. 3 The installed experimental thermal insulation on the valve 7 RC-1

4 Results and Analysis

The actual temperature of the outdoor air in the room during the measurements was 48 °C, in this regard, the actual temperature on the outer surface of the thermal insulation covers was within 65–70 °C, slightly exceeding the temperature of the installed thermal insulation using basalt mats (the sheet of measurements is presented in Table 1).

The temperature distribution diagram of the thermal insulation surface and the surface of the uncovered section of the pipeline in the study of experimental thermal insulation and basalt insulation is shown in Figs. 5 and 6. In the figures, a section of the pipeline with basalt insulation is indicated in gray, a valve with aerogel insulation in blue, and an aerogel insulated pipeline section in orange. The Y-axis shows the temperature values in degrees Celsius.

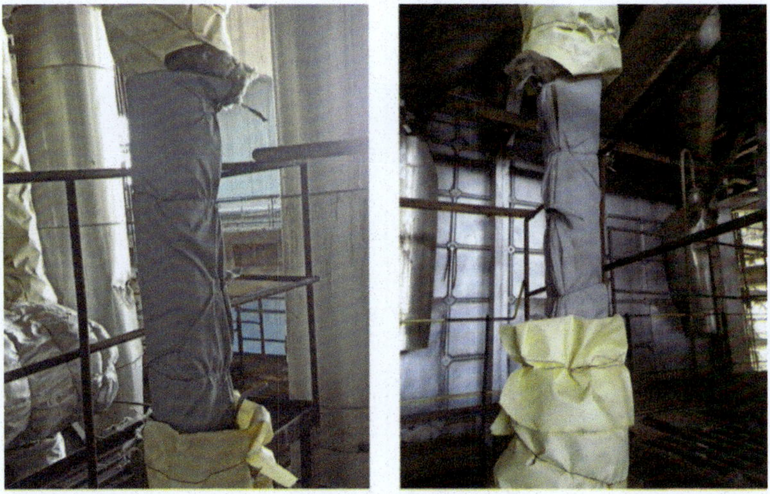

a) b)

Fig. 4 Installed experimental thermal insulation on the pipeline section

5 Conclusions

Having analyzed the temperature losses of the experimental thermal insulation under study, it is possible to form a conclusion on the study:

- The temperatures on the surface of experimental thermal insulation based on aerogel composite fiber and on the surface of thermal insulation based on BSFF (Basalt SuperFine Fiber) differ by an average of 4 °C, when the thicknesses of the compared thermal insulation materials, at similar temperatures on the surface of the insulating layer, differ by 14 times: the thickness of the aerogel fiber is 10 mm, the thickness of the basalt mat is 140 mm;
- The temperatures on the surface of the thermal insulation on the pipeline and the thermal insulation on the shut-off valves differ by 1–2 °C, showing approximately the same efficiency of thermal insulation with aerogel of both straight and complex sections, which makes it possible to simplify further simulations—by modeling a pipeline with an insulating layer, the results can be applied to shut-off valves without modeling a separate model.
- The cost of experimental thermal insulation with one layer of aerogel fiber when isolating a straight section of the pipeline exceeds the cost of thermal insulation from BSFF (the cost of experimental thermal insulation with one layer of aerogel fiber when isolating the valve is 2.97 times more expensive than installed thermal insulation from BSFF, and the direct section of the pipeline is 1.26 times more expensive).

Table 1 Statement of measurements of the surface of experimental thermal insulation on the equipment under study

	Temperature on the surface of thermal insulation (°C)											
	1 week			2 week			3 week			4 week		
	24.07	26.07	28.07	31.07	02.08	04.08	07.08	09.08	11.08	14.08	16.08	18.08
Plot without insulation	188	189	187	190	186	187	195	185	188	187	186	188
Valve with aerogel	65	63	63	69	66	68	68	66	67	68	68	67
The section of the pipeline with aerogel	66	64	69	71	69	66	69	69	69	71	69	70
A basalt-insulated site	61	61	62	67	65	64	67	65	63	61	60	63

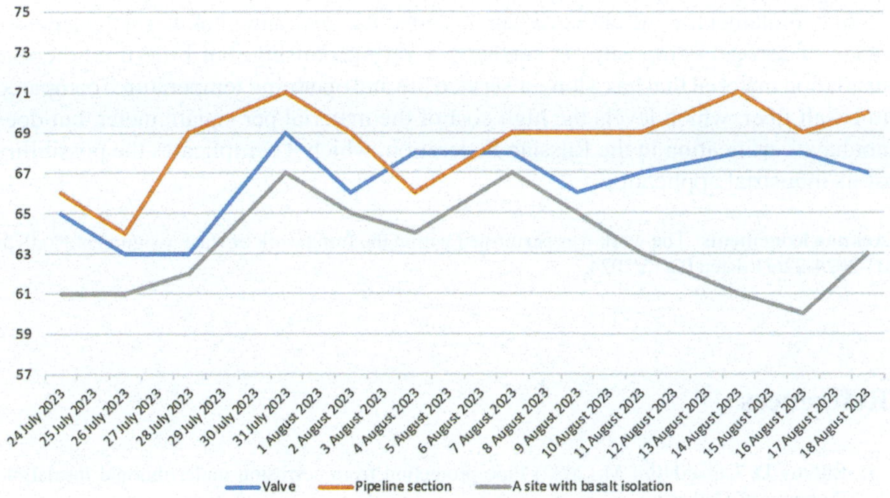

Fig. 5 Temperature distribution diagram for measurements

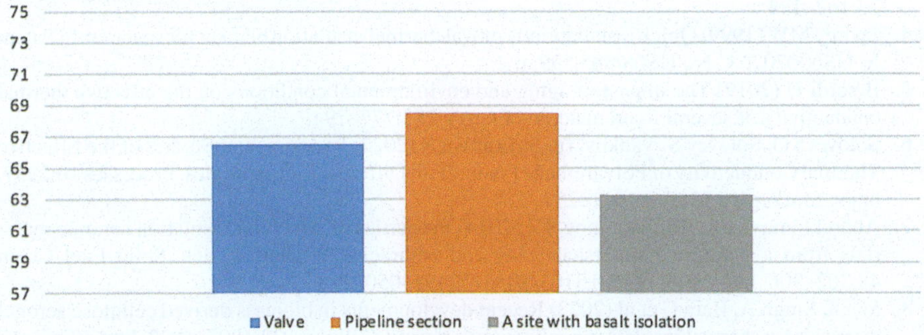

Fig. 6 Comparison of average values of temperature measurements

- The design of the developed covers requires minor refinement, taking into account better coverage of the insulated surfaces, increasing the contact of the insulating material with the heating source.
- A significant reduction in the cost of using thermal covers with aerogel insulation for pipelines and valves of both large diameters (200 mm or more) and very small diameters (less than 60 mm). In addition, the advantages of using aerogel fiber are fully revealed when operating at high temperatures (more than 500 °C), since traditional thermal insulation materials quickly lose their properties and become unusable during high-temperature operation. For small diameters of shut-off valves or pipelines, it is more efficient to use aerogel with a smaller thickness of the thermal insulation layer, which reduces the dimensions of the insulated unit.

In complement to the above, it can be added that experimental thermal insulation made of aerogel composites is currently a very promising, but little-used thermal insulation material that has a longer service life and operating temperatures compared to basalt fiber, which levels the high cost of the material per square meter, but does not have certification in the Russian Federation, which it complicates the possibility of its industrial application.

Acknowledgements The work was performed within the framework of state assignment No.075-03-2024-226/1 dated 02/15/2024.

References

1. Burkov D, Gubaidullin M (2023) Pipe protection from corrosion under thermal insulation. Problems of Gathering
2. Treatment and Transportation of Oil and Oil Products, pp 94–102. https://doi.org/10.17122/ntj-oil-2023-3-94-102
3. Gu T-S, Xie L-Y, Chen G (2006) Building energy conservation and wall thermal insulation 23:167–184
4. Shorey MW (1999) Quick installation-removal thermal insulation blanket for space craft. Patent № US6007026A. No US08/885,599, p 7
5. Berardi U (2019) The impact of aging and environmental conditions on the effective thermal conductivity of several foam materials.Energy 182:777–794
6. Soloveva O, Solovev S, Vankov Yu, Shakurova R (2022) Experimental Studies of the Effective Thermal Conductivity of Polyurethane Foams with Different Morphologies. Processes 10:2257. https://doi.org/10.3390/pr10112257
7. Abdrakhimova ES, Abdrakhimov VZ (2014) Properties of structural insulation ceramic materials from a mixture of intershale clay and anthracite flotation wastes. Solid Fuel Chem 48:302–306. https://doi.org/10.3103/S0361521914050024
8. Sen S, Singh A, Bera C et al (2022) Recent developments in biomass derived cellulose aerogel materials for thermal insulation application: a review. Cellulose 29:4805–4833. https://doi.org/10.1007/s10570-022-04586-7
9. Apkaryan AS, Kulkov SN (2021) Study of the Thermophysical Properties of a Glass-Ceramic Material for Thermal Insulation of Pipelines of Heating Pipe Mains and Utilities. Inorg. Mater. Appl. Res. 12:499–503
10. Bhuiyan MAR, Ali A, Mohebbullah M et al (2023) Recycling of cotton apparel waste and its utilization as a thermal insulation layer in high performance clothing. Fash Text 10:22. https://doi.org/10.1186/s40691-023-00342-y
11. Netsch N, Simons M, Feil A, Leibold H, Richter F, Slama J, Yogish SP, Greiff K, Stapf D (2022) Recycling of polystyrene-based external thermal insulation composite systems - Application of combined mechanical and chemical recycling. Waste Manag. 1(150):141–150. https://doi.org/10.1016/j.wasman.2022.07.001
12. Karamahmut Mermer N, Sari Yilmaz M, Dere Ozdemir O et al (2017) The synthesis of silica-based aerogel from gold mine waste for thermal insulation. J Therm Anal Calorim 129:1807–1812. https://doi.org/10.1007/s10973-017-6371-8
13. Suitcases AN, Gorinov YuA, Safin RG, Alibekov SYa, Gainullin RH (2015) Composite thermal insulation and ballast material based on wood waste. Patent No. RU 2544194 C1 IPC C04B 18/26, C04B 28/02. No. 2014103401/03, p 5
14. Mahajan UR, Emmanuel I, Sreenivasarao A et al (2023) Development of smart polyurethane foam with combined capabilities of thermal insulation and thermal energy storage by integrating

microencapsulated phase change material. Polym. Bull. 80:13099–13115. https://doi.org/10.1007/s00289-023-04695-8

15. Selyaev VP, Kupriyashkina LI, Neverov VA et al (2015) Vacuum thermal insulation panel. Utility Model Patent No. 150467 U1 Russian Federation, IPC E04C 2/02, F16L 59/06. No. 2014133568/03, N.P. Ogarev Mordovian State University
16. Berge A, Adl-Zarrabi B, Hagentoft CE (2015) Assessing the thermal performance of district heating twin pipes with vacuuminsulation panels. Energy Procedia 78:382–387
17. Cha J, Kim S, Park KW et al (2014) Improvement of window thermal performance using aerogel insulation film for building energy saving. J Therm Anal Calorim 116:219–224. https://doi.org/10.1007/s10973-013-3521-5
18. Oh KW, Kim DK, Kim SH (2009) Ultra-porous flexible PET/Aerogel blanket for sound absorption and thermal insulation. Fibers Polym 10:731–737. https://doi.org/10.1007/s12221-010-0731-3
19. Yang Jianming, Wu Huijun, Liang Yuying, Cen Jian, Zhang Xianyong (2023) A Comparative Assessment of Different Aerogel-Insulated Building Walls for Enhanced Thermal Insulation Performance. Gels 9:943. https://doi.org/10.3390/gels9120943
20. ISO 12241:2022 "Thermal insulation of buildings - Testing and rating of materials for thermal resistance and thermal transmittance - Specification used in Europe"
21. ISO 5725-1:2023 Accuracy (reliability and precision) of measurement methods and results Part 1: General principles and definitions: date of introduction 07/20/2023. TC 69/SC 6, Switzerland, 2023, p 26
22. File classification of construction products and building elements - Part 1: Classification using data from reaction to fire tests (2019) German version EN 13501-1, p 62
23. Order of the Ministry of Energy of the Russian Federation of 30 December 2008 No. 325 "On Approval of the Procedure for Determiningthe Standards for Technological Losses during the Transfer of Heat Energy, Heat Carrier" (2008) Ministry of Energy, Moscow, Russia

Environmental Perspectives on Zinc Plating Technologies for Steel

L. V. Radionova

Abstract The article discusses the main technologies for applying zinc coatings, which makes it possible to protect steel products from corrosion and thereby reduce the harmful impact on the environment. A description of the technological features of hot-dip galvanizing, electro galvanizing, thermal diffusion galvanizing, cold galvanizing, and thermal spraying galvanizing is given. A comparative analysis of the considered galvanizing technologies was carried out from the point of view of their harmful effects on the environment and human health. The article shows that surface preparation before hot-dip galvanizing, including pickling in acids, fluxing, degreasing in alkaline solutions and rinsing in water, has a negative impact on the environment and requires significant costs for disposal and neutralization of waste solutions. Electroplating production is one of the most dangerous sources of environmental pollution, due to the fact that both in the surface preparation process and during the coating itself, reagents and electrolytes are used that are harmful to both the environment and human health. The most environmentally friendly technologies for applying zinc coatings are thermal diffusion and thermal spraying galvanizing. When preparing the surface, these zinc plating methods allow mechanical cleaning of the surface using abrasive materials, which significantly reduces the burden on the environment. The use of zinc wire instead of powder during thermal spraying will improve the economic efficiency of the process compared to powder technologies, which will make this galvanizing method the most promising and environmentally friendly for protecting steel structures from corrosion.

Keywords Environment · Zinc coatings · Hot-dip galvanizing · Electro galvanizing · Thermal diffusion galvanizing · Cold galvanizing · Thermal spraying

L. V. Radionova (✉)
Moscow Polytechnic University, 38, Bolshaya Semyonovskaya Str., Moscow 107023, Russia
e-mail: radionovalv@rambler.ru

L. V. Radionova and D. V. Ulrikh (eds.), *Advances in Ecology and Environmental Engineering*, Springer Proceedings in Earth and Environmental Sciences,
https://doi.org/10.1007/978-3-031-64423-8_48

1 Introduction

Steel is the most common structural material, which, depending on its chemical composition, has a wide range of mechanical properties. It is known that the properties of steel are determined by the carbon content and the presence of alloying elements, as well as the applied mechanical and heat treatment [1]. The disadvantage of steel products is their low corrosion resistance [2]. Steel structures and products oxidize under the influence of moisture and other more aggressive environments, resulting in the formation of iron oxides (rust) on their surface [3]. Corrosion processes spoil the appearance of structures, reduce their strength and lead to premature destruction [4]. This negatively affects safety, economics and the environment [5]. Reducing the service life of steel structures due to corrosion processes requires their replacement with new ones. This requires new steel to be smelted and treated mechanically and thermally, which has a significant additional impact on the environment. In addition, oxides of iron and other elements present in steel enter the soil, water and air, causing harm to the environment [6].

The corrosion resistance of structures can be increased by alloying steel [7] and applying protective coatings [8, 9]. For steel structures for bridges, fences and others, alloying with chemical elements that increase corrosion resistance is an economically unjustified measure due to the high price of chromium and nickel. This method is applicable to increase the corrosion resistance of critical products, where a two- and three-fold increase in the price of steel is economically justified. In other cases, the application of corrosion-resistant coatings is used to protect steel structures [10].

Among anti-corrosion coatings, zinc coatings [11] are widely used, which have a high degree of protection for steel, extend the service life of products and are relatively safe for humans [12] and the environment [13]. Zinc coated steel does not emit toxic gases and does not pose a danger to humans during operation, since this chemical element is present in all living organisms [14].

Currently, there are several technologies for applying zinc to steel structures and products: hot-dip galvanizing [15, 16], electro galvanizing [17], thermal diffusion galvanizing [18], cold galvanizing [19], and thermal spraying [20]. In this article, we will consider the technological features and environmental prospects for the use of each of these coatings.

2 Technological Features of Galvanizing Steel

2.1 Hot-Dip Galvanizing

The hot-dip galvanizing method involves immersing products in a bath of molten zinc at a temperature of 450–480 °C (Fig. 1). The surface of the products is carefully prepared before immersion in molten zinc. The surface preparation process includes degreasing steel products in caustic cleaning, picking, rinsing and flux solution, most often in solutions of zinc and ammonium chlorides, followed by drying (Fig. 2).

When cleaning steel, grease stains, oils and synthetic contaminants are removed from the surface, which slow down and worsen etching, reducing the adhesion of zinc. To do this, use a solution of orthophosphoric acid and surfactants. The next step is rinsing the product in water to remove any phosphoric acid residue. An important point in surface preparation is picking in a 15% solution of hydrochloric acid. When pickling, rust and scale are removed and the steel surface is chemically activated. Residues of hydrochloric acid and iron oxides are rinsed with water. Then flux, which, being a protective film, improves the adhesion of the zinc melt to the surface of the product, is applied to the surface. Drying at a temperature of 120 °C removes residual moisture and prepares the surface for galvanizing.

As a result of hot-dip galvanizing, a protective coating with a thickness of 30–200 microns is formed on the surface of a steel product. This method is used when applying zinc coating to steel strip, wire, structures and products.

Fig. 1 Hot-dip galvanizing of steel strip: 1—support rollers; 2—zinc bath; 3—air knifes; 4—steel strip

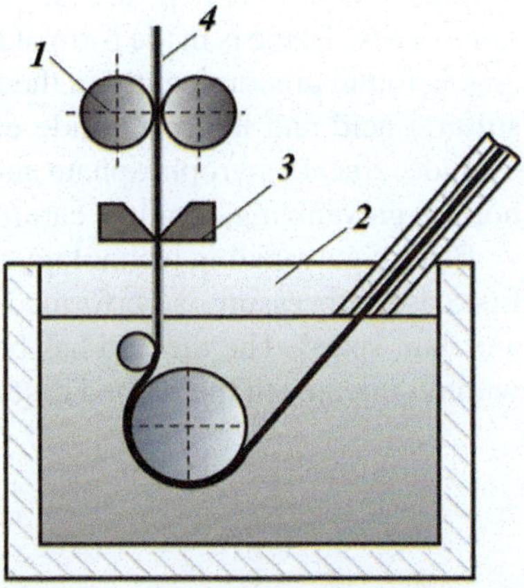

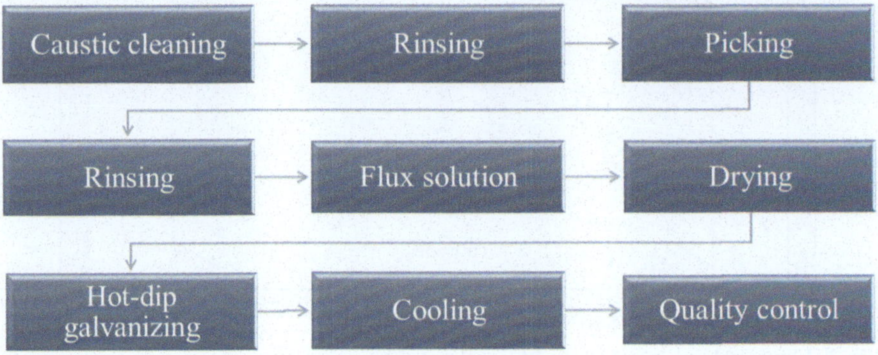

Fig. 2 Hot-dip galvanizing process

Direct application of molten zinc to the surface of steel strip, wire, structures and products is a fairly environmentally friendly process, but preparing the surface before it has a negative impact on the environment and involves high costs for the neutralization of hazardous waste, which requires significant production space.

2.2 *Electro Galvanizing*

Applying a thin layer of zinc to the surface of metal products in an electrolyte solution is called electro galvanizing. Electro galvanizing as a production technology is a chemical process—electrolysis. Two metals, steel products and pure zinc are placed in an electrolyte bath (Fig. 3). Zinc, i.e. the anode dissolves during electrolysis. It is this technology of applying zinc coating that allows you to obtain uniform and shiny surfaces. Such methods are used mainly in the production of bolts, nuts, pipes and pipe fittings.

Zinc coatings are deposited mainly from two types of electrolytes: (a) from solutions in which zinc is in the form of hydrated zinc cations; (b) from solutions in which zinc is found predominantly in the form of complex anions. The first group includes sulfuric acid and hydro fluoride electrolytes, the second group includes complex cyanide, zincate, pyrophosphate and other electrolytes. Electro galvanizing, just like hot-dip galvanizing, requires careful preparation of the product surface.

Electro galvanizing technology is produced by two technological methods. In the first case, processing is performed in special installations (drums), which rotate at a certain speed. The method has become widespread for processing parts with and without threads. In the second case, the metal structures are suspended using copper

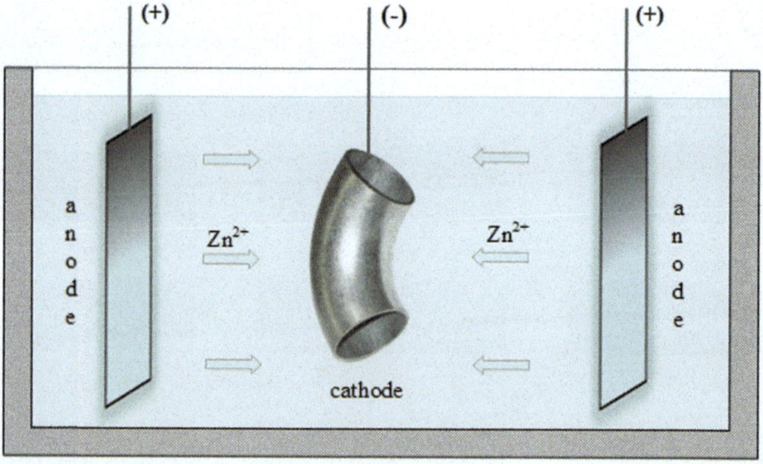

Fig. 3 Electro galvanizing of steel

wire and then lowered into the electrolyte. This method is used for electro galvanizing large-sized products and structures.

The results of corrosion tests of zinc coatings applied by various methods indicate that electro galvanized coatings are more corrosion resistant. This is due to a higher degree of purity, fine-crystalline structure, greater plastic properties, better coating adhesion and more uniform thickness. Metal savings using the electro galvanizing method compared to hot galvanizing are 50% or more.

2.3 *Thermal Diffusion Galvanizing*

Thermal diffusion galvanizing is a relatively new technology for applying zinc coating, which has advantages over traditional methods of electro galvanizing and hot-dip galvanizing. In this regard, the popularity of this method is rapidly growing. The thermal diffusion method of application is based on the process of diffusion of a metal into the surface layers of products at high temperatures. The product is placed in a container with zinc powder and heated for a set time (Fig. 4).

In a zinc coating obtained by chemical-thermal treatment in zinc-based powder mixtures at 400–500 °C for 2–5 h, mainly two phases are clearly observed: the γ-phase, which is a thin (2–10 μm) layer, located directly on the surface of the coated product, and the δ-phase following this layer (Fig. 5).

With this galvanizing method, as a result of diffusion, zinc is introduced into the crystal lattice of steel, so the bond strength of the coating with the coated material increases sharply. The higher the temperature and the longer the diffusion process, the greater the depth of penetration of zinc into the metal being coated. The adhesion of

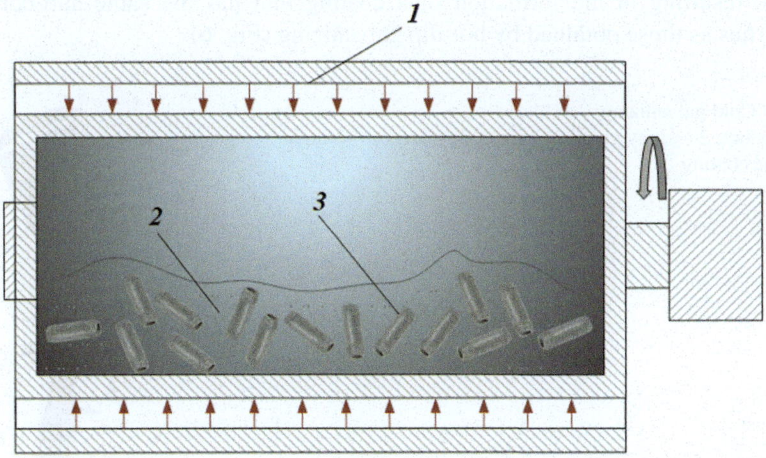

Fig. 4 Thermal diffusion galvanizing: 1—drum furnace; 2—powder mixture; 3—products

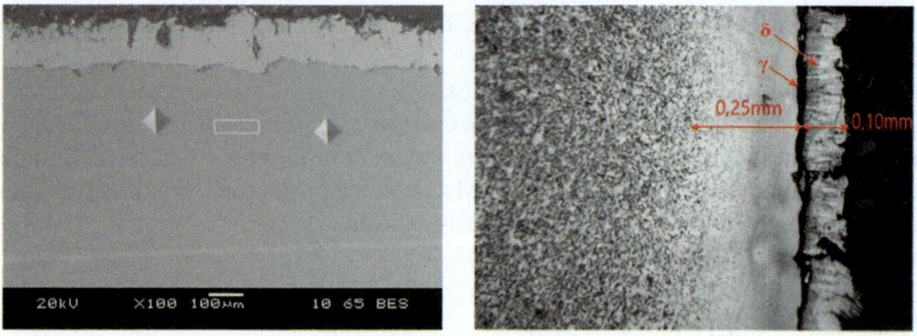

Fig. 5 Microstructure of the zinc coating

diffusion zinc coating, based on diffusion bonding, is superior to the adhesion of zinc coatings applied from molten zinc (hot-dip galvanizing) or other known methods.

Thermal diffusion galvanizing is most often used to coat small and medium-sized parts. The thin thickness, strength and uniformity of the coating make it optimal for coating threaded connections, ensuring repeated screwing. This advantage is also relevant when used for connecting fittings, pipes, fittings, and various plumbing products.

2.4 *Cold Galvanizing*

Cold galvanizing is the application of a special zinc-containing composition (more than 95%) to a prepared surface by techniques used in working with conventional paints, resulting in the formation of a coating that has the same anti-corrosion properties as those obtained by hot-dip galvanizing (Fig. 6).

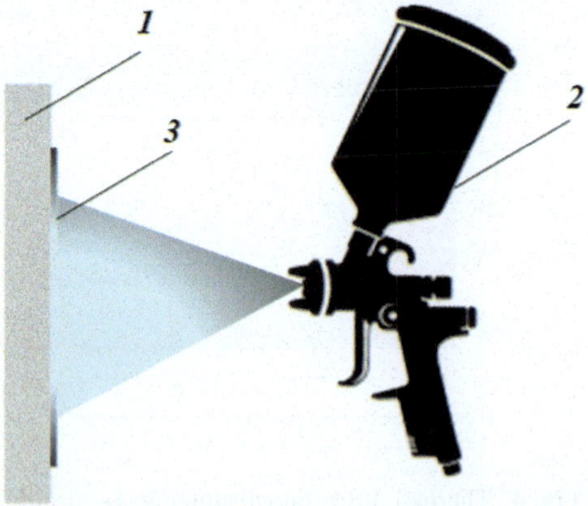

Fig. 6 Cold galvanizing: 1—product; 2—spray gun; 3—zinc coating

Just as with other zinc plating methods, surface preparation determines the quality of the coating. Unlike hot-dip galvanizing and electro galvanizing, cold galvanizing often uses mechanical methods of surface preparation. To remove dirt and rust, surface treatment with metal brushes or sandblasting is used, when an abrasive under high pressure is applied to the surface of the product without the use of chemical reagents. The abrasive particles destroy the dirt, leaving the material itself only slightly rough. Additional roughness improves the adhesion of the zinc-containing composition to the surface of the product.

The main advantage of cold galvanizing steel is its general availability. Zinc coatings can be applied to various metal structures, including large or complex shapes. The equipment and tool for applying such coatings is a spray gun, which is small in size and can be used in any condition and directly at the place where the product is used. The thickness of the coating is regulated by the number of applied layers and can range from 5 to 35 microns.

Cold galvanizing is a very promising method of protecting steel structures from corrosion since it is widely available but quite effective. Cold galvanizing has a large positive environmental effect by increasing the service life of steel products while minimizing the negative environmental impact of surface preparation methods and the coating method itself.

2.5 Zinc Thermal Spraying

The method of metallization of steel products by high-speed spraying of zinc powder onto their surface is usually called thermal spraying. As a result of heating, dispersion and transfer of condensed particles of sprayed zinc by a gas or plasma flow, a coating is formed on the surface of the part (Fig. 7).

Delivery of fine droplets of zinc to the surface to be treated can be carried out by gas-flame, detonation, plasma spraying or electric arc metallization. Sprayed zinc

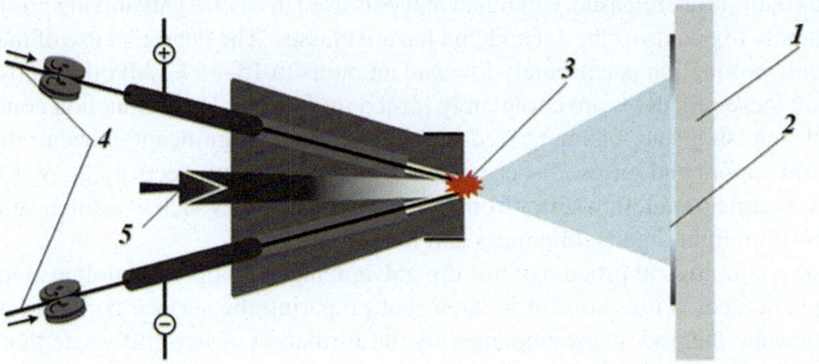

Fig. 7 Zinc thermal spraying

particles harden on the surface of a part or product, forming a protective layer 30–50 microns thick. The thickness of the coating, as with cold galvanizing, can be adjusted by the number of layers applied.

As an advantage, it is worth noting the possibility of applying the coating outside production conditions and without dismantling metal structures. Abrasive treatment is used to prepare the surface. The equipment used for thermal spraying is universal and can be used for applying various protective coatings, as well as for restoring worn parts. The impact of this zinc plating method on the environment is minimal and is today one of the most promising areas of research to improve zinc plating technologies.

The material used for coating can be zinc powder or zinc wire. The wire is obtained through the process of plastic deformation—drawing. This process is quite complex and labor-intensive, but even with such features, wire is a cheaper and more convenient material to use compared to powder. The cost of the wire determines the economic attractiveness of the thermal spraying method as one of the environmentally friendly zinc plating methods. Therefore, reducing the cost of zinc wire by increasing the efficiency of its production technology is an urgent issue that should be given attention.

3 Analysis of the Harmful Effects of Zinc Plating Technologies on the Environment

Galvanic production is the most harmful to both the environment and human health among the considered technologies for galvanizing steel products. In galvanic zinc plating units, caustic cleaning, pickling in hydrochloric acid, removal of pickling sludge, cyanide-free galvanizing, and passivation with intermediate rinsing in the hot and cold water of steel parts are carried out in baths. Picking is carried out in a bath of hydrochloric acid, which releases hydrogen chloride into the atmospheric air. During electro galvanizing, sodium hydroxide, an aerosol of sulfuric acid, and sodium sulfate are released. Chemical reagents used in electro galvanizing production are largely toxicants of the 1st and 2nd hazard classes. The degree of use of metals in galvanic production is extremely low and amounts to 15–35%. All other substances used in these processes are completely turned into waste. The production conditions in galvanic shops are characterized by high humidity, significant concentrations of harmful vapors and gases, dispersed mists and splashes of electrolytes. Workers in electroplating production suffer from occupational diseases such as asthma, allergies, ulcers of internal organs, blindness and loss of smell.

The technological process of hot-dip galvanizing, i.e. applying molten zinc to the surface of steel is not harmful in itself, but preparing the surface before it, as with the galvanic method, is accompanied by the formation of harmful waste that needs to be neutralized.

During thermal diffusion, cold and thermal spraying zinc plating, mechanical removal of rust, dirt and other contaminants is used during surface preparation. Sandblasting not only allows you to clean the surface of the product before applying the coating but also creates a roughness that leads to better adhesion of the zinc to the base metal. The processes of applying zinc coating using these methods are environmentally friendly and, when workers use personal protective equipment, do not cause harm to human health.

4 Conclusions

Analysis of the technological features of five methods of applying zinc plating allowed us to establish that:

- coating obtained by hot-dip galvanizing, electro galvanizing, thermal diffusion galvanizing, cold galvanizing, and thermal spraying allows you to effectively protect steel products and structures from corrosion and significantly extends their service life;
- to protect steel strips and wires, hot-dip galvanizing is used, which has minimal impact on the environment, but the surface preparation processes for it cause significant harm to the environment and human health;
- electro galvanizing production is one of the most dangerous sources of environmental pollution, this is due to the fact that both in the process of preparing the surface and during the application of the coating itself, reagents and electrolytes are used that are harmful to both the environment and human health;
- thermal diffusion, cold and thermal spraying galvanizing during surface preparation allows mechanical cleaning using abrasive materials, which significantly increases the environmental friendliness of these methods;
- thermal diffusion and thermal spraying galvanizing are by far the most environmentally friendly processes and allow the application of corrosion-resistant coatings to products of complex shapes;
- the use of wire in thermal spraying zinc plating will improve the economic efficiency of the process compared to powder technologies, which will make this zinc plating method the most promising and environmentally friendly for protecting steel structures from corrosion.

References

1. Callister Jr WD, Rethwisch DG (2020) Fundamentals of materials science and engineering: an integrated approach. John Wiley & Sons

2. Soleimani M, Mirzadeh H, Dehghanian C (2020) Unraveling the Effect of Martensite Volume Fraction on the Mechanical and Corrosion Properties of Low-Carbon Dual-Phase Steel. Steel research international 91(2):1900327
3. Harsimran S, Santosh K, Rakesh K (2021) Overview of corrosion and its control: A critical review. Proc. Eng. Sci 3(1):13–24
4. Abubakar SA, Mori S, Sumner J (2022) A review of factors affecting SCC initiation and propagation in pipeline carbon steels. Metals 12(8):1397
5. Panchal J, Shah D, Patel R, Shah S, Prajapati M, Shah M (2021) Comprehensive review and critical data analysis on corrosion and emphasizing on green eco-friendly corrosion inhibitors for oil and gas industries. Journal of Bio-and Tribo-Corrosion 7(3):107
6. Nortjé GP, Laker MC (2021) Factors that determine the sorption of mineral elements in soils and their impact on soil and water pollution. Minerals 11(8):821
7. Sun J, Tang H, Wang C, Han Z, Li S (2022) Effects of alloying elements and microstructure on stainless steel corrosion: a review. Steel research international 93(5):2100450
8. Lazorenko G, Kasprzhitskii A, Nazdracheva T (2021) Anti-corrosion coatings for protection of steel railway structures exposed to atmospheric environments: A review. Construction and Building Materials 288:123115
9. Cui G, Bi Z, Wang S, Liu J, Xing X, Li Z, Wang B (2020) A comprehensive review on smart anti-corrosive coatings. Progress in organic coatings 148:105821
10. Kumar SS, Tripathi V, Sharma R, Puthilibai G, Sudhakar M, Negash K (2022) Study on developments in protection coating techniques for steel. Advances in Materials Science and Engineering
11. Dan A, Bijalwan PK, Pathakm AS, Bhagat AN (2022) A review on physical vapor deposition-based metallic coatings on steel as an alternative to conventional galvanized coatings. Journal of Coatings Technology and Research 19(2):403–438
12. Vimercati L, Cavone D, Caputi A, De Maria L, Tria M, Prato E, Ferri GM (2020) Nanoparticles: An experimental study of zinc nanoparticles toxicity on marine crustaceans. General overview on the health implications in humans. Frontiers in Public Health 8:499190
13. Czyżowska A, Barbasz A (2022) A review: zinc oxide nanoparticles–friends or enemies? International journal of environmental health research 32(4):885–901
14. Gondal AH, Zafar A, Zainab D, Toor MD, Sohail S, Ameen S, Younas N (2021) A detailed review study of zinc involvement in animal, plant and human nutrition. Indian Journal of Pure & Applied Biosciences 9(2):262–271
15. Radionova LV, Subbotina YuM (2016) The Study of Aging Cold Rolled and Hot Dip Galvanized Automotive Steel. In Materials Science Forum 870:352–359
16. El Kacimi Y, Touir R, Alaoui K, Touhami ME (2023) Recent Progress and Comprehension of Corrosion of Steels in the Hot-Dip Galvanizing Industry. Corrosion Science, pp 81–106
17. Mohammed AJ, Moats M (2022). Effects of Carrier, Leveller, and Booster Concentrations on Zinc Plating from Alkaline Zincate Baths. Metals 12(4)–621
18. Galin RG, Shaburova NA, Zakharyevich DA (2016) Thermal Diffusion Galvanizing in Ferriferous Zinc Powder. In Materials Science Forum 870:129–134
19. Li J, Wang Q, Gao N, Nwokolo IK, Zhang W, Ma L, Han EH (2021) Interface characteristics and anticorrosion performances of cold galvanizing coatings incorporated with γ-chloropropyl triethoxysilane on hot-dip galvanized steel. Coatings 11(4):402
20. Kahar S, Singh A, Vala U, Desai A, Desai S (2020) Thermal sprayed coating using zinc: A review. Int. Res. J. Eng. Technol 7:6497–6503

The manufacturer's authorised representative in the EU is Springer Nature Customer Service Centre GmbH, Europaplatz 3, 69115 Heidelberg, Germany. If you have any concerns regarding our products, please contact ProductSafety@springernature.com

Printed and bound by CPI Group (UK) Ltd, Croydon, CR0 4YY
01/10/2026
02225907-0017